10TH EDITION ABRIDGED

Elementary Linear Algebra

HOWARD ANTON

Professor Emeritus, Drexel University

CHRIS RORRES

University of Pennsylvania

WILEY

John Wiley & Sons Canada, Ltd.

Library and Archives Canada Cataloguing in Publication

Anton, Howard
   Elementary linear algebra / Howard Anton, Chris Rorres. -- 10th ed. abridged

Includes index.
ISBN 978-0-470-93747-1

      1. Algebras, Linear--Textbooks.
2. Geometry--Textbooks. I. Rorres, Chris II. Title.

QA184.A57 2010   512'.5   C2010-905368-0

Printed in the United States

2 3 4 5 EB 15 1 4 13

John Wiley & Sons Canada, Ltd.
6045 Freemont Blvd.
Mississauga, Ontario L5R 4J3

www.wiley.ca

This textbook is an abridged version of *Elementary Linear Algebra, Applications Version*, tenth edition, by Howard Anton. The fifth chapter consists of eight applications of linear algebra drawn from business, economics, and computer science. The applications are largely independent of each other, and each includes a list of mathematical prerequisites. Thus, each instructor has the flexibility to choose those applications that are suitable for his or her students and to incorporate each application anywhere in the course after the mathematical prerequisites have been satisfied. Chapters 1–4 include simpler treatments of some of the applications covered in more depth in Chapter 5.

This edition gives an introductory treatment of linear algebra that is suitable for students in the CEGEP system. Its aim is to present the fundamentals of linear algebra in the clearest possible way—sound pedagogy is the main consideration. Although calculus is not a prerequisite, there is some optional material that is clearly marked for students with a calculus background. If desired, that material can be omitted without loss of continuity.

Technology is not required to use this text, but for instructors who would like to use MATLAB, *Mathematica*, Maple, or calculators with linear algebra capabilities, we have posted some supporting material that can be accessed at either of the following Web sites:

www.wiley.com/canada/anton

Summary of Changes in this Edition

This edition is a major revision of its predecessor. In addition to including some new material, some of the old material has been streamlined to ensure that the major topics can all be covered in a standard course. These are the most significant changes:

- **Vectors in 2-space, 3-space, and *n*-space** Chapters 3 and 4 of the previous edition have been combined into a single chapter. This has enabled us to eliminate some duplicate exposition and to juxtapose concepts in *n*-space with those in 2-space and 3-space, thereby conveying more clearly how *n*-space ideas generalize those already familiar to the student.

- **New Pedagogical Elements** Each section now ends with a *Concept Review* and a *Skills* mastery that provide the student a convenient reference to the main ideas in that section.

- **New Exercises** Many new exercises have been added, including a set of True/False exercises at the end of most sections.

Hallmark Features

- **Relationships Among Concepts** One of our main pedagogical goals is to convey to the student that linear algebra is a cohesive subject and not simply a collection of isolated definitions and techniques. One way in which we do this is by using a crescendo of *Equivalent Statements* theorems that continually revisit relationships among systems of equations, matrices, determinants, vectors, linear transformations, and eigenvalues. To get a general sense of how we use this technique see Theorems 1.5.3, 1.6.4, 2.3.8, 4.8.10, and then 4.10.4, for example.

- **Smooth Transition to Abstraction** Because the transition from R^n to general vector spaces is difficult for many students, considerable effort is devoted to explaining the purpose of abstraction and helping the student to "visualize" abstract ideas by drawing analogies to familiar geometric ideas.

- **Mathematical Precision** When reasonable, we try to be mathematically precise. In keeping with the level of student audience, proofs are presented in a patient style that is tailored for beginners.

- **Suitability for a Diverse Audience**   This text is designed to serve the needs of students in engineering, computer science, biology, physics, business, and economics as well as those majoring in mathematics.

- **Historical Notes**   To give the students a sense of mathematical history and to convey that real people created the mathematical theorems and equations they are studying, we have included numerous *Historical Notes* that put the topic being studied in historical perspective.

About the Exercises

- **Graded Exercise Sets**   Each exercise set begins with routine drill problems and progresses to problems with more substance.

- **True/False Exercises**   Most exercise sets end with a set of True/False exercises that are designed to check conceptual understanding and logical reasoning. To avoid pure guessing, the students are required to justify their responses in some way.

- **Supplementary Exercise Sets**   Most chapters end with a set of supplementary exercises that tend to be more challenging and force the student to draw on ideas from the *entire* chapter rather than a specific section.

Supplementary Materials for Students

- **Technology Exercises and Data Files**   The technology exercises that appeared in the previous edition have been moved to the Web site that accompanies this text. Those exercises are designed to be solved using MATLAB, *Mathematica*, or Maple and are accompanied by data files in all three formats. The exercises and data can be downloaded from either of the following Web sites:

www.wiley.com/canada/anton

CONTENTS

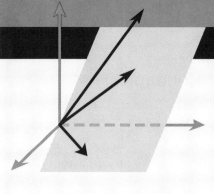

CHAPTER 1

Systems of Linear Equations and Matrices

INTRODUCTION

Information in science, business, and mathematics is often organized into rows and columns to form rectangular arrays called "matrices" (plural of "matrix"). Matrices often appear as tables of numerical data that arise from physical observations, but they occur in various mathematical contexts as well. For example, we will see in this chapter that all of the information required to solve a system of equations such as

$$5x + y = 3$$
$$2x - y = 4$$

is embodied in the matrix

$$\begin{bmatrix} 5 & 1 & 3 \\ 2 & -1 & 4 \end{bmatrix}$$

and that the solution of the system can be obtained by performing appropriate operations on this matrix. This is particularly important in developing computer programs for solving systems of equations because computers are well suited for manipulating arrays of numerical information. However, matrices are not simply a notational tool for solving systems of equations; they can be viewed as mathematical objects in their own right, and there is a rich and important theory associated with them that has a multitude of practical applications. It is the study of matrices and related topics that forms the mathematical field that we call "linear algebra." In this chapter we will begin our study of matrices.

1.1 Introduction to Systems of Linear Equations

Systems of linear equations and their solutions constitute one of the major topics that we will study in this course. In this first section we will introduce some basic terminology and discuss a method for solving such systems.

Linear Equations Recall that in two dimensions a line in a rectangular xy-coordinate system can be represented by an equation of the form

$$ax + by = c \quad (a, b \text{ not both } 0)$$

and in three dimensions a plane in a rectangular xyz-coordinate system can be represented by an equation of the form

$$ax + by + cz = d \quad (a, b, c \text{ not all } 0)$$

These are examples of "linear equations," the first being a linear equation in the variables x and y and the second a linear equation in the variables x, y, and z. More generally, we define a **linear equation** in the n variables $x_1, x_2, \ldots, x_n$ to be one that can be expressed in the form

$$a_1x_1 + a_2x_2 + \cdots + a_nx_n = b \tag{1}$$

where $a_1, a_2, \ldots, a_n$ and b are constants, and the a's are not all zero. In the special cases where $n = 2$ or $n = 3$, we will often use variables without subscripts and write linear equations as

$$a_1x + a_2y = b \quad (a_1, a_2 \text{ not both } 0) \tag{2}$$

$$a_1x + a_2y + a_3z = b \quad (a_1, a_2, a_3 \text{ not all } 0) \tag{3}$$

In the special case where $b = 0$, Equation (1) has the form

$$a_1x_1 + a_2x_2 + \cdots + a_nx_n = 0 \tag{4}$$

which is called a **homogeneous linear equation** in the variables $x_1, x_2, \ldots, x_n$.

▶ **EXAMPLE 1 Linear Equations**

Observe that a linear equation does not involve any products or roots of variables. All variables occur only to the first power and do not appear, for example, as arguments of trigonometric, logarithmic, or exponential functions. The following are linear equations:

$$x + 3y = 7 \qquad\qquad x_1 - 2x_2 - 3x_3 + x_4 = 0$$
$$\tfrac{1}{2}x - y + 3z = -1 \qquad\qquad x_1 + x_2 + \cdots + x_n = 1$$

The following are not linear equations:

$$x + 3y^2 = 4 \qquad\qquad 3x + 2y - xy = 5$$
$$\sin x + y = 0 \qquad\qquad \sqrt{x_1} + 2x_2 + x_3 = 1 \ \blacktriangleleft$$

A finite set of linear equations is called a **system of linear equations** or, more briefly, a **linear system**. The variables are called **unknowns**. For example, system (5) that follows has unknowns x and y, and system (6) has unknowns x_1, x_2, and x_3.

$$\begin{array}{ll} 5x + y = 3 & 4x_1 - x_2 + 3x_3 = -1 \\ 2x - y = 4 & 3x_1 + x_2 + 9x_3 = -4 \end{array} \tag{5–6}$$

A general linear system of m equations in the n unknowns $x_1, x_2, \ldots, x_n$ can be written as

$$
\begin{aligned}
a_{11}x_1 + a_{12}x_2 + \cdots + a_{1n}x_n &= b_1 \\
a_{21}x_1 + a_{22}x_2 + \cdots + a_{2n}x_n &= b_2 \\
\vdots \qquad \vdots \qquad\qquad \vdots \qquad &\ \ \vdots \\
a_{m1}x_1 + a_{m2}x_2 + \cdots + a_{mn}x_n &= b_m
\end{aligned}
\tag{7}
$$

A **solution** of a linear system in n unknowns $x_1, x_2, \ldots, x_n$ is a sequence of n numbers $s_1, s_2, \ldots, s_n$ for which the substitution

$$x_1 = s_1, \quad x_2 = s_2, \ldots, \quad x_n = s_n$$

makes each equation a true statement. For example, the system in (5) has the solution

$$x = 1, \quad y = -2$$

and the system in (6) has the solution

$$x_1 = 1, \quad x_2 = 2, \quad x_3 = -1$$

These solutions can be written more succinctly as

$$(1, -2) \quad \text{and} \quad (1, 2, -1)$$

in which the names of the variables are omitted. This notation allows us to interpret these solutions geometrically as points in two-dimensional and three-dimensional space. More generally, a solution

$$x_1 = s_1, \quad x_2 = s_2, \ldots, \quad x_n = s_n$$

of a linear system in n unknowns can be written as

$$(s_1, s_2, \ldots, s_n)$$

which is called an **ordered n-tuple**. With this notation it is understood that all variables appear in the same order in each equation. If $n = 2$, then the n-tuple is called an **ordered pair**, and if $n = 3$, then it is called an **ordered triple**.

Linear Systems with Two and Three Unknowns

Linear systems in two unknowns arise in connection with intersections of lines. For example, consider the linear system

$$
\begin{aligned}
a_1 x + b_1 y &= c_1 \\
a_2 x + b_2 y &= c_2
\end{aligned}
$$

in which the graphs of the equations are lines in the xy-plane. Each solution (x, y) of this system corresponds to a point of intersection of the lines, so there are three possibilities (Figure 1.1.1):

1. The lines may be parallel and distinct, in which case there is no intersection and consequently no solution.

2. The lines may intersect at only one point, in which case the system has exactly one solution.

3. The lines may coincide, in which case there are infinitely many points of intersection (the points on the common line) and consequently infinitely many solutions.

In general, we say that a linear system is **consistent** if it has at least one solution and **inconsistent** if it has no solutions. Thus, a *consistent* linear system of two equations in

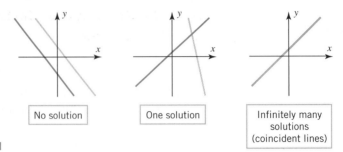

No solution   One solution   Infinitely many solutions (coincident lines)

▶ Figure 1.1.1

two unknowns has either one solution or infinitely many solutions—there are no other possibilities. The same is true for a linear system of three equations in three unknowns

$$a_1x + b_1y + c_1z = d_1$$
$$a_2x + b_2y + c_2z = d_2$$
$$a_3x + b_3y + c_3z = d_3$$

in which the graphs of the equations are planes. The solutions of the system, if any, correspond to points where all three planes intersect, so again we see that there are only three possibilities—no solutions, one solution, or infinitely many solutions (Figure 1.1.2).

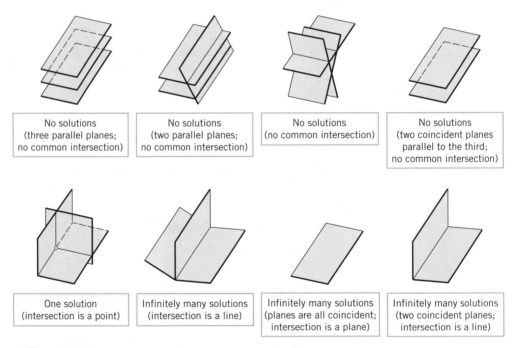

No solutions (three parallel planes; no common intersection)   No solutions (two parallel planes; no common intersection)   No solutions (no common intersection)   No solutions (two coincident planes parallel to the third; no common intersection)

One solution (intersection is a point)   Infinitely many solutions (intersection is a line)   Infinitely many solutions (planes are all coincident; intersection is a plane)   Infinitely many solutions (two coincident planes; intersection is a line)

▲ Figure 1.1.2

We will prove later that our observations about the number of solutions of linear systems of two equations in two unknowns and linear systems of three equations in three unknowns actually hold for *all* linear systems. That is:

Every system of linear equations has zero, one, or infinitely many solutions. There are no other possibilities.

▶ EXAMPLE 2 **A Linear System with One Solution**

Solve the linear system

$$x - y = 1$$
$$2x + y = 6$$

Solution We can eliminate x from the second equation by adding -2 times the first equation to the second. This yields the simplified system

$$x - y = 1$$
$$3y = 4$$

From the second equation we obtain $y = \frac{4}{3}$, and on substituting this value in the first equation we obtain $x = 1 + y = \frac{7}{3}$. Thus, the system has the unique solution

$$x = \tfrac{7}{3}, \quad y = \tfrac{4}{3}$$

Geometrically, this means that the lines represented by the equations in the system intersect at the single point $\left(\frac{7}{3}, \frac{4}{3}\right)$. We leave it for you to check this by graphing the lines.

▶ EXAMPLE 3 **A Linear System with No Solutions**

Solve the linear system

$$x + y = 4$$
$$3x + 3y = 6$$

Solution We can eliminate x from the second equation by adding -3 times the first equation to the second equation. This yields the simplified system

$$x + y = 4$$
$$0 = -6.$$

The second equation is contradictory, so the given system has no solution. Geometrically, this means that the lines corresponding to the equations in the original system are parallel and distinct. We leave it for you to check this by graphing the lines or by showing that they have the same slope but different y-intercepts.

▶ EXAMPLE 4 **A Linear System with Infinitely Many Solutions**

Solve the linear system

$$4x - 2y = 1$$
$$16x - 8y = 4$$

Solution We can eliminate x from the second equation by adding -4 times the first equation to the second. This yields the simplified system

$$4x - 2y = 1$$
$$0 = 0$$

The second equation does not impose any restrictions on x and y and hence can be omitted. Thus, the solutions of the system are those values of x and y that satisfy the single equation

$$4x - 2y = 1 \tag{8}$$

Geometrically, this means the lines corresponding to the two equations in the original system coincide. One way to describe the solution set is to solve this equation for x in terms of y to obtain $x = \frac{1}{4} + \frac{1}{2}y$ and then assign an arbitrary value t (called a ***parameter***)

In Example 4 we could have also obtained parametric equations for the solutions by solving (8) for y in terms of x, and letting $x = t$ be the parameter. The resulting parametric equations would look different but would define the same solution set.

to y. This allows us to express the solution by the pair of equations (called **parametric equations**)

$$x = \tfrac{1}{4} + \tfrac{1}{2}t, \quad y = t$$

We can obtain specific numerical solutions from these equations by substituting numerical values for the parameter. For example, $t = 0$ yields the solution $\left(\tfrac{1}{4}, 0\right)$, $t = 1$ yields the solution $\left(\tfrac{3}{4}, 1\right)$, and $t = -1$ yields the solution $\left(-\tfrac{1}{4}, -1\right)$. You can confirm that these are solutions by substituting the coordinates into the given equations.

▶ **EXAMPLE 5 A Linear System with Infinitely Many Solutions**

Solve the linear system

$$
\begin{aligned}
x - y + 2z &= 5 \\
2x - 2y + 4z &= 10 \\
3x - 3y + 6z &= 15
\end{aligned}
$$

Solution This system can be solved by inspection, since the second and third equations are multiples of the first. Geometrically, this means that the three planes coincide and that those values of x, y, and z that satisfy the equation

$$x - y + 2z = 5 \tag{9}$$

automatically satisfy all three equations. Thus, it suffices to find the solutions of (9). We can do this by first solving (9) for x in terms of y and z, then assigning arbitrary values r and s (parameters) to these two variables, and then expressing the solution by the three parametric equations

$$x = 5 + r - 2s, \quad y = r, \quad z = s$$

Specific solutions can be obtained by choosing numerical values for the parameters r and s. For example, taking $r = 1$ and $s = 0$ yields the solution $(6, 1, 0)$. ◀

Augmented Matrices and Elementary Row Operations

As the number of equations and unknowns in a linear system increases, so does the complexity of the algebra involved in finding solutions. The required computations can be made more manageable by simplifying notation and standardizing procedures. For example, by mentally keeping track of the location of the $+$'s, the x's, and the $=$'s in the linear system

$$
\begin{aligned}
a_{11}x_1 + a_{12}x_2 + \cdots + a_{1n}x_n &= b_1 \\
a_{21}x_1 + a_{22}x_2 + \cdots + a_{2n}x_n &= b_2 \\
\vdots \quad\quad \vdots \quad\quad\quad\quad \vdots \quad\quad \vdots \\
a_{m1}x_1 + a_{m2}x_2 + \cdots + a_{mn}x_n &= b_m
\end{aligned}
$$

we can abbreviate the system by writing only the rectangular array of numbers

$$
\begin{bmatrix}
a_{11} & a_{12} & \cdots & a_{1n} & b_1 \\
a_{21} & a_{22} & \cdots & a_{2n} & b_2 \\
\vdots & \vdots & & \vdots & \vdots \\
a_{m1} & a_{m2} & \cdots & a_{mn} & b_m
\end{bmatrix}
$$

As noted in the introduction to this chapter, the term "matrix" is used in mathematics to denote a rectangular array of numbers. In a later section we will study matrices in detail, but for now we will only be concerned with augmented matrices for linear systems.

This is called the **augmented matrix** for the system. For example, the augmented matrix for the system of equations

$$
\begin{aligned}
x_1 + x_2 + 2x_3 &= 9 \\
2x_1 + 4x_2 - 3x_3 &= 1 \\
3x_1 + 6x_2 - 5x_3 &= 0
\end{aligned}
\qquad \text{is} \qquad
\begin{bmatrix}
1 & 1 & 2 & 9 \\
2 & 4 & -3 & 1 \\
3 & 6 & -5 & 0
\end{bmatrix}
$$

The basic method for solving a linear system is to perform algebraic operations on the system that do not alter the solution set and that produce a succession of increasingly simpler systems, until a point is reached where it can be ascertained whether the system is consistent, and if so, what its solutions are. Typically, the algebraic operations are as follows:

1. Multiply an equation through by a nonzero constant.

2. Interchange two equations.

3. Add a constant times one equation to another.

Since the rows (horizontal lines) of an augmented matrix correspond to the equations in the associated system, these three operations correspond to the following operations on the rows of the augmented matrix:

1. Multiply a row through by a nonzero constant.

2. Interchange two rows.

3. Add a constant times one row to another.

These are called *elementary row operations* on a matrix.

In the following example we will illustrate how to use elementary row operations and an augmented matrix to solve a linear system in three unknowns. Since a systematic procedure for solving linear systems will be developed in the next section, do not worry about how the steps in the example were chosen. Your objective here should be simply to understand the computations.

▶ **EXAMPLE 6 Using Elementary Row Operations**

In the left column we solve a system of linear equations by operating on the equations in the system, and in the right column we solve the same system by operating on the rows of the augmented matrix.

$$\begin{aligned} x + y + 2z &= 9 \\ 2x + 4y - 3z &= 1 \\ 3x + 6y - 5z &= 0 \end{aligned} \qquad \begin{bmatrix} 1 & 1 & 2 & 9 \\ 2 & 4 & -3 & 1 \\ 3 & 6 & -5 & 0 \end{bmatrix}$$

Add -2 times the first equation to the second to obtain

Add -2 times the first row to the second to obtain

$$\begin{aligned} x + y + 2z &= 9 \\ 2y - 7z &= -17 \\ 3x + 6y - 5z &= 0 \end{aligned} \qquad \begin{bmatrix} 1 & 1 & 2 & 9 \\ 0 & 2 & -7 & -17 \\ 3 & 6 & -5 & 0 \end{bmatrix}$$

Historical Note The first known use of augmented matrices appeared between 200 B.C. and 100 B.C. in a Chinese manuscript entitled *Nine Chapters of Mathematical Art*. The coefficients were arranged in columns rather than in rows, as today, but remarkably the system was solved by performing a succession of operations on the columns. The actual use of the term *augmented matrix* appears to have been introduced by the American mathematician Maxime Bôcher in his book *Introduction to Higher Algebra*, published in 1907. In addition to being an outstanding research mathematician and an expert in Latin, chemistry, philosophy, zoology, geography, meteorology, art, and music, Bôcher was an outstanding expositor of mathematics whose elementary textbooks were greatly appreciated by students and are still in demand today.

[*Image: Courtesy of the American Mathematical Society*]

Maxime Bôcher
(1867–1918)

Add -3 times the first equation to the third to obtain

$$
\begin{aligned}
x + y + 2z &= 9 \\
2y - 7z &= -17 \\
3y - 11z &= -27
\end{aligned}
$$

Add -3 times the first row to the third to obtain

$$
\begin{bmatrix}
1 & 1 & 2 & 9 \\
0 & 2 & -7 & -17 \\
0 & 3 & -11 & -27
\end{bmatrix}
$$

Multiply the second equation by $\frac{1}{2}$ to obtain

$$
\begin{aligned}
x + y + 2z &= 9 \\
y - \tfrac{7}{2}z &= -\tfrac{17}{2} \\
3y - 11z &= -27
\end{aligned}
$$

Multiply the second row by $\frac{1}{2}$ to obtain

$$
\begin{bmatrix}
1 & 1 & 2 & 9 \\
0 & 1 & -\tfrac{7}{2} & -\tfrac{17}{2} \\
0 & 3 & -11 & -27
\end{bmatrix}
$$

Add -3 times the second equation to the third to obtain

$$
\begin{aligned}
x + y + 2z &= 9 \\
y - \tfrac{7}{2}z &= -\tfrac{17}{2} \\
-\tfrac{1}{2}z &= -\tfrac{3}{2}
\end{aligned}
$$

Add -3 times the second row to the third to obtain

$$
\begin{bmatrix}
1 & 1 & 2 & 9 \\
0 & 1 & -\tfrac{7}{2} & -\tfrac{17}{2} \\
0 & 0 & -\tfrac{1}{2} & -\tfrac{3}{2}
\end{bmatrix}
$$

Multiply the third equation by -2 to obtain

$$
\begin{aligned}
x + y + 2z &= 9 \\
y - \tfrac{7}{2}z &= -\tfrac{17}{2} \\
z &= 3
\end{aligned}
$$

Multiply the third row by -2 to obtain

$$
\begin{bmatrix}
1 & 1 & 2 & 9 \\
0 & 1 & -\tfrac{7}{2} & -\tfrac{17}{2} \\
0 & 0 & 1 & 3
\end{bmatrix}
$$

Add -1 times the second equation to the first to obtain

$$
\begin{aligned}
x + \tfrac{11}{2}z &= \tfrac{35}{2} \\
y - \tfrac{7}{2}z &= -\tfrac{17}{2} \\
z &= 3
\end{aligned}
$$

Add -1 times the second row to the first to obtain

$$
\begin{bmatrix}
1 & 0 & \tfrac{11}{2} & \tfrac{35}{2} \\
0 & 1 & -\tfrac{7}{2} & -\tfrac{17}{2} \\
0 & 0 & 1 & 3
\end{bmatrix}
$$

Add $-\frac{11}{2}$ times the third equation to the first and $\frac{7}{2}$ times the third equation to the second to obtain

$$
\begin{aligned}
x &= 1 \\
y &= 2 \\
z &= 3
\end{aligned}
$$

Add $-\frac{11}{2}$ times the third row to the first and $\frac{7}{2}$ times the third row to the second to obtain

$$
\begin{bmatrix}
1 & 0 & 0 & 1 \\
0 & 1 & 0 & 2 \\
0 & 0 & 1 & 3
\end{bmatrix}
$$

The solution $x = 1$, $y = 2$, $z = 3$ is now evident. ◀

Concept Review

- Linear equation
- Homogeneous linear equation
- System of linear equations
- Solution of a linear system

- Ordered n-tuple
- Consistent linear system
- Inconsistent linear system
- Parameter

- Parametric equations
- Augmented matrix
- Elementary row operations

Skills

- Determine whether a given equation is linear.
- Determine whether a given n-tuple is a solution of a linear system.
- Find the augmented matrix of a linear system.
- Find the linear system corresponding to a given augmented matrix.
- Perform elementary row operations on a linear system and on its corresponding augmented matrix.
- Determine whether a linear system is consistent or inconsistent.
- Find the set of solutions to a consistent linear system.

Exercise Set 1.1

1. In each part, determine whether the equation is linear in x_1, x_2, and x_3.

   (a) $x_1 + 5x_2 - \sqrt{2}x_3 = 1$ (b) $x_1 + 3x_2 + x_1x_3 = 2$

   (c) $x_1 = -7x_2 + 3x_3$ (d) $x_1^{-2} + x_2 + 8x_3 = 5$

   (e) $x_1^{3/5} - 2x_2 + x_3 = 4$

   (f) $\pi x_1 - \sqrt{2}x_2 + \frac{1}{3}x_3 = 7^{1/3}$

2. In each part, determine whether the equations form a linear system.

   (a) $-2x + 4y + z = 2$ (b) $x = 4$

   $3x - \dfrac{2}{y} = 0$ $2x = 8$

   (c) $4x - y + 2z = -1$

   $-x + (\ln 2)y - 3z = 0$

   (d) $3z + x = -4$

   $y + 5z = 1$

   $6x + 2z = 3$

   $-x - y - z = 4$

3. In each part, determine whether the equations form a linear system.

   (a) $2x_1$ $- x_4 = 5$

   $-x_1 + 5x_2 + 3x_3 - 2x_4 = -1$

   (b) $\sin(2x_1 + x_3) = \sqrt{5}$

   $e^{-2x_2 - 2x_4} = \dfrac{1}{x_2}$

   $4x_4 = 4$

   (c) $7x_1 - x_2 + 2x_3 = 0$ (d) $x_1 + x_2 = x_3 + x_4$

   $2x_1 + x_2 - x_3x_4 = 3$

   $-x_1 + 5x_2 - x_4 = -1$

4. For each system in Exercise 2 that is linear, determine whether it is consistent.

5. For each system in Exercise 3 that is linear, determine whether it is consistent.

6. Write a system of linear equations consisting of three equations in three unknowns with

   (a) no solutions.

   (b) exactly one solution.   3 leading 1

   (c) infinitely many solutions.

7. In each part, determine whether the given vector is a solution of the linear system

   $$2x_1 - 4x_2 - x_3 = 1$$
   $$x_1 - 3x_2 + x_3 = 1$$
   $$3x_1 - 5x_2 - 3x_3 = 1$$

   (a) $(3, 1, 1)$ (b) $(3, -1, 1)$ (c) $(13, 5, 2)$

   (d) $\left(\frac{13}{2}, \frac{5}{2}, 2\right)$ (e) $(17, 7, 5)$

8. In each part, determine whether the given vector is a solution of the linear system

   $$x_1 + 2x_2 - 2x_3 = 3$$
   $$3x_1 - x_2 + x_3 = 1$$
   $$-x_1 + 5x_2 - 5x_3 = 5$$

   (a) $\left(\frac{5}{7}, \frac{8}{7}, 1\right)$ (b) $\left(\frac{5}{7}, \frac{8}{7}, 0\right)$ (c) $(5, 8, 1)$

   (d) $\left(\frac{5}{7}, \frac{10}{7}, \frac{2}{7}\right)$ (e) $\left(\frac{5}{7}, \frac{22}{7}, 2\right)$

9. In each part, find the solution set of the linear equation by using parameters as necessary.

   (a) $7x - 5y = 3$

   (b) $-8x_1 + 2x_2 - 5x_3 + 6x_4 = 1$

10. In each part, find the solution set of the linear equation by using parameters as necessary.

    (a) $3x_1 - 5x_2 + 4x_3 = 7$

    (b) $3v - 8w + 2x - y + 4z = 0$

11. In each part, find a system of linear equations corresponding to the given augmented matrix.

(a) $\begin{bmatrix} 2 & 0 & 0 \\ 3 & -4 & 0 \\ 0 & 1 & 1 \end{bmatrix}$

(b) $\begin{bmatrix} 3 & 0 & -2 & 5 \\ 7 & 1 & 4 & -3 \\ 0 & -2 & 1 & 7 \end{bmatrix}$

(c) $\begin{bmatrix} 7 & 2 & 1 & -3 & 5 \\ 1 & 2 & 4 & 0 & 1 \end{bmatrix}$

(d) $\begin{bmatrix} 1 & 0 & 0 & 0 & 7 \\ 0 & 1 & 0 & 0 & -2 \\ 0 & 0 & 1 & 0 & 3 \\ 0 & 0 & 0 & 1 & 4 \end{bmatrix}$

12. In each part, find a system of linear equations corresponding to the given augmented matrix.

(a) $\begin{bmatrix} 2 & -1 \\ -4 & -6 \\ 1 & -1 \\ 3 & 0 \end{bmatrix}$

(b) $\begin{bmatrix} 0 & 3 & -1 & -1 & -1 \\ 5 & 2 & 0 & -3 & -6 \end{bmatrix}$

(c) $\begin{bmatrix} 1 & 2 & 3 & 4 \\ -4 & -3 & -2 & -1 \\ 5 & -6 & 1 & 1 \\ -8 & 0 & 0 & 3 \end{bmatrix}$

(d) $\begin{bmatrix} 3 & 0 & 1 & -4 & 3 \\ -4 & 0 & 4 & 1 & -3 \\ -1 & 3 & 0 & -2 & -9 \\ 0 & 0 & 0 & -1 & -2 \end{bmatrix}$

13. In each part, find the augmented matrix for the given system of linear equations.

(a) $-2x_1 = 6$
$3x_1 = 8$
$9x_1 = -3$

(b) $6x_1 - x_2 + 3x_3 = 4$
$5x_2 - x_3 = 1$

(c) $2x_2 - 3x_4 + x_5 = 0$
$-3x_1 - x_2 + x_3 = -1$
$6x_1 + 2x_2 - x_3 + 2x_4 - 3x_5 = 6$

(d) $x_1 - x_5 = 7$

14. In each part, find the augmented matrix for the given system of linear equations.

(a) $3x_1 - 2x_2 = -1$
$4x_1 + 5x_2 = 3$
$7x_1 + 3x_2 = 2$

(b) $2x_1 + 2x_3 = 1$
$3x_1 - x_2 + 4x_3 = 7$
$6x_1 + x_2 - x_3 = 0$

(c) $x_1 + 2x_2 - x_4 + x_5 = 1$
$3x_2 + x_3 - x_5 = 2$
$x_3 + 7x_4 = 1$

(d) $x_1 = 1$
$x_2 = 2$
$x_3 = 3$

15. The curve $y = ax^2 + bx + c$ shown in the accompanying figure passes through the points (x_1, y_1), (x_2, y_2), and (x_3, y_3).

Show that the coefficients a, b, and c are a solution of the system of linear equations whose augmented matrix is

$$\begin{bmatrix} x_1^2 & x_1 & 1 & y_1 \\ x_2^2 & x_2 & 1 & y_2 \\ x_3^2 & x_3 & 1 & y_3 \end{bmatrix}$$

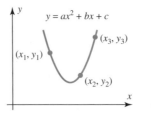

◀ **Figure Ex-15**

16. Explain why each of the three elementary row operations does not affect the solution set of a linear system.

17. Show that if the linear equations

$$x_1 + kx_2 = c \quad \text{and} \quad x_1 + lx_2 = d$$

have the same solution set, then the two equations are identical (i.e., $k = l$ and $c = d$).

True-False Exercises

In parts (a)–(h) determine whether the statement is true or false, and justify your answer.

(a) A linear system whose equations are all homogeneous must be consistent.

(b) Multiplying a linear equation through by zero is an acceptable elementary row operation.

(c) The linear system

$$x - y = 3$$
$$2x - 2y = k$$

cannot have a unique solution, regardless of the value of k.

(d) A single linear equation with two or more unknowns must always have infinitely many solutions.

(e) If the number of equations in a linear system exceeds the number of unknowns, then the system must be inconsistent.

(f) If each equation in a consistent linear system is multiplied through by a constant c, then all solutions to the new system can be obtained by multiplying solutions from the original system by c.

(g) Elementary row operations permit one equation in a linear system to be subtracted from another.

(h) The linear system with corresponding augmented matrix

$$\begin{bmatrix} 2 & -1 & 4 \\ 0 & 0 & -1 \end{bmatrix}$$

is consistent.

1.2 Gaussian Elimination

In this section we will develop a systematic procedure for solving systems of linear equations. The procedure is based on the idea of performing certain operations on the rows of the augmented matrix that simplifies it to a form from which the solution of the system can be ascertained by inspection.

Considerations in Solving Linear Systems

When considering methods for solving systems of linear equations, it is important to distinguish between large systems that must be solved by computer and small systems that can be solved by hand. For example, there are many applications that lead to linear systems in thousands or even millions of unknowns. Large systems require special techniques to deal with issues of memory size, roundoff errors, solution time, and so forth. Such techniques are studied in the field of *numerical analysis* and will only be touched on in this text. However, almost all of the methods that are used for large systems are based on the ideas that we will develop in this section.

Echelon Forms

In Example 6 of the last section, we solved a linear system in the unknowns x, y, and z by reducing the augmented matrix to the form

$$\begin{bmatrix} 1 & 0 & 0 & 1 \\ 0 & 1 & 0 & 2 \\ 0 & 0 & 1 & 3 \end{bmatrix}$$

from which the solution $x = 1$, $y = 2$, $z = 3$ became evident. This is an example of a matrix that is in *reduced row echelon form*. To be of this form, a matrix must have the following properties:

1. If a row does not consist entirely of zeros, then the first nonzero number in the row is a 1. We call this a *leading 1*.

2. If there are any rows that consist entirely of zeros, then they are grouped together at the bottom of the matrix.

3. In any two successive rows that do not consist entirely of zeros, the leading 1 in the lower row occurs farther to the right than the leading 1 in the higher row.

4. Each column that contains a leading 1 has zeros everywhere else in that column.

A matrix that has the first three properties is said to be in *row echelon form*. (Thus, a matrix in reduced row echelon form is of necessity in row echelon form, but not conversely.)

▶ **EXAMPLE 1 Row Echelon and Reduced Row Echelon Form**

The following matrices are in reduced row echelon form.

$$\begin{bmatrix} 1 & 0 & 0 & 4 \\ 0 & 1 & 0 & 7 \\ 0 & 0 & 1 & -1 \end{bmatrix}, \quad \begin{bmatrix} 1 & 0 & 0 \\ 0 & 1 & 0 \\ 0 & 0 & 1 \end{bmatrix}, \quad \begin{bmatrix} 0 & 1 & -2 & 0 & 1 \\ 0 & 0 & 0 & 1 & 3 \\ 0 & 0 & 0 & 0 & 0 \\ 0 & 0 & 0 & 0 & 0 \end{bmatrix}, \quad \begin{bmatrix} 0 & 0 \\ 0 & 0 \end{bmatrix}$$

The following matrices are in row echelon form but not reduced row echelon form.

$$\begin{bmatrix} 1 & 4 & -3 & 7 \\ 0 & 1 & 6 & 2 \\ 0 & 0 & 1 & 5 \end{bmatrix}, \quad \begin{bmatrix} 1 & 1 & 0 \\ 0 & 1 & 0 \\ 0 & 0 & 0 \end{bmatrix}, \quad \begin{bmatrix} 0 & 1 & 2 & 6 & 0 \\ 0 & 0 & 1 & -1 & 0 \\ 0 & 0 & 0 & 0 & 1 \end{bmatrix}$$

▶ **EXAMPLE 2 More on Row Echelon and Reduced Row Echelon Form**

As Example 1 illustrates, a matrix in row echelon form has zeros below each leading 1, whereas a matrix in reduced row echelon form has zeros below *and above* each leading 1. Thus, with any real numbers substituted for the $*$'s, all matrices of the following types are in row echelon form:

$$\begin{bmatrix} 1 & * & * & * \\ 0 & 1 & * & * \\ 0 & 0 & 1 & * \\ 0 & 0 & 0 & 1 \end{bmatrix}, \begin{bmatrix} 1 & * & * & * \\ 0 & 1 & * & * \\ 0 & 0 & 1 & * \\ 0 & 0 & 0 & 0 \end{bmatrix}, \begin{bmatrix} 1 & * & * & * \\ 0 & 1 & * & * \\ 0 & 0 & 0 & 0 \\ 0 & 0 & 0 & 0 \end{bmatrix}, \begin{bmatrix} 0 & 1 & * & * & * & * & * & * & * \\ 0 & 0 & 0 & 1 & * & * & * & * & * \\ 0 & 0 & 0 & 0 & 1 & * & * & * & * \\ 0 & 0 & 0 & 0 & 0 & 1 & * & * & * \\ 0 & 0 & 0 & 0 & 0 & 0 & 0 & 1 & * \end{bmatrix}$$

All matrices of the following types are in reduced row echelon form:

$$\begin{bmatrix} 1 & 0 & 0 & 0 \\ 0 & 1 & 0 & 0 \\ 0 & 0 & 1 & 0 \\ 0 & 0 & 0 & 1 \end{bmatrix}, \begin{bmatrix} 1 & 0 & 0 & * \\ 0 & 1 & 0 & * \\ 0 & 0 & 1 & * \\ 0 & 0 & 0 & 0 \end{bmatrix}, \begin{bmatrix} 1 & 0 & * & * \\ 0 & 1 & * & * \\ 0 & 0 & 0 & 0 \\ 0 & 0 & 0 & 0 \end{bmatrix}, \begin{bmatrix} 0 & 1 & * & 0 & 0 & 0 & * & * & 0 & * \\ 0 & 0 & 0 & 1 & 0 & 0 & * & * & 0 & * \\ 0 & 0 & 0 & 0 & 1 & 0 & * & * & 0 & * \\ 0 & 0 & 0 & 0 & 0 & 1 & * & * & 0 & * \\ 0 & 0 & 0 & 0 & 0 & 0 & 0 & 0 & 1 & * \end{bmatrix} ◀$$

If, by a sequence of elementary row operations, the augmented matrix for a system of linear equations is put in *reduced* row echelon form, then the solution set can be obtained either by inspection or by converting certain linear equations to parametric form. Here are some examples.

▶ **EXAMPLE 3 Unique Solution**

Suppose that the augmented matrix for a linear system in the unknowns x_1, x_2, x_3, and x_4 has been reduced by elementary row operations to

$$\begin{bmatrix} 1 & 0 & 0 & 0 & 3 \\ 0 & 1 & 0 & 0 & -1 \\ 0 & 0 & 1 & 0 & 0 \\ 0 & 0 & 0 & 1 & 5 \end{bmatrix}$$

This matrix is in reduced row echelon form and corresponds to the equations

$$\begin{aligned} x_1 & & & &= 3 \\ & x_2 & & &= -1 \\ & & x_3 & &= 0 \\ & & & x_4 &= 5 \end{aligned}$$

In Example 3 we could, if desired, express the solution more succinctly as the 4-tuple $(3, -1, 0, 5)$.

Thus, the system has a unique solution, namely, $x_1 = 3$, $x_2 = -1$, $x_3 = 0$, $x_4 = 5$.

▶ **EXAMPLE 4 Linear Systems in Three Unknowns**

In each part, suppose that the augmented matrix for a linear system in the unknowns x, y, and z has been reduced by elementary row operations to the given reduced row echelon form. Solve the system.

$$(a)\begin{bmatrix} 1 & 0 & 0 & 0 \\ 0 & 1 & 2 & 0 \\ 0 & 0 & 0 & 1 \end{bmatrix} \quad (b)\begin{bmatrix} 1 & 0 & 3 & -1 \\ 0 & 1 & -4 & 2 \\ 0 & 0 & 0 & 0 \end{bmatrix} \quad (c)\begin{bmatrix} 1 & -5 & 1 & 4 \\ 0 & 0 & 0 & 0 \\ 0 & 0 & 0 & 0 \end{bmatrix}$$

Solution (a) The equation that corresponds to the last row of the augmented matrix is

$$0x + 0y + 0z = 1$$

Since this equation is not satisfied by any values of x, y, and z, the system is inconsistent.

Solution (b) The equation that corresponds to the last row of the augmented matrix is

$$0x + 0y + 0z = 0$$

This equation can be omitted since it imposes no restrictions on x, y, and z; hence, the linear system corresponding to the augmented matrix is

$$
\begin{aligned}
x \quad\quad + 3z &= -1 \\
y - 4z &= 2
\end{aligned}
$$

Since x and y correspond to the leading 1's in the augmented matrix, we call these the *leading variables*. The remaining variables (in this case z) are called *free variables*. Solving for the leading variables in terms of the free variables gives

$$
\begin{aligned}
x &= -1 - 3z \\
y &= 2 + 4z
\end{aligned}
$$

From these equations we see that the free variable z can be treated as a parameter and assigned an arbitrary value t, which then determines values for x and y. Thus, the solution set can be represented by the parametric equations

$$x = -1 - 3t, \quad y = 2 + 4t, \quad z = t$$

By substituting various values for t in these equations we can obtain various solutions of the system. For example, setting $t = 0$ yields the solution

$$x = -1, \quad y = 2, \quad z = 0$$

and setting $t = 1$ yields the solution

$$x = -4, \quad y = 6, \quad z = 1$$

Solution (c) As explained in part (b), we can omit the equations corresponding to the zero rows, in which case the linear system associated with the augmented matrix consists of the single equation

$$x - 5y + z = 4 \tag{1}$$

from which we see that the solution set is a plane in three-dimensional space. Although (1) is a valid form of the solution set, there are many applications in which it is preferable to express the solution set in parametric form. We can convert (1) to parametric form by solving for the leading variable x in terms of the free variables y and z to obtain

$$x = 4 + 5y - z$$

From this equation we see that the free variables can be assigned arbitrary values, say $y = s$ and $z = t$, which then determine the value of x. Thus, the solution set can be expressed parametrically as

$$x = 4 + 5s - t, \quad y = s, \quad z = t \quad \blacktriangleleft \tag{2}$$

We will usually denote parameters in a general solution by the letters $r, s, t, \ldots$, but any letters that do not conflict with the names of the unknowns can be used. For systems with more than three unknowns, subscripted letters such as $t_1, t_2, t_3, \ldots$ are convenient.

Formulas, such as (2), that express the solution set of a linear system parametrically have some associated terminology.

DEFINITION 1 If a linear system has infinitely many solutions, then a set of parametric equations from which all solutions can be obtained by assigning numerical values to the parameters is called a *general solution* of the system.

Elimination Methods   We have just seen how easy it is to solve a system of linear equations once its augmented matrix is in reduced row echelon form. Now we will give a step-by-step *elimination procedure* that can be used to reduce any matrix to reduced row echelon form. As we state each step in the procedure, we illustrate the idea by reducing the following matrix to reduced row echelon form.

$$\begin{bmatrix} 0 & 0 & -2 & 0 & 7 & 12 \\ 2 & 4 & -10 & 6 & 12 & 28 \\ 2 & 4 & -5 & 6 & -5 & -1 \end{bmatrix}$$

Step 1.   Locate the leftmost column that does not consist entirely of zeros.

$$\begin{bmatrix} 0 & 0 & -2 & 0 & 7 & 12 \\ 2 & 4 & -10 & 6 & 12 & 28 \\ 2 & 4 & -5 & 6 & -5 & -1 \end{bmatrix}$$

$\uparrow$
Leftmost nonzero column

Step 2.   Interchange the top row with another row, if necessary, to bring a nonzero entry to the top of the column found in Step 1.

$$\begin{bmatrix} 2 & 4 & -10 & 6 & 12 & 28 \\ 0 & 0 & -2 & 0 & 7 & 12 \\ 2 & 4 & -5 & 6 & -5 & -1 \end{bmatrix}$$ ⟵ The first and second rows in the preceding matrix were interchanged.

Step 3.   If the entry that is now at the top of the column found in Step 1 is a, multiply the first row by $1/a$ in order to introduce a leading 1.

$$\begin{bmatrix} 1 & 2 & -5 & 3 & 6 & 14 \\ 0 & 0 & -2 & 0 & 7 & 12 \\ 2 & 4 & -5 & 6 & -5 & -1 \end{bmatrix}$$ ⟵ The first row of the preceding matrix was multiplied by $\frac{1}{2}$.

Step 4.   Add suitable multiples of the top row to the rows below so that all entries below the leading 1 become zeros.

$$\begin{bmatrix} 1 & 2 & -5 & 3 & 6 & 14 \\ 0 & 0 & -2 & 0 & 7 & 12 \\ 0 & 0 & 5 & 0 & -17 & -29 \end{bmatrix}$$ ⟵ -2 times the first row of the preceding matrix was added to the third row.

Step 5.   Now cover the top row in the matrix and begin again with Step 1 applied to the submatrix that remains. Continue in this way until the *entire* matrix is in row echelon form.

$$\begin{bmatrix} 1 & 2 & -5 & 3 & 6 & 14 \\ 0 & 0 & -2 & 0 & 7 & 12 \\ 0 & 0 & 5 & 0 & -17 & -29 \end{bmatrix}$$

$\uparrow$
**Leftmost nonzero column
in the submatrix**

$$\begin{bmatrix} 1 & 2 & -5 & 3 & 6 & 14 \\ 0 & 0 & 1 & 0 & -\frac{7}{2} & -6 \\ 0 & 0 & 5 & 0 & -17 & -29 \end{bmatrix}$$ ⟵ The first row in the submatrix was multiplied by $-\frac{1}{2}$ to introduce a leading 1.

$$\begin{bmatrix} 1 & 2 & -5 & 3 & 6 & 14 \\ 0 & 0 & 1 & 0 & -\frac{7}{2} & -6 \\ 0 & 0 & 0 & 0 & \frac{1}{2} & 1 \end{bmatrix}$$

← −5 times the first row of the submatrix was added to the second row of the submatrix to introduce a zero below the leading 1.

$$\begin{bmatrix} 1 & 2 & -5 & 3 & 6 & 14 \\ 0 & 0 & 1 & 0 & -\frac{7}{2} & -6 \\ 0 & 0 & 0 & 0 & \frac{1}{2} & 1 \end{bmatrix}$$

← The top row in the submatrix was covered, and we returned again to Step 1.

**⌐ Leftmost nonzero column
in the new submatrix**

$$\begin{bmatrix} 1 & 2 & -5 & 3 & 6 & 14 \\ 0 & 0 & 1 & 0 & -\frac{7}{2} & -6 \\ 0 & 0 & 0 & 0 & 1 & 2 \end{bmatrix}$$

← The first (and only) row in the new submatrix was multiplied by 2 to introduce a leading 1.

The *entire* matrix is now in row echelon form. To find the reduced row echelon form we need the following additional step.

Step 6. Beginning with the last nonzero row and working upward, add suitable multiples of each row to the rows above to introduce zeros above the leading 1's.

$$\begin{bmatrix} 1 & 2 & -5 & 3 & 6 & 14 \\ 0 & 0 & 1 & 0 & 0 & 1 \\ 0 & 0 & 0 & 0 & 1 & 2 \end{bmatrix}$$

← $\frac{7}{2}$ times the third row of the preceding matrix was added to the second row.

$$\begin{bmatrix} 1 & 2 & -5 & 3 & 0 & 2 \\ 0 & 0 & 1 & 0 & 0 & 1 \\ 0 & 0 & 0 & 0 & 1 & 2 \end{bmatrix}$$

← −6 times the third row was added to the first row.

$$\begin{bmatrix} 1 & 2 & 0 & 3 & 0 & 7 \\ 0 & 0 & 1 & 0 & 0 & 1 \\ 0 & 0 & 0 & 0 & 1 & 2 \end{bmatrix}$$

← 5 times the second row was added to the first row.

The last matrix is in reduced row echelon form.

The procedure (or algorithm) we have just described for reducing a matrix to reduced row echelon form is called ***Gauss–Jordan elimination***. This algorithm consists of two parts, a ***forward phase*** in which zeros are introduced below the leading 1's and a ***backward phase*** in which zeros are introduced above the leading 1's. If only the forward phase

**Carl Friedrich Gauss
(1777–1855)**

**Wilhelm Jordan
(1842–1899)**

Historical Note Although versions of Gaussian elimination were known much earlier, the power of the method was not recognized until the great German mathematician Carl Friedrich Gauss used it to compute the orbit of the asteroid Ceres from limited data. What happened was this: On January 1, 1801 the Sicilian astronomer Giuseppe Piazzi (1746–1826) noticed a dim celestial object that he believed might be a "missing planet." He named the object Ceres and made a limited number of positional observations but then lost the object as it neared the Sun. Gauss undertook the problem of computing the orbit from the limited data and the procedure that we now call Gaussian elimination. The work of Gauss caused a sensation when Ceres reappeared a year later in the constellation Virgo at almost the precise position that Gauss predicted! The method was further popularized by the German engineer Wilhelm Jordan in his handbook on geodesy (the science of measuring Earth shapes) entitled *Handbuch der Vermessungskunde* and published in 1888.
[*Images: Granger Collection (Gauss); wikipedia (Jordan)*]

is used, then the procedure produces a row echelon form only and is called *Gaussian elimination*. For example, in the preceding computations a row echelon form was obtained at the end of Step 5.

▶ **EXAMPLE 5 Gauss–Jordan Elimination**

Solve by Gauss–Jordan elimination.

$$\begin{array}{rcrcrcrcrcrcr} x_1 &+& 3x_2 &-& 2x_3 & & & +& 2x_5 & & &=& 0 \\ 2x_1 &+& 6x_2 &-& 5x_3 &-& 2x_4 &+& 4x_5 &-& 3x_6 &=& -1 \\ & & & & 5x_3 &+& 10x_4 & & &+& 15x_6 &=& 5 \\ 2x_1 &+& 6x_2 & & &+& 8x_4 &+& 4x_5 &+& 18x_6 &=& 6 \end{array}$$

Solution The augmented matrix for the system is

$$\begin{bmatrix} 1 & 3 & -2 & 0 & 2 & 0 & 0 \\ 2 & 6 & -5 & -2 & 4 & -3 & -1 \\ 0 & 0 & 5 & 10 & 0 & 15 & 5 \\ 2 & 6 & 0 & 8 & 4 & 18 & 6 \end{bmatrix}$$

Adding −2 times the first row to the second and fourth rows gives

$$\begin{bmatrix} 1 & 3 & -2 & 0 & 2 & 0 & 0 \\ 0 & 0 & -1 & -2 & 0 & -3 & -1 \\ 0 & 0 & 5 & 10 & 0 & 15 & 5 \\ 0 & 0 & 4 & 8 & 0 & 18 & 6 \end{bmatrix}$$

Multiplying the second row by −1 and then adding −5 times the new second row to the third row and −4 times the new second row to the fourth row gives

$$\begin{bmatrix} 1 & 3 & -2 & 0 & 2 & 0 & 0 \\ 0 & 0 & 1 & 2 & 0 & 3 & 1 \\ 0 & 0 & 0 & 0 & 0 & 0 & 0 \\ 0 & 0 & 0 & 0 & 0 & 6 & 2 \end{bmatrix}$$

Interchanging the third and fourth rows and then multiplying the third row of the resulting matrix by $\frac{1}{6}$ gives the row echelon form

$$\begin{bmatrix} 1 & 3 & -2 & 0 & 2 & 0 & 0 \\ 0 & 0 & 1 & 2 & 0 & 3 & 1 \\ 0 & 0 & 0 & 0 & 0 & 1 & \frac{1}{3} \\ 0 & 0 & 0 & 0 & 0 & 0 & 0 \end{bmatrix}$$

This completes the forward phase since there are zeros below the leading 1's.

Adding −3 times the third row to the second row and then adding 2 times the second row of the resulting matrix to the first row yields the reduced row echelon form

$$\begin{bmatrix} 1 & 3 & 0 & 4 & 2 & 0 & 0 \\ 0 & 0 & 1 & 2 & 0 & 0 & 0 \\ 0 & 0 & 0 & 0 & 0 & 1 & \frac{1}{3} \\ 0 & 0 & 0 & 0 & 0 & 0 & 0 \end{bmatrix}$$

This completes the backward phase since there are zeros above the leading 1's.

The corresponding system of equations is

$$x_1 + 3x_2 \qquad + 4x_4 + 2x_5 \qquad = 0$$
$$x_3 + 2x_4 \qquad\qquad = 0 \qquad (3)$$
$$x_6 = \tfrac{1}{3}$$

Note that in constructing the linear system in (3) we ignored the row of zeros in the corresponding augmented matrix. Why is this justified?

Solving for the leading variables we obtain

$$x_1 = -3x_2 - 4x_4 - 2x_5$$
$$x_3 = -2x_4$$
$$x_6 = \tfrac{1}{3}$$

Finally, we express the general solution of the system parametrically by assigning the free variables x_2, x_4, and x_5 arbitrary values r, s, and t, respectively. This yields

$$x_1 = -3r - 4s - 2t, \quad x_2 = r, \quad x_3 = -2s, \quad x_4 = s, \quad x_5 = t, \quad x_6 = \tfrac{1}{3} \;\blacktriangleleft$$

Homogeneous Linear Systems

A system of linear equations is said to be **homogeneous** if the constant terms are all zero; that is, the system has the form

$$a_{11}x_1 + a_{12}x_2 + \cdots + a_{1n}x_n = 0$$
$$a_{21}x_1 + a_{22}x_2 + \cdots + a_{2n}x_n = 0$$
$$\vdots \qquad \vdots \qquad\qquad \vdots \qquad \vdots$$
$$a_{m1}x_1 + a_{m2}x_2 + \cdots + a_{mn}x_n = 0$$

Every homogeneous system of linear equations is consistent because all such systems have $x_1 = 0, x_2 = 0, \ldots, x_n = 0$ as a solution. This solution is called the **trivial solution**; if there are other solutions, they are called **nontrivial solutions**.

Because a homogeneous linear system always has the trivial solution, there are only two possibilities for its solutions:

• The system has only the trivial solution.

• The system has infinitely many solutions in addition to the trivial solution.

In the special case of a homogeneous linear system of two equations in two unknowns, say

$$a_1 x + b_1 y = 0 \quad (a_1, b_1 \text{ not both zero})$$
$$a_2 x + b_2 y = 0 \quad (a_2, b_2 \text{ not both zero})$$

the graphs of the equations are lines through the origin, and the trivial solution corresponds to the point of intersection at the origin (Figure 1.2.1).

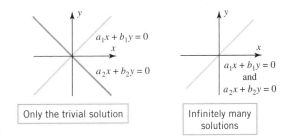

▶ Figure 1.2.1

There is one case in which a homogeneous system is assured of having nontrivial solutions—namely, whenever the system involves more unknowns than equations. To see why, consider the following example of four equations in six unknowns.

▶ **EXAMPLE 6 A Homogeneous System**

Use Gauss–Jordan elimination to solve the homogeneous linear system

$$\begin{array}{r}
x_1 + 3x_2 - 2x_3 + 2x_5 = 0 \\
2x_1 + 6x_2 - 5x_3 - 2x_4 + 4x_5 - 3x_6 = 0 \\
5x_3 + 10x_4 + 15x_6 = 0 \\
2x_1 + 6x_2 + 8x_4 + 4x_5 + 18x_6 = 0
\end{array} \tag{4}$$

Solution Observe first that the coefficients of the unknowns in this system are the same as those in Example 5; that is, the two systems differ only in the constants on the right side. The augmented matrix for the given homogeneous system is

$$\begin{bmatrix}
1 & 3 & -2 & 0 & 2 & 0 & 0 \\
2 & 6 & -5 & -2 & 4 & -3 & 0 \\
0 & 0 & 5 & 10 & 0 & 15 & 0 \\
2 & 6 & 0 & 8 & 4 & 18 & 0
\end{bmatrix} \tag{5}$$

which is the same as the augmented matrix for the system in Example 5, except for zeros in the last column. Thus, the reduced row echelon form of this matrix will be the same as that of the augmented matrix in Example 5, except for the last column. However, a moment's reflection will make it evident that a column of zeros is not changed by an elementary row operation, so the reduced row echelon form of (5) is

$$\begin{bmatrix}
1 & 3 & 0 & 4 & 2 & 0 & 0 \\
0 & 0 & 1 & 2 & 0 & 0 & 0 \\
0 & 0 & 0 & 0 & 0 & 1 & 0 \\
0 & 0 & 0 & 0 & 0 & 0 & 0
\end{bmatrix} \tag{6}$$

The corresponding system of equations is

$$\begin{array}{r}
x_1 + 3x_2 + 4x_4 + 2x_5 = 0 \\
x_3 + 2x_4 = 0 \\
x_6 = 0
\end{array}$$

Solving for the leading variables we obtain

$$\begin{array}{l}
x_1 = -3x_2 - 4x_4 - 2x_5 \\
x_3 = -2x_4 \\
x_6 = 0
\end{array} \tag{7}$$

If we now assign the free variables x_2, x_4, and x_5 arbitrary values r, s, and t, respectively, then we can express the solution set parametrically as

$$x_1 = -3r - 4s - 2t, \quad x_2 = r, \quad x_3 = -2s, \quad x_4 = s, \quad x_5 = t, \quad x_6 = 0$$

Note that the trivial solution results when $r = s = t = 0$. ◀

Free Variables in Homogeneous Linear Systems

Example 6 illustrates two important points about solving homogeneous linear systems:

1. Elementary row operations do not alter columns of zeros in a matrix, so the reduced row echelon form of the augmented matrix for a homogeneous linear system has a final column of zeros. This implies that the linear system corresponding to the reduced row echelon form is homogeneous, just like the original system.

2. When we constructed the homogeneous linear system corresponding to augmented matrix (6), we ignored the row of zeros because the corresponding equation

$$0x_1 + 0x_2 + 0x_3 + 0x_4 + 0x_5 + 0x_6 = 0$$

does not impose any conditions on the unknowns. Thus, depending on whether or not the reduced row echelon form of the augmented matrix for a homogeneous linear system has any rows of zero, the linear system corresponding to that reduced row echelon form will either have the same number of equations as the original system or it will have fewer.

Now consider a general homogeneous linear system with n unknowns, and suppose that the reduced row echelon form of the augmented matrix has r nonzero rows. Since each nonzero row has a leading 1, and since each leading 1 corresponds to a leading variable, the homogeneous system corresponding to the reduced row echelon form of the augmented matrix must have r leading variables and $n - r$ free variables. Thus, this system is of the form

$$
\begin{aligned}
x_{k_1} \quad\quad\quad\quad + \textstyle\sum(\ ) &= 0 \\
x_{k_2} \quad\quad\quad + \textstyle\sum(\ ) &= 0 \\
\ddots \quad\quad \vdots \quad\quad & \\
x_{k_r} + \textstyle\sum(\ ) &= 0
\end{aligned}
\tag{8}
$$

where in each equation the expression $\sum(\ )$ denotes a sum that involves the free variables, if any [see (7), for example]. In summary, we have the following result.

THEOREM 1.2.1 Free Variable Theorem for Homogeneous Systems

If a homogeneous linear system has n unknowns, and if the reduced row echelon form of its augmented matrix has r nonzero rows, then the system has n − r free variables.

Theorem 1.2.1 has an important implication for homogeneous linear systems with more unknowns than equations. Specifically, if a homogeneous linear system has m equations in n unknowns, and if $m < n$, then it must also be true that $r < n$ (why?). This being the case, the theorem implies that there is at least one free variable, and this implies that the system has infinitely many solutions. Thus, we have the following result.

Note that Theorem 1.2.2 applies only to homogeneous systems—a *nonhomogeneous* system with more unknowns than equations need not be consistent. However, we will prove later that if a nonhomogeneous system with more unknowns then equations is consistent, then it has infinitely many solutions.

THEOREM 1.2.2 *A homogeneous linear system with more unknowns than equations has infinitely many solutions.*

In retrospect, we could have anticipated that the homogeneous system in Example 6 would have infinitely many solutions since it has four equations in six unknowns.

Gaussian Elimination and
Back-Substitution

For small linear systems that are solved by hand (such as most of those in this text), Gauss–Jordan elimination (reduction to reduced row echelon form) is a good procedure to use. However, for large linear systems that require a computer solution, it is generally more efficient to use Gaussian elimination (reduction to row echelon form) followed by a technique known as **back-substitution** to complete the process of solving the system. The next example illustrates this technique.

▶ **EXAMPLE 7 Example 5 Solved by Back-Substitution**

From the computations in Example 5, a row echelon form of the augmented matrix is

$$
\begin{bmatrix}
1 & 3 & -2 & 0 & 2 & 0 & 0 \\
0 & 0 & 1 & 2 & 0 & 3 & 1 \\
0 & 0 & 0 & 0 & 0 & 1 & \frac{1}{3} \\
0 & 0 & 0 & 0 & 0 & 0 & 0
\end{bmatrix}
$$

To solve the corresponding system of equations

$$x_1 + 3x_2 - 2x_3 \qquad\quad + 2x_5 \qquad\qquad = 0$$
$$x_3 + 2x_4 \qquad\quad + 3x_6 = 1$$
$$x_6 = \tfrac{1}{3}$$

we proceed as follows:

Step 1.   Solve the equations for the leading variables.

$$x_1 = -3x_2 + 2x_3 - 2x_5$$
$$x_3 = 1 - 2x_4 - 3x_6$$
$$x_6 = \tfrac{1}{3}$$

Step 2.   Beginning with the bottom equation and working upward, successively substitute each equation into all the equations above it.
   Substituting $x_6 = \tfrac{1}{3}$ into the second equation yield

$$x_1 = -3x_2 + 2x_3 - 2x_5$$
$$x_3 = -2x_4$$
$$x_6 = \tfrac{1}{3}$$

Substituting $x_3 = -2x_4$ into the first equation yields

$$x_1 = -3x_2 - 4x_4 - 2x_5$$
$$x_3 = -2x_4$$
$$x_6 = \tfrac{1}{3}$$

Step 3.   Assign arbitrary values to the free variables, if any.
   If we now assign x_2, x_4, and x_5 the arbitrary values r, s, and t, respectively, the general solution is given by the formulas

$$x_1 = -3r - 4s - 2t, \quad x_2 = r, \quad x_3 = -2s, \quad x_4 = s, \quad x_5 = t, \quad x_6 = \tfrac{1}{3}$$

This agrees with the solution obtained in Example 5.

▶ **EXAMPLE 8**

Suppose that the matrices below are augmented matrices for linear systems in the unknowns x_1, x_2, x_3, and x_4. These matrices are all in row echelon form but not reduced row echelon form. Discuss the existence and uniqueness of solutions to the corresponding linear systems

$$(a) \begin{bmatrix} 1 & -3 & 7 & 2 & 5 \\ 0 & 1 & 2 & -4 & 1 \\ 0 & 0 & 1 & 6 & 9 \\ 0 & 0 & 0 & 0 & 1 \end{bmatrix} \quad (b) \begin{bmatrix} 1 & -3 & 7 & 2 & 5 \\ 0 & 1 & 2 & -4 & 1 \\ 0 & 0 & 1 & 6 & 9 \\ 0 & 0 & 0 & 0 & 0 \end{bmatrix} \quad (c) \begin{bmatrix} 1 & -3 & 7 & 2 & 5 \\ 0 & 1 & 2 & -4 & 1 \\ 0 & 0 & 1 & 6 & 9 \\ 0 & 0 & 0 & 1 & 0 \end{bmatrix}$$

Solution (a)   The last row corresponds to the equation

$$0x_1 + 0x_2 + 0x_3 + 0x_4 = 1$$

from which it is evident that the system is inconsistent.

Solution (b)   The last row corresponds to the equation

$$0x_1 + 0x_2 + 0x_3 + 0x_4 = 0$$

which has no effect on the solution set. In the remaining three equations the variables x_1, x_2, and x_3 correspond to leading 1's and hence are leading variables. The variable x_4

is a free variable. With a little algebra, the leading variables can be expressed in terms of the free variable, and the free variable can be assigned an arbitrary value. Thus, the system must have infinitely many solutions.

***Solution* (c)** The last row corresponds to the equation

$$x_4 = 0$$

which gives us a numerical value for x_4. If we substitute this value into the third equation, namely,

$$x_3 + 6x_4 = 9$$

we obtain $x_3 = 9$. You should now be able to see that if we continue this process and substitute the known values of x_3 and x_4 into the equation corresponding to the second row, we will obtain a unique numerical value for x_2; and if, finally, we substitute the known values of x_4, x_3, and x_2 into the equation corresponding to the first row, we will produce a unique numerical value for x_1. Thus, the system has a unique solution. ◀

Some Facts About Echelon Forms

There are three facts about row echelon forms and reduced row echelon forms that are important to know but we will not prove:

1. Every matrix has a unique reduced row echelon form; that is, regardless of whether you use Gauss–Jordan elimination or some other sequence of elementary row operations, the same reduced row echelon form will result in the end.[*]

2. Row echelon forms are not unique; that is, different sequences of elementary row operations can result in different row echelon forms.

3. Although row echelon forms are not unique, all row echelon forms of a matrix A have the same number of zero rows, and the leading 1's always occur in the same positions in the row echelon forms of A. Those are called the ***pivot positions*** of A. A column that contains a pivot position is called a ***pivot column*** of A.

▶ **EXAMPLE 9 Pivot Positions and Columns**

Earlier in this section (immediately after Definition 1) we found a row echelon form of

$$A = \begin{bmatrix} 0 & 0 & -2 & 0 & 7 & 12 \\ 2 & 4 & -10 & 6 & 12 & 28 \\ 2 & 4 & -5 & 6 & -5 & -1 \end{bmatrix}$$

to be

$$\begin{bmatrix} 1 & 2 & -5 & 3 & 6 & 14 \\ 0 & 0 & 1 & 0 & -\frac{7}{2} & -6 \\ 0 & 0 & 0 & 0 & 1 & 2 \end{bmatrix}$$

The leading 1's occur in positions (row 1, column 1), (row 2, column 3), and (row 3, column 5). These are the pivot positions. The pivot columns are columns 1, 3, and 5. ◀

Roundoff Error and Instability

There is often a gap between mathematical theory and its practical implementation—Gauss–Jordan elimination and Gaussian elimination being good examples. The problem is that computers generally approximate numbers, thereby introducing ***roundoff*** errors,

---

[*]A proof of this result can be found in the article "The Reduced Row Echelon Form of a Matrix Is Unique: A Simple Proof," by Thomas Yuster, *Mathematics Magazine*, Vol. 57, No. 2, 1984, pp. 93–94.

so unless precautions are taken, successive calculations may degrade an answer to a degree that makes it useless. Algorithms (procedures) in which this happens are called **unstable**. There are various techniques for minimizing roundoff error and instability. For example, it can be shown that for large linear systems Gauss–Jordan elimination involves roughly 50% more operations than Gaussian elimination, so most computer algorithms are based on the latter method. Some of these matters will be considered in Chapter 9.

Concept Review

- Reduced row echelon form
- Row echelon form
- Leading 1
- Leading variables
- Free variables
- General solution to a linear system

- Gaussian elimination
- Gauss–Jordan elimination
- Forward phase
- Backward phase
- Homogeneous linear system
- Trivial solution

- Nontrivial solution
- Dimension Theorem for Homogeneous Systems
- Back-substitution

Skills

- Recognize whether a given matrix is in row echelon form, reduced row echelon form, or neither.
- Construct solutions to linear systems whose corresponding augmented matrices that are in row echelon form or reduced row echelon form.
- Use Gaussian elimination to find the general solution of a linear system.

- Use Gauss–Jordan elimination to find the general solution of a linear system.
- Analyze homogeneous linear systems using the Free Variable Theorem for Homogeneous Systems.

Exercise Set 1.2

1. In each part, determine whether the matrix is in row echelon form, reduced row echelon form, both, or neither.

(a) $\begin{bmatrix} 1 & 0 & 0 \\ 0 & 1 & 0 \\ 0 & 0 & 1 \end{bmatrix}$
(b) $\begin{bmatrix} 1 & 0 & 0 \\ 0 & 1 & 0 \\ 0 & 0 & 0 \end{bmatrix}$
(c) $\begin{bmatrix} 0 & 1 & 0 \\ 0 & 0 & 1 \\ 0 & 0 & 0 \end{bmatrix}$

(d) $\begin{bmatrix} 1 & 0 & 3 & 1 \\ 0 & 1 & 2 & 4 \end{bmatrix}$
(e) $\begin{bmatrix} 1 & 2 & 0 & 3 & 0 \\ 0 & 0 & 1 & 1 & 0 \\ 0 & 0 & 0 & 0 & 1 \\ 0 & 0 & 0 & 0 & 0 \end{bmatrix}$

(f) $\begin{bmatrix} 0 & 0 \\ 0 & 0 \\ 0 & 0 \end{bmatrix}$
(g) $\begin{bmatrix} 1 & -7 & 5 & 5 \\ 0 & 1 & 3 & 2 \end{bmatrix}$

2. In each part, determine whether the matrix is in row echelon form, reduced row echelon form, both, or neither.

(a) $\begin{bmatrix} 1 & 2 & 0 \\ 0 & 1 & 0 \\ 0 & 0 & 0 \end{bmatrix}$
(b) $\begin{bmatrix} 1 & 0 & 0 \\ 0 & 1 & 0 \\ 0 & 2 & 0 \end{bmatrix}$
(c) $\begin{bmatrix} 1 & 3 & 4 \\ 0 & 0 & 1 \\ 0 & 0 & 0 \end{bmatrix}$

(d) $\begin{bmatrix} 1 & 5 & -3 \\ 0 & 1 & 1 \\ 0 & 0 & 0 \end{bmatrix}$
(e) $\begin{bmatrix} 1 & 2 & 3 \\ 0 & 0 & 0 \\ 0 & 0 & 1 \end{bmatrix}$

(f) $\begin{bmatrix} 1 & 2 & 3 & 4 & 5 \\ 1 & 0 & 7 & 1 & 3 \\ 0 & 0 & 0 & 0 & 1 \\ 0 & 0 & 0 & 0 & 0 \end{bmatrix}$
(g) $\begin{bmatrix} 1 & -2 & 0 & 1 \\ 0 & 0 & 1 & -2 \end{bmatrix}$

3. In each part, suppose that the augmented matrix for a system of linear equations has been reduced by row operations to the given row echelon form. Solve the system.

(a) $\begin{bmatrix} 1 & -3 & 4 & 7 \\ 0 & 1 & 2 & 2 \\ 0 & 0 & 1 & 5 \end{bmatrix}$

(b) $\begin{bmatrix} 1 & 0 & 8 & -5 & 6 \\ 0 & 1 & 4 & -9 & 3 \\ 0 & 0 & 1 & 1 & 2 \end{bmatrix}$

$$(c) \begin{bmatrix} 1 & 7 & -2 & 0 & -8 & -3 \\ 0 & 0 & 1 & 1 & 6 & 5 \\ 0 & 0 & 0 & 1 & 3 & 9 \\ 0 & 0 & 0 & 0 & 0 & 0 \end{bmatrix}$$

$$(d) \begin{bmatrix} 1 & -3 & 7 & 1 \\ 0 & 1 & 4 & 0 \\ 0 & 0 & 0 & 1 \end{bmatrix}$$

4. In each part, suppose that the augmented matrix for a system of linear equations has been reduced by row operations to the given row echelon form. Solve the system.

$$(a) \begin{bmatrix} 1 & 0 & 0 & -3 \\ 0 & 1 & 0 & 0 \\ 0 & 0 & 1 & 7 \end{bmatrix}$$

$$(b) \begin{bmatrix} 1 & 0 & 0 & -7 & 8 \\ 0 & 1 & 0 & 3 & 2 \\ 0 & 0 & 1 & 1 & -5 \end{bmatrix}$$

$$(c) \begin{bmatrix} 1 & -6 & 0 & 0 & 3 & -2 \\ 0 & 0 & 1 & 0 & 4 & 7 \\ 0 & 0 & 0 & 1 & 5 & 8 \\ 0 & 0 & 0 & 0 & 0 & 0 \end{bmatrix}$$

$$(d) \begin{bmatrix} 1 & -3 & 0 & 0 \\ 0 & 0 & 1 & 0 \\ 0 & 0 & 0 & 1 \end{bmatrix}$$

In Exercises 5–8, solve the linear system by Gauss–Jordan elimination.

5.
$$\begin{aligned} x_1 + x_2 + 2x_3 &= 8 \\ -x_1 - 2x_2 + 3x_3 &= 1 \\ 3x_1 - 7x_2 + 4x_3 &= 10 \end{aligned}$$

6.
$$\begin{aligned} 2x_1 + 2x_2 + 2x_3 &= 0 \\ -2x_1 + 5x_2 + 2x_3 &= 1 \\ 8x_1 + x_2 + 4x_3 &= -1 \end{aligned}$$

7.
$$\begin{aligned} x - y + 2z - w &= -1 \\ 2x + y - 2z - 2w &= -2 \\ -x + 2y - 4z + w &= 1 \\ 3x \qquad\quad - 3w &= -3 \end{aligned}$$

8.
$$\begin{aligned} -2b + 3c &= 1 \\ 3a + 6b - 3c &= -2 \\ 6a + 6b + 3c &= 5 \end{aligned}$$

In Exercises 9–12, solve the linear system by Gaussian elimination.

9. Exercise 5

10. Exercise 6

11. Exercise 7

12. Exercise 8

In Exercises 13–16, determine whether the homogeneous system has nontrivial solutions by inspection (without pencil and paper).

13.
$$\begin{aligned} 2x_1 - 3x_2 + 4x_3 - x_4 &= 0 \\ 7x_1 + x_2 - 8x_3 + 9x_4 &= 0 \\ 2x_1 + 8x_2 + x_3 - x_4 &= 0 \end{aligned}$$

14.
$$\begin{aligned} x_1 + 3x_2 - x_3 &= 0 \\ x_2 - 8x_3 &= 0 \\ 4x_3 &= 0 \end{aligned}$$

15.
$$\begin{aligned} a_{11}x_1 + a_{12}x_2 + a_{13}x_3 &= 0 \\ a_{21}x_1 + a_{22}x_2 + a_{23}x_3 &= 0 \end{aligned}$$

16.
$$\begin{aligned} 3x_1 - 2x_2 &= 0 \\ 6x_1 - 4x_2 &= 0 \end{aligned}$$

In Exercises 17–24, solve the given linear system by any method.

17.
$$\begin{aligned} 2x_1 + x_2 + 3x_3 &= 0 \\ x_1 + 2x_2 &= 0 \\ x_2 + x_3 &= 0 \end{aligned}$$

18.
$$\begin{aligned} 2x - y - 3z &= 0 \\ -x + 2y - 3z &= 0 \\ x + y + 4z &= 0 \end{aligned}$$

19.
$$\begin{aligned} 3x_1 + x_2 + x_3 + x_4 &= 0 \\ 5x_1 - x_2 + x_3 - x_4 &= 0 \end{aligned}$$

20.
$$\begin{aligned} v + 3w - 2x &= 0 \\ 2u + v - 4w + 3x &= 0 \\ 2u + 3v + 2w - x &= 0 \\ -4u - 3v + 5w - 4x &= 0 \end{aligned}$$

21.
$$\begin{aligned} 2x + 2y + 4z &= 0 \\ w - y - 3z &= 0 \\ 2w + 3x + y + z &= 0 \\ -2w + x + 3y - 2z &= 0 \end{aligned}$$

22.
$$\begin{aligned} x_1 + 3x_2 + x_4 &= 0 \\ x_1 + 4x_2 + 2x_3 &= 0 \\ -2x_2 - 2x_3 - x_4 &= 0 \\ 2x_1 - 4x_2 + x_3 + x_4 &= 0 \\ x_1 - 2x_2 - x_3 + x_4 &= 0 \end{aligned}$$

23.
$$\begin{aligned} 2I_1 - I_2 + 3I_3 + 4I_4 &= 9 \\ I_1 - 2I_3 + 7I_4 &= 11 \\ 3I_1 - 3I_2 + I_3 + 5I_4 &= 8 \\ 2I_1 + I_2 + 4I_3 + 4I_4 &= 10 \end{aligned}$$

24.
$$\begin{aligned} Z_3 + Z_4 + Z_5 &= 0 \\ -Z_1 - Z_2 + 2Z_3 - 3Z_4 + Z_5 &= 0 \\ Z_1 + Z_2 - 2Z_3 - Z_5 &= 0 \\ 2Z_1 + 2Z_2 - Z_3 + Z_5 &= 0 \end{aligned}$$

In Exercises 25–28, determine the values of a for which the system has no solutions, exactly one solution, or infinitely many solutions.

25.
$$\begin{aligned} x + 2y - 3z &= 4 \\ 3x - y + 5z &= 2 \\ 4x + y + (a^2 - 14)z &= a + 2 \end{aligned}$$

26.
$$\begin{aligned} x + 2y + z &= 2 \\ 2x - 2y + 3z &= 1 \\ x + 2y - (a^2 - 3)z &= a \end{aligned}$$

27.
$$\begin{aligned} x + 2y &= 1 \\ 2x + (a^2 - 5)y &= a - 1 \end{aligned}$$

28.
$$\begin{aligned} x + y + 7z &= -7 \\ 2x + 3y + 17z &= -16 \\ x + 2y + (a^2 + 1)z &= 3a \end{aligned}$$

▷ In Exercises 29–30, solve the following systems, where a, b, and c are constants. ◁

29. $2x + y = a$
$3x + 6y = b$

30. $x_1 + x_2 + x_3 = a$
$2x_1 \qquad + 2x_3 = b$
$3x_2 + 3x_3 = c$

31. Find two different row echelon forms of

$$\begin{bmatrix} 1 & 3 \\ 2 & 7 \end{bmatrix}$$

This exercise shows that a matrix can have multiple row echelon forms.

32. Reduce

$$\begin{bmatrix} 2 & 1 & 3 \\ 0 & -2 & -29 \\ 3 & 4 & 5 \end{bmatrix}$$

to reduced row echelon form without introducing fractions at any intermediate stage.

33. Show that the following nonlinear system has 18 solutions if $0 \le \alpha \le 2\pi$, $0 \le \beta \le 2\pi$, and $0 \le \gamma \le 2\pi$.

$$\sin \alpha + 2 \cos \beta + 3 \tan \gamma = 0$$
$$2 \sin \alpha + 5 \cos \beta + 3 \tan \gamma = 0$$
$$-\sin \alpha - 5 \cos \beta + 5 \tan \gamma = 0$$

[*Hint:* Begin by making the substitutions $x = \sin \alpha$, $y = \cos \beta$, and $z = \tan \gamma$.]

34. Solve the following system of nonlinear equations for the unknown angles α, β, and γ, where $0 \le \alpha \le 2\pi$, $0 \le \beta \le 2\pi$, and $0 \le \gamma < \pi$.

$$2 \sin \alpha - \cos \beta + 3 \tan \gamma = 3$$
$$4 \sin \alpha + 2 \cos \beta - 2 \tan \gamma = 2$$
$$6 \sin \alpha - 3 \cos \beta + \tan \gamma = 9$$

35. Solve the following system of nonlinear equations for x, y, and z.

$$x^2 + y^2 + z^2 = 6$$
$$x^2 - y^2 + 2z^2 = 2$$
$$2x^2 + y^2 - z^2 = 3$$

[*Hint:* Begin by making the substitutions $X = x^2$, $Y = y^2$, $Z = z^2$.]

36. Solve the following system for x, y, and z.

$$\frac{1}{x} + \frac{2}{y} - \frac{4}{z} = 1$$
$$\frac{2}{x} + \frac{3}{y} + \frac{8}{z} = 0$$
$$-\frac{1}{x} + \frac{9}{y} + \frac{10}{z} = 5$$

37. Find the coefficients a, b, c, and d so that the curve shown in the accompanying figure is the graph of the equation $y = ax^3 + bx^2 + cx + d$.

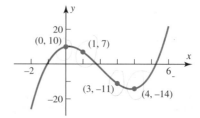

$(0, 10)$ $(1, 7)$ $(3, -11)$ $(4, -14)$

◁ **Figure Ex-37**

38. Find the coefficients a, b, c, and d so that the curve shown in the accompanying figure is given by the equation $ax^2 + ay^2 + bx + cy + d = 0$.

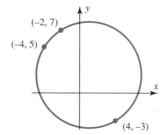

$(-2, 7)$ $(-4, 5)$ $(4, -3)$

◁ **Figure Ex-38**

39. If the linear system

$$a_1x + b_1y + c_1z = 0$$
$$a_2x - b_2y + c_2z = 0$$
$$a_3x + b_3y - c_3z = 0$$

has only the trivial solution, what can be said about the solutions of the following system?

$$a_1x + b_1y + c_1z = 3$$
$$a_2x - b_2y + c_2z = 7$$
$$a_3x + b_3y - c_3z = 11$$

40. (a) If A is a 3×5 matrix, then what is the maximum possible number of leading 1's in its reduced row echelon form?

(b) If B is a 3×6 matrix whose last column has all zeros, then what is the maximum possible number of parameters in the general solution of the linear system with augmented matrix B?

(c) If C is a 5×3 matrix, then what is the minimum possible number of rows of zeros in any row echelon form of C?

41. (a) Prove that if $ad - bc \ne 0$, then the reduced row echelon form of

$$\begin{bmatrix} a & b \\ c & d \end{bmatrix} \quad \text{is} \quad \begin{bmatrix} 1 & 0 \\ 0 & 1 \end{bmatrix}$$

(b) Use the result in part (a) to prove that if $ad - bc \ne 0$, then the linear system

$$ax + by = k$$
$$cx + dy = l$$

has exactly one solution.

42. Consider the system of equations

$$ax + by = 0$$
$$cx + dy = 0$$
$$ex + fy = 0$$

Discuss the relative positions of the lines $ax + by = 0$, $cx + dy = 0$, and $ex + fy = 0$ when (a) the system has only the trivial solution, and (b) the system has nontrivial solutions.

43. Describe all possible reduced row echelon forms of

(a) $\begin{bmatrix} a & b & c \\ d & e & f \\ g & h & i \end{bmatrix}$

(b) $\begin{bmatrix} a & b & c & d \\ e & f & g & h \\ i & j & k & l \\ m & n & p & q \end{bmatrix}$

True-False Exercises

In parts (a)–(i) determine whether the statement is true or false, and justify your answer.

(a) If a matrix is in reduced row echelon form, then it is also in row echelon form.

(b) If an elementary row operation is applied to a matrix that is in row echelon form, the resulting matrix will still be in row echelon form.

(c) Every matrix has a unique row echelon form.

(d) A homogeneous linear system in n unknowns whose corresponding augmented matrix has a reduced row echelon form with r leading 1's has $n - r$ free variables.

(e) All leading 1's in a matrix in row echelon form must occur in different columns.

(f) If every column of a matrix in row echelon form has a leading 1 then all entries that are not leading 1's are zero.

(g) If a homogeneous linear system of n equations in n unknowns has a corresponding augmented matrix with a reduced row echelon form containing n leading 1's, then the linear system has only the trivial solution.

(h) If the reduced row echelon form of the augmented matrix for a linear system has a row of zeros, then the system must have infinitely many solutions.

(i) If a linear system has more unknowns than equations, then it must have infinitely many solutions.

1.3 Matrices and Matrix Operations

Rectangular arrays of real numbers arise in contexts other than as augmented matrices for linear systems. In this section we will begin to study matrices as objects in their own right by defining operations of addition, subtraction, and multiplication on them.

Matrix Notation and Terminology

In Section 1.2 we used rectangular arrays of numbers, called *augmented matrices*, to abbreviate systems of linear equations. However, rectangular arrays of numbers occur in other contexts as well. For example, the following rectangular array with three rows and seven columns might describe the number of hours that a student spent studying three subjects during a certain week:

	Mon.	Tues.	Wed.	Thurs.	Fri.	Sat.	Sun.
Math	2	3	2	4	1	4	2
History	0	3	1	4	3	2	2
Language	4	1	3	1	0	0	2

If we suppress the headings, then we are left with the following rectangular array of numbers with three rows and seven columns, called a "matrix":

$$\begin{bmatrix} 2 & 3 & 2 & 4 & 1 & 4 & 2 \\ 0 & 3 & 1 & 4 & 3 & 2 & 2 \\ 4 & 1 & 3 & 1 & 0 & 0 & 2 \end{bmatrix}$$

More generally, we make the following definition.

> **DEFINITION 1**   A *matrix* is a rectangular array of numbers. The numbers in the array are called the *entries* in the matrix.

A matrix with only one column is called a *column vector* or a *column matrix*, and a matrix with only one row is called a *row vector* or a *row matrix*. In Example 1, the 2×1 matrix is a column vector, the 1×4 matrix is a row vector, and the 1×1 matrix is both a row vector and a column vector.

▶ **EXAMPLE 1** **Examples of Matrices**

Some examples of matrices are

$$\begin{bmatrix} 1 & 2 \\ 3 & 0 \\ -1 & 4 \end{bmatrix}, \quad [2 \quad 1 \quad 0 \quad -3], \quad \begin{bmatrix} e & \pi & -\sqrt{2} \\ 0 & \frac{1}{2} & 1 \\ 0 & 0 & 0 \end{bmatrix}, \quad \begin{bmatrix} 1 \\ 3 \end{bmatrix}, \quad [4] \quad ◀$$

The *size* of a matrix is described in terms of the number of rows (horizontal lines) and columns (vertical lines) it contains. For example, the first matrix in Example 1 has three rows and two columns, so its size is 3 by 2 (written 3×2). In a size description, the first number always denotes the number of rows, and the second denotes the number of columns. The remaining matrices in Example 1 have sizes 1×4, 3×3, 2×1, and 1×1, respectively.

We will use capital letters to denote matrices and lowercase letters to denote numerical quantities; thus we might write

$$A = \begin{bmatrix} 2 & 1 & 7 \\ 3 & 4 & 2 \end{bmatrix} \quad \text{or} \quad C = \begin{bmatrix} a & b & c \\ d & e & f \end{bmatrix}$$

When discussing matrices, it is common to refer to numerical quantities as *scalars*. Unless stated otherwise, *scalars will be real numbers*; complex scalars will be considered later in the text.

The entry that occurs in row i and column j of a matrix A will be denoted by a_{ij}. Thus a general 3×4 matrix might be written as

Matrix brackets are often omitted from 1×1 matrices, making it impossible to tell, for example, whether the symbol 4 denotes the number "four" or the matrix [4]. This rarely causes problems because it is usually possible to tell which is meant from the context.

$$A = \begin{bmatrix} a_{11} & a_{12} & a_{13} & a_{14} \\ a_{21} & a_{22} & a_{23} & a_{24} \\ a_{31} & a_{32} & a_{33} & a_{34} \end{bmatrix}$$

and a general $m \times n$ matrix as

$$A = \begin{bmatrix} a_{11} & a_{12} & \cdots & a_{1n} \\ a_{21} & a_{22} & \cdots & a_{2n} \\ \vdots & \vdots & & \vdots \\ a_{m1} & a_{m2} & \cdots & a_{mn} \end{bmatrix} \tag{1}$$

When a compact notation is desired, the preceding matrix can be written as

$$[a_{ij}]_{m \times n} \quad \text{or} \quad [a_{ij}]$$

the first notation being used when it is important in the discussion to know the size, and the second when the size need not be emphasized. Usually, we will match the letter denoting a matrix with the letter denoting its entries; thus, for a matrix B we would generally use b_{ij} for the entry in row i and column j, and for a matrix C we would use the notation c_{ij}.

The entry in row i and column j of a matrix A is also commonly denoted by the symbol $(A)_{ij}$. Thus, for matrix (1) above, we have

$$(A)_{ij} = a_{ij}$$

and for the matrix

$$A = \begin{bmatrix} 2 & -3 \\ 7 & 0 \end{bmatrix}$$

we have $(A)_{11} = 2$, $(A)_{12} = -3$, $(A)_{21} = 7$, and $(A)_{22} = 0$.

Row and column vectors are of special importance, and it is common practice to denote them by boldface lowercase letters rather than capital letters. For such matrices, double subscripting of the entries is unnecessary. Thus a general $1 \times n$ row vector $\mathbf{a}$ and a general $m \times 1$ column vector $\mathbf{b}$ would be written as

$$\mathbf{a} = \begin{bmatrix} a_1 & a_2 & \cdots & a_n \end{bmatrix} \quad \text{and} \quad \mathbf{b} = \begin{bmatrix} b_1 \\ b_2 \\ \vdots \\ b_m \end{bmatrix}$$

A matrix A with n rows and n columns is called a **square matrix of order n**, and the shaded entries $a_{11}, a_{22}, \ldots, a_{nn}$ in (2) are said to be on the **main diagonal** of A.

$$\begin{bmatrix} a_{11} & a_{12} & \cdots & a_{1n} \\ a_{21} & a_{22} & \cdots & a_{2n} \\ \vdots & \vdots & & \vdots \\ a_{n1} & a_{n2} & \cdots & a_{nn} \end{bmatrix} \tag{2}$$

Operations on Matrices

So far, we have used matrices to abbreviate the work in solving systems of linear equations. For other applications, however, it is desirable to develop an "arithmetic of matrices" in which matrices can be added, subtracted, and multiplied in a useful way. The remainder of this section will be devoted to developing this arithmetic.

> **DEFINITION 2** Two matrices are defined to be **equal** if they have the same size and their corresponding entries are equal.

The equality of two matrices

$$A = [a_{ij}] \quad \text{and} \quad B = [b_{ij}]$$

of the same size can be expressed either by writing

$$(A)_{ij} = (B)_{ij}$$

or by writing

$$a_{ij} = b_{ij}$$

where it is understood that the equalities hold for all values of i and j.

▶ **EXAMPLE 2 Equality of Matrices**

Consider the matrices

$$A = \begin{bmatrix} 2 & 1 \\ 3 & x \end{bmatrix}, \quad B = \begin{bmatrix} 2 & 1 \\ 3 & 5 \end{bmatrix}, \quad C = \begin{bmatrix} 2 & 1 & 0 \\ 3 & 4 & 0 \end{bmatrix}$$

If $x = 5$, then $A = B$, but for all other values of x the matrices A and B are not equal, since not all of their corresponding entries are equal. There is no value of x for which $A = C$ since A and C have different sizes. ◀

> **DEFINITION 3** If A and B are matrices of the same size, then the **sum $A + B$** is the matrix obtained by adding the entries of B to the corresponding entries of A, and the **difference $A - B$** is the matrix obtained by subtracting the entries of B from the corresponding entries of A. Matrices of different sizes cannot be added or subtracted.

In matrix notation, if $A = [a_{ij}]$ and $B = [b_{ij}]$ have the same size, then

$$(A + B)_{ij} = (A)_{ij} + (B)_{ij} = a_{ij} + b_{ij} \quad \text{and} \quad (A - B)_{ij} = (A)_{ij} - (B)_{ij} = a_{ij} - b_{ij}$$

▶ **EXAMPLE 3** **Addition and Subtraction**

Consider the matrices

$$A = \begin{bmatrix} 2 & 1 & 0 & 3 \\ -1 & 0 & 2 & 4 \\ 4 & -2 & 7 & 0 \end{bmatrix}, \quad B = \begin{bmatrix} -4 & 3 & 5 & 1 \\ 2 & 2 & 0 & -1 \\ 3 & 2 & -4 & 5 \end{bmatrix}, \quad C = \begin{bmatrix} 1 & 1 \\ 2 & 2 \end{bmatrix}$$

Then

$$A + B = \begin{bmatrix} -2 & 4 & 5 & 4 \\ 1 & 2 & 2 & 3 \\ 7 & 0 & 3 & 5 \end{bmatrix} \quad \text{and} \quad A - B = \begin{bmatrix} 6 & -2 & -5 & 2 \\ -3 & -2 & 2 & 5 \\ 1 & -4 & 11 & -5 \end{bmatrix}$$

The expressions $A + C$, $B + C$, $A - C$, and $B - C$ are undefined. ◀

DEFINITION 4 If A is any matrix and c is any scalar, then the **product** cA is the matrix obtained by multiplying each entry of the matrix A by c. The matrix cA is said to be a **scalar multiple** of A.

In matrix notation, if $A = [a_{ij}]$, then

$$(cA)_{ij} = c(A)_{ij} = ca_{ij}$$

▶ **EXAMPLE 4** **Scalar Multiples**

For the matrices

$$A = \begin{bmatrix} 2 & 3 & 4 \\ 1 & 3 & 1 \end{bmatrix}, \quad B = \begin{bmatrix} 0 & 2 & 7 \\ -1 & 3 & -5 \end{bmatrix}, \quad C = \begin{bmatrix} 9 & -6 & 3 \\ 3 & 0 & 12 \end{bmatrix}$$

we have

$$2A = \begin{bmatrix} 4 & 6 & 8 \\ 2 & 6 & 2 \end{bmatrix}, \quad (-1)B = \begin{bmatrix} 0 & -2 & -7 \\ 1 & -3 & 5 \end{bmatrix}, \quad \tfrac{1}{3}C = \begin{bmatrix} 3 & -2 & 1 \\ 1 & 0 & 4 \end{bmatrix}$$

It is common practice to denote $(-1)B$ by $-B$. ◀

Thus far we have defined multiplication of a matrix by a scalar but not the multiplication of two matrices. Since matrices are added by adding corresponding entries and subtracted by subtracting corresponding entries, it would seem natural to define multiplication of matrices by multiplying corresponding entries. However, it turns out that such a definition would not be very useful for most problems. Experience has led mathematicians to the following more useful definition of matrix multiplication.

DEFINITION 5 If A is an $m \times r$ matrix and B is an $r \times n$ matrix, then the **product** AB is the $m \times n$ matrix whose entries are determined as follows: To find the entry in row i and column j of AB, single out row i from the matrix A and column j from the matrix B. Multiply the corresponding entries from the row and column together, and then add up the resulting products.

▶ **EXAMPLE 5** **Multiplying Matrices**

Consider the matrices

$$A = \begin{bmatrix} 1 & 2 & 4 \\ 2 & 6 & 0 \end{bmatrix}, \quad B = \begin{bmatrix} 4 & 1 & 4 & 3 \\ 0 & -1 & 3 & 1 \\ 2 & 7 & 5 & 2 \end{bmatrix}$$

Since A is a 2×3 matrix and B is a 3×4 matrix, the product AB is a 2×4 matrix. To determine, for example, the entry in row 2 and column 3 of AB, we single out row 2 from A and column 3 from B. Then, as illustrated below, we multiply corresponding entries together and add up these products.

$$\begin{bmatrix} 1 & 2 & 4 \\ 2 & 6 & 0 \end{bmatrix} \begin{bmatrix} 4 & 1 & 4 & 3 \\ 0 & -1 & 3 & 1 \\ 2 & 7 & 5 & 2 \end{bmatrix} = \begin{bmatrix} \square & \square & \square & \square \\ \square & \square & \boxed{26} & \square \end{bmatrix}$$

$$(2 \cdot 4) + (6 \cdot 3) + (0 \cdot 5) = 26$$

The entry in row 1 and column 4 of AB is computed as follows:

$$\begin{bmatrix} 1 & 2 & 4 \\ 2 & 6 & 0 \end{bmatrix} \begin{bmatrix} 4 & 1 & 4 & 3 \\ 0 & -1 & 3 & 1 \\ 2 & 7 & 5 & 2 \end{bmatrix} = \begin{bmatrix} \square & \square & \square & \boxed{13} \\ \square & \square & \square & \square \end{bmatrix}$$

$$(1 \cdot 3) + (2 \cdot 1) + (4 \cdot 2) = 13$$

The computations for the remaining entries are

$$\begin{aligned}
(1 \cdot 4) + (2 \cdot 0) + (4 \cdot 2) &= 12 \\
(1 \cdot 1) - (2 \cdot 1) + (4 \cdot 7) &= 27 \\
(1 \cdot 4) + (2 \cdot 3) + (4 \cdot 5) &= 30 \\
(2 \cdot 4) + (6 \cdot 0) + (0 \cdot 2) &= 8 \\
(2 \cdot 1) - (6 \cdot 1) + (0 \cdot 7) &= -4 \\
(2 \cdot 3) + (6 \cdot 1) + (0 \cdot 2) &= 12
\end{aligned} \qquad AB = \begin{bmatrix} 12 & 27 & 30 & 13 \\ 8 & -4 & 26 & 12 \end{bmatrix} \blacktriangleleft$$

The definition of matrix multiplication requires that the number of columns of the first factor A be the same as the number of rows of the second factor B in order to form the product AB. If this condition is not satisfied, the product is undefined. A convenient way to determine whether a product of two matrices is defined is to write down the size of the first factor and, to the right of it, write down the size of the second factor. If, as in (3), the inside numbers are the same, then the product is defined. The outside numbers then give the size of the product.

$$\underset{m \times r}{\overset{A}{}} \quad \underset{r \times n}{\overset{B}{}} = \underset{m \times n}{\overset{AB}{}}$$

Inside

Outside

(3)

**Gotthold Eisenstein
(1823–1852)**

Historical Note The concept of matrix multiplication is due to the German mathematician Gotthold Eisenstein, who introduced the idea around 1844 to simplify the process of making substitutions in linear systems. The idea was then expanded on and formalized by Cayley in his *Memoir on the Theory of Matrices* that was published in 1858. Eisenstein was a pupil of Gauss, who ranked him as the equal of Isaac Newton and Archimedes. However, Eisenstein, suffering from bad health his entire life, died at age 30, so his potential was never realized.

[*Image: wikipedia*]

▶ **EXAMPLE 6** **Determining Whether a Product Is Defined**

Suppose that A, B, and C are matrices with the following sizes:

$$
\begin{array}{ccc}
A & B & C \\
3 \times 4 & 4 \times 7 & 7 \times 3
\end{array}
$$

Then by (3), AB is defined and is a 3×7 matrix; BC is defined and is a 4×3 matrix; and CA is defined and is a 7×4 matrix. The products AC, CB, and BA are all undefined.

◀

In general, if $A = [a_{ij}]$ is an $m \times r$ matrix and $B = [b_{ij}]$ is an $r \times n$ matrix, then, as illustrated by the shading in (4),

$$
AB =
\begin{bmatrix}
a_{11} & a_{12} & \cdots & a_{1r} \\
a_{21} & a_{22} & \cdots & a_{2r} \\
\vdots & \vdots & & \vdots \\
a_{i1} & a_{i2} & \cdots & a_{ir} \\
\vdots & \vdots & & \vdots \\
a_{m1} & a_{m2} & \cdots & a_{mr}
\end{bmatrix}
\begin{bmatrix}
b_{11} & b_{12} & \cdots & b_{1j} & \cdots & b_{1n} \\
b_{21} & b_{22} & \cdots & b_{2j} & \cdots & b_{2n} \\
\vdots & \vdots & & \vdots & & \vdots \\
b_{r1} & b_{r2} & \cdots & b_{rj} & \cdots & b_{rn}
\end{bmatrix}
\tag{4}
$$

the entry $(AB)_{ij}$ in row i and column j of AB is given by

$$
(AB)_{ij} = a_{i1}b_{1j} + a_{i2}b_{2j} + a_{i3}b_{3j} + \cdots + a_{ir}b_{rj}
\tag{5}
$$

Partitioned Matrices   A matrix can be subdivided or **partitioned** into smaller matrices by inserting horizontal and vertical rules between selected rows and columns. For example, the following are three possible partitions of a general 3×4 matrix A—the first is a partition of A into four **submatrices** A_{11}, A_{12}, A_{21}, and A_{22}; the second is a partition of A into its row vectors $\mathbf{r}_1$, $\mathbf{r}_2$, and $\mathbf{r}_3$; and the third is a partition of A into its column vectors $\mathbf{c}_1$, $\mathbf{c}_2$, $\mathbf{c}_3$, and $\mathbf{c}_4$:

$$
A =
\left[
\begin{array}{ccc|c}
a_{11} & a_{12} & a_{13} & a_{14} \\
a_{21} & a_{22} & a_{23} & a_{24} \\
\hline
a_{31} & a_{32} & a_{33} & a_{34}
\end{array}
\right]
=
\begin{bmatrix}
A_{11} & A_{12} \\
A_{21} & A_{22}
\end{bmatrix}
$$

$$
A =
\left[
\begin{array}{cccc}
a_{11} & a_{12} & a_{13} & a_{14} \\
\hline
a_{21} & a_{22} & a_{23} & a_{24} \\
\hline
a_{31} & a_{32} & a_{33} & a_{34}
\end{array}
\right]
=
\begin{bmatrix}
\mathbf{r}_1 \\
\mathbf{r}_2 \\
\mathbf{r}_3
\end{bmatrix}
$$

$$
A =
\left[
\begin{array}{c|c|c|c}
a_{11} & a_{12} & a_{13} & a_{14} \\
a_{21} & a_{22} & a_{23} & a_{24} \\
a_{31} & a_{32} & a_{33} & a_{34}
\end{array}
\right]
=
\begin{bmatrix}
\mathbf{c}_1 & \mathbf{c}_2 & \mathbf{c}_3 & \mathbf{c}_4
\end{bmatrix}
$$

Matrix Multiplication by
Columns and by Rows   Partitioning has many uses, one of which is for finding particular rows or columns of a matrix product AB without computing the entire product. Specifically, the following formulas, whose proofs are left as exercises, show how individual column vectors

of AB can be obtained by partitioning B into column vectors and how individual row vectors of AB can be obtained by partitioning A into row vectors.

$$AB = A[\mathbf{b}_1 \quad \mathbf{b}_2 \quad \cdots \quad \mathbf{b}_n] = [A\mathbf{b}_1 \quad A\mathbf{b}_2 \quad \cdots \quad A\mathbf{b}_n] \tag{6}$$

(AB computed column by column)

$$AB = \begin{bmatrix} \mathbf{a}_1 \\ \mathbf{a}_2 \\ \vdots \\ \mathbf{a}_m \end{bmatrix} B = \begin{bmatrix} \mathbf{a}_1 B \\ \mathbf{a}_2 B \\ \vdots \\ \mathbf{a}_m B \end{bmatrix} \tag{7}$$

(AB computed row by row)

In words, these formulas state that

$$j\text{th column vector of } AB = A[\,j\text{th column vector of } B] \tag{8}$$

$$i\text{th row vector of } AB = [i\text{th row vector of } A]B \tag{9}$$

► **EXAMPLE 7** **Example 5 Revisited**

If A and B are the matrices in Example 5, then from (8) the second column vector of AB can be obtained by the computation

$$\begin{bmatrix} 1 & 2 & 4 \\ 2 & 6 & 0 \end{bmatrix} \begin{bmatrix} 1 \\ -1 \\ 7 \end{bmatrix} = \begin{bmatrix} 27 \\ -4 \end{bmatrix}$$

Second column   Second column
of B   of AB

and from (9) the first row vector of AB can be obtained by the computation

$$[1 \quad 2 \quad 4] \begin{bmatrix} 4 & 1 & 4 & 3 \\ 0 & -1 & 3 & 1 \\ 2 & 7 & 5 & 2 \end{bmatrix} = [12 \quad 27 \quad 30 \quad 13] \quad ◄$$

First row of A   First row of AB

Matrix Products as Linear Combinations

We have discussed three methods for computing a matrix product AB—entry by entry, column by column, and row by row. The following definition provides yet another way of thinking about matrix multiplication.

> **DEFINITION 6** If $A_1, A_2, \ldots, A_r$ are matrices of the same size, and if $c_1, c_2, \ldots, c_r$ are scalars, then an expression of the form
>
> $$c_1 A_1 + c_2 A_2 + \cdots + c_r A_r$$
>
> is called a ***linear combination*** of $A_1, A_2, \ldots, A_r$ with ***coefficients*** $c_1, c_2, \ldots, c_r$.

To see how matrix products can be viewed as linear combinations, let A be an $m \times n$ matrix and $\mathbf{x}$ an $n \times 1$ column vector, say

$$A = \begin{bmatrix} a_{11} & a_{12} & \cdots & a_{1n} \\ a_{21} & a_{22} & \cdots & a_{2n} \\ \vdots & \vdots & & \vdots \\ a_{m1} & a_{m2} & \cdots & a_{mn} \end{bmatrix} \quad \text{and} \quad \mathbf{x} = \begin{bmatrix} x_1 \\ x_2 \\ \vdots \\ x_n \end{bmatrix}$$

Then

$$A\mathbf{x} = \begin{bmatrix} a_{11}x_1 + a_{12}x_2 + \cdots + a_{1n}x_n \\ a_{21}x_1 + a_{22}x_2 + \cdots + a_{2n}x_n \\ \vdots & \vdots & \vdots \\ a_{m1}x_1 + a_{m2}x_2 + \cdots + a_{mn}x_n \end{bmatrix} = x_1 \begin{bmatrix} a_{11} \\ a_{21} \\ \vdots \\ a_{m1} \end{bmatrix} + x_2 \begin{bmatrix} a_{12} \\ a_{22} \\ \vdots \\ a_{m2} \end{bmatrix} + \cdots + x_n \begin{bmatrix} a_{1n} \\ a_{2n} \\ \vdots \\ a_{mn} \end{bmatrix}$$

(10)

This proves the following theorem.

THEOREM 1.3.1 *If A is an $m \times n$ matrix, and if $\mathbf{x}$ is an $n \times 1$ column vector, then the product $A\mathbf{x}$ can be expressed as a linear combination of the column vectors of A in which the coefficients are the entries of $\mathbf{x}$.*

▶ **EXAMPLE 8 Matrix Products as Linear Combinations**

The matrix product

$$\begin{bmatrix} -1 & 3 & 2 \\ 1 & 2 & -3 \\ 2 & 1 & -2 \end{bmatrix} \begin{bmatrix} 2 \\ -1 \\ 3 \end{bmatrix} = \begin{bmatrix} 1 \\ -9 \\ -3 \end{bmatrix}$$

can be written as the following linear combination of column vectors

$$2 \begin{bmatrix} -1 \\ 1 \\ 2 \end{bmatrix} - 1 \begin{bmatrix} 3 \\ 2 \\ 1 \end{bmatrix} + 3 \begin{bmatrix} 2 \\ -3 \\ -2 \end{bmatrix} = \begin{bmatrix} 1 \\ -9 \\ -3 \end{bmatrix}$$

▶ **EXAMPLE 9 Columns of a Product AB as Linear Combinations**

We showed in Example 5 that

$$AB = \begin{bmatrix} 1 & 2 & 4 \\ 2 & 6 & 0 \end{bmatrix} \begin{bmatrix} 4 & 1 & 4 & 3 \\ 0 & -1 & 3 & 1 \\ 2 & 7 & 5 & 2 \end{bmatrix} = \begin{bmatrix} 12 & 27 & 30 & 13 \\ 8 & -4 & 26 & 12 \end{bmatrix}$$

It follows from Formula (6) and Theorem 1.3.1 that the jth column vector of AB can be expressed as a linear combination of the column vectors of A in which the coefficients in the linear combination are the entries from the jth column of B. The computations are as follows:

$$\begin{bmatrix} 12 \\ 8 \end{bmatrix} = 4 \begin{bmatrix} 1 \\ 2 \end{bmatrix} + 0 \begin{bmatrix} 2 \\ 6 \end{bmatrix} + 2 \begin{bmatrix} 4 \\ 0 \end{bmatrix}$$

$$\begin{bmatrix} 27 \\ -4 \end{bmatrix} = \begin{bmatrix} 1 \\ 2 \end{bmatrix} - \begin{bmatrix} 2 \\ 6 \end{bmatrix} + 7 \begin{bmatrix} 4 \\ 0 \end{bmatrix}$$

$$\begin{bmatrix} 30 \\ 26 \end{bmatrix} = 4 \begin{bmatrix} 1 \\ 2 \end{bmatrix} + 3 \begin{bmatrix} 2 \\ 6 \end{bmatrix} + 5 \begin{bmatrix} 4 \\ 0 \end{bmatrix}$$

$$\begin{bmatrix} 13 \\ 12 \end{bmatrix} = 3 \begin{bmatrix} 1 \\ 2 \end{bmatrix} + \begin{bmatrix} 2 \\ 6 \end{bmatrix} + 2 \begin{bmatrix} 4 \\ 0 \end{bmatrix} \quad \blacktriangleleft$$

Matrix Form of a Linear System

Matrix multiplication has an important application to systems of linear equations. Consider a system of m linear equations in n unknowns:

$$\begin{aligned}
a_{11}x_1 + a_{12}x_2 + \cdots + a_{1n}x_n &= b_1 \\
a_{21}x_1 + a_{22}x_2 + \cdots + a_{2n}x_n &= b_2 \\
\vdots \qquad \vdots \qquad \qquad \vdots \quad &\;\; \vdots \\
a_{m1}x_1 + a_{m2}x_2 + \cdots + a_{mn}x_n &= b_m
\end{aligned}$$

Since two matrices are equal if and only if their corresponding entries are equal, we can replace the m equations in this system by the single matrix equation

$$\begin{bmatrix} a_{11}x_1 + a_{12}x_2 + \cdots + a_{1n}x_n \\ a_{21}x_1 + a_{22}x_2 + \cdots + a_{2n}x_n \\ \vdots \qquad \vdots \qquad \qquad \vdots \\ a_{m1}x_1 + a_{m2}x_2 + \cdots + a_{mn}x_n \end{bmatrix} = \begin{bmatrix} b_1 \\ b_2 \\ \vdots \\ b_m \end{bmatrix}$$

The $m \times 1$ matrix on the left side of this equation can be written as a product to give

$$\begin{bmatrix} a_{11} & a_{12} & \cdots & a_{1n} \\ a_{21} & a_{22} & \cdots & a_{2n} \\ \vdots & \vdots & & \vdots \\ a_{m1} & a_{m2} & \cdots & a_{mn} \end{bmatrix} \begin{bmatrix} x_1 \\ x_2 \\ \vdots \\ x_n \end{bmatrix} = \begin{bmatrix} b_1 \\ b_2 \\ \vdots \\ b_m \end{bmatrix}$$

If we designate these matrices by A, $\mathbf{x}$, and $\mathbf{b}$, respectively, then we can replace the original system of m equations in n unknowns by the single matrix equation

$$A\mathbf{x} = \mathbf{b}$$

The matrix A in this equation is called the ***coefficient matrix*** of the system. The ***augmented matrix*** for the system is obtained by adjoining $\mathbf{b}$ to A as the last column; thus the augmented matrix is

The vertical bar in $[A \mid \mathbf{b}]$ is a convenient way to separate A from $\mathbf{b}$ visually; it has no mathematical significance.

$$[A \mid \mathbf{b}] = \left[\begin{array}{cccc|c} a_{11} & a_{12} & \cdots & a_{1n} & b_1 \\ a_{21} & a_{22} & \cdots & a_{2n} & b_2 \\ \vdots & \vdots & & \vdots & \vdots \\ a_{m1} & a_{m2} & \cdots & a_{mn} & b_m \end{array} \right]$$

Transpose of a Matrix

We conclude this section by defining two matrix operations that have no analogs in the arithmetic of real numbers.

> **DEFINITION 7** If A is any $m \times n$ matrix, then the **transpose of A**, denoted by A^T, is defined to be the $n \times m$ matrix that results by interchanging the rows and columns of A; that is, the first column of A^T is the first row of A, the second column of A^T is the second row of A, and so forth.

▶ **EXAMPLE 10** **Some Transposes**

The following are some examples of matrices and their transposes.

$$A = \begin{bmatrix} a_{11} & a_{12} & a_{13} & a_{14} \\ a_{21} & a_{22} & a_{23} & a_{24} \\ a_{31} & a_{32} & a_{33} & a_{34} \end{bmatrix}, \quad B = \begin{bmatrix} 2 & 3 \\ 1 & 4 \\ 5 & 6 \end{bmatrix}, \quad C = \begin{bmatrix} 1 & 3 & 5 \end{bmatrix}, \quad D = [4]$$

$$A^T = \begin{bmatrix} a_{11} & a_{21} & a_{31} \\ a_{12} & a_{22} & a_{32} \\ a_{13} & a_{23} & a_{33} \\ a_{14} & a_{24} & a_{34} \end{bmatrix}, \quad B^T = \begin{bmatrix} 2 & 1 & 5 \\ 3 & 4 & 6 \end{bmatrix}, \quad C^T = \begin{bmatrix} 1 \\ 3 \\ 5 \end{bmatrix}, \quad D^T = [4] \quad ◀$$

Observe that not only are the columns of A^T the rows of A, but the rows of A^T are the columns of A. Thus the entry in row i and column j of A^T is the entry in row j and column i of A; that is,

$$(A^T)_{ij} = (A)_{ji} \tag{11}$$

Note the reversal of the subscripts.

In the special case where A is a square matrix, the transpose of A can be obtained by interchanging entries that are symmetrically positioned about the main diagonal. In (12) we see that A^T can also be obtained by "reflecting" A about its main diagonal.

$$A = \begin{bmatrix} 1 & -2 & 4 \\ 3 & 7 & 0 \\ -5 & 8 & 6 \end{bmatrix} \rightarrow \begin{bmatrix} 1 & -2 & 4 \\ 3 & 7 & 0 \\ -5 & 8 & 6 \end{bmatrix} \rightarrow A^T = \begin{bmatrix} 1 & 3 & -5 \\ -2 & 7 & 8 \\ 4 & 0 & 6 \end{bmatrix} \tag{12}$$

Interchange entries that are symmetrically positioned about the main diagonal.

James Sylvester
(1814–1897)

Arthur Cayley
(1821–1895)

Historical Note The term *matrix* was first used by the English mathematician James Sylvester, who defined the term in 1850 to be an "oblong arrangement of terms." Sylvester communicated his work on matrices to a fellow English mathematician and lawyer named Arthur Cayley, who then introduced some of the basic operations on matrices in a book entitled *Memoir on the Theory of Matrices* that was published in 1858. As a matter of interest, Sylvester, who was Jewish, did not get his college degree because he refused to sign a required oath to the Church of England. He was appointed to a chair at the University of Virginia in the United States but resigned after swatting a student with a stick because he was reading a newspaper in class. Sylvester, thinking he had killed the student, fled back to England on the first available ship. Fortunately, the student was not dead, just in shock!

[Images: The Granger Collection, New York]

> **DEFINITION 8** If A is a square matrix, then the ***trace of*** A, denoted by $\text{tr}(A)$, is defined to be the sum of the entries on the main diagonal of A. The trace of A is undefined if A is not a square matrix.

▶ **EXAMPLE 11 Trace of a Matrix**

The following are examples of matrices and their traces.

$$A = \begin{bmatrix} a_{11} & a_{12} & a_{13} \\ a_{21} & a_{22} & a_{23} \\ a_{31} & a_{32} & a_{33} \end{bmatrix}, \quad B = \begin{bmatrix} -1 & 2 & 7 & 0 \\ 3 & 5 & -8 & 4 \\ 1 & 2 & 7 & -3 \\ 4 & -2 & 1 & 0 \end{bmatrix}$$

$$\text{tr}(A) = a_{11} + a_{22} + a_{33} \qquad \text{tr}(B) = -1 + 5 + 7 + 0 = 11 \quad ◀$$

In the exercises you will have some practice working with the transpose and trace operations.

Concept Review

- Matrix
- Entries
- Column vector (or column matrix)
- Row vector (or row matrix)
- Square matrix
- Main diagonal
- Equal matrices

- Matrix operations: sum, difference, scalar multiplication
- Linear combination of matrices
- Product of matrices (matrix multiplication)
- Partitioned matrices
- Submatrices

- Row-column method
- Column method
- Row method
- Coefficient matrix of a linear system
- Transpose
- Trace

Skills

- Determine the size of a given matrix.
- Identify the row vectors and column vectors of a given matrix.
- Perform the arithmetic operations of matrix addition, subtraction, scalar multiplication, and multiplication.
- Determine whether the product of two given matrices is defined.
- Compute matrix products using the row-column method, the column method, and the row method.

- Express the product of a matrix and a column vector as a linear combination of the columns of the matrix.
- Express a linear system as a matrix equation, and identify the coefficient matrix.
- Compute the transpose of a matrix.
- Compute the trace of a square matrix.

Exercise Set 1.3

1. Suppose that A, B, C, D, and E are matrices with the following sizes:

A	B	C	D	E
(4×5)	(4×5)	(5×2)	(4×2)	(5×4)

In each part, determine whether the given matrix expression is defined. For those that are defined, give the size of the resulting matrix.

(a) BA

(b) $AC + D$

(c) $AE + B$

(d) $AB + B$

(e) $E(A + B)$

(f) $E(AC)$

(g) $E^T A$

(h) $(A^T + E)D$

2. Suppose that $A, B, C, D,$ and E are matrices with the following sizes:

A	B	C	D	E
(3×1)	(3×6)	(6×2)	(2×6)	(1×3)

In each part, determine whether the given matrix expression is defined. For those that are defined, give the size of the resulting matrix.

(a) EA

(b) AB^T

(c) $B^T(A + E^T)$

(d) $2A + C$

(e) $(C^T + D)B^T$

(f) $CD + B^T E^T$

(g) $(BD^T)C^T$

(h) $DC + EA$

3. Consider the matrices

$$A = \begin{bmatrix} 3 & 0 \\ -1 & 2 \\ 1 & 1 \end{bmatrix}, \quad B = \begin{bmatrix} 4 & -1 \\ 0 & 2 \end{bmatrix}, \quad C = \begin{bmatrix} 1 & 4 & 2 \\ 3 & 1 & 5 \end{bmatrix},$$

$$D = \begin{bmatrix} 1 & 5 & 2 \\ -1 & 0 & 1 \\ 3 & 2 & 4 \end{bmatrix}, \quad E = \begin{bmatrix} 6 & 1 & 3 \\ -1 & 1 & 2 \\ 4 & 1 & 3 \end{bmatrix}$$

In each part, compute the given expression (where possible).

(a) $D + E$

(b) $D - E$

(c) $5A$

(d) $-7C$

(e) $2B - C$

(f) $4E - 2D$

(g) $-3(D + 2E)$

(h) $A - A$

(i) $\operatorname{tr}(D)$

(j) $\operatorname{tr}(D - 3E)$

(k) $4\operatorname{tr}(7B)$

(l) $\operatorname{tr}(A)$

4. Using the matrices in Exercise 3, in each part compute the given expression (where possible).

(a) $2A^T + C$

(b) $D^T - E^T$

(c) $(D - E)^T$

(d) $B^T + 5C^T$

(e) $\frac{1}{2}C^T - \frac{1}{4}A$

(f) $B - B^T$

(g) $2E^T - 3D^T$

(h) $(2E^T - 3D^T)^T$

(i) $(CD)E$

(j) $C(BA)$

(k) $\operatorname{tr}(DE^T)$

(l) $\operatorname{tr}(BC)$

5. Using the matrices in Exercise 3, in each part compute the given expression (where possible).

(a) AB

(b) BA

(c) $(3E)D$

(d) $(AB)C$

(e) $A(BC)$

(f) CC^T

(g) $(DA)^T$

(h) $(C^TB)A^T$

(i) $\operatorname{tr}(DD^T)$

(j) $\operatorname{tr}(4E^T - D)$

(k) $\operatorname{tr}(C^TA^T + 2E^T)$

(l) $\operatorname{tr}((EC^T)^TA)$

6. Using the matrices in Exercise 3, in each part compute the given expression (where possible).

(a) $(2D^T - E)A$

(b) $(4B)C + 2B$

(c) $(-AC)^T + 5D^T$

(d) $(BA^T - 2C)^T$

(e) $B^T(CC^T - A^TA)$

(f) $D^TE^T - (ED)^T$

7. Let

$$A = \begin{bmatrix} 3 & -2 & 7 \\ 6 & 5 & 4 \\ 0 & 4 & 9 \end{bmatrix} \quad \text{and} \quad B = \begin{bmatrix} 6 & -2 & 4 \\ 0 & 1 & 3 \\ 7 & 7 & 5 \end{bmatrix}$$

Use the row method or column method (as appropriate) to find

(a) the first row of AB.

(b) the third row of AB.

(c) the second column of AB.

(d) the first column of BA.

(e) the third row of AA.

(f) the third column of AA.

8. Referring to the matrices in Exercise 7, use the row method or column method (as appropriate) to find

(a) the first column of AB.

(b) the third column of BB.

(c) the second row of BB.

(d) the first column of AA.

(e) the third column of AB.

(f) the first row of BA.

9. Referring to the matrices in Exercise 7 and Example 9,

(a) express each column vector of AA as a linear combination of the column vectors of A.

(b) express each column vector of BB as a linear combination of the column vectors of B.

10. Referring to the matrices in Exercise 7 and Example 9,

(a) express each column vector of AB as a linear combination of the column vectors of A.

(b) express each column vector of BA as a linear combination of the column vectors of B.

11. In each part, find matrices A, $\mathbf{x}$, and $\mathbf{b}$ that express the given system of linear equations as a single matrix equation $A\mathbf{x} = \mathbf{b}$, and write out this matrix equation.

(a)
$$\begin{aligned} 2x_1 - 3x_2 + 5x_3 &= 7 \\ 9x_1 - x_2 + x_3 &= -1 \\ x_1 + 5x_2 + 4x_3 &= 0 \end{aligned}$$

(b)
$$\begin{aligned} 4x_1 \qquad - 3x_3 + x_4 &= 1 \\ 5x_1 + x_2 \qquad - 8x_4 &= 3 \\ 2x_1 - 5x_2 + 9x_3 - x_4 &= 0 \\ 3x_2 - x_3 + 7x_4 &= 2 \end{aligned}$$

12. In each part, find matrices A, $\mathbf{x}$, and $\mathbf{b}$ that express the given system of linear equations as a single matrix equation $A\mathbf{x} = \mathbf{b}$, and write out this matrix equation.

(a)
$$\begin{aligned} x_1 - 2x_2 + 3x_3 &= -3 \\ 2x_1 + x_2 \qquad &= 0 \\ -3x_2 + 4x_3 &= 1 \\ x_1 \qquad + x_3 &= 5 \end{aligned}$$

(b)
$$\begin{aligned} 3x_1 + 3x_2 + 3x_3 &= -3 \\ -x_1 - 5x_2 - 2x_3 &= 3 \\ -4x_2 + x_3 &= 0 \end{aligned}$$

13. In each part, express the matrix equation as a system of linear equations.

(a)
$$\begin{bmatrix} 5 & 6 & -7 \\ -1 & -2 & 3 \\ 0 & 4 & -1 \end{bmatrix} \begin{bmatrix} x_1 \\ x_2 \\ x_3 \end{bmatrix} = \begin{bmatrix} 2 \\ 0 \\ 3 \end{bmatrix}$$

(b) $\begin{bmatrix} 1 & 1 & 1 \\ 2 & 3 & 0 \\ 5 & -3 & -6 \end{bmatrix} \begin{bmatrix} x_1 \\ x_2 \\ x_3 \end{bmatrix} = \begin{bmatrix} 2 \\ 2 \\ -9 \end{bmatrix}$

14. In each part, express the matrix equation as a system of linear equations.

(a) $\begin{bmatrix} 3 & -1 & 2 \\ 4 & 3 & 7 \\ -2 & 1 & 5 \end{bmatrix} \begin{bmatrix} x_1 \\ x_2 \\ x_3 \end{bmatrix} = \begin{bmatrix} 2 \\ -1 \\ 4 \end{bmatrix}$

(b) $\begin{bmatrix} 3 & -2 & 0 & 1 \\ 5 & 0 & 2 & -2 \\ 3 & 1 & 4 & 7 \\ -2 & 5 & 1 & 6 \end{bmatrix} \begin{bmatrix} w \\ x \\ y \\ z \end{bmatrix} = \begin{bmatrix} 0 \\ 0 \\ 0 \\ 0 \end{bmatrix}$

In Exercises 15–16, find all values of k, if any, that satisfy the equation.

15. $\begin{bmatrix} k & 1 & 1 \end{bmatrix} \begin{bmatrix} 1 & 1 & 0 \\ 1 & 0 & 2 \\ 0 & 2 & -3 \end{bmatrix} \begin{bmatrix} k \\ 1 \\ 1 \end{bmatrix} = 0$

16. $\begin{bmatrix} 2 & 2 & k \end{bmatrix} \begin{bmatrix} 1 & 2 & 0 \\ 2 & 0 & 3 \\ 0 & 3 & 1 \end{bmatrix} \begin{bmatrix} 2 \\ 2 \\ k \end{bmatrix} = 0$

In Exercises 17–18, solve the matrix equation for a, b, c, and d.

17. $\begin{bmatrix} a & 3 \\ -1 & a+b \end{bmatrix} = \begin{bmatrix} 4 & d-2c \\ d+2c & -2 \end{bmatrix}$

18. $\begin{bmatrix} a-b & b+a \\ 3d+c & 2d-c \end{bmatrix} = \begin{bmatrix} 8 & 1 \\ 7 & 6 \end{bmatrix}$

19. Let A be any $m \times n$ matrix and let 0 be the $m \times n$ matrix each of whose entries is zero. Show that if $kA = 0$, then $k = 0$ or $A = 0$.

20. (a) Show that if AB and BA are both defined, then AB and BA are square matrices.

(b) Show that if A is an $m \times n$ matrix and $A(BA)$ is defined, then B is an $n \times m$ matrix.

21. Prove: If A and B are $n \times n$ matrices, then

$$\text{tr}(A+B) = \text{tr}(A) + \text{tr}(B)$$

22. (a) Show that if A has a row of zeros and B is any matrix for which AB is defined, then AB also has a row of zeros.

(b) Find a similar result involving a column of zeros.

23. In each part, find a 6×6 matrix $[a_{ij}]$ that satisfies the stated condition. Make your answers as general as possible by using letters rather than specific numbers for the nonzero entries.

(a) $a_{ij} = 0$ if $i \neq j$ (b) $a_{ij} = 0$ if $i > j$

(c) $a_{ij} = 0$ if $i < j$

(d) $a_{ij} = 0$ if $|i - j| > 1$

24. Find the 4×4 matrix $A = [a_{ij}]$ whose entries satisfy the stated condition.

(a) $a_{ij} = i + j$ (b) $a_{ij} = i^{j-1}$

(c) $a_{ij} = \begin{cases} 1 & \text{if } |i-j| > 1 \\ -1 & \text{if } |i-j| \leq 1 \end{cases}$

25. Consider the function $y = f(x)$ defined for 2×1 matrices x by $y = Ax$, where

$$A = \begin{bmatrix} 1 & 1 \\ 0 & 1 \end{bmatrix}$$

Plot $f(x)$ together with x in each case below. How would you describe the action of f?

(a) $x = \begin{pmatrix} 1 \\ 1 \end{pmatrix}$ (b) $x = \begin{pmatrix} 2 \\ 0 \end{pmatrix}$

(c) $x = \begin{pmatrix} 4 \\ 3 \end{pmatrix}$ (d) $x = \begin{pmatrix} 2 \\ -2 \end{pmatrix}$

26. Let I be the $n \times n$ matrix whose entry in row i and column j is

$$\begin{cases} 1 & \text{if } i = j \\ 0 & \text{if } i \neq j \end{cases}$$

Show that $AI = IA = A$ for every $n \times n$ matrix A.

27. How many 3×3 matrices A can you find such that

$$A \begin{bmatrix} x \\ y \\ z \end{bmatrix} = \begin{bmatrix} x + y \\ x - y \\ 0 \end{bmatrix}$$

for all choices of x, y, and z?

28. How many 3×3 matrices A can you find such that

$$A \begin{bmatrix} x \\ y \\ z \end{bmatrix} = \begin{bmatrix} xy \\ 0 \\ 0 \end{bmatrix}$$

for all choices of x, y, and z?

29. A matrix B is said to be a *square root* of a matrix A if $BB = A$.

(a) Find two square roots of $A = \begin{bmatrix} 2 & 2 \\ 2 & 2 \end{bmatrix}$.

(b) How many different square roots can you find of $A = \begin{bmatrix} 5 & 0 \\ 0 & 9 \end{bmatrix}$?

(c) Do you think that every 2×2 matrix has at least one square root? Explain your reasoning.

30. Let 0 denote a 2×2 matrix, each of whose entries is zero.

(a) Is there a 2×2 matrix A such that $A \neq 0$ and $AA = 0$? Justify your answer.

(b) Is there a 2×2 matrix A such that $A \neq 0$ and $AA = A$? Justify your answer.

True-False Exercises

In parts (a)–(o) determine whether the statement is true or false, and justify your answer.

(a) The matrix $\begin{bmatrix} 1 & 2 & 3 \\ 4 & 5 & 6 \end{bmatrix}$ has no main diagonal.

(b) An $m \times n$ matrix has m column vectors and n row vectors.

(c) If A and B are 2×2 matrices, then $AB = BA$.

(d) The ith row vector of a matrix product AB can be computed by multiplying A by the ith row vector of B.

(e) For every matrix A, it is true that $(A^T)^T = A$.

(f) If A and B are square matrices of the same order, then $\text{tr}(AB) = \text{tr}(A)\text{tr}(B)$.

(g) If A and B are square matrices of the same order, then $(AB)^T = A^T B^T$.

(h) For every square matrix A, it is true that $\text{tr}(A^T) = \text{tr}(A)$.

(i) If A is a 6×4 matrix and B is an $m \times n$ matrix such that $B^T A^T$ is a 2×6 matrix, then $m = 4$ and $n = 2$.

(j) If A is an $n \times n$ matrix and c is a scalar, then $\text{tr}(cA) = c\,\text{tr}(A)$.

(k) If A, B, and C are matrices of the same size such that $A - C = B - C$, then $A = B$.

(l) If A, B, and C are square matrices of the same order such that $AC = BC$, then $A = B$.

(m) If $AB + BA$ is defined, then A and B are square matrices of the same size.

(n) If B has a column of zeros, then so does AB if this product is defined.

(o) If B has a column of zeros, then so does BA if this product is defined.

1.4 Inverses; Algebraic Properties of Matrices

In this section we will discuss some of the algebraic properties of matrix operations. We will see that many of the basic rules of arithmetic for real numbers hold for matrices, but we will also see that some do not.

Properties of Matrix Addition and Scalar Multiplication

The following theorem lists the basic algebraic properties of the matrix operations.

THEOREM 1.4.1 **Properties of Matrix Arithmetic**

Assuming that the sizes of the matrices are such that the indicated operations can be performed, the following rules of matrix arithmetic are valid.

(a) $A + B = B + A$ **(Commutative law for addition)**

(b) $A + (B + C) = (A + B) + C$ **(Associative law for addition)**

(c) $A(BC) = (AB)C$ **(Associative law for multiplication)**

(d) $A(B + C) = AB + AC$ **(Left distributive law)**

(e) $(B + C)A = BA + CA$ **(Right distributive law)**

(f) $A(B - C) = AB - AC$

(g) $(B - C)A = BA - CA$

(h) $a(B + C) = aB + aC$

(i) $a(B - C) = aB - aC$

(j) $(a + b)C = aC + bC$

(k) $(a - b)C = aC - bC$

(l) $a(bC) = (ab)C$

(m) $a(BC) = (aB)C = B(aC)$

To prove any of the equalities in this theorem we must show that the matrix on the left side has the same size as that on the right and that the corresponding entries on the two sides are the same. Most of the proofs follow the same pattern, so we will prove part (d) as a sample. The proof of the associative law for multiplication is more complicated than the rest and is outlined in the exercises.

Proof (d) We must show that $A(B + C)$ and $AB + AC$ have the same size and that corresponding entries are equal. To form $A(B + C)$, the matrices B and C must have the same size, say $m \times n$, and the matrix A must then have m columns, so its size must be of the form $r \times m$. This makes $A(B + C)$ an $r \times n$ matrix. It follows that $AB + AC$ is also an $r \times n$ matrix and, consequently, $A(B + C)$ and $AB + AC$ have the same size.

Suppose that $A = [a_{ij}]$, $B = [b_{ij}]$, and $C = [c_{ij}]$. We want to show that corresponding entries of $A(B + C)$ and $AB + AC$ are equal; that is,

$$[A(B + C)]_{ij} = [AB + AC]_{ij}$$

for all values of i and j. But from the definitions of matrix addition and matrix multiplication, we have

$$[A(B + C)]_{ij} = a_{i1}(b_{1j} + c_{1j}) + a_{i2}(b_{2j} + c_{2j}) + \cdots + a_{im}(b_{mj} + c_{mj})$$
$$= (a_{i1}b_{1j} + a_{i2}b_{2j} + \cdots + a_{im}b_{mj}) + (a_{i1}c_{1j} + a_{i2}c_{2j} + \cdots + a_{im}c_{mj})$$
$$= [AB]_{ij} + [AC]_{ij} = [AB + AC]_{ij} \blacktriangleleft$$

> There are three basic ways to prove that two matrices of the same size are equal—prove that corresponding entries are the same, prove that corresponding row vectors are the same, or prove that corresponding column vectors are the same.

Remark Although the operations of matrix addition and matrix multiplication were defined for pairs of matrices, associative laws (b) and (c) enable us to denote sums and products of three matrices as $A + B + C$ and ABC without inserting any parentheses. This is justified by the fact that no matter how parentheses are inserted, the associative laws guarantee that the same end result will be obtained. In general, *given any sum or any product of matrices, pairs of parentheses can be inserted or deleted anywhere within the expression without affecting the end result.*

▶ **EXAMPLE 1 Associativity of Matrix Multiplication**

As an illustration of the associative law for matrix multiplication, consider

$$A = \begin{bmatrix} 1 & 2 \\ 3 & 4 \\ 0 & 1 \end{bmatrix}, \quad B = \begin{bmatrix} 4 & 3 \\ 2 & 1 \end{bmatrix}, \quad C = \begin{bmatrix} 1 & 0 \\ 2 & 3 \end{bmatrix}$$

Then

$$AB = \begin{bmatrix} 1 & 2 \\ 3 & 4 \\ 0 & 1 \end{bmatrix} \begin{bmatrix} 4 & 3 \\ 2 & 1 \end{bmatrix} = \begin{bmatrix} 8 & 5 \\ 20 & 13 \\ 2 & 1 \end{bmatrix} \quad \text{and} \quad BC = \begin{bmatrix} 4 & 3 \\ 2 & 1 \end{bmatrix} \begin{bmatrix} 1 & 0 \\ 2 & 3 \end{bmatrix} = \begin{bmatrix} 10 & 9 \\ 4 & 3 \end{bmatrix}$$

Thus

$$(AB)C = \begin{bmatrix} 8 & 5 \\ 20 & 13 \\ 2 & 1 \end{bmatrix} \begin{bmatrix} 1 & 0 \\ 2 & 3 \end{bmatrix} = \begin{bmatrix} 18 & 15 \\ 46 & 39 \\ 4 & 3 \end{bmatrix}$$

and

$$A(BC) = \begin{bmatrix} 1 & 2 \\ 3 & 4 \\ 0 & 1 \end{bmatrix} \begin{bmatrix} 10 & 9 \\ 4 & 3 \end{bmatrix} = \begin{bmatrix} 18 & 15 \\ 46 & 39 \\ 4 & 3 \end{bmatrix}$$

so $(AB)C = A(BC)$, as guaranteed by Theorem 1.4.1(c). ◀

Properties of Matrix Multiplication

Do not let Theorem 1.4.1 lull you into believing that *all* laws of real arithmetic carry over to matrix arithmetic. For example, you know that in real arithmetic it is always true that $ab = ba$, which is called the *commutative law for multiplication*. In matrix arithmetic, however, the equality of AB and BA can fail for three possible reasons:

1. AB may be defined and BA may not (for example, if A is 2×3 and B is 3×4).

2. AB and BA may both be defined, but they may have different sizes (for example, if A is 2×3 and B is 3×2).

3. AB and BA may both be defined and have the same size, but the two matrices may be different (as illustrated in the next example).

▶ **EXAMPLE 2 Order Matters in Matrix Multiplication**

Consider the matrices

$$A = \begin{bmatrix} -1 & 0 \\ 2 & 3 \end{bmatrix} \quad \text{and} \quad B = \begin{bmatrix} 1 & 2 \\ 3 & 0 \end{bmatrix}$$

Multiplying gives

$$AB = \begin{bmatrix} -1 & -2 \\ 11 & 4 \end{bmatrix} \quad \text{and} \quad BA = \begin{bmatrix} 3 & 6 \\ -3 & 0 \end{bmatrix}$$

Thus, $AB \neq BA$. ◀

> Do not read too much into Example 2—it does not rule out the possibility that AB and BA may be equal in *certain* cases, just that they are not equal in *all* cases. If it so happens that $AB = BA$, then we say that AB and BA **commute**.

Zero Matrices

A matrix whose entries are all zero is called a **zero matrix**. Some examples are

$$\begin{bmatrix} 0 & 0 \\ 0 & 0 \end{bmatrix}, \quad \begin{bmatrix} 0 & 0 & 0 \\ 0 & 0 & 0 \\ 0 & 0 & 0 \end{bmatrix}, \quad \begin{bmatrix} 0 & 0 & 0 & 0 \\ 0 & 0 & 0 & 0 \end{bmatrix}, \quad \begin{bmatrix} 0 \\ 0 \\ 0 \\ 0 \end{bmatrix}, \quad [0]$$

We will denote a zero matrix by 0 unless it is important to specify its size, in which case we will denote the $m \times n$ zero matrix by $0_{m \times n}$.

It should be evident that if A and 0 are matrices with the same size, then

$$A + 0 = 0 + A = A$$

Thus, 0 plays the same role in this matrix equation that the number 0 plays in the numerical equation $a + 0 = 0 + a = a$.

The following theorem lists the basic properties of zero matrices. Since the results should be self-evident, we will omit the formal proofs.

THEOREM 1.4.2 Properties of Zero Matrices

If c is a scalar, and if the sizes of the matrices are such that the operations can be perfomed, then:

(a) $A + 0 = 0 + A = A$

(b) $A - 0 = A$

(c) $A - A = A + (-A) = 0$

(d) $0A = 0$

(e) *If $cA = 0$, then $c = 0$ or $A = 0$.*

Since we know that the commutative law of real arithmetic is not valid in matrix arithmetic, it should not be surprising that there are other rules that fail as well. For example, consider the following two laws of real arithmetic:

- If $ab = bc$ and $a \neq 0$, then $b = c$. [**The cancellation law**]
- If $ab = 0$, then at least one of the factors on the left is 0.

The next two examples show that these laws are not universally true in matrix arithmetic.

▶ **EXAMPLE 3** **Failure of the Cancellation Law**

Consider the matrices

$$A = \begin{bmatrix} 0 & 1 \\ 0 & 2 \end{bmatrix}, \quad B = \begin{bmatrix} 1 & 1 \\ 3 & 4 \end{bmatrix}, \quad C = \begin{bmatrix} 2 & 5 \\ 3 & 4 \end{bmatrix}$$

We leave it for you to confirm that

$$AB = AC = \begin{bmatrix} 3 & 4 \\ 6 & 8 \end{bmatrix}$$

Although $A \neq 0$, canceling A from both sides of the equation $AB = AC$ would lead to the incorrect conclusion that $B = C$. Thus, the cancellation law does not hold, in general, for matrix multiplication.

▶ **EXAMPLE 4** **A Zero Product with Nonzero Factors**

Here are two matrices for which $AB = 0$, but $A \neq 0$ and $B \neq 0$:

$$A = \begin{bmatrix} 0 & 1 \\ 0 & 2 \end{bmatrix}, \quad B = \begin{bmatrix} 3 & 7 \\ 0 & 0 \end{bmatrix} \quad ◀$$

Identity Matrices A square matrix with 1's on the main diagonal and zeros elsewhere is called an ***identity matrix***. Some examples are

$$[1], \quad \begin{bmatrix} 1 & 0 \\ 0 & 1 \end{bmatrix}, \quad \begin{bmatrix} 1 & 0 & 0 \\ 0 & 1 & 0 \\ 0 & 0 & 1 \end{bmatrix}, \quad \begin{bmatrix} 1 & 0 & 0 & 0 \\ 0 & 1 & 0 & 0 \\ 0 & 0 & 1 & 0 \\ 0 & 0 & 0 & 1 \end{bmatrix}$$

An identity matrix is denoted by the letter I. If it is important to emphasize the size, we will write I_n for the $n \times n$ identity matrix.

To explain the role of identity matrices in matrix arithmetic, let us consider the effect of multiplying a general 2×3 matrix A on each side by an identity matrix. Multiplying on the right by the 3×3 identity matrix yields

$$AI_3 = \begin{bmatrix} a_{11} & a_{12} & a_{13} \\ a_{21} & a_{22} & a_{23} \end{bmatrix} \begin{bmatrix} 1 & 0 & 0 \\ 0 & 1 & 0 \\ 0 & 0 & 1 \end{bmatrix} = \begin{bmatrix} a_{11} & a_{12} & a_{13} \\ a_{21} & a_{22} & a_{23} \end{bmatrix} = A$$

and multiplying on the left by the 2×2 identity matrix yields

$$I_2 A = \begin{bmatrix} 1 & 0 \\ 0 & 1 \end{bmatrix} \begin{bmatrix} a_{11} & a_{12} & a_{13} \\ a_{21} & a_{22} & a_{23} \end{bmatrix} = \begin{bmatrix} a_{11} & a_{12} & a_{13} \\ a_{21} & a_{22} & a_{23} \end{bmatrix} = A$$

The same result holds in general; that is, if A is any $m \times n$ matrix, then

$$AI_n = A \quad \text{and} \quad I_m A = A$$

Thus, the identity matrices play the same role in these matrix equations that the number 1 plays in the numerical equation $a \cdot 1 = 1 \cdot a = a$.

As the next theorem shows, identity matrices arise naturally in studying reduced row echelon forms of *square* matrices.

THEOREM 1.4.3 *If R is the reduced row echelon form of an $n \times n$ matrix A, then either R has a row of zeros or R is the identity matrix I_n.*

Proof   Suppose that the reduced row echelon form of A is

$$R = \begin{bmatrix} r_{11} & r_{12} & \cdots & r_{1n} \\ r_{21} & r_{22} & \cdots & r_{2n} \\ \vdots & \vdots & & \vdots \\ r_{n1} & r_{n2} & \cdots & r_{nn} \end{bmatrix}$$

Either the last row in this matrix consists entirely of zeros or it does not. If not, the matrix contains no zero rows, and consequently each of the n rows has a leading entry of 1. Since these leading 1's occur progressively farther to the right as we move down the matrix, each of these 1's must occur on the main diagonal. Since the other entries in the same column as one of these 1's are zero, R must be I_n. Thus, either R has a row of zeros or $R = I_n$. ◄

Inverse of a Matrix   In real arithmetic every nonzero number a has a reciprocal $a^{-1}(= 1/a)$ with the property

$$a \cdot a^{-1} = a^{-1} \cdot a = 1$$

The number a^{-1} is sometimes called the *multiplicative inverse* of a. Our next objective is to develop an analog of this result for matrix arithmetic. For this purpose we make the following definition.

DEFINITION 1   If A is a square matrix, and if a matrix B of the same size can be found such that $AB = BA = I$, then A is said to be **invertible** (or **nonsingular**) and B is called an **inverse** of A. If no such matrix B can be found, then A is said to be **singular**.

Remark   The relationship $AB = BA = I$ is not changed by interchanging A and B, so if A is invertible and B is an inverse of A, then it is also true that B is invertible, and A is an inverse of B. Thus, when

$$AB = BA = I$$

we say that A and B are *inverses of one another*.

► **EXAMPLE 5 An Invertible Matrix**

Let

$$A = \begin{bmatrix} 2 & -5 \\ -1 & 3 \end{bmatrix} \quad \text{and} \quad B = \begin{bmatrix} 3 & 5 \\ 1 & 2 \end{bmatrix}$$

Then

$$AB = \begin{bmatrix} 2 & -5 \\ -1 & 3 \end{bmatrix}\begin{bmatrix} 3 & 5 \\ 1 & 2 \end{bmatrix} = \begin{bmatrix} 1 & 0 \\ 0 & 1 \end{bmatrix} = I$$

$$BA = \begin{bmatrix} 3 & 5 \\ 1 & 2 \end{bmatrix}\begin{bmatrix} 2 & -5 \\ -1 & 3 \end{bmatrix} = \begin{bmatrix} 1 & 0 \\ 0 & 1 \end{bmatrix} = I$$

Thus, A and B are invertible and each is an inverse of the other. ◄

▶ **EXAMPLE 6 A Class of Singular Matrices**

In general, a square matrix with a row or column of zeros is singular. To help understand why this is so, consider the matrix

$$A = \begin{bmatrix} 1 & 4 & 0 \\ 2 & 5 & 0 \\ 3 & 6 & 0 \end{bmatrix}$$

To prove that A is singular we must show that there is no 3×3 matrix B such that $AB = BA = I$. For this purpose let $\mathbf{c}_1$, $\mathbf{c}_2$, $\mathbf{0}$ be the column vectors of A. Thus, for any 3×3 matrix B we can express the product BA as

$$BA = B[\mathbf{c}_1 \quad \mathbf{c}_2 \quad \mathbf{0}] = [B\mathbf{c}_1 \quad B\mathbf{c}_2 \quad \mathbf{0}] \quad \text{[Formula (6) of Section 1.3]}$$

The column of zeros shows that $BA \neq I$ and hence that A is singular. ◀

Properties of Inverses

It is reasonable to ask whether an invertible matrix can have more than one inverse. The next theorem shows that the answer is no—*an invertible matrix has exactly one inverse.*

THEOREM 1.4.4 *If B and C are both inverses of the matrix A, then B = C.*

Proof Since B is an inverse of A, we have $BA = I$. Multiplying both sides on the right by C gives $(BA)C = IC = C$. But it is also true that $(BA)C = B(AC) = BI = B$, so $C = B$. ◀

As a consequence of this important result, we can now speak of "the" inverse of an invertible matrix. If A is invertible, then its inverse will be denoted by the symbol A^{-1}. Thus,

$$AA^{-1} = I \quad \text{and} \quad A^{-1}A = I \tag{1}$$

The inverse of A plays much the same role in matrix arithmetic that the reciprocal a^{-1} plays in the numerical relationships $aa^{-1} = 1$ and $a^{-1}a = 1$.

In the next section we will develop a method for computing the inverse of an invertible matrix of any size. For now we give the following theorem that specifies conditions under which a 2×2 matrix is invertible and provides a simple formula for its inverse.

THEOREM 1.4.5 *The matrix*

$$A = \begin{bmatrix} a & b \\ c & d \end{bmatrix}$$

is invertible if and only if $ad - bc \neq 0$, in which case the inverse is given by the formula

$$A^{-1} = \frac{1}{ad - bc} \begin{bmatrix} d & -b \\ -c & a \end{bmatrix} \tag{2}$$

The quantity $ad - bc$ in Theorem 1.4.5 is called the ***determinant*** of the 2×2 matrix A and is denoted by

$$\det(A) = ad - bc$$

or alternatively by

$$\begin{vmatrix} a & b \\ c & d \end{vmatrix} = ad - bc$$

We will omit the proof, because we will study a more general version of this theorem later. For now, you should at least confirm the validity of Formula (2) by showing that $AA^{-1} = A^{-1}A = I$.

Historical Note The formula for A^{-1} given in Theorem 1.4.5 first appeared (in a more general form) in Arthur Cayley's 1858 *Memoir on the Theory of Matrices*. The more general result that Cayley discovered will be studied later.

$$\det(A) = \begin{vmatrix} a & b \\ c & d \end{vmatrix} = ad - bc$$

▲ Figure 1.4.1

Remark Figure 1.4.1 illustrates that the determinant of a 2×2 matrix A is the product of the entries on its main diagonal minus the product of the entries *off* its main diagonal. In words, Theorem 1.4.5 states that a 2×2 matrix A is invertible if and only if its determinant is nonzero, and if invertible, then its inverse can be obtained by interchanging its diagonal entries, reversing the signs of its off-diagonal entries, and multiplying the entries by the reciprocal of the determinant of A.

▶ **EXAMPLE 7** Calculating the Inverse of a 2 × 2 Matrix

In each part, determine whether the matrix is invertible. If so, find its inverse.

$$\text{(a) } A = \begin{bmatrix} 6 & 1 \\ 5 & 2 \end{bmatrix} \qquad \text{(b) } A = \begin{bmatrix} -1 & 2 \\ 3 & -6 \end{bmatrix}$$

Solution (a) The determinant of A is $\det(A) = (6)(2) - (1)(5) = 7$, which is nonzero. Thus, A is invertible, and its inverse is

$$A^{-1} = \frac{1}{7} \begin{bmatrix} 2 & -1 \\ -5 & 6 \end{bmatrix} = \begin{bmatrix} \frac{2}{7} & -\frac{1}{7} \\ -\frac{5}{7} & \frac{6}{7} \end{bmatrix}$$

We leave it for you to confirm that $AA^{-1} = A^{-1}A = I$.

Solution (b) The matrix is not invertible since $\det(A) = (-1)(-6) - (2)(3) = 0$.

▶ **EXAMPLE 8** Solution of a Linear System by Matrix Inversion

A problem that arises in many applications is to solve a pair of equations of the form

$$u = ax + by$$
$$v = cx + dy$$

for x and y in terms of u and v. One approach is to treat this as a linear system of two equations in the unknowns x and y and use Gauss–Jordan elimination to solve for x and y. However, because the coefficients of the unknowns are *literal* rather than *numerical*, this procedure is a little clumsy. As an alternative approach, let us replace the two equations by the single matrix equation

$$\begin{bmatrix} u \\ v \end{bmatrix} = \begin{bmatrix} ax + by \\ cx + dy \end{bmatrix}$$

which we can rewrite as

$$\begin{bmatrix} u \\ v \end{bmatrix} = \begin{bmatrix} a & b \\ c & d \end{bmatrix} \begin{bmatrix} x \\ y \end{bmatrix}$$

If we assume that the 2×2 matrix is invertible (i.e., $ad - bc \neq 0$), then we can multiply through on the left by the inverse and rewrite the equation as

$$\begin{bmatrix} a & b \\ c & d \end{bmatrix}^{-1} \begin{bmatrix} u \\ v \end{bmatrix} = \begin{bmatrix} a & b \\ c & d \end{bmatrix}^{-1} \begin{bmatrix} a & b \\ c & d \end{bmatrix} \begin{bmatrix} x \\ y \end{bmatrix}$$

which simplifies to

$$\begin{bmatrix} a & b \\ c & d \end{bmatrix}^{-1} \begin{bmatrix} u \\ v \end{bmatrix} = \begin{bmatrix} x \\ y \end{bmatrix}$$

Using Theorem 1.4.5, we can rewrite this equation as

$$\frac{1}{ad - bc} \begin{bmatrix} d & -b \\ -c & a \end{bmatrix} \begin{bmatrix} u \\ v \end{bmatrix} = \begin{bmatrix} x \\ y \end{bmatrix}$$

from which we obtain

$$x = \frac{du - bv}{ad - bc}, \quad y = \frac{av - cu}{ad - bc} \blacktriangleleft$$

The next theorem is concerned with inverses of matrix products.

THEOREM 1.4.6 *If A and B are invertible matrices with the same size, then AB is invertible and*

$$(AB)^{-1} = B^{-1}A^{-1}$$

Proof We can establish the invertibility and obtain the stated formula at the same time by showing that

$$(AB)(B^{-1}A^{-1}) = (B^{-1}A^{-1})(AB) = I$$

But

$$(AB)(B^{-1}A^{-1}) = A(BB^{-1})A^{-1} = AIA^{-1} = AA^{-1} = I$$

and similarly, $(B^{-1}A^{-1})(AB) = I$. ◀

Although we will not prove it, this result can be extended to three or more factors:

A product of any number of invertible matrices is invertible, and the inverse of the product is the product of the inverses in the reverse order.

▶ **EXAMPLE 9** **The Inverse of a Product**

Consider the matrices

$$A = \begin{bmatrix} 1 & 2 \\ 1 & 3 \end{bmatrix}, \quad B = \begin{bmatrix} 3 & 2 \\ 2 & 2 \end{bmatrix}$$

We leave it for you to show that

| If a product of matrices is singular, then at least one of the factors must be singular. Why? |

$$AB = \begin{bmatrix} 7 & 6 \\ 9 & 8 \end{bmatrix}, \quad (AB)^{-1} = \begin{bmatrix} 4 & -3 \\ -\frac{9}{2} & \frac{7}{2} \end{bmatrix}$$

and also that

$$A^{-1} = \begin{bmatrix} 3 & -2 \\ -1 & 1 \end{bmatrix}, \quad B^{-1} = \begin{bmatrix} 1 & -1 \\ -1 & \frac{3}{2} \end{bmatrix}, \quad B^{-1}A^{-1} = \begin{bmatrix} 1 & -1 \\ -1 & \frac{3}{2} \end{bmatrix}\begin{bmatrix} 3 & -2 \\ -1 & 1 \end{bmatrix} = \begin{bmatrix} 4 & -3 \\ -\frac{9}{2} & \frac{7}{2} \end{bmatrix}$$

Thus, $(AB)^{-1} = B^{-1}A^{-1}$ as guaranteed by Theorem 1.4.6. ◀

Powers of a Matrix If A is a *square* matrix, then we define the nonnegative integer powers of A to be

$$A^0 = I \quad \text{and} \quad A^n = AA \cdots A \quad \text{[\textit{n} factors]}$$

and if A is invertible, then we define the negative integer powers of A to be

$$A^{-n} = (A^{-1})^n = A^{-1}A^{-1} \cdots A^{-1} \quad \text{[\textit{n} factors]}$$

Because these definitions parallel those for real numbers, the usual laws of nonnegative exponents hold; for example,

$$A^r A^s = A^{r+s} \quad \text{and} \quad (A^r)^s = A^{rs}$$

In addition, we have the following properties of negative exponents.

THEOREM 1.4.7 *If A is invertible and n is a nonnegative integer, then*:

(a) A^{-1} *is invertible and* $(A^{-1})^{-1} = A$.

(b) A^n *is invertible and* $(A^n)^{-1} = A^{-n} = (A^{-1})^n$.

(c) kA *is invertible for any nonzero scalar* k, *and* $(kA)^{-1} = k^{-1}A^{-1}$.

We will prove part (c) and leave the proofs of parts (a) and (b) as exercises.

Proof (c) Property (c) in Theorem 1.4.1 and property (f) in Theorem 1.4.2 imply that

$$(kA)(k^{-1}A^{-1}) = k^{-1}(kA)A^{-1} = (k^{-1}k)AA^{-1} = (1)I = I$$

and similarly, $(k^{-1}A^{-1})(kA) = I$. Thus, kA is invertible and $(kA)^{-1} = k^{-1}A^{-1}$. ◄

▶ **EXAMPLE 10 Properties of Exponents**

Let A and A^{-1} be the matrices in Example 9; that is,

$$A = \begin{bmatrix} 1 & 2 \\ 1 & 3 \end{bmatrix} \quad \text{and} \quad A^{-1} = \begin{bmatrix} 3 & -2 \\ -1 & 1 \end{bmatrix}$$

Then

$$A^{-3} = (A^{-1})^3 = \begin{bmatrix} 3 & -2 \\ -1 & 1 \end{bmatrix}\begin{bmatrix} 3 & -2 \\ -1 & 1 \end{bmatrix}\begin{bmatrix} 3 & -2 \\ -1 & 1 \end{bmatrix} = \begin{bmatrix} 41 & -30 \\ -15 & 11 \end{bmatrix}$$

Also,

$$A^3 = \begin{bmatrix} 1 & 2 \\ 1 & 3 \end{bmatrix}\begin{bmatrix} 1 & 2 \\ 1 & 3 \end{bmatrix}\begin{bmatrix} 1 & 2 \\ 1 & 3 \end{bmatrix} = \begin{bmatrix} 11 & 30 \\ 15 & 41 \end{bmatrix}$$

so, as expected from Theorem 1.4.7(b),

$$(A^3)^{-1} = \frac{1}{(11)(41) - (30)(15)}\begin{bmatrix} 41 & -30 \\ -15 & 11 \end{bmatrix} = \begin{bmatrix} 41 & -30 \\ -15 & 11 \end{bmatrix} = (A^{-1})^3$$

▶ **EXAMPLE 11 The Square of a Matrix Sum**

In real arithmetic, where we have a commutative law for multiplication, we can write

$$(a + b)^2 = a^2 + ab + ba + b^2 = a^2 + ab + ab + b^2 = a^2 + 2ab + b^2$$

However, in matrix arithmetic, where we have no commutative law for multiplication, the best we can do is to write

$$(A + B)^2 = A^2 + AB + BA + B^2$$

It is only in the special case where A and B *commute* (i.e., $AB = BA$) that we can go a step further and write

$$(A + B)^2 = A^2 + 2AB + B^2 \quad ◄$$

Matrix Polynomials   If A is a square matrix, say $n \times n$, and if

$$p(x) = a_0 + a_1x + a_2x^2 + \cdots + a_mx^m$$

is any polynomial, then we define the $n \times n$ matrix $p(A)$ to be

$$p(A) = a_0I + a_1A + a_2A^2 + \cdots + a_mA^m \tag{3}$$

where I is the $n \times n$ identity matrix; that is, $p(A)$ is obtained by substituting A for x and replacing the constant term a_0 by the matrix $a_0 I$. An expression of form (3) is called a *matrix polynomial in A*.

▶ **EXAMPLE 12 A Matrix Polynomial**

Find $p(A)$ for

$$p(x) = x^2 - 2x - 3 \quad \text{and} \quad A = \begin{bmatrix} -1 & 2 \\ 0 & 3 \end{bmatrix}$$

Solution

$$p(A) = A^2 - 2A - 3I$$

$$= \begin{bmatrix} -1 & 2 \\ 0 & 3 \end{bmatrix}^2 - 2 \begin{bmatrix} -1 & 2 \\ 0 & 3 \end{bmatrix} - 3 \begin{bmatrix} 1 & 0 \\ 0 & 1 \end{bmatrix}$$

$$= \begin{bmatrix} 1 & 4 \\ 0 & 9 \end{bmatrix} - \begin{bmatrix} -2 & 4 \\ 0 & 6 \end{bmatrix} - \begin{bmatrix} 3 & 0 \\ 0 & 3 \end{bmatrix} = \begin{bmatrix} 0 & 0 \\ 0 & 0 \end{bmatrix}$$

or more briefly, $p(A) = 0$. ◀

Remark It follows from the fact that $A^r A^s = A^{r+s} = A^{s+r} = A^s A^r$ that powers of a square matrix commute, and since a matrix polynomial in A is built up from powers of A, any two matrix polynomials in A also commute; that is, for any polynomials p_1 and p_2 we have

$$p_1(A)p_2(A) = p_2(A)p_1(A) \tag{4}$$

Properties of the Transpose   The following theorem lists the main properties of the transpose.

THEOREM 1.4.8 *If the sizes of the matrices are such that the stated operations can be performed, then:*

(a) $(A^T)^T = A$

(b) $(A + B)^T = A^T + B^T$

(c) $(A - B)^T = A^T - B^T$

(d) $(kA)^T = kA^T$

(e) $(AB)^T = B^T A^T$

If you keep in mind that transposing a matrix interchanges its rows and columns, then you should have little trouble visualizing the results in parts (*a*)–(*d*). For example, part (*a*) states the obvious fact that interchanging rows and columns twice leaves a matrix unchanged; and part (*b*) states that adding two matrices and then interchanging the rows and columns produces the same result as interchanging the rows and columns before adding. We will omit the formal proofs. Part (*e*) is a less obvious, but for brevity we will omit its proof as well. The result in that part can be extended to three or more factors and restated as:

The transpose of a product of any number of matrices is the product of the transposes in the reverse order.

The following theorem establishes a relationship between the inverse of a matrix and the inverse of its transpose.

> **THEOREM 1.4.9** *If A is an invertible matrix, then A^T is also invertible and*
> $$(A^T)^{-1} = (A^{-1})^T$$

Proof   We can establish the invertibility and obtain the formula at the same time by showing that
$$A^T(A^{-1})^T = (A^{-1})^T A^T = I$$

But from part (*e*) of Theorem 1.4.8 and the fact that $I^T = I$, we have
$$A^T(A^{-1})^T = (A^{-1}A)^T = I^T = I$$
$$(A^{-1})^T A^T = (AA^{-1})^T = I^T = I$$

which completes the proof. ◄

▶ **EXAMPLE 13   Inverse of a Transpose**

Consider a general 2×2 invertible matrix and its transpose:
$$A = \begin{bmatrix} a & b \\ c & d \end{bmatrix} \quad \text{and} \quad A^T = \begin{bmatrix} a & c \\ b & d \end{bmatrix}$$

Since A is invertible, its determinant $ad - bc$ is nonzero. But the determinant of A^T is also $ad - bc$ (verify), so A^T is also invertible. It follows from Theorem 1.4.5 that
$$(A^T)^{-1} = \begin{bmatrix} \dfrac{d}{ad - bc} & -\dfrac{c}{ad - bc} \\ -\dfrac{b}{ad - bc} & \dfrac{a}{ad - bc} \end{bmatrix}$$

which is the same matrix that results if A^{-1} is transposed (verify). Thus,
$$(A^T)^{-1} = (A^{-1})^T$$

as guaranteed by Theorem 1.4.9. ◄

Concept Review

- Commutative law for matrix addition
- Associative law for matrix addition
- Associative law for matrix multiplication

- Left and right distributive laws
- Zero matrix
- Identity matrix
- Inverse of a matrix
- Invertible matrix

- Nonsingular matrix
- Singular matrix
- Determinant
- Power of a matrix
- Matrix polynomial

Skills

- Know the arithmetic properties of matrix operations.
- Be able to prove arithmetic properties of matrices.
- Know the properties of zero matrices.
- Know the properties of identity matrices.
- Be able to recognize when two square matrices are inverses of each other.
- Be able to determine whether a 2×2 matrix is invertible.

- Be able to solve a linear system of two equations in two unknowns whose coefficient matrix is invertible.
- Be able to prove basic properties involving invertible matrices.
- Know the properties of the matrix transpose and its relationship with invertible matrices.

Exercise Set 1.4

1. Let

$$A = \begin{bmatrix} 2 & -1 & 3 \\ 0 & 4 & 5 \\ -2 & 1 & 4 \end{bmatrix}, \quad B = \begin{bmatrix} 8 & -3 & -5 \\ 0 & 1 & 2 \\ 4 & -7 & 6 \end{bmatrix},$$

$$C = \begin{bmatrix} 0 & -2 & 3 \\ 1 & 7 & 4 \\ 3 & 5 & 9 \end{bmatrix}, \quad a = 4, \quad b = -7$$

Show that

(a) $A + (B + C) = (A + B) + C$

(b) $(AB)C = A(BC)$ (c) $(a + b)C = aC + bC$

(d) $a(B - C) = aB - aC$

2. Using the matrices and scalars in Exercise 1, verify that

(a) $a(BC) = (aB)C = B(aC)$

(b) $A(B - C) = AB - AC$ (c) $(B + C)A = BA + CA$

(d) $a(bC) = (ab)C$

3. Using the matrices and scalars in Exercise 1, verify that

(a) $(A^T)^T = A$ (b) $(A + B)^T = A^T + B^T$

(c) $(aC)^T = aC^T$ (d) $(AB)^T = B^T A^T$

In Exercises 4–7, use Theorem 1.4.5 to compute the inverses of the following matrices.

4. $A = \begin{bmatrix} 3 & 1 \\ 5 & 2 \end{bmatrix}$ **5.** $B = \begin{bmatrix} 2 & -3 \\ 4 & 4 \end{bmatrix}$

6. $C = \begin{bmatrix} 6 & 4 \\ -2 & -1 \end{bmatrix}$ **7.** $D = \begin{bmatrix} 2 & 0 \\ 0 & 3 \end{bmatrix}$

8. Find the inverse of

$$\begin{bmatrix} \cos\theta & \sin\theta \\ -\sin\theta & \cos\theta \end{bmatrix}$$

9. Find the inverse of

$$\begin{bmatrix} \frac{1}{2}(e^x + e^{-x}) & \frac{1}{2}(e^x - e^{-x}) \\ \frac{1}{2}(e^x - e^{-x}) & \frac{1}{2}(e^x + e^{-x}) \end{bmatrix}$$

10. Use the matrix A in Exercise 4 to verify that $(A^T)^{-1} = (A^{-1})^T$.

11. Use the matrix B in Exercise 5 to verify that $(B^T)^{-1} = (B^{-1})^T$.

12. Use the matrices A and B in Exercises 4 and 5 to verify that $(AB)^{-1} = B^{-1}A^{-1}$.

13. Use the matrices A, B, and C in Exercises 4–6 to verify that $(ABC)^{-1} = C^{-1}B^{-1}A^{-1}$.

In Exercises 14–17, use the given information to find A.

14. $A^{-1} = \begin{bmatrix} 2 & -1 \\ 3 & 5 \end{bmatrix}$ **15.** $(7A)^{-1} = \begin{bmatrix} -3 & 7 \\ 1 & -2 \end{bmatrix}$

16. $(5A^T)^{-1} = \begin{bmatrix} -3 & -1 \\ 5 & 2 \end{bmatrix}$ **17.** $(I + 2A)^{-1} = \begin{bmatrix} -1 & 2 \\ 4 & 5 \end{bmatrix}$

18. Let A be the matrix

$$\begin{bmatrix} 2 & 0 \\ 4 & 1 \end{bmatrix}$$

In each part, compute the given quantity.

(a) A^3 (b) A^{-3} (c) $A^2 - 2A + I$

(d) $p(A)$, where $p(x) = x - 2$

(e) $p(A)$, where $p(x) = 2x^2 - x + 1$

(f) $p(A)$, where $p(x) = x^3 - 2x + 4$

19. Repeat Exercise 18 for the matrix

$$A = \begin{bmatrix} 3 & 1 \\ 2 & 1 \end{bmatrix}$$

20. Repeat Exercise 18 for the matrix

$$A = \begin{bmatrix} 3 & 0 & -1 \\ 0 & -2 & 0 \\ 5 & 0 & 2 \end{bmatrix}$$

21. Repeat Exercise 18 for the matrix

$$A = \begin{bmatrix} 3 & 0 & 0 \\ 0 & -1 & 3 \\ 0 & -3 & -1 \end{bmatrix}$$

In Exercises 22–24, let $p_1(x) = x^2 - 9$, $p_2(x) = x + 3$, and $p_3(x) = x - 3$. Show that $p_1(A) = p_2(A)p_3(A)$ for the given matrix.

22. The matrix A in Exercise 18.

23. The matrix A in Exercise 21.

24. An arbitrary square matrix A.

25. Show that if $p(x) = x^2 - (a + d)x + (ad - bc)$ and

$$A = \begin{bmatrix} a & b \\ c & d \end{bmatrix}$$

then $p(A) = 0$.

26. Show that if $p(x) = x^3 - (a + b + c)x^2 + (ab + ae + be - cd)x - a(be - cd)$ and

$$A = \begin{bmatrix} a & 0 & 0 \\ 0 & b & c \\ 0 & d & e \end{bmatrix}$$

then $p(A) = 0$.

27. Consider the matrix

$$A = \begin{bmatrix} a_{11} & 0 & \cdots & 0 \\ 0 & a_{22} & \cdots & 0 \\ \vdots & \vdots & & \vdots \\ 0 & 0 & \cdots & a_{nn} \end{bmatrix}$$

where $a_{11}a_{22}\cdots a_{nn} \neq 0$. Show that A is invertible and find its inverse.

28. Show that if a square matrix A satisfies $A^2 - 3A + I = 0$, then $A^{-1} = 3I - A$.

29. (a) Show that a matrix with a row of zeros cannot have an inverse.

(b) Show that a matrix with a column of zeros cannot have an inverse.

30. Assuming that all matrices are $n \times n$ and invertible, solve for D.
$$ABC^T DBA^T C = AB^T$$

31. Assuming that all matrices are $n \times n$ and invertible, solve for D.
$$C^T B^{-1} A^2 BAC^{-1} DA^{-2} B^T C^{-2} = C^T$$

32. If A is a square matrix and n is a positive integer, is it true that $(A^n)^T = (A^T)^n$? Justify your answer.

33. Simplify:
$$(AB)^{-1}(AC^{-1})(D^{-1}C^{-1})^{-1}D^{-1}$$

34. Simplify:
$$(AC^{-1})^{-1}(AC^{-1})(AC^{-1})^{-1}AD^{-1}$$

In Exercises 35–37, determine whether A is invertible, and if so, find the inverse. [*Hint:* Solve $AX = I$ for X by equating corresponding entries on the two sides.]

35. $A = \begin{bmatrix} 1 & 0 & 1 \\ 1 & 1 & 0 \\ 0 & 1 & 1 \end{bmatrix}$

36. $A = \begin{bmatrix} 1 & 1 & 1 \\ 1 & 0 & 0 \\ 0 & 1 & 1 \end{bmatrix}$ **37.** $A = \begin{bmatrix} 0 & 0 & 1 \\ 1 & 1 & 0 \\ -1 & 1 & 1 \end{bmatrix}$

38. Prove Theorem 1.4.2.

In Exercises 39–42, use the method of Example 8 to find the unique solution of the given linear system.

39. $3x_1 - 2x_2 = -1$
$\quad\ 4x_1 + 5x_2 = \ \ 3$

40. $-x_1 + 5x_2 = 4$
$\quad\ \ -x_1 - 3x_2 = 1$

41. $6x_1 + \ x_2 = \ \ 0$
$\quad\ 4x_1 - 3x_2 = -2$

42. $2x_1 - 2x_2 = 4$
$\quad\ \ x_1 + 4x_2 = 4$

43. Prove part (a) of Theorem 1.4.1.

44. Prove part (c) of Theorem 1.4.1.

45. Prove part (f) of Theorem 1.4.1.

46. Prove part (b) of Theorem 1.4.2.

47. Prove part (c) of Theorem 1.4.2.

48. Verify Formula (4) in the text by a direct calculation.

49. Prove part (d) of Theorem 1.4.8.

50. Prove part (e) of Theorem 1.4.8.

51. (a) Show that if A is invertible and $AB = AC$, then $B = C$.

(b) Explain why part (a) and Example 3 do not contradict one another.

52. Show that if A is invertible and k is any nonzero scalar, then $(kA)^n = k^n A^n$ for all integer values of n.

53. (a) Show that if A, B, and $A + B$ are invertible matrices with the same size, then
$$A(A^{-1} + B^{-1})B(A + B)^{-1} = I$$

(b) What does the result in part (a) tell you about the matrix $A^{-1} + B^{-1}$?

54. A square matrix A is said to be ***idempotent*** if $A^2 = A$.

(a) Show that if A is idempotent, then so is $I - A$.

(b) Show that if A is idempotent, then $2A - I$ is invertible and is its own inverse.

55. Show that if A is a square matrix such that $A^k = 0$ for some positive integer k, then the matrix A is invertible and
$$(I - A)^{-1} = I + A + A^2 + \cdots + A^{k-1}$$

True-False Exercises

In parts (a)–(k) determine whether the statement is true or false, and justify your answer.

(a) Two $n \times n$ matrices, A and B, are inverses of one another if and only if $AB = BA = 0$.

(b) For all square matrices A and B of the same size, it is true that $(A + B)^2 = A^2 + 2AB + B^2$.

(c) For all square matrices A and B of the same size, it is true that $A^2 - B^2 = (A - B)(A + B)$.

(d) If A and B are invertible matrices of the same size, then AB is invertible and $(AB)^{-1} = A^{-1}B^{-1}$.

(e) If A and B are matrices such that AB is defined, then it is true that $(AB)^T = A^T B^T$.

(f) The matrix
$$A = \begin{bmatrix} a & b \\ c & d \end{bmatrix}$$
is invertible if and only if $ad - bc \neq 0$.

(g) If A and B are matrices of the same size and k is a constant, then $(kA + B)^T = kA^T + B^T$.

(h) If A is an invertible matrix, then so is A^T.

(i) If $p(x) = a_0 + a_1x + a_2x^2 + \cdots + a_mx^m$ and I is an identity matrix, then $p(I) = a_0 + a_1 + a_2 + \cdots + a_m$.

(j) A square matrix containing a row or column of zeros cannot be invertible.

(k) The sum of two invertible matrices of the same size must be invertible.

1.5 Elementary Matrices and a Method for Finding A^{-1}

In this section we will develop an algorithm for finding the inverse of a matrix, and we will discuss some of the basic properties of invertible matrices.

In Section 1.1 we defined three elementary row operations on a matrix A:

1. Multiply a row by a nonzero constant c.
2. Interchange two rows.
3. Add a constant c times one row to another.

It should be evident that if we let B be the matrix that results from A by performing one of the operations in this list, then the matrix A can be recovered from B by performing the corresponding operation in the following list:

1. Multiply the same row by $1/c$.
2. Interchange the same two rows.
3. If B resulted by adding c times row r_1 of A to row r_2, then add $-c$ times r_1 to row r_2.

It follows that if B is obtained from A by performing a sequence of elementary row operations, then there is a second sequence of elementary row operations, which when applied to B recovers A (Exercise 43). Accordingly, we make the following definition.

> **DEFINITION 1** Matrices A and B are said to be *row equivalent* if either (hence each) can be obtained from the other by a sequence of elementary row operations.

Our next goal is to show how matrix multiplication can be used to carry out an elementary row operation.

> **DEFINITION 2** An $n \times n$ matrix is called an *elementary matrix* if it can be obtained from the $n \times n$ identity matrix I_n by performing a *single* elementary row operation.

▶ **EXAMPLE 1 Elementary Matrices and Row Operations**

Listed below are four elementary matrices and the operations that produce them.

$$\begin{bmatrix} 1 & 0 \\ 0 & -3 \end{bmatrix} \qquad \begin{bmatrix} 1 & 0 & 0 & 0 \\ 0 & 0 & 0 & 1 \\ 0 & 0 & 1 & 0 \\ 0 & 1 & 0 & 0 \end{bmatrix} \qquad \begin{bmatrix} 1 & 0 & 3 \\ 0 & 1 & 0 \\ 0 & 0 & 1 \end{bmatrix} \qquad \begin{bmatrix} 1 & 0 & 0 \\ 0 & 1 & 0 \\ 0 & 0 & 1 \end{bmatrix}$$

↑ Multiply the second row of I_2 by -3.

↑ Interchange the second and fourth rows of I_4.

↑ Add 3 times the third row of I_3 to the first row.

↑ Multiply the first row of I_3 by 1.

The following theorem, whose proof is left as an exercise, shows that when a matrix A is multiplied on the *left* by an elementary matrix E, the effect is to perform an elementary row operation on A.

THEOREM 1.5.1 Row Operations by Matrix Multiplication

If the elementary matrix E results from performing a certain row operation on I_m and if A is an $m \times n$ matrix, then the product EA is the matrix that results when this same row operation is performed on A.

▶ **EXAMPLE 2 Using Elementary Matrices**

Consider the matrix

$$A = \begin{bmatrix} 1 & 0 & 2 & 3 \\ 2 & -1 & 3 & 6 \\ 1 & 4 & 4 & 0 \end{bmatrix}$$

and consider the elementary matrix

$$E = \begin{bmatrix} 1 & 0 & 0 \\ 0 & 1 & 0 \\ 3 & 0 & 1 \end{bmatrix}$$

which results from adding 3 times the first row of I_3 to the third row. The product EA is

$$EA = \begin{bmatrix} 1 & 0 & 2 & 3 \\ 2 & -1 & 3 & 6 \\ 4 & 4 & 10 & 9 \end{bmatrix}$$

which is precisely the matrix that results when we add 3 times the first row of A to the third row. ◀

Theorem 1.5.1 will be a useful tool for developing new results about matrices, but as a practical matter it is usually preferable to perform row operations directly.

We know from the discussion at the beginning of this section that if E is an elementary matrix that results from performing an elementary row operation on an identity matrix I, then there is a second elementary row operation, which when applied to E, produces I back again. Table 1 lists these operations. The operations on the right side of the table are called the ***inverse operations*** of the corresponding operations on the left.

Table 1

Row Operation on I That Produces E	Row Operation on E That Reproduces I
Multiply row i by $c \neq 0$	Multiply row i by $1/c$
Interchange rows i and j	Interchange rows i and j
Add c times row i to row j	Add $-c$ times row i to row j

▶ **EXAMPLE 3 Row Operations and Inverse Row Operations**

In each of the following, an elementary row operation is applied to the 2×2 identity matrix to obtain an elementary matrix E, then E is restored to the identity matrix by applying the inverse row operation.

$$\begin{bmatrix} 1 & 0 \\ 0 & 1 \end{bmatrix} \longrightarrow \begin{bmatrix} 1 & 0 \\ 0 & 7 \end{bmatrix} \longrightarrow \begin{bmatrix} 1 & 0 \\ 0 & 1 \end{bmatrix}$$

Multiply the second   Multiply the second
row by 7.   row by $\frac{1}{7}$.

$$\begin{bmatrix} 1 & 0 \\ 0 & 1 \end{bmatrix} \longrightarrow \begin{bmatrix} 0 & 1 \\ 1 & 0 \end{bmatrix} \longrightarrow \begin{bmatrix} 1 & 0 \\ 0 & 1 \end{bmatrix}$$

Interchange the first   Interchange the first
and second rows.   and second rows.

$$\begin{bmatrix} 1 & 0 \\ 0 & 1 \end{bmatrix} \longrightarrow \begin{bmatrix} 1 & 5 \\ 0 & 1 \end{bmatrix} \longrightarrow \begin{bmatrix} 1 & 0 \\ 0 & 1 \end{bmatrix}$$

Add 5 times the   Add -5 times the
second row to the   second row to the
first.   first. ◀

The next theorem is a key result about invertibility of elementary matrices. It will be a building block for many results that follow.

> **THEOREM 1.5.2** *Every elementary matrix is invertible, and the inverse is also an elementary matrix.*

Proof If E is an elementary matrix, then E results by performing some row operation on I. Let E_0 be the matrix that results when the inverse of this operation is performed on I. Applying Theorem 1.5.1 and using the fact that inverse row operations cancel the effect of each other, it follows that

$$E_0 E = I \quad \text{and} \quad E E_0 = I$$

Thus, the elementary matrix E_0 is the inverse of E. ◀

Equivalence Theorem One of our objectives as we progress through this text is to show how seemingly diverse ideas in linear algebra are related. The following theorem, which relates results we have obtained about invertibility of matrices, homogeneous linear systems, reduced row echelon forms, and elementary matrices, is our first step in that direction. As we study new topics, more statements will be added to this theorem.

> **THEOREM 1.5.3 Equivalent Statements**
>
> *If A is an $n \times n$ matrix, then the following statements are equivalent, that is, all true or all false.*
>
> (a) *A is invertible.*
> (b) *$A\mathbf{x} = \mathbf{0}$ has only the trivial solution.*
> (c) *The reduced row echelon form of A is I_n.*
> (d) *A is expressible as a product of elementary matrices.*

It may make the logic of our proof of Theorem 1.5.3 more apparent by writing the implications

$$(a) \Rightarrow (b) \Rightarrow (c) \Rightarrow (d) \Rightarrow (a)$$

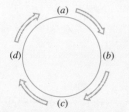

This makes it evident visually that the validity of any one statement implies the validity of all the others, and hence that the falsity of any one implies the falsity of the others.

Proof   We will prove the equivalence by establishing the chain of implications: $(a) \Rightarrow (b) \Rightarrow (c) \Rightarrow (d) \Rightarrow (a)$.

(a) $\Rightarrow$ (b)   Assume A is invertible and let $\mathbf{x}_0$ be any solution of $A\mathbf{x} = \mathbf{0}$. Multiplying both sides of this equation by the matrix A^{-1} gives $A^{-1}(A\mathbf{x}_0) = A^{-1}\mathbf{0}$, or $(A^{-1}A)\mathbf{x}_0 = \mathbf{0}$, or $I\mathbf{x}_0 = \mathbf{0}$, or $\mathbf{x}_0 = \mathbf{0}$. Thus, $A\mathbf{x} = \mathbf{0}$ has only the trivial solution.

(b) $\Rightarrow$ (c)   Let $A\mathbf{x} = \mathbf{0}$ be the matrix form of the system

$$\begin{aligned} a_{11}x_1 + a_{12}x_2 + \cdots + a_{1n}x_n &= 0 \\ a_{21}x_1 + a_{22}x_2 + \cdots + a_{2n}x_n &= 0 \\ \vdots \qquad \vdots \qquad\qquad \vdots \quad\ \ \vdots \\ a_{n1}x_1 + a_{n2}x_2 + \cdots + a_{nn}x_n &= 0 \end{aligned} \tag{1}$$

and assume that the system has only the trivial solution. If we solve by Gauss–Jordan elimination, then the system of equations corresponding to the reduced row echelon form of the augmented matrix will be

$$\begin{aligned} x_1 \qquad\qquad\qquad &= 0 \\ x_2 \qquad\qquad &= 0 \\ \ddots \qquad \\ x_n &= 0 \end{aligned} \tag{2}$$

Thus the augmented matrix

$$\begin{bmatrix} a_{11} & a_{12} & \cdots & a_{1n} & 0 \\ a_{21} & a_{22} & \cdots & a_{2n} & 0 \\ \vdots & \vdots & & \vdots & \vdots \\ a_{n1} & a_{n2} & \cdots & a_{nn} & 0 \end{bmatrix}$$

for (1) can be reduced to the augmented matrix

$$\begin{bmatrix} 1 & 0 & 0 & \cdots & 0 & 0 \\ 0 & 1 & 0 & \cdots & 0 & 0 \\ 0 & 0 & 1 & \cdots & 0 & 0 \\ \vdots & \vdots & \vdots & & \vdots & \vdots \\ 0 & 0 & 0 & \cdots & 1 & 0 \end{bmatrix}$$

for (2) by a sequence of elementary row operations. If we disregard the last column (all zeros) in each of these matrices, we can conclude that the reduced row echelon form of A is I_n.

(c) $\Rightarrow$ (d)   Assume that the reduced row echelon form of A is I_n, so that A can be reduced to I_n by a finite sequence of elementary row operations. By Theorem 1.5.1, each of these operations can be accomplished by multiplying on the left by an appropriate elementary matrix. Thus we can find elementary matrices $E_1, E_2, \ldots, E_k$ such that

$$E_k \cdots E_2 E_1 A = I_n \tag{3}$$

By Theorem 1.5.2, $E_1, E_2, \ldots, E_k$ are invertible. Multiplying both sides of Equation (3) on the left successively by $E_k^{-1}, \ldots, E_2^{-1}, E_1^{-1}$ we obtain

$$A = E_1^{-1}E_2^{-1} \cdots E_k^{-1}I_n = E_1^{-1}E_2^{-1} \cdots E_k^{-1} \tag{4}$$

By Theorem 1.5.2, this equation expresses A as a product of elementary matrices.

(d) $\Rightarrow$ **(a)**   If A is a product of elementary matrices, then from Theorems 1.4.7 and 1.5.2, the matrix A is a product of invertible matrices and hence is invertible. ◀

A Method for Inverting Matrices   As a first application of Theorem 1.5.3, we will develop a procedure (or algorithm) that can be used to tell whether a given matrix is invertible, and if so, produce its inverse. To derive this algorithm, assume for the moment, that A is an invertible $n \times n$ matrix. In Equation (3), the elementary matrices execute a sequence of row operations that reduce A to I_n. If we multiply both sides of this equation on the right by A^{-1} and simplify, we obtain

$$A^{-1} = E_k \cdots E_2 E_1 I_n$$

But this equation tells us that *the same sequence of row operations that reduces A to I_n will transform I_n to A^{-1}*. Thus, we have established the following result.

> **Inversion Algorithm** To find the inverse of an invertible matrix A, find a sequence of elementary row operations that reduces A to the identity and then perform that same sequence of operations on I_n to obtain A^{-1}.

A simple method for carrying out this procedure is given in the following example.

▶ **EXAMPLE 4 Using Row Operations to Find A^{-1}**

Find the inverse of

$$A = \begin{bmatrix} 1 & 2 & 3 \\ 2 & 5 & 3 \\ 1 & 0 & 8 \end{bmatrix}$$

Solution We want to reduce A to the identity matrix by row operations and simultaneously apply these operations to I to produce A^{-1}. To accomplish this we will adjoin the identity matrix to the right side of A, thereby producing a partitioned matrix of the form

$$[A \mid I]$$

Then we will apply row operations to this matrix until the left side is reduced to I; these operations will convert the right side to A^{-1}, so the final matrix will have the form

$$[I \mid A^{-1}]$$

The computations are as follows:

$$\left[\begin{array}{ccc|ccc} 1 & 2 & 3 & 1 & 0 & 0 \\ 2 & 5 & 3 & 0 & 1 & 0 \\ 1 & 0 & 8 & 0 & 0 & 1 \end{array}\right]$$

$$\left[\begin{array}{ccc|ccc} 1 & 2 & 3 & 1 & 0 & 0 \\ 0 & 1 & -3 & -2 & 1 & 0 \\ 0 & -2 & 5 & -1 & 0 & 1 \end{array}\right]$$ We added -2 times the first row to the second and -1 times the first row to the third.

$$\left[\begin{array}{ccc|ccc} 1 & 2 & 3 & 1 & 0 & 0 \\ 0 & 1 & -3 & -2 & 1 & 0 \\ 0 & 0 & -1 & -5 & 2 & 1 \end{array}\right]$$ We added 2 times the second row to the third.

$$\left[\begin{array}{ccc|ccc} 1 & 2 & 3 & 1 & 0 & 0 \\ 0 & 1 & -3 & -2 & 1 & 0 \\ 0 & 0 & 1 & 5 & -2 & -1 \end{array}\right]$$ We multiplied the third row by -1.

$$\left[\begin{array}{ccc|ccc} 1 & 2 & 0 & -14 & 6 & 3 \\ 0 & 1 & 0 & 13 & -5 & -3 \\ 0 & 0 & 1 & 5 & -2 & -1 \end{array}\right]$$ We added 3 times the third row to the second and -3 times the third row to the first.

$$\left[\begin{array}{ccc|ccc} 1 & 0 & 0 & -40 & 16 & 9 \\ 0 & 1 & 0 & 13 & -5 & -3 \\ 0 & 0 & 1 & 5 & -2 & -1 \end{array}\right]$$ We added -2 times the second row to the first.

Thus,

$$A^{-1} = \left[\begin{array}{ccc} -40 & 16 & 9 \\ 13 & -5 & -3 \\ 5 & -2 & -1 \end{array}\right] \blacktriangleleft$$

Often it will not be known in advance if a given $n \times n$ matrix A is invertible. However, if it is not, then by parts (a) and (c) of Theorem 1.5.3 it will be impossible to reduce A to I_n by elementary row operations. This will be signaled by a row of zeros appearing on the *left side* of the partition at some stage of the inversion algorithm. If this occurs, then you can stop the computations and conclude that A is not invertible.

▶ **EXAMPLE 5** **Showing That a Matrix Is Not Invertible**

Consider the matrix

$$A = \left[\begin{array}{ccc} 1 & 6 & 4 \\ 2 & 4 & -1 \\ -1 & 2 & 5 \end{array}\right]$$

Applying the procedure of Example 4 yields

$$\left[\begin{array}{ccc|ccc} 1 & 6 & 4 & 1 & 0 & 0 \\ 2 & 4 & -1 & 0 & 1 & 0 \\ -1 & 2 & 5 & 0 & 0 & 1 \end{array}\right]$$

$$\left[\begin{array}{ccc|ccc} 1 & 6 & 4 & 1 & 0 & 0 \\ 0 & -8 & -9 & -2 & 1 & 0 \\ 0 & 8 & 9 & 1 & 0 & 1 \end{array}\right] \longleftarrow \begin{array}{l}\text{We added } -2 \text{ times the first} \\ \text{row to the second and added} \\ \text{the first row to the third.}\end{array}$$

$$\left[\begin{array}{ccc|ccc} 1 & 6 & 4 & 1 & 0 & 0 \\ 0 & -8 & -9 & -2 & 1 & 0 \\ 0 & 0 & 0 & -1 & 1 & 1 \end{array}\right] \longleftarrow \begin{array}{l}\text{We added the second} \\ \text{row to the third.}\end{array}$$

Since we have obtained a row of zeros on the left side, A is not invertible.

▶ **EXAMPLE 6 Analyzing Homogeneous Systems**

Use Theorem 1.5.3 to determine whether the given homogeneous system has nontrivial solutions.

$$
\begin{array}{ll}
\begin{aligned}
x_1 + 2x_2 + 3x_3 &= 0 \\
\text{(a)} \quad 2x_1 + 5x_2 + 3x_3 &= 0 \\
x_1 \qquad\quad + 8x_3 &= 0
\end{aligned}
&
\begin{aligned}
x_1 + 6x_2 + 4x_3 &= 0 \\
\text{(b)} \quad 2x_1 + 4x_2 - \quad x_3 &= 0 \\
-x_1 + 2x_2 + 5x_3 &= 0
\end{aligned}
\end{array}
$$

Solution From parts (*a*) and (*b*) of Theorem 1.5.3 a homogeneous linear system has only the trivial solution if and only if its coefficient matrix is invertible. From Examples 4 and 5 the coefficient matrix of system (a) is invertible and that of system (b) is not. Thus, system (a) has only the trivial solution whereas system (b) has nontrivial solutions.

◀

Concept Review

- Row equivalent matrices
- Elementary matrix

- Inverse operations
- Inversion algorithm

Skills

- Determine whether a given square matrix is an elementary.
- Determine whether two square matrices are row equivalent.
- Apply the inverse of a given elementary row operation to a matrix.
- Apply elementary row operations to reduce a given square matrix to the identity matrix.

- Understand the relationships between statements that are equivalent to the invertibility of a square matrix (Theorem 1.5.3).
- Use the inversion algorithm to find the inverse of an invertible matrix.
- Express an invertible matrix as a product of elementary matrices.

Exercise Set 1.5

1. Decide whether each matrix below is an elementary matrix.

(a) $\begin{bmatrix} 1 & 0 \\ -5 & 1 \end{bmatrix}$

(b) $\begin{bmatrix} -5 & 1 \\ 1 & 0 \end{bmatrix}$

(c) $\begin{bmatrix} 1 & 1 & 0 \\ 0 & 0 & 1 \\ 0 & 0 & 0 \end{bmatrix}$

(d) $\begin{bmatrix} 2 & 0 & 0 & 2 \\ 0 & 1 & 0 & 0 \\ 0 & 0 & 1 & 0 \\ 0 & 0 & 0 & 1 \end{bmatrix}$

2. Decide whether each matrix below is an elementary matrix.

(a) $\begin{bmatrix} 1 & 0 \\ 0 & \sqrt{3} \end{bmatrix}$

(b) $\begin{bmatrix} 0 & 0 & 1 \\ 0 & 1 & 0 \\ 1 & 0 & 0 \end{bmatrix}$

(c) $\begin{bmatrix} 1 & 0 & 0 \\ 0 & 1 & 9 \\ 0 & 0 & 1 \end{bmatrix}$

(d) $\begin{bmatrix} -1 & 0 & 0 \\ 0 & 0 & 1 \\ 0 & 1 & 0 \end{bmatrix}$

3. Find a row operation and the corresponding elementary matrix that will restore the given elementary matrix to the identity matrix.

(a) $\begin{bmatrix} 1 & -3 \\ 0 & 1 \end{bmatrix}$

(b) $\begin{bmatrix} -7 & 0 & 0 \\ 0 & 1 & 0 \\ 0 & 0 & 1 \end{bmatrix}$

(c) $\begin{bmatrix} 1 & 0 & 0 \\ 0 & 1 & 0 \\ -5 & 0 & 1 \end{bmatrix}$

(d) $\begin{bmatrix} 0 & 0 & 1 & 0 \\ 0 & 1 & 0 & 0 \\ 1 & 0 & 0 & 0 \\ 0 & 0 & 0 & 1 \end{bmatrix}$

4. Find a row operation and the corresponding elementary matrix that will restore the given elementary matrix to the identity matrix.

(a) $\begin{bmatrix} 1 & 0 \\ -3 & 1 \end{bmatrix}$

(b) $\begin{bmatrix} 1 & 0 & 0 \\ 0 & 1 & 0 \\ 0 & 0 & 3 \end{bmatrix}$

(c) $\begin{bmatrix} 0 & 0 & 0 & 1 \\ 0 & 1 & 0 & 0 \\ 0 & 0 & 1 & 0 \\ 1 & 0 & 0 & 0 \end{bmatrix}$

(d) $\begin{bmatrix} 1 & 0 & -\frac{1}{7} & 0 \\ 0 & 1 & 0 & 0 \\ 0 & 0 & 1 & 0 \\ 0 & 0 & 0 & 1 \end{bmatrix}$

5. In each part, an elementary matrix E and a matrix A are given. Write down the row operation corresponding to E and show that the product EA results from applying the row operation to A.

(a) $E = \begin{bmatrix} 0 & 1 \\ 1 & 0 \end{bmatrix}$, $A = \begin{bmatrix} -1 & -2 & 5 & -1 \\ 3 & -6 & -6 & -6 \end{bmatrix}$

(b) $E = \begin{bmatrix} 1 & 0 & 0 \\ 0 & 1 & 0 \\ 0 & -3 & 1 \end{bmatrix}$, $A = \begin{bmatrix} 2 & -1 & 0 & -4 & -4 \\ 1 & -3 & -1 & 5 & 3 \\ 2 & 0 & 1 & 3 & -1 \end{bmatrix}$

(c) $E = \begin{bmatrix} 1 & 0 & 4 \\ 0 & 1 & 0 \\ 0 & 0 & 1 \end{bmatrix}$, $A = \begin{bmatrix} 1 & 4 \\ 2 & 5 \\ 3 & 6 \end{bmatrix}$

6. In each part, an elementary matrix E and a matrix A are given. Write down the row operation corresponding to E and show that the product EA results from applying the row operation to A.

(a) $E = \begin{bmatrix} -6 & 0 \\ 0 & 1 \end{bmatrix}$, $A = \begin{bmatrix} -1 & -2 & 5 & -1 \\ 3 & -6 & -6 & -6 \end{bmatrix}$

(b) $E = \begin{bmatrix} 1 & 0 & 0 \\ -4 & 1 & 0 \\ 0 & 0 & 1 \end{bmatrix}$, $A = \begin{bmatrix} 2 & -1 & 0 & -4 & -4 \\ 1 & -3 & -1 & 5 & 3 \\ 2 & 0 & 1 & 3 & -1 \end{bmatrix}$

(c) $E = \begin{bmatrix} 1 & 0 & 0 \\ 0 & 5 & 0 \\ 0 & 0 & 1 \end{bmatrix}$, $A = \begin{bmatrix} 1 & 4 \\ 2 & 5 \\ 3 & 6 \end{bmatrix}$

In Exercises 7–8, use the following matrices.

$$A = \begin{bmatrix} 3 & 4 & 1 \\ 2 & -7 & -1 \\ 8 & 1 & 5 \end{bmatrix}, \quad B = \begin{bmatrix} 8 & 1 & 5 \\ 2 & -7 & -1 \\ 3 & 4 & 1 \end{bmatrix}$$

$$C = \begin{bmatrix} 3 & 4 & 1 \\ 2 & -7 & -1 \\ 2 & -7 & 3 \end{bmatrix}, \quad D = \begin{bmatrix} 8 & 1 & 5 \\ -6 & 21 & 3 \\ 3 & 4 & 1 \end{bmatrix}$$

$$F = \begin{bmatrix} 8 & 1 & 5 \\ 8 & 1 & 1 \\ 3 & 4 & 1 \end{bmatrix}$$

7. Find an elementary matrix E that satisfies the equation.

(a) $EA = B$ (b) $EB = A$

(c) $EA = C$ (d) $EC = A$

8. Find an elementary matrix E that satisfies the equation.

(a) $EB = D$ (b) $ED = B$

(c) $EB = F$ (d) $EF = B$

In Exercises 9–24, use the inversion algorithm to find the inverse of the given matrix, if the inverse exists.

9. $\begin{bmatrix} 1 & 4 \\ 2 & 7 \end{bmatrix}$

10. $\begin{bmatrix} -3 & 6 \\ 4 & 5 \end{bmatrix}$

11. $\begin{bmatrix} -1 & 3 \\ 3 & -2 \end{bmatrix}$

12. $\begin{bmatrix} 6 & -4 \\ -3 & 2 \end{bmatrix}$

13. $\begin{bmatrix} 3 & 4 & -1 \\ 1 & 0 & 3 \\ 2 & 5 & -4 \end{bmatrix}$

14. $\begin{bmatrix} 1 & 2 & 0 \\ 2 & 1 & 2 \\ 0 & 2 & 1 \end{bmatrix}$

15. $\begin{bmatrix} -1 & 3 & -4 \\ 2 & 4 & 1 \\ -4 & 2 & -9 \end{bmatrix}$

16. $\begin{bmatrix} \frac{1}{5} & \frac{1}{5} & -\frac{2}{5} \\ \frac{1}{5} & \frac{1}{5} & \frac{1}{10} \\ \frac{1}{5} & -\frac{4}{5} & \frac{1}{10} \end{bmatrix}$

17. $\begin{bmatrix} 1 & 0 & 1 \\ 0 & 1 & 1 \\ 1 & 1 & 0 \end{bmatrix}$

18. $\begin{bmatrix} \sqrt{2} & 3\sqrt{2} & 0 \\ -4\sqrt{2} & \sqrt{2} & 0 \\ 0 & 0 & 1 \end{bmatrix}$

19. $\begin{bmatrix} 2 & 6 & 6 \\ 2 & 7 & 6 \\ 2 & 7 & 7 \end{bmatrix}$

20. $\begin{bmatrix} 1 & 0 & 0 & 0 \\ 1 & 3 & 0 & 0 \\ 1 & 3 & 5 & 0 \\ 1 & 3 & 5 & 7 \end{bmatrix}$

21. $\begin{bmatrix} 2 & -4 & 0 & 0 \\ 1 & 2 & 12 & 0 \\ 0 & 0 & 2 & 0 \\ 0 & -1 & -4 & -5 \end{bmatrix}$

22. $\begin{bmatrix} -8 & 17 & 2 & \frac{1}{3} \\ 4 & 0 & \frac{2}{5} & -9 \\ 0 & 0 & 0 & 0 \\ -1 & 13 & 4 & 2 \end{bmatrix}$

23. $\begin{bmatrix} -1 & 0 & 1 & 0 \\ 2 & 3 & -2 & 6 \\ 0 & -1 & 2 & 0 \\ 0 & 0 & 1 & 5 \end{bmatrix}$

24. $\begin{bmatrix} 0 & 0 & 2 & 0 \\ 1 & 0 & 0 & 1 \\ 0 & -1 & 3 & 0 \\ 2 & 1 & 5 & -3 \end{bmatrix}$

In Exercises 25–26, find the inverse of each of the following 4×4 matrices, where k_1, k_2, k_3, k_4, and k are all nonzero.

25. (a) $\begin{bmatrix} k_1 & 0 & 0 & 0 \\ 0 & k_2 & 0 & 0 \\ 0 & 0 & k_3 & 0 \\ 0 & 0 & 0 & k_4 \end{bmatrix}$ (b) $\begin{bmatrix} k & 1 & 0 & 0 \\ 0 & 1 & 0 & 0 \\ 0 & 0 & k & 1 \\ 0 & 0 & 0 & 1 \end{bmatrix}$

26. (a) $\begin{bmatrix} 0 & 0 & 0 & k_1 \\ 0 & 0 & k_2 & 0 \\ 0 & k_3 & 0 & 0 \\ k_4 & 0 & 0 & 0 \end{bmatrix}$ (b) $\begin{bmatrix} k & 0 & 0 & 0 \\ 1 & k & 0 & 0 \\ 0 & 1 & k & 0 \\ 0 & 0 & 1 & k \end{bmatrix}$

In Exercises 27–28, find all values of c, if any, for which the given matrix is invertible.

27. $\begin{bmatrix} c & c & c \\ 1 & c & c \\ 1 & 1 & c \end{bmatrix}$

28. $\begin{bmatrix} c & 1 & 0 \\ 1 & c & 1 \\ 0 & 1 & c \end{bmatrix}$

In Exercises 29–32, write the given matrix as a product of elementary matrices.

29. $\begin{bmatrix} -3 & 1 \\ 2 & 2 \end{bmatrix}$

30. $\begin{bmatrix} 1 & 0 \\ -5 & 2 \end{bmatrix}$

31. $\begin{bmatrix} 1 & 0 & -2 \\ 0 & 4 & 3 \\ 0 & 0 & 1 \end{bmatrix}$

32. $\begin{bmatrix} 1 & 1 & 0 \\ 1 & 1 & 1 \\ 0 & 1 & 1 \end{bmatrix}$

In Exercises 33–36, write the *inverse* of the given matrix as a product of elementary matrices.

33. The matrix in Exercise 29.

34. The matrix in Exercise 30.

35. The matrix in Exercise 31.

36. The matrix in Exercise 32.

In Exercises 37–38, show that the given matrices A and B are row equivalent, and find a sequence of elementary row operations that produces B from A.

37. $A = \begin{bmatrix} 1 & 2 & 3 \\ 1 & 4 & 1 \\ 2 & 1 & 9 \end{bmatrix}$, $B = \begin{bmatrix} 1 & 0 & 5 \\ 0 & 2 & -2 \\ 1 & 1 & 4 \end{bmatrix}$

38. $A = \begin{bmatrix} 2 & 1 & 0 \\ -1 & 1 & 0 \\ 3 & 0 & -1 \end{bmatrix}$, $B = \begin{bmatrix} 6 & 9 & 4 \\ -5 & -1 & 0 \\ -1 & -2 & -1 \end{bmatrix}$

39. Show that if

$$A = \begin{bmatrix} 1 & 0 & 0 \\ 0 & 1 & 0 \\ a & b & c \end{bmatrix}$$

is an elementary matrix, then at least one entry in the third row must be zero.

40. Show that

$$A = \begin{bmatrix} 0 & a & 0 & 0 & 0 \\ b & 0 & c & 0 & 0 \\ 0 & d & 0 & e & 0 \\ 0 & 0 & f & 0 & g \\ 0 & 0 & 0 & h & 0 \end{bmatrix}$$

is not invertible for any values of the entries.

41. Prove that if A and B are $m \times n$ matrices, then A and B are row equivalent if and only if A and B have the same reduced row echelon form.

42. Prove that if A is an invertible matrix and B is row equivalent to A, then B is also invertible.

43. Show that if B is obtained from A by performing a sequence of elementary row operations, then there is a second sequence of elementary row operations, which when applied to B recovers A.

True-False Exercises

In parts (a)–(g) determine whether the statement is true or false, and justify your answer.

(a) The product of two elementary matrices of the same size must be an elementary matrix.

(b) Every elementary matrix is invertible.

(c) If A and B are row equivalent, and if B and C are row equivalent, then A and C are row equivalent.

(d) If A is an $n \times n$ matrix that is not invertible, then the linear system $A\mathbf{x} = 0$ has infinitely many solutions.

(e) If A is an $n \times n$ matrix that is not invertible, then the matrix obtained by interchanging two rows of A cannot be invertible.

(f) If A is invertible and a multiple of the first row of A is added to the second row, then the resulting matrix is invertible.

(g) An expression of the invertible matrix A as a product of elementary matrices is unique.

1.6   More on Linear Systems and Invertible Matrices

In this section we will show how the inverse of a matrix can be used to solve a linear system and we will develop some more results about invertible matrices.

Number of Solutions of a Linear System

In Section 1.1 we made the statement (based on Figures 1.1.1 and 1.1.2) that every linear system has either no solutions, has exactly one solution, or has infinitely many solutions. We are now in a position to prove this fundamental result.

> **THEOREM 1.6.1** *A system of linear equations has zero, one, or infinitely many solutions. There are no other possibilities.*

Proof   If $A\mathbf{x} = \mathbf{b}$ is a system of linear equations, exactly one of the following is true: (a) the system has no solutions, (b) the system has exactly one solution, or (c) the system has more than one solution. The proof will be complete if we can show that the system has infinitely many solutions in case (c).

Assume that $A\mathbf{x} = \mathbf{b}$ has more than one solution, and let $\mathbf{x}_0 = \mathbf{x}_1 - \mathbf{x}_2$, where $\mathbf{x}_1$ and $\mathbf{x}_2$ are any two distinct solutions. Because $\mathbf{x}_1$ and $\mathbf{x}_2$ are distinct, the matrix $\mathbf{x}_0$ is nonzero; moreover,

$$A\mathbf{x}_0 = A(\mathbf{x}_1 - \mathbf{x}_2) = A\mathbf{x}_1 - A\mathbf{x}_2 = \mathbf{b} - \mathbf{b} = 0$$

If we now let k be any scalar, then

$$A(\mathbf{x}_1 + k\mathbf{x}_0) = A\mathbf{x}_1 + A(k\mathbf{x}_0) = A\mathbf{x}_1 + k(A\mathbf{x}_0)$$
$$= \mathbf{b} + k0 = \mathbf{b} + 0 = \mathbf{b}$$

But this says that $\mathbf{x}_1 + k\mathbf{x}_0$ is a solution of $A\mathbf{x} = \mathbf{b}$. Since $\mathbf{x}_0$ is nonzero and there are infinitely many choices for k, the system $A\mathbf{x} = \mathbf{b}$ has infinitely many solutions. ◄

Solving Linear Systems by Matrix Inversion

Thus far we have studied two *procedures* for solving linear systems—Gauss–Jordan elimination and Gaussian elimination. The following theorem provides an actual *formula* for the solution of a linear system of n equations in n unknowns in the case where the coefficient matrix is invertible.

> **THEOREM 1.6.2** *If A is an invertible $n \times n$ matrix, then for each $n \times 1$ matrix $\mathbf{b}$, the system of equations $A\mathbf{x} = \mathbf{b}$ has exactly one solution, namely, $\mathbf{x} = A^{-1}\mathbf{b}$.*

Proof Since $A(A^{-1}\mathbf{b}) = \mathbf{b}$, it follows that $\mathbf{x} = A^{-1}\mathbf{b}$ is a solution of $A\mathbf{x} = \mathbf{b}$. To show that this is the only solution, we will assume that $\mathbf{x}_0$ is an arbitrary solution and then show that $\mathbf{x}_0$ must be the solution $A^{-1}\mathbf{b}$.

If $\mathbf{x}_0$ is any solution of $A\mathbf{x} = \mathbf{b}$, then $A\mathbf{x}_0 = \mathbf{b}$. Multiplying both sides of this equation by A^{-1}, we obtain $\mathbf{x}_0 = A^{-1}\mathbf{b}$. ◄

▶ **EXAMPLE 1 Solution of a Linear System Using A^{-1}**

Consider the system of linear equations

$$
\begin{aligned}
x_1 + 2x_2 + 3x_3 &= 5 \\
2x_1 + 5x_2 + 3x_3 &= 3 \\
x_1 \qquad\quad + 8x_3 &= 17
\end{aligned}
$$

In matrix form this system can be written as $A\mathbf{x} = \mathbf{b}$, where

$$
A = \begin{bmatrix} 1 & 2 & 3 \\ 2 & 5 & 3 \\ 1 & 0 & 8 \end{bmatrix}, \quad \mathbf{x} = \begin{bmatrix} x_1 \\ x_2 \\ x_3 \end{bmatrix}, \quad \mathbf{b} = \begin{bmatrix} 5 \\ 3 \\ 17 \end{bmatrix}
$$

In Example 4 of the preceding section, we showed that A is invertible and

$$
A^{-1} = \begin{bmatrix} -40 & 16 & 9 \\ 13 & -5 & -3 \\ 5 & -2 & -1 \end{bmatrix}
$$

By Theorem 1.6.2, the solution of the system is

$$
\mathbf{x} = A^{-1}\mathbf{b} = \begin{bmatrix} -40 & 16 & 9 \\ 13 & -5 & -3 \\ 5 & -2 & -1 \end{bmatrix} \begin{bmatrix} 5 \\ 3 \\ 17 \end{bmatrix} = \begin{bmatrix} 1 \\ -1 \\ 2 \end{bmatrix}
$$

Keep in mind that the method of Example 1 only applies when the system has as many equations as unknowns and the coefficient matrix is invertible.

or $x_1 = 1$, $x_2 = -1$, $x_3 = 2$. ◄

Linear Systems with a Common Coefficient Matrix

Frequently, one is concerned with solving a sequence of systems

$$
A\mathbf{x} = \mathbf{b}_1, \quad A\mathbf{x} = \mathbf{b}_2, \quad A\mathbf{x} = \mathbf{b}_3, \ldots, \quad A\mathbf{x} = \mathbf{b}_k
$$

each of which has the same square coefficient matrix A. If A is invertible, then the solutions

$$
\mathbf{x}_1 = A^{-1}\mathbf{b}_1, \quad \mathbf{x}_2 = A^{-1}\mathbf{b}_2, \quad \mathbf{x}_3 = A^{-1}\mathbf{b}_3, \ldots, \quad \mathbf{x}_k = A^{-1}\mathbf{b}_k
$$

can be obtained with one matrix inversion and k matrix multiplications. An efficient way to do this is to form the partitioned matrix

$$
[A \mid \mathbf{b}_1 \mid \mathbf{b}_2 \mid \cdots \mid \mathbf{b}_k] \tag{1}
$$

in which the coefficient matrix A is "augmented" by all k of the matrices $\mathbf{b}_1, \mathbf{b}_2, \ldots, \mathbf{b}_k$, and then reduce (1) to reduced row echelon form by Gauss–Jordan elimination. In this way we can solve all k systems at once. This method has the added advantage that it applies even when A is not invertible.

▶ **EXAMPLE 2** Solving Two Linear Systems at Once

Solve the systems

(a)
$$\begin{aligned} x_1 + 2x_2 + 3x_3 &= 4 \\ 2x_1 + 5x_2 + 3x_3 &= 5 \\ x_1 \qquad\quad + 8x_3 &= 9 \end{aligned}$$

(b)
$$\begin{aligned} x_1 + 2x_2 + 3x_3 &= 1 \\ 2x_1 + 5x_2 + 3x_3 &= 6 \\ x_1 \qquad\quad + 8x_3 &= -6 \end{aligned}$$

Solution The two systems have the same coefficient matrix. If we augment this co-efficient matrix with the columns of constants on the right sides of these systems, we obtain

$$\left[\begin{array}{ccc|cc} 1 & 2 & 3 & 4 & 1 \\ 2 & 5 & 3 & 5 & 6 \\ 1 & 0 & 8 & 9 & -6 \end{array}\right]$$

Reducing this matrix to reduced row echelon form yields (verify)

$$\left[\begin{array}{ccc|cc} 1 & 0 & 0 & 1 & 2 \\ 0 & 1 & 0 & 0 & 1 \\ 0 & 0 & 1 & 1 & -1 \end{array}\right]$$

It follows from the last two columns that the solution of system (a) is $x_1 = 1$, $x_2 = 0$, $x_3 = 1$ and the solution of system (b) is $x_1 = 2$, $x_2 = 1$, $x_3 = -1$. ◀

Properties of Invertible Matrices   Up to now, to show that an $n \times n$ matrix A is invertible, it has been necessary to find an $n \times n$ matrix B such that

$$AB = I \quad\text{and}\quad BA = I$$

The next theorem shows that if we produce an $n \times n$ matrix B satisfying *either* condition, then the other condition holds automatically.

> **THEOREM 1.6.3** *Let A be a square matrix.*
>
> (*a*) *If B is a square matrix satisfying BA = I, then B = A^{-1}.*
> (*b*) *If B is a square matrix satisfying AB = I, then B = A^{-1}.*

We will prove part (*a*) and leave part (*b*) as an exercise.

Proof (a)   Assume that $BA = I$. If we can show that A is invertible, the proof can be completed by multiplying $BA = I$ on both sides by A^{-1} to obtain

$$BAA^{-1} = IA^{-1} \quad\text{or}\quad BI = IA^{-1} \quad\text{or}\quad B = A^{-1}$$

To show that A is invertible, it suffices to show that the system $A\mathbf{x} = \mathbf{0}$ has only the trivial solution (see Theorem 1.5.3). Let $\mathbf{x}_0$ be any solution of this system. If we multiply both sides of $A\mathbf{x}_0 = \mathbf{0}$ on the left by B, we obtain $BA\mathbf{x}_0 = B\mathbf{0}$ or $I\mathbf{x}_0 = \mathbf{0}$ or $\mathbf{x}_0 = \mathbf{0}$. Thus, the system of equations $A\mathbf{x} = \mathbf{0}$ has only the trivial solution. ◀

Equivalence Theorem   We are now in a position to add two more statements to the four given in Theorem 1.5.3.

THEOREM 1.6.4 Equivalent Statements

If A is an n × n matrix, then the following are equivalent.

(a) *A is invertible.*

(b) $A\mathbf{x} = \mathbf{0}$ *has only the trivial solution.*

(c) *The reduced row echelon form of A is I_n.*

(d) *A is expressible as a product of elementary matrices.*

(e) $A\mathbf{x} = \mathbf{b}$ *is consistent for every n × 1 matrix $\mathbf{b}$.*

(f) $A\mathbf{x} = \mathbf{b}$ *has exactly one solution for every n × 1 matrix $\mathbf{b}$.*

Proof Since we proved in Theorem 1.5.3 that (*a*), (*b*), (*c*), and (*d*) are equivalent, it will be sufficient to prove that $(a) \Rightarrow (f) \Rightarrow (e) \Rightarrow (a)$.

(a) ⇒ (f) This was already proved in Theorem 1.6.2.

(f) ⇒ (e) This is almost self-evident, for if $A\mathbf{x} = \mathbf{b}$ has exactly one solution for every $n \times 1$ matrix $\mathbf{b}$, then $A\mathbf{x} = \mathbf{b}$ is consistent for every $n \times 1$ matrix $\mathbf{b}$.

(e) ⇒ (a) If the system $A\mathbf{x} = \mathbf{b}$ is consistent for every $n \times 1$ matrix $\mathbf{b}$, then, in particular, this is so for the systems

$$A\mathbf{x} = \begin{bmatrix} 1 \\ 0 \\ 0 \\ \vdots \\ 0 \end{bmatrix}, \quad A\mathbf{x} = \begin{bmatrix} 0 \\ 1 \\ 0 \\ \vdots \\ 0 \end{bmatrix}, \dots, \quad A\mathbf{x} = \begin{bmatrix} 0 \\ 0 \\ 0 \\ \vdots \\ 1 \end{bmatrix}$$

Let $\mathbf{x}_1, \mathbf{x}_2, \dots, \mathbf{x}_n$ be solutions of the respective systems, and let us form an $n \times n$ matrix C having these solutions as columns. Thus C has the form

$$C = [\mathbf{x}_1 \mid \mathbf{x}_2 \mid \cdots \mid \mathbf{x}_n]$$

As discussed in Section 1.3, the successive columns of the product AC will be

$$A\mathbf{x}_1, A\mathbf{x}_2, \dots, A\mathbf{x}_n$$

[see Formula (8) of Section 1.3]. Thus,

It follows from the equivalency of parts (*e*) and (*f*) that if you can show that $A\mathbf{x} = \mathbf{b}$ has at *least one* solution for every $n \times 1$ matrix $\mathbf{b}$, then you can conclude that it has *exactly one* solution for every $n \times 1$ matrix $\mathbf{b}$.

$$AC = [A\mathbf{x}_1 \mid A\mathbf{x}_2 \mid \cdots \mid A\mathbf{x}_n] = \begin{bmatrix} 1 & 0 & \cdots & 0 \\ 0 & 1 & \cdots & 0 \\ 0 & 0 & \cdots & 0 \\ \vdots & \vdots & & \vdots \\ 0 & 0 & \cdots & 1 \end{bmatrix} = I$$

By part (*b*) of Theorem 1.6.3, it follows that $C = A^{-1}$. Thus, A is invertible. ◀

We know from earlier work that invertible matrix factors produce an invertible product. Conversely, the following theorem shows that if the product of square matrices is invertible, then the factors themselves must be invertible.

THEOREM 1.6.5 *Let A and B be square matrices of the same size. If AB is invertible, then A and B must also be invertible.*

In our later work the following fundamental problem will occur frequently in various contexts.

> **A Fundamental Problem** Let A be a fixed $m \times n$ matrix. Find all $m \times 1$ matrices $\mathbf{b}$ such that the system of equations $A\mathbf{x} = \mathbf{b}$ is consistent.

If A is an invertible matrix, Theorem 1.6.2 completely solves this problem by asserting that for *every* $m \times 1$ matrix $\mathbf{b}$, the linear system $A\mathbf{x} = \mathbf{b}$ has the unique solution $\mathbf{x} = A^{-1}\mathbf{b}$. If A is not square, or if A is square but not invertible, then Theorem 1.6.2 does not apply. In these cases the matrix $\mathbf{b}$ must usually satisfy certain conditions in order for $A\mathbf{x} = \mathbf{b}$ to be consistent. The following example illustrates how the methods of Section 1.2 can be used to determine such conditions.

▶ **EXAMPLE 3 Determining Consistency by Elimination**

What conditions must b_1, b_2, and b_3 satisfy in order for the system of equations

$$
\begin{array}{rcl}
x_1 + x_2 + 2x_3 &=& b_1 \\
x_1 \quad\;\;\; + \; x_3 &=& b_2 \\
2x_1 + x_2 + 3x_3 &=& b_3
\end{array}
$$

to be consistent?

Solution The augmented matrix is

$$
\begin{bmatrix}
1 & 1 & 2 & b_1 \\
1 & 0 & 1 & b_2 \\
2 & 1 & 3 & b_3
\end{bmatrix}
$$

which can be reduced to row echelon form as follows:

$$
\begin{bmatrix}
1 & 1 & 2 & b_1 \\
0 & -1 & -1 & b_2 - b_1 \\
0 & -1 & -1 & b_3 - 2b_1
\end{bmatrix}
$$

    ⟵ -1 times the first row was added to the second and -2 times the first row was added to the third.

$$
\begin{bmatrix}
1 & 1 & 2 & b_1 \\
0 & 1 & 1 & b_1 - b_2 \\
0 & -1 & -1 & b_3 - 2b_1
\end{bmatrix}
$$

    ⟵ The second row was multiplied by -1.

$$
\begin{bmatrix}
1 & 1 & 2 & b_1 \\
0 & 1 & 1 & b_1 - b_2 \\
0 & 0 & 0 & b_3 - b_2 - b_1
\end{bmatrix}
$$

    ⟵ The second row was added to the third.

It is now evident from the third row in the matrix that the system has a solution if and only if b_1, b_2, and b_3 satisfy the condition

$$
b_3 - b_2 - b_1 = 0 \quad \text{or} \quad b_3 = b_1 + b_2
$$

To express this condition another way, $A\mathbf{x} = \mathbf{b}$ is consistent if and only if $\mathbf{b}$ is a matrix of the form

$$
\mathbf{b} = \begin{bmatrix} b_1 \\ b_2 \\ b_1 + b_2 \end{bmatrix}
$$

where b_1 and b_2 are arbitrary.

▶ **EXAMPLE 4** Determining Consistency by Elimination

What conditions must b_1, b_2, and b_3 satisfy in order for the system of equations

$$\begin{aligned}
x_1 + 2x_2 + 3x_3 &= b_1 \\
2x_1 + 5x_2 + 3x_3 &= b_2 \\
x_1 \qquad\quad + 8x_3 &= b_3
\end{aligned}$$

to be consistent?

Solution The augmented matrix is

$$\begin{bmatrix} 1 & 2 & 3 & b_1 \\ 2 & 5 & 3 & b_2 \\ 1 & 0 & 8 & b_3 \end{bmatrix}$$

Reducing this to reduced row echelon form yields (verify)

$$\begin{bmatrix} 1 & 0 & 0 & -40b_1 + 16b_2 + 9b_3 \\ 0 & 1 & 0 & 13b_1 - 5b_2 - 3b_3 \\ 0 & 0 & 1 & 5b_1 - 2b_2 - b_3 \end{bmatrix} \qquad (2)$$

In this case there are no restrictions on b_1, b_2, and b_3, so the system has the unique solution

$$x_1 = -40b_1 + 16b_2 + 9b_3, \quad x_2 = 13b_1 - 5b_2 - 3b_3, \quad x_3 = 5b_1 - 2b_2 - b_3 \quad (3)$$

for all values of b_1, b_2, and b_3. ◀

What does the result in Example 4 tell you about the coefficient matrix of the system?

Skills

- Determine whether a linear system of equations has no solutions, exactly one solution, or infinitely many solutions.
- Solve linear systems by inverting its coefficient matrix.
- Solve multiple linear systems with the same coefficient matrix simultaneously.
- Be familiar with the additional conditions of invertibility stated in the Equivalence Theorem.

Exercise Set 1.6

In Exercises 1–8, solve the system by inverting the coefficient matrix and using Theorem 1.6.2.

1. $\begin{aligned} x_1 + x_2 &= 2 \\ 5x_1 + 6x_2 &= 9 \end{aligned}$

2. $\begin{aligned} 4x_1 - 3x_2 &= -3 \\ 2x_1 - 5x_2 &= 9 \end{aligned}$

3. $\begin{aligned} x_1 + 3x_2 + x_3 &= 4 \\ 2x_1 + 2x_2 + x_3 &= -1 \\ 2x_1 + 3x_2 + x_3 &= 3 \end{aligned}$

4. $\begin{aligned} 5x_1 + 3x_2 + 2x_3 &= 4 \\ 3x_1 + 3x_2 + 2x_3 &= 2 \\ x_2 + x_3 &= 5 \end{aligned}$

5. $\begin{aligned} x + y + z &= 5 \\ x + y - 4z &= 10 \\ -4x + y + z &= 0 \end{aligned}$

6. $\begin{aligned} -x - 2y - 3z &= 0 \\ w + x + 4y + 4z &= 7 \\ w + 3x + 7y + 9z &= 4 \\ -w - 2x - 4y - 6z &= 6 \end{aligned}$

7. $\begin{aligned} 3x_1 + 5x_2 &= b_1 \\ x_1 + 2x_2 &= b_2 \end{aligned}$

8. $\begin{aligned} x_1 + 2x_2 + 3x_3 &= b_1 \\ 2x_1 + 5x_2 + 5x_3 &= b_2 \\ 3x_1 + 5x_2 + 8x_3 &= b_3 \end{aligned}$

In Exercises 9–12, solve the linear systems together by reducing the appropriate augmented matrix.

9. $\begin{aligned} x_1 - 5x_2 &= b_1 \\ 3x_1 + 2x_2 &= b_2 \end{aligned}$
   (i) $b_1 = 1$, $b_2 = 4$ (ii) $b_1 = -2$, $b_2 = 5$

10. $\begin{aligned} -x_1 + 4x_2 + x_3 &= b_1 \\ x_1 + 9x_2 - 2x_3 &= b_2 \\ 6x_1 + 4x_2 - 8x_3 &= b_3 \end{aligned}$
   (i) $b_1 = 0$, $b_2 = 1$, $b_3 = 0$
   (ii) $b_1 = -3$, $b_2 = 4$, $b_3 = -5$

11. $\begin{aligned} 4x_1 - 7x_2 &= b_1 \\ x_1 + 2x_2 &= b_2 \end{aligned}$
   (i) $b_1 = 0$, $b_2 = 1$ (ii) $b_1 = -4$, $b_2 = 6$
   (iii) $b_1 = -1$, $b_2 = 3$ (iv) $b_1 = -5$, $b_2 = 1$

12.
$$x_1 + 3x_2 + 5x_3 = b_1$$
$$-x_1 - 2x_2 \qquad = b_2$$
$$2x_1 + 5x_2 + 4x_3 = b_3$$

(i) $b_1 = 1, \quad b_2 = 0, \quad b_3 = -1$

(ii) $b_1 = 0, \quad b_2 = 1, \quad b_3 = 1$

(iii) $b_1 = -1, \quad b_2 = -1, \quad b_3 = 0$

▶ In Exercises 13–17, determine conditions on the b_i's, if any, in order to guarantee that the linear system is consistent. ◀

13.
$$x_1 + 3x_2 = b_1$$
$$-2x_1 + x_2 = b_2$$

14.
$$6x_1 - 4x_2 = b_1$$
$$3x_1 - 2x_2 = b_2$$

15.
$$x_1 - 2x_2 + 5x_3 = b_1$$
$$4x_1 - 5x_2 + 8x_3 = b_2$$
$$-3x_1 + 3x_2 - 3x_3 = b_3$$

16.
$$x_1 - 2x_2 - x_3 = b_1$$
$$-4x_1 + 5x_2 + 2x_3 = b_2$$
$$-4x_1 + 7x_2 + 4x_3 = b_3$$

17.
$$x_1 - x_2 + 3x_3 + 2x_4 = b_1$$
$$-2x_1 + x_2 + 5x_3 + x_4 = b_2$$
$$-3x_1 + 2x_2 + 2x_3 - x_4 = b_3$$
$$4x_1 - 3x_2 + x_3 + 3x_4 = b_4$$

18. Consider the matrices

$$A = \begin{bmatrix} 2 & 1 & 2 \\ 2 & 2 & -2 \\ 3 & 1 & 1 \end{bmatrix} \quad \text{and} \quad \mathbf{x} = \begin{bmatrix} x_1 \\ x_2 \\ x_3 \end{bmatrix}$$

(a) Show that the equation $A\mathbf{x} = \mathbf{x}$ can be rewritten as $(A - I)\mathbf{x} = \mathbf{0}$ and use this result to solve $A\mathbf{x} = \mathbf{x}$ for $\mathbf{x}$.

(b) Solve $A\mathbf{x} = 4\mathbf{x}$.

▶ In Exercises 19–20, solve the given matrix equation for X. ◀

19.
$$\begin{bmatrix} 1 & -1 & 1 \\ 2 & 3 & 0 \\ 0 & 2 & -1 \end{bmatrix} X = \begin{bmatrix} 2 & -1 & 5 & 7 & 8 \\ 4 & 0 & -3 & 0 & 1 \\ 3 & 5 & -7 & 2 & 1 \end{bmatrix}$$

20.
$$\begin{bmatrix} -2 & 0 & 1 \\ 0 & -1 & -1 \\ 1 & 1 & -4 \end{bmatrix} X = \begin{bmatrix} 4 & 3 & 2 & 1 \\ 6 & 7 & 8 & 9 \\ 1 & 3 & 7 & 9 \end{bmatrix}$$

21. Let $A\mathbf{x} = \mathbf{0}$ be a homogeneous system of n linear equations in n unknowns that has only the trivial solution. Show that if k is any positive integer, then the system $A^k\mathbf{x} = \mathbf{0}$ also has only the trivial solution.

22. Let $A\mathbf{x} = \mathbf{0}$ be a homogeneous system of n linear equations in n unknowns, and let Q be an invertible $n \times n$ matrix. Show that $A\mathbf{x} = \mathbf{0}$ has just the trivial solution if and only if $(QA)\mathbf{x} = \mathbf{0}$ has just the trivial solution.

23. Let $A\mathbf{x} = \mathbf{b}$ be any consistent system of linear equations, and let $\mathbf{x}_1$ be a fixed solution. Show that every solution to the system can be written in the form $\mathbf{x} = \mathbf{x}_1 + \mathbf{x}_0$, where $\mathbf{x}_0$ is a solution to $A\mathbf{x} = \mathbf{0}$. Show also that every matrix of this form is a solution.

24. Use part (*a*) of Theorem 1.6.3 to prove part (*b*).

True-False Exercises

In parts (a)–(g) determine whether the statement is true or false, and justify your answer.

(a) It is impossible for a system of linear equations to have exactly two solutions.

(b) If the linear system $A\mathbf{x} = \mathbf{b}$ has a unique solution, then the linear system $A\mathbf{x} = \mathbf{c}$ also must have a unique solution.

(c) If A and B are $n \times n$ matrices such that $AB = I_n$, then $BA = I_n$.

(d) If A and B are row equivalent matrices, then the linear systems $A\mathbf{x} = \mathbf{0}$ and $B\mathbf{x} = \mathbf{0}$ have the same solution set.

(e) If A is an $n \times n$ matrix and S is an $n \times n$ invertible matrix, then if $\mathbf{x}$ is a solution to the linear system $(S^{-1}AS)\mathbf{x} = \mathbf{b}$, then $S\mathbf{x}$ is a solution to the linear system $A\mathbf{y} = S\mathbf{b}$.

(f) Let A be an $n \times n$ matrix. The linear system $A\mathbf{x} = 4\mathbf{x}$ has a unique solution if and only if $A - 4I$ is an invertible matrix.

(g) Let A and B be $n \times n$ matrices. If A or B (or both) are not invertible, then neither is AB.

1.7 Diagonal, Triangular, and Symmetric Matrices

In this section we will discuss matrices that have various special forms. These matrices arise in a wide variety of applications and will play an important role in our subsequent work.

Diagonal Matrices   A square matrix in which all the entries off the main diagonal are zero is called a ***diagonal matrix***. Here are some examples:

$$\begin{bmatrix} 0 & 0 \\ 0 & 0 \end{bmatrix}, \quad \begin{bmatrix} 2 & 0 \\ 0 & -5 \end{bmatrix}, \quad \begin{bmatrix} 1 & 0 & 0 \\ 0 & 1 & 0 \\ 0 & 0 & 1 \end{bmatrix}, \quad \begin{bmatrix} 6 & 0 & 0 & 0 \\ 0 & -4 & 0 & 0 \\ 0 & 0 & 0 & 0 \\ 0 & 0 & 0 & 8 \end{bmatrix}$$

A general $n \times n$ diagonal matrix D can be written as

$$D = \begin{bmatrix} d_1 & 0 & \cdots & 0 \\ 0 & d_2 & \cdots & 0 \\ \vdots & \vdots & & \vdots \\ 0 & 0 & \cdots & d_n \end{bmatrix} \tag{1}$$

A diagonal matrix is invertible if and only if all of its diagonal entries are nonzero; in this case the inverse of (1) is

Confirm Formula (2) by showing that

$$DD^{-1} = D^{-1}D = I$$

$$D^{-1} = \begin{bmatrix} 1/d_1 & 0 & \cdots & 0 \\ 0 & 1/d_2 & \cdots & 0 \\ \vdots & \vdots & & \vdots \\ 0 & 0 & \cdots & 1/d_n \end{bmatrix} \tag{2}$$

Powers of diagonal matrices are easy to compute; we leave it for you to verify that if D is the diagonal matrix (1) and k is a positive integer, then

$$D^k = \begin{bmatrix} d_1^k & 0 & \cdots & 0 \\ 0 & d_2^k & \cdots & 0 \\ \vdots & \vdots & & \vdots \\ 0 & 0 & \cdots & d_n^k \end{bmatrix} \tag{3}$$

▶ **EXAMPLE 1** **Inverses and Powers of Diagonal Matrices**

If

$$A = \begin{bmatrix} 1 & 0 & 0 \\ 0 & -3 & 0 \\ 0 & 0 & 2 \end{bmatrix}$$

then

$$A^{-1} = \begin{bmatrix} 1 & 0 & 0 \\ 0 & -\frac{1}{3} & 0 \\ 0 & 0 & \frac{1}{2} \end{bmatrix}, \quad A^5 = \begin{bmatrix} 1 & 0 & 0 \\ 0 & -243 & 0 \\ 0 & 0 & 32 \end{bmatrix}, \quad A^{-5} = \begin{bmatrix} 1 & 0 & 0 \\ 0 & -\frac{1}{243} & 0 \\ 0 & 0 & \frac{1}{32} \end{bmatrix}$$

◀

Matrix products that involve diagonal factors are especially easy to compute. For example,

$$\begin{bmatrix} d_1 & 0 & 0 \\ 0 & d_2 & 0 \\ 0 & 0 & d_3 \end{bmatrix} \begin{bmatrix} a_{11} & a_{12} & a_{13} & a_{14} \\ a_{21} & a_{22} & a_{23} & a_{24} \\ a_{31} & a_{32} & a_{33} & a_{34} \end{bmatrix} = \begin{bmatrix} d_1 a_{11} & d_1 a_{12} & d_1 a_{13} & d_1 a_{14} \\ d_2 a_{21} & d_2 a_{22} & d_2 a_{23} & d_2 a_{24} \\ d_3 a_{31} & d_3 a_{32} & d_3 a_{33} & d_3 a_{34} \end{bmatrix}$$

$$\begin{bmatrix} a_{11} & a_{12} & a_{13} \\ a_{21} & a_{22} & a_{23} \\ a_{31} & a_{32} & a_{33} \\ a_{41} & a_{42} & a_{43} \end{bmatrix} \begin{bmatrix} d_1 & 0 & 0 \\ 0 & d_2 & 0 \\ 0 & 0 & d_3 \end{bmatrix} = \begin{bmatrix} d_1 a_{11} & d_2 a_{12} & d_3 a_{13} \\ d_1 a_{21} & d_2 a_{22} & d_3 a_{23} \\ d_1 a_{31} & d_2 a_{32} & d_3 a_{33} \\ d_1 a_{41} & d_2 a_{42} & d_3 a_{43} \end{bmatrix}$$

In words, *to multiply a matrix A on the left by a diagonal matrix D, one can multiply successive rows of A by the successive diagonal entries of D, and to multiply A on the right by D, one can multiply successive columns of A by the successive diagonal entries of D.*

Triangular Matrices   A square matrix in which all the entries above the main diagonal are zero is called **lower triangular**, and a square matrix in which all the entries below the main diagonal are zero is called **upper triangular**. A matrix that is either upper triangular or lower triangular is called **triangular**.

▶ **EXAMPLE 2 Upper and Lower Triangular Matrices**

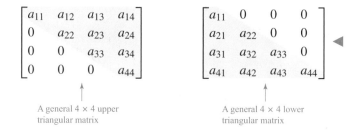

A general 4 × 4 upper
triangular matrix

A general 4 × 4 lower
triangular matrix

Remark   Observe that diagonal matrices are both upper triangular and lower triangular since they have zeros below and above the main diagonal. Observe also that a *square* matrix in row echelon form is upper triangular since it has zeros below the main diagonal.

Properties of Triangular Matrices   Example 2 illustrates the following four facts about triangular matrices that we will state without formal proof:

- A square matrix $A = [a_{ij}]$ is upper triangular if and only if all entries to the left of the main diagonal are zero; that is, $a_{ij} = 0$ if $i > j$ (Figure 1.7.1).

- A square matrix $A = [a_{ij}]$ is lower triangular if and only if all entries to the right of the main diagonal are zero; that is, $a_{ij} = 0$ if $i < j$ (Figure 1.7.1).

- A square matrix $A = [a_{ij}]$ is upper triangular if and only if the ith row starts with at least $i - 1$ zeros for every i.

- A square matrix $A = [a_{ij}]$ is lower triangular if and only if the jth column starts with at least $j - 1$ zeros for every j.

▲ Figure 1.7.1

The following theorem lists some of the basic properties of triangular matrices.

> **THEOREM 1.7.1**
>
> (*a*) *The transpose of a lower triangular matrix is upper triangular, and the transpose of an upper triangular matrix is lower triangular.*
>
> (*b*) *The product of lower triangular matrices is lower triangular, and the product of upper triangular matrices is upper triangular.*
>
> (*c*) *A triangular matrix is invertible if and only if its diagonal entries are all nonzero.*
>
> (*d*) *The inverse of an invertible lower triangular matrix is lower triangular, and the inverse of an invertible upper triangular matrix is upper triangular.*

Part (*a*) is evident from the fact that transposing a square matrix can be accomplished by reflecting the entries about the main diagonal; we omit the formal proof. We will prove (*b*), but we will defer the proofs of (*c*) and (*d*) to the next chapter, where we will have the tools to prove those results more efficiently.

Proof (b)   We will prove the result for lower triangular matrices; the proof for upper triangular matrices is similar. Let $A = [a_{ij}]$ and $B = [b_{ij}]$ be lower triangular $n \times n$ matrices, and let $C = [c_{ij}]$ be the product $C = AB$. We can prove that C is lower triangular by showing that $c_{ij} = 0$ for $i < j$. But from the definition of matrix multiplication,

$$c_{ij} = a_{i1}b_{1j} + a_{i2}b_{2j} + \cdots + a_{in}b_{nj}$$

If we assume that $i < j$, then the terms in this expression can be grouped as follows:

$$c_{ij} = \underbrace{a_{i1}b_{1j} + a_{i2}b_{2j} + \cdots + a_{i(j-1)}b_{(j-1)j}}_{\substack{\text{Terms in which the row} \\ \text{number of } b \text{ is less than} \\ \text{the column number of } b}} + \underbrace{a_{ij}b_j + \cdots + a_{in}b_{nj}}_{\substack{\text{Terms in which the row} \\ \text{number of } a \text{ is less than} \\ \text{the column number of } a}}$$

In the first grouping all of the b factors are zero since B is lower triangular, and in the second grouping all of the a factors are zero since A is lower triangular. Thus, $c_{ij} = 0$, which is what we wanted to prove. ◄

▶ **EXAMPLE 3 Computations with Triangular Matrices**

Consider the upper triangular matrices

$$A = \begin{bmatrix} 1 & 3 & -1 \\ 0 & 2 & 4 \\ 0 & 0 & 5 \end{bmatrix}, \quad B = \begin{bmatrix} 3 & -2 & 2 \\ 0 & 0 & -1 \\ 0 & 0 & 1 \end{bmatrix}$$

It follows from part (c) of Theorem 1.7.1 that the matrix A is invertible but the matrix B is not. Moreover, the theorem also tells us that A^{-1}, AB, and BA must be upper triangular. We leave it for you to confirm these three statements by showing that

$$A^{-1} = \begin{bmatrix} 1 & -\frac{3}{2} & \frac{7}{5} \\ 0 & \frac{1}{2} & -\frac{2}{5} \\ 0 & 0 & \frac{1}{5} \end{bmatrix}, \quad AB = \begin{bmatrix} 3 & -2 & -2 \\ 0 & 0 & 2 \\ 0 & 0 & 5 \end{bmatrix}, \quad BA = \begin{bmatrix} 3 & 5 & -1 \\ 0 & 0 & -5 \\ 0 & 0 & 5 \end{bmatrix} \quad ◄$$

Symmetric Matrices

DEFINITION 1   A square matrix A is said to be **symmetric** if $A = A^T$.

It is easy to recognize a symmetric matrix by inspection: The entries on the main diagonal have no restrictions, but mirror images of entries *across* the main diagonal must be equal. Here is a picture using the second matrix in Example 4:

$$\begin{bmatrix} 1 & 4 & 5 \\ 4 & -3 & 0 \\ 5 & 0 & 7 \end{bmatrix}$$

All diagonal matrices, such as the third matrix in Example 4, have this property.

▶ **EXAMPLE 4 Symmetric Matrices**

The following matrices are symmetric, since each is equal to its own transpose (verify).

$$\begin{bmatrix} 7 & -3 \\ -3 & 5 \end{bmatrix}, \quad \begin{bmatrix} 1 & 4 & 5 \\ 4 & -3 & 0 \\ 5 & 0 & 7 \end{bmatrix}, \quad \begin{bmatrix} d_1 & 0 & 0 & 0 \\ 0 & d_2 & 0 & 0 \\ 0 & 0 & d_3 & 0 \\ 0 & 0 & 0 & d_4 \end{bmatrix} \quad ◄$$

Remark   It follows from Formula (11) of Section 1.3 that a square matrix $A = [a_{ij}]$ is symmetric if and only if

$$(A)_{ij} = (A)_{ji} \tag{4}$$

for all values of i and j.

The following theorem lists the main algebraic properties of symmetric matrices. The proofs are direct consequences of Theorem 1.4.8 and are omitted.

> **THEOREM 1.7.2** *If A and B are symmetric matrices with the same size, and if k is any scalar, then:*
>
> (*a*) A^T *is symmetric.*
> (*b*) $A + B$ *and* $A - B$ *are symmetric.*
> (*c*) kA *is symmetric.*

It is not true, in general, that the product of symmetric matrices is symmetric. To see why this is so, let A and B be symmetric matrices with the same size. Then it follows from part (*e*) of Theorem 1.4.8 and the symmetry of A and B that

$$(AB)^T = B^T A^T = BA$$

Thus, $(AB)^T = AB$ if and only if $AB = BA$, that is, if and only if A and B commute. In summary, we have the following result.

Use the result in Theorem 1.4.5 to confirm Theorem 1.7.3 in the case where the symmetric matrix

$$\begin{bmatrix} a & b \\ b & d \end{bmatrix}$$

is invertible.

> **THEOREM 1.7.3** *The product of two symmetric matrices is symmetric if and only if the matrices commute.*

▶ **EXAMPLE 5 Products of Symmetric Matrices**

The first of the following equations shows a product of symmetric matrices that *is not* symmetric, and the second shows a product of symmetric matrices that *is* symmetric. We conclude that the factors in the first equation do not commute, but those in the second equation do. We leave it for you to verify that this is so.

$$\begin{bmatrix} 1 & 2 \\ 2 & 3 \end{bmatrix} \begin{bmatrix} -4 & 1 \\ 1 & 0 \end{bmatrix} = \begin{bmatrix} -2 & 1 \\ -5 & 2 \end{bmatrix}$$

$$\begin{bmatrix} 1 & 2 \\ 2 & 3 \end{bmatrix} \begin{bmatrix} -4 & 3 \\ 3 & -1 \end{bmatrix} = \begin{bmatrix} 2 & 1 \\ 1 & 3 \end{bmatrix} \blacktriangleleft$$

Invertibility of Symmetric Matrices

In general, a symmetric matrix need not be invertible. For example, a diagonal matrix with a zero on the main diagonal is symmetric but not invertible. However, the following theorem shows that if a symmetric matrix happens to be invertible, then its inverse must also be symmetric.

> **THEOREM 1.7.4** *If A is an invertible symmetric matrix, then* A^{-1} *is symmetric.*

Proof Assume that A is symmetric and invertible. From Theorem 1.4.9 and the fact that $A = A^T$, we have

$$(A^{-1})^T = (A^T)^{-1} = A^{-1}$$

which proves that A^{-1} is symmetric. ◀

Products AA^T and A^TA

Matrix products of the form AA^T and A^TA arise in a variety of applications. If A is an $m \times n$ matrix, then A^T is an $n \times m$ matrix, so the products AA^T and A^TA are both square matrices—the matrix AA^T has size $m \times m$, and the matrix A^TA has size $n \times n$. Such products are always symmetric since

$$(AA^T)^T = (A^T)^T A^T = AA^T \quad \text{and} \quad (A^TA)^T = A^T(A^T)^T = A^TA$$

▶ **EXAMPLE 6** **The Product of a Matrix and Its Transpose Is Symmetric**

Let A be the 2×3 matrix

$$A = \begin{bmatrix} 1 & -2 & 4 \\ 3 & 0 & -5 \end{bmatrix}$$

Then

$$A^T A = \begin{bmatrix} 1 & 3 \\ -2 & 0 \\ 4 & -5 \end{bmatrix} \begin{bmatrix} 1 & -2 & 4 \\ 3 & 0 & -5 \end{bmatrix} = \begin{bmatrix} 10 & -2 & -11 \\ -2 & 4 & -8 \\ -11 & -8 & 41 \end{bmatrix}$$

$$A A^T = \begin{bmatrix} 1 & -2 & 4 \\ 3 & 0 & -5 \end{bmatrix} \begin{bmatrix} 1 & 3 \\ -2 & 0 \\ 4 & -5 \end{bmatrix} = \begin{bmatrix} 21 & -17 \\ -17 & 34 \end{bmatrix}$$

Observe that $A^T A$ and $A A^T$ are symmetric as expected. ◀

Later in this text, we will obtain general conditions on A under which $A A^T$ and $A^T A$ are invertible. However, in the special case where A is *square*, we have the following result.

THEOREM 1.7.5 *If A is an invertible matrix, then $A A^T$ and $A^T A$ are also invertible.*

Proof Since A is invertible, so is A^T by Theorem 1.4.9. Thus $A A^T$ and $A^T A$ are invertible, since they are the products of invertible matrices. ◀

Concept Review

- Diagonal matrix
- Lower triangular matrix
- Upper triangular matrix
- Triangular matrix
- Symmetric matrix

Skills

- Determine whether a diagonal matrix is invertible with no computations.
- Compute matrix products involving diagonal matrices by inspection.
- Determine whether a matrix is triangular.
- Understand how the transpose operation affects diagonal and triangular matrices.
- Understand how inversion affects diagonal and triangular matrices.
- Determine whether a matrix is a symmetric matrix.

Exercise Set 1.7

In Exercises 1–4, determine whether the given matrix is invertible.

1. $\begin{bmatrix} 2 & 0 \\ 0 & -5 \end{bmatrix}$

2. $\begin{bmatrix} 4 & 0 & 0 \\ 0 & 0 & 0 \\ 0 & 0 & 5 \end{bmatrix}$

3. $\begin{bmatrix} -1 & 0 & 0 \\ 0 & 2 & 0 \\ 0 & 0 & \frac{1}{3} \end{bmatrix}$

4. $\begin{bmatrix} -1 & 0 & 0 & 0 \\ 0 & 3 & 0 & 0 \\ 0 & 0 & -3 & 0 \\ 0 & 0 & 0 & -2 \end{bmatrix}$

In Exercises 5–8, determine the product by inspection.

5. $\begin{bmatrix} 3 & 0 & 0 \\ 0 & -1 & 0 \\ 0 & 0 & 2 \end{bmatrix} \begin{bmatrix} 2 & 1 \\ -4 & 1 \\ 2 & 5 \end{bmatrix}$

6. $\begin{bmatrix} 1 & 2 & -5 \\ -3 & -1 & 0 \end{bmatrix} \begin{bmatrix} -4 & 0 & 0 \\ 0 & 3 & 0 \\ 0 & 0 & 2 \end{bmatrix}$

7. $\begin{bmatrix} 5 & 0 & 0 \\ 0 & 2 & 0 \\ 0 & 0 & -3 \end{bmatrix} \begin{bmatrix} -3 & 2 & 0 & 4 & -4 \\ 1 & -5 & 3 & 0 & 3 \\ -6 & 2 & 2 & 2 & 2 \end{bmatrix}$

8. $\begin{bmatrix} 2 & 0 & 0 \\ 0 & -1 & 0 \\ 0 & 0 & 4 \end{bmatrix} \begin{bmatrix} 4 & -1 & 3 \\ 1 & 2 & 0 \\ -5 & 1 & -2 \end{bmatrix} \begin{bmatrix} -3 & 0 & 0 \\ 0 & 5 & 0 \\ 0 & 0 & 2 \end{bmatrix}$

▶ In Exercises 9–12, find A^2, A^{-2}, and A^{-k} (where k is any integer) by inspection. ◀

9. $A = \begin{bmatrix} 1 & 0 \\ 0 & -2 \end{bmatrix}$

10. $A = \begin{bmatrix} -6 & 0 & 0 \\ 0 & 3 & 0 \\ 0 & 0 & 5 \end{bmatrix}$

11. $A = \begin{bmatrix} \frac{1}{2} & 0 & 0 \\ 0 & \frac{1}{3} & 0 \\ 0 & 0 & \frac{1}{4} \end{bmatrix}$

12. $A = \begin{bmatrix} -2 & 0 & 0 & 0 \\ 0 & -4 & 0 & 0 \\ 0 & 0 & -3 & 0 \\ 0 & 0 & 0 & 2 \end{bmatrix}$

▶ In Exercises 13–19, decide whether the given matrix is symmetric. ◀

13. $\begin{bmatrix} -8 & -8 \\ 0 & 0 \end{bmatrix}$ **14.** $\begin{bmatrix} 2 & -1 \\ 1 & 2 \end{bmatrix}$ **15.** $\begin{bmatrix} 0 & -7 \\ -7 & 7 \end{bmatrix}$

16. $\begin{bmatrix} 3 & 4 \\ 4 & 0 \end{bmatrix}$ **17.** $\begin{bmatrix} 0 & 1 & 2 \\ 1 & 5 & -6 \\ 2 & 6 & 6 \end{bmatrix}$

18. $\begin{bmatrix} 2 & -1 & 3 \\ -1 & 5 & 1 \\ 3 & 1 & 7 \end{bmatrix}$ **19.** $\begin{bmatrix} 0 & 0 & 1 \\ 0 & 2 & 0 \\ 3 & 0 & 0 \end{bmatrix}$

▶ In Exercises 20–22, decide by inspection whether the given matrix is invertible. ◀

20. $\begin{bmatrix} -1 & 2 & 4 \\ 0 & 3 & 0 \\ 0 & 0 & 5 \end{bmatrix}$ **21.** $\begin{bmatrix} 0 & 1 & -2 & 5 \\ 0 & 1 & 5 & 6 \\ 0 & 0 & -3 & 1 \\ 0 & 0 & 0 & 5 \end{bmatrix}$

22. $\begin{bmatrix} 2 & 0 & 0 & 0 \\ -3 & -1 & 0 & 0 \\ -4 & -6 & 0 & 0 \\ 0 & 3 & 8 & -5 \end{bmatrix}$

▶ In Exercises 23–24, find all values of the unknown constant(s) in order for A to be symmetric. ◀

23. $A = \begin{bmatrix} 4 & -3 \\ a+5 & -1 \end{bmatrix}$

24. $A = \begin{bmatrix} 2 & a-2b+2c & 2a+b+c \\ 3 & 5 & a+c \\ 0 & -2 & 7 \end{bmatrix}$

▶ In Exercises 25–26, find all values of x in order for A to be invertible. ◀

25. $A = \begin{bmatrix} x-1 & x^2 & x^4 \\ 0 & x+2 & x^3 \\ 0 & 0 & x-4 \end{bmatrix}$

26. $A = \begin{bmatrix} x-\frac{1}{2} & 0 & 0 \\ x & x-\frac{1}{3} & 0 \\ x^2 & x^3 & x+\frac{1}{4} \end{bmatrix}$

▶ In Exercises 27–28, find a diagonal matrix A that satisfies the given condition. ◀

27. $A^5 = \begin{bmatrix} 1 & 0 & 0 \\ 0 & -1 & 0 \\ 0 & 0 & -1 \end{bmatrix}$ **28.** $A^{-2} = \begin{bmatrix} 9 & 0 & 0 \\ 0 & 4 & 0 \\ 0 & 0 & 1 \end{bmatrix}$

29. Verify Theorem 1.7.1(*b*) for the product AB, where

$$A = \begin{bmatrix} -1 & 2 & 5 \\ 0 & 1 & 3 \\ 0 & 0 & -4 \end{bmatrix}, \quad B = \begin{bmatrix} 2 & -8 & 0 \\ 0 & 2 & 1 \\ 0 & 0 & 3 \end{bmatrix}$$

30. Verify Theorem 1.7.1(*d*) for the matrices A and B in Exercise 29.

31. Verify Theorem 1.7.4 for the given matrix A.

(a) $A = \begin{bmatrix} 2 & -1 \\ -1 & 3 \end{bmatrix}$ (b) $A = \begin{bmatrix} 1 & -2 & 3 \\ -2 & 1 & -7 \\ 3 & -7 & 4 \end{bmatrix}$

32. Let A be an $n \times n$ symmetric matrix.

(a) Show that A^2 is symmetric.

(b) Show that $2A^2 - 3A + I$ is symmetric.

33. Prove: If $A^T A = A$, then A is symmetric and $A = A^2$.

34. Find all 3×3 diagonal matrices A that satisfy $A^2 - 3A - 4I = 0$.

35. Let $A = [a_{ij}]$ be an $n \times n$ matrix. Determine whether A is symmetric.

(a) $a_{ij} = i^2 + j^2$ (b) $a_{ij} = i^2 - j^2$

(c) $a_{ij} = 2i + 2j$ (d) $a_{ij} = 2i^2 + 2j^3$

36. On the basis of your experience with Exercise 35, devise a general test that can be applied to a formula for a_{ij} to determine whether $A = [a_{ij}]$ is symmetric.

37. A square matrix A is called ***skew-symmetric*** if $A^T = -A$. Prove:

(a) If A is an invertible skew-symmetric matrix, then A^{-1} is skew-symmetric.

(b) If A and B are skew-symmetric matrices, then so are A^T, $A + B$, $A - B$, and kA for any scalar k.

(c) Every square matrix A can be expressed as the sum of a symmetric matrix and a skew-symmetric matrix. [*Hint:* Note the identity $A = \frac{1}{2}(A + A^T) + \frac{1}{2}(A - A^T)$.]

In Exercises 38–39, fill in the missing entries (marked with ×) to produce a skew-symmetric matrix.

38. $A = \begin{bmatrix} \times & \times & 4 \\ 0 & \times & \times \\ \times & -1 & \times \end{bmatrix}$ **39.** $A = \begin{bmatrix} \times & 0 & \times \\ \times & \times & -4 \\ 8 & \times & \times \end{bmatrix}$

40. Find all values of $a, b, c,$ and d for which A is skew-symmetric.

$$A = \begin{bmatrix} 0 & 2a - 3b + c & 3a - 5b + 5c \\ -2 & 0 & 5a - 8b + 6c \\ -3 & -5 & d \end{bmatrix}$$

41. We showed in the text that the product of symmetric matrices is symmetric if and only if the matrices commute. Is the product of commuting skew-symmetric matrices skew-symmetric? Explain. [*Note:* See Exercise 37 for the definition of *skew-symmetric*.]

42. If the $n \times n$ matrix A can be expressed as $A = LU$, where L is a lower triangular matrix and U is an upper triangular matrix, then the linear system $A\mathbf{x} = \mathbf{b}$ can be expressed as $LU\mathbf{x} = \mathbf{b}$ and can be solved in two steps:

Step 1. Let $U\mathbf{x} = \mathbf{y}$, so that $LU\mathbf{x} = \mathbf{b}$ can be expressed as $L\mathbf{y} = \mathbf{b}$. Solve this system.

Step 2. Solve the system $U\mathbf{x} = \mathbf{y}$ for $\mathbf{x}$.

In each part, use this two-step method to solve the given system.

(a) $\begin{bmatrix} 1 & 0 & 0 \\ -2 & 3 & 0 \\ 2 & 4 & 1 \end{bmatrix} \begin{bmatrix} 2 & -1 & 3 \\ 0 & 1 & 2 \\ 0 & 0 & 4 \end{bmatrix} \begin{bmatrix} x_1 \\ x_2 \\ x_3 \end{bmatrix} = \begin{bmatrix} 1 \\ -2 \\ 0 \end{bmatrix}$

(b) $\begin{bmatrix} 2 & 0 & 0 \\ 4 & 1 & 0 \\ -3 & -2 & 3 \end{bmatrix} \begin{bmatrix} 3 & -5 & 2 \\ 0 & 4 & 1 \\ 0 & 0 & 2 \end{bmatrix} \begin{bmatrix} x_1 \\ x_2 \\ x_3 \end{bmatrix} = \begin{bmatrix} 4 \\ -5 \\ 2 \end{bmatrix}$

43. Find an upper triangular matrix that satisfies

$$A^3 = \begin{bmatrix} 1 & 30 \\ 0 & -8 \end{bmatrix}$$

True-False Exercises

In parts (a)–(m) determine whether the statement is true or false, and justify your answer.

(a) The transpose of a diagonal matrix is a diagonal matrix.

(b) The transpose of an upper triangular matrix is an upper triangular matrix.

(c) The sum of an upper triangular matrix and a lower triangular matrix is a diagonal matrix.

(d) All entries of a symmetric matrix are determined by the entries occurring on and above the main diagonal.

(e) All entries of an upper triangular matrix are determined by the entries occurring on and above the main diagonal.

(f) The inverse of an invertible lower triangular matrix is an upper triangular matrix.

(g) A diagonal matrix is invertible if and only if all of its diagonal entries are positive.

(h) The sum of a diagonal matrix and a lower triangular matrix is a lower triangular matrix.

(i) A matrix that is both symmetric and upper triangular must be a diagonal matrix.

(j) If A and B are $n \times n$ matrices such that $A + B$ is symmetric, then A and B are symmetric.

(k) If A and B are $n \times n$ matrices such that $A + B$ is upper triangular, then A and B are upper triangular.

(l) If A^2 is a symmetric matrix, then A is a symmetric matrix.

(m) If kA is a symmetric matrix for some $k \neq 0$, then A is a symmetric matrix.

1.8 Applications of Linear Systems

In this section we will discuss some brief applications of linear systems. These are but a small sample of the wide variety of real-world problems to which our study of linear systems is applicable.

Network Analysis   The concept of a *network* appears in a variety of applications. Loosely stated, a **network** is a set of **branches** through which something "flows." For example, the branches might be electrical wires through which electricity flows, pipes through which water or oil flows, traffic lanes through which vehicular traffic flows, or economic linkages through which money flows, to name a few possibilities.

In most networks, the branches meet at points, called **nodes** or **junctions**, where the flow divides. For example, in an electrical network, nodes occur where three or more wires join, in a traffic network they occur at street intersections, and in a financial network they occur at banking centers where incoming money is distributed to individuals or other institutions.

In the study of networks, there is generally some numerical measure of the rate at which the medium flows through a branch. For example, the flow rate of electricity is often measured in amperes, the flow rate of water or oil in gallons per minute, the flow rate of traffic in vehicles per hour, and the flow rate of European currency in millions of Euros per day. We will restrict our attention to networks in which there is *flow conservation* at each node, by which we mean that *the rate of flow into any node is equal to the rate of flow out of that node.* This ensures that the flow medium does not build up at the nodes and block the free movement of the medium through the network.

A common problem in network analysis is to use known flow rates in certain branches to find the flow rates in all of the branches. Here is an example.

▶ **EXAMPLE 1** **Network Analysis Using Linear Systems**

Figure 1.8.1 shows a network with four nodes in which the flow rate and direction of flow in certain branches are known. Find the flow rates and directions of flow in the remaining branches.

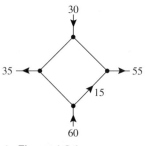
▲ Figure 1.8.1

Solution As illustrated in Figure 1.8.2, we have assigned arbitrary directions to the unknown flow rates x_1, x_2, and x_3. We need not be concerned if some of the directions are incorrect, since an incorrect direction will be signaled by a negative value for the flow rate when we solve for the unknowns.

It follows from the conservation of flow at node A that

$$x_1 + x_2 = 30$$

Similarly, at the other nodes we have

$$x_2 + x_3 = 35 \quad (\text{node } B)$$
$$x_3 + 15 = 60 \quad (\text{node } C)$$
$$x_1 + 15 = 55 \quad (\text{node } D)$$

These four conditions produce the linear system

$$
\begin{aligned}
x_1 + x_2 \quad\ \ &= 30 \\
x_2 + x_3 &= 35 \\
x_3 &= 45 \\
x_1 \qquad\qquad &= 40
\end{aligned}
$$

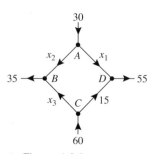
▲ Figure 1.8.2

which we can now try to solve for the unknown flow rates. In this particular case the system is sufficiently simple that it can be solved by inspection (work from the bottom up). We leave it for you to confirm that the solution is

$$x_1 = 40, \quad x_2 = -10, \quad x_3 = 45$$

The fact that x_2 is negative tells us that the direction assigned to that flow in Figure 1.8.2 is incorrect; that is, the flow in that branch is *into* node A.

▶ **EXAMPLE 2** **Design of Traffic Patterns**

The network in Figure 1.8.3 shows a proposed plan for the traffic flow around a new park that will house the Liberty Bell in Philadelphia, Pennsylvania. The plan calls for a computerized traffic light at the north exit on Fifth Street, and the diagram indicates the average number of vehicles per hour that are expected to flow in and out of the streets that border the complex. All streets are one-way.

(a) How many vehicles per hour should the traffic light let through to ensure that the average number of vehicles per hour flowing into the complex is the same as the average number of vehicles flowing out?

(b) Assuming that the traffic light has been set to balance the total flow in and out of the complex, what can you say about the average number of vehicles per hour that will flow along the streets that border the complex?

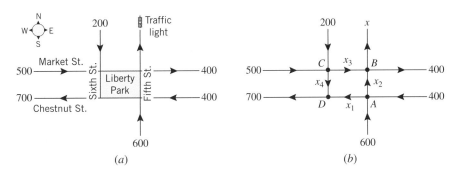

▶ Figure 1.8.3

(a)   *(b)*

Solution (a) If, as indicated in Figure 1.8.3*b*, we let x denote the number of vehicles per hour that the traffic light must let through, then the total number of vehicles per hour that flow in and out of the complex will be

Flowing in: $500 + 400 + 600 + 200 = 1700$

Flowing out: $x + 700 + 400$

Equating the flows in and out shows that the traffic light should let $x = 600$ vehicles per hour pass through.

Solution (b) To avoid traffic congestion, the flow in must equal the flow out at each intersection. For this to happen, the following conditions must be satisfied:

Intersection	Flow In		Flow Out
A	$400 + 600$	$=$	$x_1 + x_2$
B	$x_2 + x_3$	$=$	$400 + x$
C	$500 + 200$	$=$	$x_3 + x_4$
D	$x_1 + x_4$	$=$	700

Thus, with $x = 600$, as computed in part (a), we obtain the following linear system:

$$
\begin{aligned}
x_1 + x_2 \phantom{{}+{} x_3 + x_4} &= 1000 \\
x_2 + x_3 \phantom{{}+{} x_4} &= 1000 \\
x_3 + x_4 &= 700 \\
x_1 \phantom{{}+{} x_2 x_3} + x_4 &= 700
\end{aligned}
$$

We leave it for you to show that the system has infinitely many solutions and that these are given by the parametric equations

$$x_1 = 700 - t, \quad x_2 = 300 + t, \quad x_3 = 700 - t, \quad x_4 = t \tag{1}$$

However, the parameter t is not completely arbitrary here, since there are physical constraints to be considered. For example, the average flow rates must be nonnegative since we have assumed the streets to be one-way, and a negative flow rate would indicate a flow in the wrong direction. This being the case, we see from (1) that t can be any real number that satisfies $0 \le t \le 700$, which implies that the average flow rates along the streets will fall in the ranges

$$0 \le x_1 \le 700, \quad 300 \le x_2 \le 1000, \quad 0 \le x_3 \le 700, \quad 0 \le x_4 \le 700 \quad \blacktriangleleft$$

Electrical Circuits

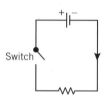

▲ Figure 1.8.4

Next, we will show how network analysis can be used to analyze electrical circuits consisting of batteries and resistors. A **battery** is a source of electric energy, and a **resistor**, such as a lightbulb, is an element that dissipates electric energy. Figure 1.8.4 shows a schematic diagram of a circuit with one battery (represented by the symbol ─┤├─), one resistor (represented by the symbol ─w─), and a switch. The battery has a **positive pole** ($+$) and a **negative pole** ($-$). When the switch is closed, electrical current is considered to flow from the positive pole of the battery, through the resistor, and back to the negative pole (indicated by the arrowhead in the figure).

Electrical current, which is a flow of electrons through wires, behaves much like the flow of water through pipes. A battery acts like a pump that creates "electrical pressure" to increase the flow rate of electrons, and a resistor acts like a restriction in a pipe that reduces the flow rate of electrons. The technical term for electrical pressure is **electrical potential**; it is commonly measured in **volts** (V). The degree to which a resistor reduces the electrical potential is called its **resistance** and is commonly measured in **ohms** (Ω). The rate of flow of electrons in a wire is called **current** and is commonly measured in **amperes** (also called **amps**) (A). The precise effect of a resistor is given by the following law:

> **Ohm's Law**   If a current of I amperes passes through a resistor with a resistance of R ohms, then there is a resulting drop of E volts in electrical potential that is the product of the current and resistance; that is,
>
> $$E = IR$$

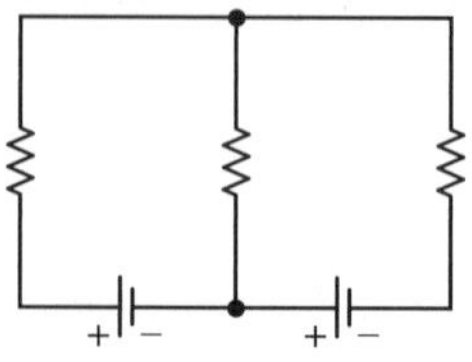

▲ Figure 1.8.5

A typical electrical network will have multiple batteries and resistors joined by some configuration of wires. A point at which three or more wires in a network are joined is called a **node** (or **junction point**). A **branch** is a wire connecting two nodes, and a **closed loop** is a succession of connected branches that begin and end at the same node. For example, the electrical network in Figure 1.8.5 has two nodes and three closed loops—two inner loops and one outer loop. As current flows through an electrical network, it undergoes increases and decreases in electrical potential, called **voltage rises** and **voltage drops**, respectively. The behavior of the current at the nodes and around closed loops is governed by two fundamental laws:

> **Kirchhoff's Current Law**   The sum of the currents flowing into any node is equal to the sum of the currents flowing out.

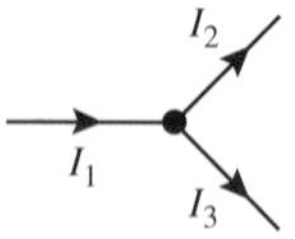

▲ Figure 1.8.6

> **Kirchhoff's Voltage Law**   In one traversal of any closed loop, the sum of the voltage rises equals the sum of the voltage drops.

Kirchhoff's current law is a restatement of the principle of flow conservation at a node that was stated for general networks. Thus, for example, the currents at the top node in Figure 1.8.6 satisfy the equation $I_1 = I_2 + I_3$.

In circuits with multiple loops and batteries there is usually no way to tell in advance which way the currents are flowing, so the usual procedure in circuit analysis is to assign *arbitrary* directions to the current flows in the branches and let the mathematical computations determine whether the assignments are correct. In addition to assigning directions to the current flows, Kirchhoff's voltage law requires a direction of travel for each closed loop. The choice is arbitrary, but for consistency we will always take this direction to be *clockwise* (Figure 1.8.7). We also make the following conventions:

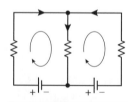

Clockwise closed-loop convention with arbitrary direction assignments to currents in the branches

▲ Figure 1.8.7

- A voltage drop occurs at a resistor if the direction assigned to the current through the resistor is the same as the direction assigned to the loop, and a voltage rise occurs at a resistor if the direction assigned to the current through the resistor is the opposite to that assigned to the loop.

- A voltage rise occurs at a battery if the direction assigned to the loop is from $-$ to $+$ through the battery, and a voltage drop occurs at a battery if the direction assigned to the loop is from $+$ to $-$ through the battery.

If you follow these conventions when calculating currents, then those currents whose directions were assigned correctly will have positive values and those whose directions were assigned incorrectly will have negative values.

▶ **EXAMPLE 3** **A Circuit with One Closed Loop**

Determine the current I in the circuit shown in Figure 1.8.8.

Solution Since the direction assigned to the current through the resistor is the same as the direction of the loop, there is a voltage drop at the resistor. By Ohm's law this voltage drop is $E = IR = 3I$. Also, since the direction assigned to the loop is from $-$ to $+$ through the battery, there is a voltage rise of 6 volts at the battery. Thus, it follows from Kirchhoff's voltage law that

$$3I = 6$$

from which we conclude that the current is $I = 2$ A. Since I is positive, the direction assigned to the current flow is correct.

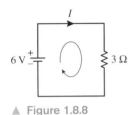

▲ Figure 1.8.8

▶ **EXAMPLE 4** **A Circuit with Three Closed Loops**

Determine the currents I_1, I_2, and I_3 in the circuit shown in Figure 1.8.9.

Solution Using the assigned directions for the currents, Kirchhoff's current law provides one equation for each node:

Node	Current In		Current Out
A	$I_1 + I_2$	$=$	I_3
B	I_3	$=$	$I_1 + I_2$

However, these equations are really the same, since both can be expressed as

$$I_1 + I_2 - I_3 = 0 \tag{2}$$

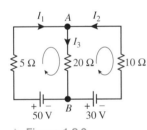

▲ Figure 1.8.9

Gustav Kirchhoff
(1824–1887)

Historical Note The German physicist Gustav Kirchhoff was a student of Gauss. His work on Kirchhoff's laws, announced in 1854, was a major advance in the calculation of currents, voltages, and resistances of electrical circuits. Kirchhoff was severely disabled and spent most of his life on crutches or in a wheelchair.
[*Image:* ©*SSPL/The Image Works*]

To find unique values for the currents we will need two more equations, which we will obtain from Kirchhoff's voltage law. We can see from the network diagram that there are three closed loops, a left inner loop containing the 50 V battery, a right inner loop containing the 30 V battery, and an outer loop that contains both batteries. Thus, Kirchhoff's voltage law will actually produce three equations. With a clockwise traversal of the loops, the voltage rises and drops in these loops are as follows:

	Voltage Rises	**Voltage Drops**
Left Inside Loop	50	$5I_1 + 20I_3$
Right Inside Loop	$30 + 10I_2 + 20I_3$	0
Outside Loop	$30 + 50 + 10I_2$	$5I_1$

These conditions can be rewritten as

$$\begin{aligned} 5I_1 \qquad\quad + 20I_3 &= 50 \\ 10I_2 + 20I_3 &= -30 \\ 5I_1 - 10I_2 \qquad\quad &= 80 \end{aligned} \qquad (3)$$

However, the last equation is superfluous, since it is the difference of the first two. Thus, if we combine (2) and the first two equations in (3), we obtain the following linear system of three equations in the three unknown currents:

$$\begin{aligned} I_1 + I_2 - I_3 &= 0 \\ 5I_1 \qquad\quad + 20I_3 &= 50 \\ 10I_2 + 20I_3 &= -30 \end{aligned}$$

We leave it for you to solve this system and show that $I_1 = 6$ A, $I_2 = -5$ A, and $I_3 = 1$ A. The fact that I_2 is negative tells us that the direction of this current is opposite to that indicated in Figure 1.8.9. ◀

Balancing Chemical Equations

Chemical compounds are represented by **chemical formulas** that describe the atomic makeup of their molecules. For example, water is composed of two hydrogen atoms and one oxygen atom, so its chemical formula is H_2O; and stable oxygen is composed of two oxygen atoms, so its chemical formula is O_2.

When chemical compounds are combined under the right conditions, the atoms in their molecules rearrange to form new compounds. For example, when methane burns, the methane (CH_4) and stable oxygen (O_2) react to form carbon dioxide (CO_2) and water (H_2O). This is indicated by the **chemical equation**

$$CH_4 + O_2 \longrightarrow CO_2 + H_2O \qquad (4)$$

The molecules to the left of the arrow are called the **reactants** and those to the right the **products**. In this equation the plus signs serve to separate the molecules and are not intended as algebraic operations. However, this equation does not tell the whole story, since it fails to account for the proportions of molecules required for a **complete reaction** (no reactants left over). For example, we can see from the right side of (4) that to produce one molecule of carbon dioxide and one molecule of water, one needs *three* oxygen atoms for each carbon atom. However, from the left side of (4) we see that one molecule of methane and one molecule of stable oxygen have only *two* oxygen atoms for each carbon atom. Thus, on the reactant side the ratio of methane to stable oxygen cannot be one-to-one in a complete reaction.

A chemical equation is said to be ***balanced*** if for each type of atom in the reaction, the same number of atoms appears on each side of the arrow. For example, the balanced version of Equation (4) is

$$CH_4 + 2O_2 \longrightarrow CO_2 + 2H_2O \tag{5}$$

by which we mean that one methane molecule combines with two stable oxygen molecules to produce one carbon dioxide molecule and two water molecules. In theory, one could multiply this equation through by any positive integer. For example, multiplying through by 2 yields the balanced chemical equation

$$2CH_4 + 4O_2 \longrightarrow 2CO_2 + 4H_2O$$

However, the standard convention is to use the smallest positive integers that will balance the equation.

Equation (4) is sufficiently simple that it could have been balanced by trial and error, but for more complicated chemical equations we will need a systematic method. There are various methods that can be used, but we will give one that uses systems of linear equations. To illustrate the method let us reexamine Equation (4). To balance this equation we must find positive integers, x_1, x_2, x_3, and x_4 such that

$$x_1 (CH_4) + x_2 (O_2) \longrightarrow x_3 (CO_2) + x_4 (H_2O) \tag{6}$$

For each of the atoms in the equation, the number of atoms on the left must be equal to the number of atoms on the right. Expressing this in tabular form we have

	Left Side		**Right Side**
Carbon	x_1	$=$	x_3
Hydrogen	$4x_1$	$=$	$2x_4$
Oxygen	$2x_2$	$=$	$2x_3 + x_4$

from which we obtain the homogeneous linear system

$$
\begin{aligned}
x_1 \quad\quad - \quad x_3 \quad\quad &= 0 \\
4x_1 \quad\quad\quad\quad - 2x_4 &= 0 \\
2x_2 - 2x_3 - \quad x_4 &= 0
\end{aligned}
$$

The augmented matrix for this system is

$$
\begin{bmatrix}
1 & 0 & -1 & 0 & 0 \\
4 & 0 & 0 & -2 & 0 \\
0 & 2 & -2 & -1 & 0
\end{bmatrix}
$$

We leave it for you to show that the reduced row echelon form of this matrix is

$$
\begin{bmatrix}
1 & 0 & 0 & -\frac{1}{2} & 0 \\
0 & 1 & 0 & -1 & 0 \\
0 & 0 & 1 & -\frac{1}{2} & 0
\end{bmatrix}
$$

from which we conclude that the general solution of the system is

$$x_1 = t/2, \quad x_2 = t, \quad x_3 = t/2, \quad x_4 = t$$

where t is arbitrary. The smallest positive integer values for the unknowns occur when we let $t = 2$, so the equation can be balanced by letting $x_1 = 1, x_2 = 2, x_3 = 1, x_4 = 2$. This agrees with our earlier conclusions, since substituting these values into Equation (6) yields Equation (5).

▶ **EXAMPLE 5** **Balancing Chemical Equations Using Linear Systems**

Balance the chemical equation

$$\text{HCl} \quad + \quad \text{Na}_3\text{PO}_4 \quad \longrightarrow \quad \text{H}_3\text{PO}_4 \quad + \quad \text{NaCl}$$

[hydrochloric acid] + [sodium phosphate] ⟶ [phosphoric acid] + [sodium chloride]

Solution Let $x_1, x_2, x_3,$ and x_4 be positive integers that balance the equation

$$x_1 \, (\text{HCl}) + x_2 \, (\text{Na}_3\text{PO}_4) \longrightarrow x_3 \, (\text{H}_3\text{PO}_4) + x_4 \, (\text{NaCl}) \tag{7}$$

Equating the number of atoms of each type on the two sides yields

$$\begin{aligned}
1x_1 &= 3x_3 \quad &\textbf{Hydrogen (H)} \\
1x_1 &= 1x_4 \quad &\textbf{Chlorine (Cl)} \\
3x_2 &= 1x_4 \quad &\textbf{Sodium (Na)} \\
1x_2 &= 1x_3 \quad &\textbf{Phosphorous (P)} \\
4x_2 &= 4x_3 \quad &\textbf{Oxygen (O)}
\end{aligned}$$

from which we obtain the homogeneous linear system

$$\begin{aligned}
x_1 \quad\quad - 3x_3 \quad\quad &= 0 \\
x_1 \quad\quad\quad\quad - x_4 &= 0 \\
3x_2 \quad\quad - x_4 &= 0 \\
x_2 - \; x_3 \quad\quad &= 0 \\
4x_2 - 4x_3 \quad\quad &= 0
\end{aligned}$$

We leave it for you to show that the reduced row echelon form of the augmented matrix for this system is

$$\begin{bmatrix}
1 & 0 & 0 & -1 & 0 \\
0 & 1 & 0 & -\frac{1}{3} & 0 \\
0 & 0 & 1 & -\frac{1}{3} & 0 \\
0 & 0 & 0 & 0 & 0 \\
0 & 0 & 0 & 0 & 0
\end{bmatrix}$$

from which we conclude that the general solution of the system is

$$x_1 = t, \quad x_2 = t/3, \quad x_3 = t/3, \quad x_4 = t$$

where t is arbitrary. To obtain the smallest positive integers that balance the equation, we let $t = 3$, in which case we obtain $x_1 = 3, x_2 = 1, x_3 = 1,$ and $x_4 = 3$. Substituting these values in (7) produces the balanced equation

$$3\text{HCl} + \text{Na}_3\text{PO}_4 \longrightarrow \text{H}_3\text{PO}_4 + 3\text{NaCl} \; \blacktriangleleft$$

Polynomial Interpolation   An important problem in various applications is to find a polynomial whose graph passes through a specified set of points in the plane; this is called an ***interpolating polynomial*** for the points. The simplest example of such a problem is to find a linear polynomial

$$p(x) = ax + b \tag{8}$$

whose graph passes through two known distinct points, (x_1, y_1) and (x_2, y_2), in the xy-plane (Figure 1.8.10). You have probably encountered various methods in analytic geometry for finding the equation of a line through two points, but here we will give a method based on linear systems that can be adapted to general polynomial interpolation.

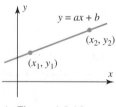

▲ Figure 1.8.10

The graph of (8) is the line $y = ax + b$, and for this line to pass through the points (x_1, y_1) and (x_2, y_2), we must have

$$y_1 = ax_1 + b \quad \text{and} \quad y_2 = ax_2 + b$$

Therefore, the unknown coefficients a and b can be obtained by solving the linear system

$$ax_1 + b = y_1$$
$$ax_2 + b = y_2$$

We don't need any fancy methods to solve this system—the value of a can be obtained by subtracting the equations to eliminate b, and then the value of a can be substituted into either equation to find b. We leave it as an exercise for you to find a and b and then show that they can be expressed in the form

$$a = \frac{y_2 - y_1}{x_2 - x_1} \quad \text{and} \quad b = \frac{y_1 x_2 - y_2 x_1}{x_2 - x_1} \tag{9}$$

provided $x_1 \neq x_2$. Thus, for example, the line $y = ax + b$ that passes through the points

$$(2, 1) \quad \text{and} \quad (5, 4)$$

can be obtained by taking $(x_1, y_1) = (2, 1)$ and $(x_2, y_2) = (5, 4)$, in which case (9) yields

$$a = \frac{4 - 1}{5 - 2} = 1 \quad \text{and} \quad b = \frac{(1)(5) - (4)(2)}{5 - 2} = -1$$

Therefore, the equation of the line is

$$y = x - 1$$

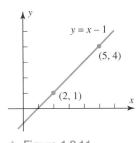

▲ Figure 1.8.11

(Figure 1.8.11).

Now let us consider the more general problem of finding a polynomial whose graph passes through n points with distinct x-coordinates

$$(x_1, y_1), \quad (x_2, y_2), \quad (x_3, y_3), \ldots, \quad (x_n, y_n) \tag{10}$$

Since there are n conditions to be satisfied, intuition suggests that we should begin by looking for a polynomial of the form

$$p(x) = a_0 + a_1 x + a_2 x^2 + \cdots + a_{n-1} x^{n-1} \tag{11}$$

since a polynomial of this form has n coefficients that are at our disposal to satisfy the n conditions. However, we want to allow for cases where the points may lie on a line or have some other configuration that would make it possible to use a polynomial whose degree is less than $n - 1$; thus, we allow for the possibility that a_{n-1} and other coefficients in (11) may be zero.

The following theorem, which we will prove later in the text, is the basic result on polynomial interpolation.

THEOREM 1.8.1 Polynomial Interpolation

Given any n points in the xy-plane that have distinct x-coordinates, there is a unique polynomial of degree $n - 1$ or less whose graph passes through those points.

Let us now consider how we might go about finding the interpolating polynomial (11) whose graph passes through the points in (10). Since the graph of this polynomial is the graph of the equation

$$y = a_0 + a_1 x + a_2 x^2 + \cdots + a_{n-1} x^{n-1} \tag{12}$$

it follows that the coordinates of the points must satisfy

$$a_0 + a_1x_1 + a_2x_1^2 + \cdots + a_{n-1}x_1^{n-1} = y_1$$
$$a_0 + a_1x_2 + a_2x_2^2 + \cdots + a_{n-1}x_2^{n-1} = y_2$$
$$\vdots \qquad \vdots \qquad \vdots \qquad\qquad \vdots \qquad \vdots \qquad\qquad (13)$$
$$a_0 + a_1x_n + a_2x_n^2 + \cdots + a_{n-1}x_n^{n-1} = y_n$$

In these equations the values of x's and y's are assumed to be known, so we can view this as a linear system in the unknowns $a_0, a_1, \ldots, a_{n-1}$. From this point of view the augmented matrix for the system is

$$\begin{bmatrix} 1 & x_1 & x_1^2 & \cdots & x_1^{n-1} & y_1 \\ 1 & x_2 & x_2^2 & \cdots & x_2^{n-1} & y_2 \\ \vdots & \vdots & \vdots & & \vdots & \vdots \\ 1 & x_n & x_n^2 & \cdots & x_n^{n-1} & y_n \end{bmatrix} \qquad (14)$$

and hence the interpolating polynomial can be found by reducing this matrix to reduced row echelon form (Gauss–Jordan elimination).

▶ **EXAMPLE 6 Polynomial Interpolation by Gauss–Jordan Elimination**

Find a cubic polynomial whose graph passes through the points

$$(1, 3), \quad (2, -2), \quad (3, -5), \quad (4, 0)$$

Solution Since there are four points, we will use an interpolating polynomial of degree $n = 3$. Denote this polynomial by

$$p(x) = a_0 + a_1x + a_2x^2 + a_3x^3$$

and denote the x- and y-coordinates of the given points by

$$x_1 = 1, \quad x_2 = 2, \quad x_3 = 3, \quad x_4 = 4 \quad \text{and} \quad y_1 = 3, \quad y_2 = -2, \quad y_3 = -5, \quad y_4 = 0$$

Thus, it follows from (14) that the augmented matrix for the linear system in the unknowns a_0, a_1, a_2, and a_3 is

$$\begin{bmatrix} 1 & x_1 & x_1^2 & x_1^3 & y_1 \\ 1 & x_2 & x_2^2 & x_2^3 & y_2 \\ 1 & x_3 & x_3^2 & x_3^3 & y_3 \\ 1 & x_4 & x_4^2 & x_4^3 & y_4 \end{bmatrix} = \begin{bmatrix} 1 & 1 & 1 & 1 & 3 \\ 1 & 2 & 4 & 8 & -2 \\ 1 & 3 & 9 & 27 & -5 \\ 1 & 4 & 16 & 64 & 0 \end{bmatrix}$$

We leave it for you to confirm that the reduced row echelon form of this matrix is

$$\begin{bmatrix} 1 & 0 & 0 & 0 & 4 \\ 0 & 1 & 0 & 0 & 3 \\ 0 & 0 & 1 & 0 & -5 \\ 0 & 0 & 0 & 1 & 1 \end{bmatrix}$$

from which it follows that $a_0 = 4, a_1 = 3, a_2 = -5, a_3 = 1$. Thus, the interpolating polynomial is

$$p(x) = 4 + 3x - 5x^2 + x^3$$

The graph of this polynomial and the given points are shown in Figure 1.8.12. ◀

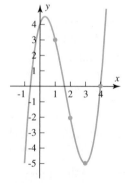

▲ Figure 1.8.12

Remark Later we will give a more efficient method for finding interpolating polynomials that is better suited for problems in which the number of data points is large.

▶ **EXAMPLE 7 Approximate Integration**

There is no way to evaluate the integral

$$\int_0^1 \sin\left(\frac{\pi x^2}{2}\right) dx$$

directly since there is no way to express an antiderivative of the integrand in terms of elementary functions. This integral could be approximated by Simpson's rule or some comparable method, but an alternative approach is to approximate the integrand by an interpolating polynomial and integrate the approximating polynomial. For example, let us consider the five points

$$x_0 = 0, \quad x_1 = 0.25, \quad x_2 = 0.5, \quad x_3 = 0.75, \quad x_4 = 1$$

that divide the interval $[0, 1]$ into four equally spaced subintervals. The values of

$$f(x) = \sin\left(\frac{\pi x^2}{2}\right)$$

at these points are approximately

$$f(0) = 0, \quad f(0.25) = 0.098017, \quad f(0.5) = 0.382683,$$
$$f(0.75) = 0.77301, \quad f(1) = 1$$

The interpolating polynomial is (verify)

$$p(x) = 0.098796x + 0.762356x^2 + 2.14429x^3 - 2.00544x^4 \tag{15}$$

and

$$\int_0^1 p(x)\, dx \approx 0.438501 \tag{16}$$

As shown in Figure 1.8.13, the graphs of f and p match very closely over the interval $[0, 1]$, so the approximation is quite good. ◀

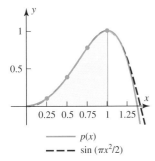

$$\underline{\hspace{1.2cm}}\ p(x)$$
$$\text{-- -- --}\ \sin(\pi x^2/2)$$

▲ Figure 1.8.13

Concept Review

- Network
- Branches
- Nodes
- Flow conservation

- Electrical circuits: battery, resistor, poles (positive and negative), electrical potential, Ohm's law, Kirchhoff's current law, Kirchhoff's voltage law

- Chemical equations: reactants, products, balanced equation
- Interpolating polynomial

Skills

- Find the flow rates and directions of flow in branches of a network.
- Find the amount of current flowing through parts of an electrical circuit.

- Write a balanced chemical equation for a given chemical reaction.
- Find an interpolating polynomial for a graph passing through a given collection of points.

Exercise Set 1.8

1. The accompanying figure shows a network in which the flow rate and direction of flow in certain branches are known. Find the flow rates and directions of flow in the remaining branches.

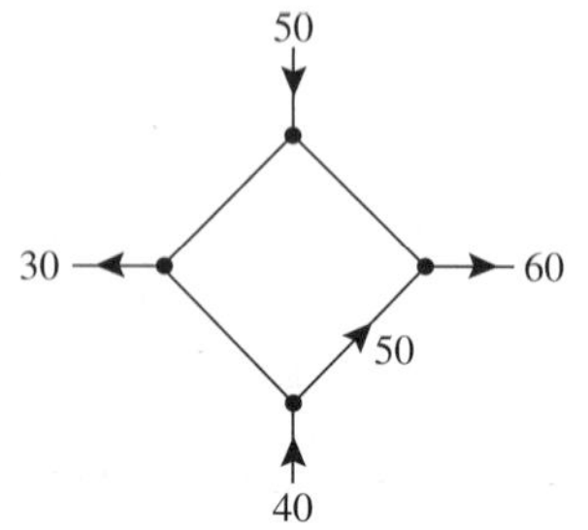

◀ Figure Ex-1

2. The accompanying figure shows known flow rates of hydrocarbons into and out of a network of pipes at an oil refinery.

(a) Set up a linear system whose solution provides the unknown flow rates.

(b) Solve the system for the unknown flow rates.

(c) Find the flow rates and directions of flow if $x_4 = 50$ and $x_6 = 0$.

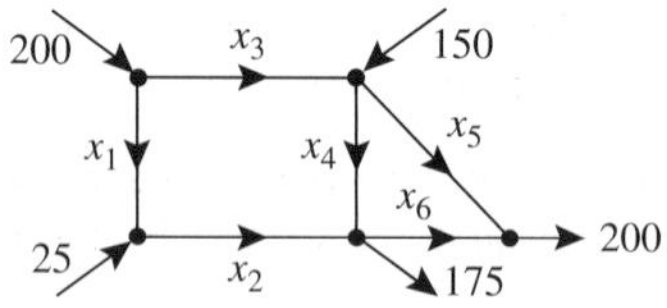

◀ Figure Ex-2

3. The accompanying figure shows a network of one-way streets with traffic flowing in the directions indicated. The flow rates along the streets are measured as the average number of vehicles per hour.

(a) Set up a linear system whose solution provides the unknown flow rates.

(b) Solve the system for the unknown flow rates.

(c) If the flow along the road from A to B must be reduced for construction, what is the minimum flow that is required to keep traffic flowing on all roads?

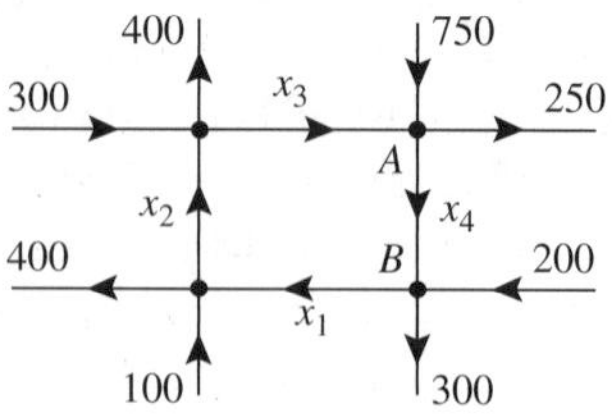

◀ Figure Ex-3

4. The accompanying figure shows a network of one-way streets with traffic flowing in the directions indicated. The flow rates along the streets are measured as the average number of vehicles per hour.

(a) Set up a linear system whose solution provides the unknown flow rates.

(b) Solve the system for the unknown flow rates.

(c) Is it possible to close the road from A to B for construction and keep traffic flowing on the other streets? Explain.

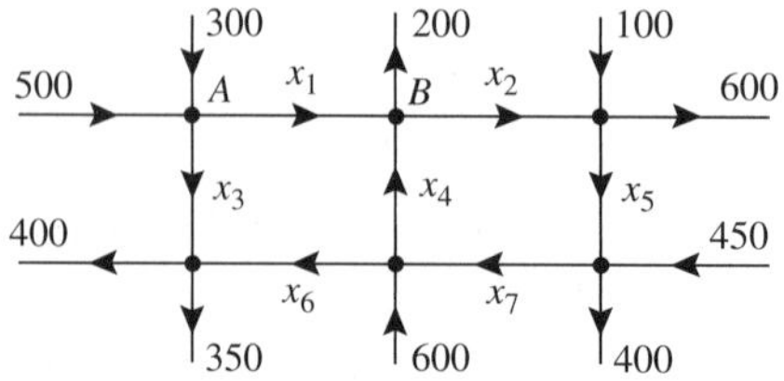

◀ Figure Ex-4

▶ In Exercises 5–8, analyze the given electrical circuits by finding the unknown currents. ◀

5.

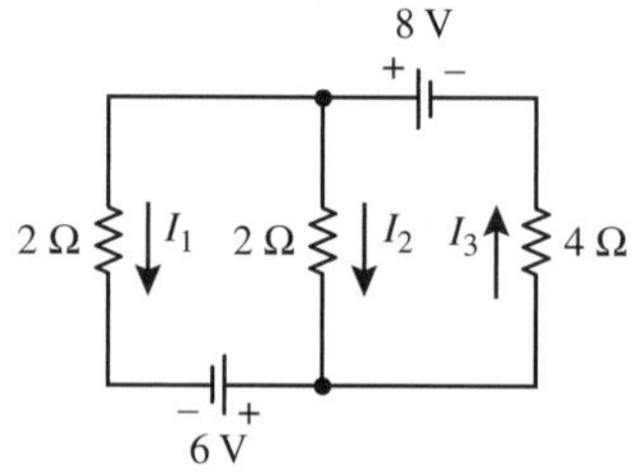

6.

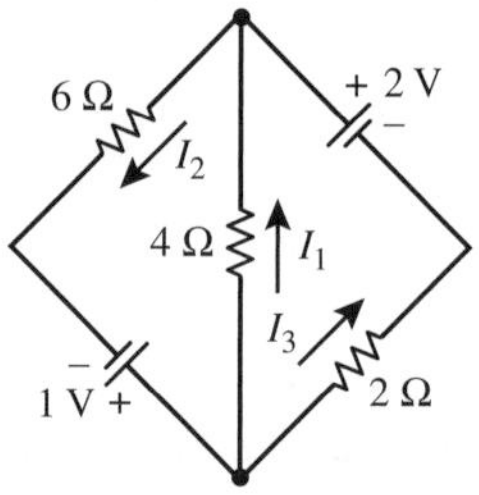

7.

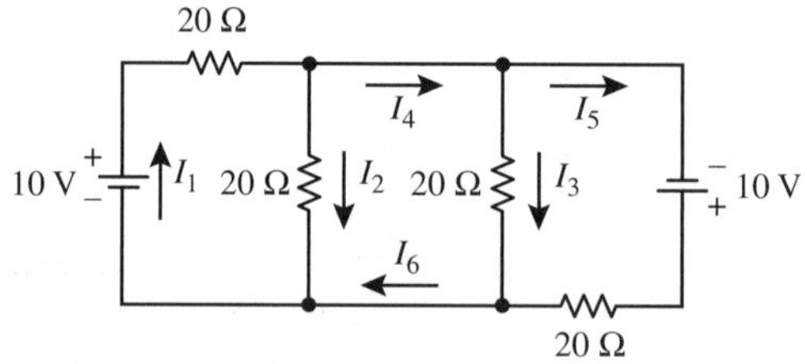

8.

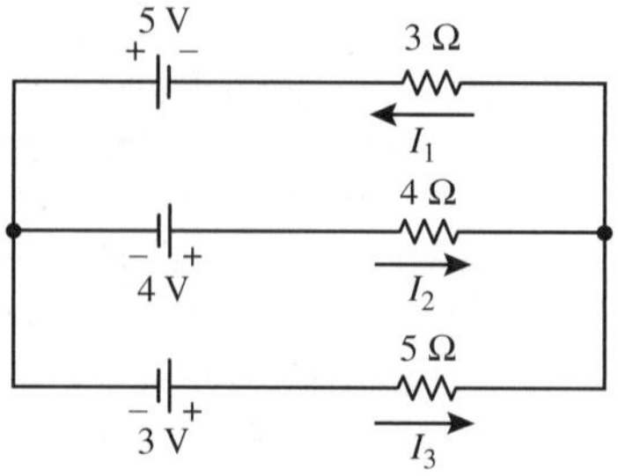

▶ In Exercises 9–12, write a balanced equation for the given chemical reaction. ◀

9. $C_3H_8 + O_2 \rightarrow CO_2 + H_2O$ (propane combustion)

10. $C_6H_{12}O_6 \rightarrow CO_2 + C_2H_5OH$ (fermentation of sugar)

11. $CH_3COF + H_2O \rightarrow CH_3COOH + HF$

12. $CO_2 + H_2O \rightarrow C_6H_{12}O_6 + O_2$ (photosynthesis)

13. Find the quadratic polynomial whose graph passes through the points $(1, 1)$, $(2, 2)$, and $(3, 5)$.

14. Find the quadratic polynomial whose graph passes through the points $(0, 0)$, $(-1, 1)$, and $(1, 1)$.

15. Find the cubic polynomial whose graph passes through the points $(-1, -1)$, $(0, 1)$, $(1, 3)$, $(4, -1)$.

16. The accompanying figure shows the graph of a cubic polynomial. Find the polynomial.

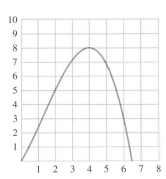

◄ **Figure Ex-16**

17. (a) Find an equation that represents the family of all second-degree polynomials that pass through the points $(0, 1)$

and $(1, 2)$. [*Hint:* The equation will involve one arbitrary parameter that produces the members of the family when varied.]

(b) By hand, or with the help of a graphing utility, sketch four curves in the family.

18. In this section we have selected only a few applications of linear systems. Using the Internet as a search tool, try to find some more real-world applications of such systems. Select one that is of interest to you, and write a paragraph about it.

True-False Exercises

In parts (a)–(e) determine whether the statement is true or false, and justify your answer.

(a) In any network, the sum of the flows out of a node must equal the sum of the flows into a node.

(b) When a current passes through a resistor, there is an increase in the electrical potential in a circuit.

(c) Kirchhoff's current law states that the sum of the currents flowing into a node equals the sum of the currents flowing out of the node.

(d) A chemical equation is called balanced if the total number of atoms on each side of the equation is the same.

(e) Given any n points in the xy-plane, there is a unique polynomial of degree $n - 1$ or less whose graph passes through those points.

1.9 Leontief Input-Output Models

In 1973 the economist Wassily Leontief was awarded the Nobel prize for his work on economic modeling in which he used matrix methods to study the relationships between different sectors in an economy. In this section we will discuss some of the ideas developed by Leontief.

Inputs and Outputs in an Economy

One way to analyze an economy is to divide it into **sectors** and study how the sectors interact with one another. For example, a simple economy might be divided into three sectors—manufacturing, agriculture, and utilities. Typically, a sector will produce certain **outputs** but will require **inputs** from the other sectors and itself. For example, the agricultural sector may produce wheat as an output but will require inputs of farm machinery from the manufacturing sector, electrical power from the utilities sector, and food from its own sector to feed its workers. Thus, we can imagine an economy to be a network in which inputs and outputs flow in and out of the sectors; the study of such flows is called **input-output analysis**. Inputs and outputs are commonly measured in monetary units (dollars or millions of dollars, for example) but other units of measurement are also possible.

The flows between sectors of a real economy are not always obvious. For example, in World War II the United States had a demand for 50,000 new airplanes that required the construction of many new aluminum manufacturing plants. This produced an unexpectedly large demand for certain copper electrical components, which in turn produced a copper shortage. The problem was eventually resolved by using silver borrowed from Fort Knox as a copper substitute. In all likelihood modern input-output analysis would have anticipated the copper shortage.

Most sectors of an economy will produce outputs, but there may exist sectors that consume outputs without producing anything themselves (the consumer market, for example). Those sectors that do not produce outputs are called *open sectors*. Economies with no open sectors are called *closed economies*, and economies with one or more open sectors are called *open economies* (Figure 1.9.1). In this section we will be concerned with economies with one open sector, and our primary goal will be to determine the output levels that are required for the productive sectors to sustain themselves and satisfy the demand of the open sector.

Leontief Model of an Open Economy

Let us consider a simple open economy with one open sector and three product-producing sectors: manufacturing, agriculture, and utilities. Assume that inputs and outputs are measured in dollars and that the inputs required by the productive sectors to produce one dollar's worth of output are in accordance with Table 1.

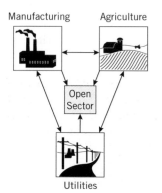

Manufacturing   Agriculture

Utilities

▲ Figure 1.9.1

Table 1

		Input Required per Dollar Output		
		Manufacturing	**Agriculture**	**Utilities**
	Manufacturing	$ 0.50	$ 0.10	$ 0.10
Provider	**Agriculture**	$ 0.20	$ 0.50	$ 0.30
	Utilities	$ 0.10	$ 0.30	$ 0.40

Usually, one would suppress the labeling and express this matrix as

$$C = \begin{bmatrix} 0.5 & 0.1 & 0.1 \\ 0.2 & 0.5 & 0.3 \\ 0.1 & 0.3 & 0.4 \end{bmatrix} \quad (1)$$

This is called the *consumption matrix* (or sometimes the *technology matrix*) for the economy. The column vectors

$$\mathbf{c}_1 = \begin{bmatrix} 0.5 \\ 0.2 \\ 0.1 \end{bmatrix}, \quad \mathbf{c}_2 = \begin{bmatrix} 0.1 \\ 0.5 \\ 0.3 \end{bmatrix}, \quad \mathbf{c}_3 = \begin{bmatrix} 0.1 \\ 0.3 \\ 0.4 \end{bmatrix}$$

Historical Note   It is somewhat ironic that it was the Russian-born Wassily Leontief who won the Nobel prize in 1973 for pioneering the modern methods for analyzing free-market economies. Leontief was a precocious student who entered the University of Leningrad at age 15. Bothered by the intellectual restrictions of the Soviet system, he was put in jail for anti-Communist activities, after which he headed for the University of Berlin, receiving his Ph.D. there in 1928. He came to the United States in 1931, where he held professorships at Harvard and then New York University.

[Image: ©Bettmann/©Corbis]

**Wassily Leontief
(1906–1999)**

in C list the inputs required by the manufacturing, agricultural, and utilities sectors, respectively, to produce $1.00 worth of output. These are called the ***consumption vectors*** of the sectors. For example, c_1 tells us that to produce $1.00 worth of output the manufacturing sector needs $0.50 worth of manufacturing output, $0.20 worth of agricultural output, and $0.10 worth of utilities output.

What is the economic significance of the row sums of the consumption matrix?

Continuing with the above example, suppose that the open sector wants the economy to supply it manufactured goods, agricultural products, and utilities with dollar values:

$$d_1 \text{ dollars of manufactured goods}$$
$$d_2 \text{ dollars of agricultural products}$$
$$d_3 \text{ dollars of utilities}$$

The column vector $\mathbf{d}$ that has these numbers as successive components is called the ***outside demand vector***. Since the product-producing sectors consume some of their own output, the dollar value of their output must cover their own needs plus the outside demand. Suppose that the dollar values required to do this are

$$x_1 \text{ dollars of manufactured goods}$$
$$x_2 \text{ dollars of agricultural products}$$
$$x_3 \text{ dollars of utilities}$$

The column vector $\mathbf{x}$ that has these numbers as successive components is called the ***production vector*** for the economy. For the economy with consumption matrix (1), that portion of the production vector $\mathbf{x}$ that will be consumed by the three productive sectors is

$$x_1 \begin{bmatrix} 0.5 \\ 0.2 \\ 0.1 \end{bmatrix} + x_2 \begin{bmatrix} 0.1 \\ 0.5 \\ 0.3 \end{bmatrix} + x_3 \begin{bmatrix} 0.1 \\ 0.3 \\ 0.4 \end{bmatrix} = \begin{bmatrix} 0.5 & 0.1 & 0.1 \\ 0.2 & 0.5 & 0.3 \\ 0.1 & 0.3 & 0.4 \end{bmatrix} \begin{bmatrix} x_1 \\ x_2 \\ x_3 \end{bmatrix} = C\mathbf{x}$$

Fractions consumed by manufacturing	Fractions consumed by agriculture	Fractions consumed by utilities

The vector $C\mathbf{x}$ is called the ***intermediate demand vector*** for the economy. Once the intermediate demand is met, the portion of the production that is left to satisfy the outside demand is $\mathbf{x} - C\mathbf{x}$. Thus, if the outside demand vector is $\mathbf{d}$, then $\mathbf{x}$ must satisfy the equation

$$\underset{\substack{\text{Amount} \\ \text{produced}}}{\mathbf{x}} - \underset{\substack{\text{Intermediate} \\ \text{demand}}}{C\mathbf{x}} = \underset{\substack{\text{Outside} \\ \text{demand}}}{\mathbf{d}}$$

which we will find convenient to rewrite as

$$(I - C)\mathbf{x} = \mathbf{d} \tag{2}$$

The matrix $I - C$ is called the ***Leontief matrix*** and (2) is called the ***Leontief equation***.

▶ **EXAMPLE 1 Satisfying Outside Demand**

Consider the economy described in Table 1. Suppose that the open sector has a demand for $7900 worth of manufacturing products, $3950 worth of agricultural products, and $1975 worth of utilities.

(a)   Can the economy meet this demand?

(b)   If so, find a production vector $\mathbf{x}$ that will meet it exactly.

Solution   The consumption matrix, production vector, and outside demand vector are

$$C = \begin{bmatrix} 0.5 & 0.1 & 0.1 \\ 0.2 & 0.5 & 0.3 \\ 0.1 & 0.3 & 0.4 \end{bmatrix}, \quad \mathbf{x} = \begin{bmatrix} x_1 \\ x_2 \\ x_3 \end{bmatrix}, \quad \mathbf{d} = \begin{bmatrix} 7900 \\ 3950 \\ 1975 \end{bmatrix} \tag{3}$$

To meet the outside demand, the vector $\mathbf{x}$ must satisfy the Leontief equation (2), so the problem reduces to solving the linear system

$$\underbrace{\begin{bmatrix} 0.5 & -0.1 & -0.1 \\ -0.2 & 0.5 & -0.3 \\ -0.1 & -0.3 & 0.6 \end{bmatrix}}_{I - C} \underbrace{\begin{bmatrix} x_1 \\ x_2 \\ x_3 \end{bmatrix}}_{\mathbf{x}} = \underbrace{\begin{bmatrix} 7900 \\ 3950 \\ 1975 \end{bmatrix}}_{\mathbf{d}} \tag{4}$$

(if consistent). We leave it for you to show that the reduced row echelon form of the augmented matrix for this system is

$$\begin{bmatrix} 1 & 0 & 0 & | & 27{,}500 \\ 0 & 1 & 0 & | & 33{,}750 \\ 0 & 0 & 1 & | & 24{,}750 \end{bmatrix}$$

This tells us that (4) is consistent, and the economy can satisfy the demand of the open sector exactly by producing \$27,500 worth of manufacturing output, \$33,750 worth of agricultural output, and \$24,750 worth of utilities output. ◀

Productive Open Economies   In the preceding discussion we considered an open economy with three product-producing sectors; the same ideas apply to an open economy with n product-producing sectors. In this case, the consumption matrix, production vector, and outside demand vector have the form

$$C = \begin{bmatrix} c_{11} & c_{12} & \cdots & c_{1n} \\ c_{21} & c_{22} & \cdots & c_{2n} \\ \vdots & \vdots & & \vdots \\ c_{n1} & c_{n2} & \cdots & c_{nn} \end{bmatrix}, \quad \mathbf{x} = \begin{bmatrix} x_1 \\ x_2 \\ \vdots \\ x_n \end{bmatrix}, \quad \mathbf{d} = \begin{bmatrix} d_1 \\ d_2 \\ \vdots \\ d_n \end{bmatrix}$$

where all entries are nonnegative and

$c_{ij} = $ the monetary value of the output of the ith sector that is needed by the jth sector to produce one unit of output

$x_i = $ the monetary value of the output of the ith sector

$d_i = $ the monetary value of the output of the ith sector that is required to meet the demand of the open sector

Remark   Note that the jth column vector of C contains the monetary values that the jth sector requires of the other sectors to produce one monetary unit of output, and the ith row vector of C contains the monetary values required of the ith sector by the other sectors for each of them to produce one monetary unit of output.

As discussed in our example above, a production vector $\mathbf{x}$ that meets the demand $\mathbf{d}$ of the outside sector must satisfy the Leontief equation

$$(I - C)\mathbf{x} = \mathbf{d}$$

If the matrix $I - C$ is invertible, then this equation has the unique solution

$$\mathbf{x} = (I - C)^{-1}\mathbf{d} \tag{5}$$

for every demand vector **d**. However, for **x** to be a valid production vector it must have nonnegative entries, so the problem of importance in economics is to determine conditions under which the Leontief equation has a solution with nonnegative entries.

It is evident from the form of (5) that if $I - C$ is invertible, and if $(I - C)^{-1}$ has nonnegative entries, then for every demand vector **d** the corresponding **x** will also have nonnegative entries, and hence will be a valid production vector for the economy. Economies for which $(I - C)^{-1}$ has nonnegative entries are said to be *productive*. Such economies are desirable because demand can always be met by some level of production. The following theorem, whose proof can be found in many books on economics, gives conditions under which open economies are productive.

THEOREM 1.9.1 *If C is the consumption matrix for an open economy, and if all of the column sums are less than* 1, *then the matrix* $I - C$ *is invertible, the entries of* $(I - C)^{-1}$ *are nonnegative, and the economy is productive.*

Remark The jth column sum of C represents the total dollar value of input that the jth sector requires to produce $1 of output, so if the jth column sum is less than 1, then the jth sector requires less than $1 of input to produce $1 of output; in this case we say that the jth sector is *profitable*. Thus, Theorem 1.9.1 states that if all product-producing sectors of an open economy are profitable, then the economy is productive. In the exercises we will ask you to show that an open economy is productive if all of the row sums of C are less than 1 (Exercise 11). Thus, an open economy is productive if *either* all of the column sums or all of the row sums of C are less than 1.

▶ **EXAMPLE 2 An Open Economy Whose Sectors Are All Profitable**

The column sums of the consumption matrix C in (1) are less than 1, so $(I - C)^{-1}$ exists and has nonnegative entries. Use a calculating utility to confirm this, and use this inverse to solve Equation (4) in Example 1.

Solution We leave it for you to show that

$$(I - C)^{-1} \approx \begin{bmatrix} 2.65823 & 1.13924 & 1.01266 \\ 1.89873 & 3.67089 & 2.15190 \\ 1.39241 & 2.02532 & 2.91139 \end{bmatrix}$$

This matrix has nonnegative entries, and

$$\mathbf{x} = (I - C)^{-1}\mathbf{d} \approx \begin{bmatrix} 2.65823 & 1.13924 & 1.01266 \\ 1.89873 & 3.67089 & 2.15190 \\ 1.39241 & 2.02532 & 2.91139 \end{bmatrix} \begin{bmatrix} 7900 \\ 3950 \\ 1975 \end{bmatrix} \approx \begin{bmatrix} 27,500 \\ 33,750 \\ 24,750 \end{bmatrix}$$

which is consistent with the solution in Example 1. ◀

Concept Review

- Sectors
- Inputs
- Outputs
- Input-output analysis
- Open sector

- Economies: open, closed
- Consumption (technology) matrix
- Consumption vector
- Outside demand vector
- Production vector

- Intermediate demand vector
- Leontief matrix
- Leontief equation

Skills

- Construct a consumption matrix for an economy.

- Understand the relationships among the vectors of a sector of an economy: consumption, outside demand, production, and intermediate demand.

Exercise Set 1.9

1. An automobile mechanic (M) and a body shop (B) use each other's services. For each $1.00 of business that M does, it uses $0.50 of its own services and $0.25 of B's services, and for each $1.00 of business that B does it uses $0.10 of its own services and $0.25 of M's services.

   (a) Construct a consumption matrix for this economy.

   (b) How much must M and B each produce to provide customers with $7000 worth of mechanical work and $14,000 worth of body work?

2. A simple economy produces food (F) and housing (H). The production of $1.00 worth of food requires $0.30 worth of food and $0.10 worth of housing, and the production of $1.00 worth of housing requires $0.20 worth of food and $0.60 worth of housing.

   (a) Construct a consumption matrix for this economy.

   (b) What dollar value of food and housing must be produced for the economy to provide consumers $130,000 worth of food and $130,000 worth of housing?

3. Consider the open economy described by the accompanying table, where the input is in dollars needed for $1.00 of output.

   (a) Find the consumption matrix for the economy.

   (b) Suppose that the open sector has a demand for $1930 worth of housing, $3860 worth of food, and $5790 worth of utilities. Use row reduction to find a production vector that will meet this demand exactly.

Table Ex-3

Input Required per Dollar Output

	Housing	Food	Utilities
Housing	$ 0.10	$ 0.60	$ 0.40
Food	$ 0.30	$ 0.20	$ 0.30
Utilities	$ 0.40	$ 0.10	$ 0.20

Provider

4. A company produces Web design, software, and networking services. View the company as an open economy described by the accompanying table, where input is in dollars needed for $1.00 of output.

   (a) Find the consumption matrix for the company.

   (b) Suppose that the customers (the open sector) have a demand for $5400 worth of Web design, $2700 worth of software, and $900 worth of networking. Use row reduction to find a production vector that will meet this demand exactly.

Table Ex-4

Input Required per Dollar Output

	Web Design	Software	Networking
Web Design	$ 0.40	$ 0.20	$ 0.45
Software	$ 0.30	$ 0.35	$ 0.30
Networking	$ 0.15	$ 0.10	$ 0.20

Provider

In Exercises 5–6, use matrix inversion to find the production vector $\mathbf{x}$ that meets the demand $\mathbf{d}$ for the consumption matrix C.

5. $C = \begin{bmatrix} 0.1 & 0.3 \\ 0.5 & 0.4 \end{bmatrix}$; $\mathbf{d} = \begin{bmatrix} 50 \\ 60 \end{bmatrix}$

6. $C = \begin{bmatrix} 0.3 & 0.1 \\ 0.3 & 0.7 \end{bmatrix}$; $\mathbf{d} = \begin{bmatrix} 22 \\ 14 \end{bmatrix}$

7. Consider an open economy with consumption matrix

$$C = \begin{bmatrix} \frac{1}{2} & 0 \\ 0 & 1 \end{bmatrix}$$

   (a) Show that the economy can meet a demand of $d_1 = 2$ units from the first sector and $d_2 = 0$ units from the second sector, but it cannot meet a demand of $d_1 = 2$ units from the first sector and $d_2 = 1$ unit from the second sector.

(b) Give both a mathematical and an economic explanation of the result in part (a).

8. Consider an open economy with consumption matrix

$$C = \begin{bmatrix} \frac{1}{2} & \frac{1}{4} & \frac{1}{4} \\ \frac{1}{2} & \frac{1}{8} & \frac{1}{4} \\ \frac{1}{2} & \frac{1}{4} & \frac{1}{8} \end{bmatrix}$$

If the open sector demands the same dollar value from each product-producing sector, which such sector must produce the greatest dollar value to meet the demand?

9. Consider an open economy with consumption matrix

$$C = \begin{bmatrix} c_{11} & c_{12} \\ c_{21} & 0 \end{bmatrix}$$

Show that the Leontief equation $\mathbf{x} - C\mathbf{x} = \mathbf{d}$ has a unique solution for every demand vector $\mathbf{d}$ if $c_{21}c_{12} < 1 - c_{11}$.

10. (a) Consider an open economy with a consumption matrix C whose column sums are less than 1, and let $\mathbf{x}$ be the production vector that satisfies an outside demand $\mathbf{d}$; that is, $(I - C)^{-1}\mathbf{d} = \mathbf{x}$. Let $\mathbf{d}_j$ be the demand vector that is obtained by increasing the jth entry of $\mathbf{d}$ by 1 and leaving the other entries fixed. Prove that the production vector

$\mathbf{x}_j$ that meets this demand is

$$\mathbf{x}_j = \mathbf{x} + j\text{th column vector of } (I - C)^{-1}$$

(b) In words, what is the economic significance of the jth column vector of $(I - C)^{-1}$? [*Hint:* Look at $\mathbf{x}_j - \mathbf{x}$.]

11. Prove: If C is an $n \times n$ matrix whose entries are nonnegative and whose row sums are less than 1, then $I - C$ is invertible and has nonnegative entries. [*Hint:* $(A^T)^{-1} = (A^{-1})^T$ for any invertible matrix A.]

True-False Exercises

In parts (a)–(e) determine whether the statement is true or false, and justify your answer.

(a) Sectors of an economy that produce outputs are called open sectors.

(b) A closed economy is an economy that has no open sectors.

(c) The rows of a consumption matrix represent the outputs in a sector of an economy.

(d) If the column sums of the consumption matrix are all less than 1, then the Leontif matrix is invertible.

(e) The Leontif equation relates the production vector for an economy to the outside demand vector.

Chapter 1   Supplementary Exercises

In Exercises 1–4 the given matrix represents an augmented matrix for a linear system. Write the corresponding set of linear equations for the system, and use Gaussian elimination to solve the linear system. Introduce free parameters as necessary.

1. $\begin{bmatrix} 3 & -1 & 0 & 4 & 1 \\ 2 & 0 & 3 & 3 & -1 \end{bmatrix}$

2. $\begin{bmatrix} 1 & 4 & -1 \\ -2 & -8 & 2 \\ 3 & 12 & -3 \\ 0 & 0 & 0 \end{bmatrix}$

3. $\begin{bmatrix} 2 & -4 & 1 & 6 \\ -4 & 0 & 3 & -1 \\ 0 & 1 & -1 & 3 \end{bmatrix}$

4. $\begin{bmatrix} 3 & 1 & -2 \\ -9 & -3 & 6 \\ 6 & 2 & 1 \end{bmatrix}$

5. Use Gauss–Jordan elimination to solve for x' and y' in terms of x and y.

$$x = \tfrac{3}{5}x' - \tfrac{4}{5}y'$$
$$y = \tfrac{4}{5}x' + \tfrac{3}{5}y'$$

6. Use Gauss–Jordan elimination to solve for x' and y' in terms of x and y.

$$x = x'\cos\theta - y'\sin\theta$$
$$y = x'\sin\theta + y'\cos\theta$$

7. Find positive integers that satisfy

$$\begin{aligned} x + y + z &= 9 \\ x + 5y + 10z &= 44 \end{aligned}$$

8. A box containing pennies, nickels, and dimes has 13 coins with a total value of 83 cents. How many coins of each type are in the box?

9. Let

$$\begin{bmatrix} a & 0 & b & 2 \\ a & a & 4 & 4 \\ 0 & a & 2 & b \end{bmatrix}$$

be the augmented matrix for a linear system. Find for what values of a and b the system has

(a) a unique solution.

(b) a one-parameter solution.

(c) a two-parameter solution.   (d) no solution.

10. For which value(s) of a does the following system have zero solutions? One solution? Infinitely many solutions?

$$\begin{aligned} x_1 + x_2 + x_3 &= 4 \\ x_3 &= 2 \\ (a^2 - 4)x_3 &= a - 2 \end{aligned}$$

11. Find a matrix K such that $AKB = C$ given that

$$A = \begin{bmatrix} 1 & 4 \\ -2 & 3 \\ 1 & -2 \end{bmatrix}, \quad B = \begin{bmatrix} 2 & 0 & 0 \\ 0 & 1 & -1 \end{bmatrix},$$

$$C = \begin{bmatrix} 8 & 6 & -6 \\ 6 & -1 & 1 \\ -4 & 0 & 0 \end{bmatrix},$$

12. How should the coefficients a, b, and c be chosen so that the system

$$ax + by - 3z = -3$$
$$-2x - by + cz = -1$$
$$ax + 3y - cz = -3$$

has the solution $x = 1$, $y = -1$, and $z = 2$?

13. In each part, solve the matrix equation for X.

(a) $X \begin{bmatrix} -1 & 0 & 1 \\ 1 & 1 & 0 \\ 3 & 1 & -1 \end{bmatrix} = \begin{bmatrix} 1 & 2 & 0 \\ -3 & 1 & 5 \end{bmatrix}$

(b) $X \begin{bmatrix} 1 & -1 & 2 \\ 3 & 0 & 1 \end{bmatrix} = \begin{bmatrix} -5 & -1 & 0 \\ 6 & -3 & 7 \end{bmatrix}$

(c) $\begin{bmatrix} 3 & 1 \\ -1 & 2 \end{bmatrix} X - X \begin{bmatrix} 1 & 4 \\ 2 & 0 \end{bmatrix} = \begin{bmatrix} 2 & -2 \\ 5 & 4 \end{bmatrix}$

14. Let A be a square matrix.

(a) Show that $(I - A)^{-1} = I + A + A^2 + A^3$ if $A^4 = 0$.

(b) Show that

$$(I - A)^{-1} = I + A + A^2 + \cdots + A^n$$

if $A^{n+1} = 0$.

15. Find values of a, b, and c such that the graph of the polynomial $p(x) = ax^2 + bx + c$ passes through the points $(1, 2)$, $(-1, 6)$, and $(2, 3)$.

16. (*Calculus required*) Find values of a, b, and c such that the graph of $p(x) = ax^2 + bx + c$ passes through the point $(-1, 0)$ and has a horizontal tangent at $(2, -9)$.

17. Let J_n be the $n \times n$ matrix each of whose entries is 1. Show that if $n > 1$, then

$$(I - J_n)^{-1} = I - \frac{1}{n - 1} J_n$$

18. Show that if a square matrix A satisfies

$$A^3 + 4A^2 - 2A + 7I = 0$$

then so does A^T.

19. Prove: If B is invertible, then $AB^{-1} = B^{-1}A$ if and only if $AB = BA$.

20. Prove: If A is invertible, then $A + B$ and $I + BA^{-1}$ are both invertible or both not invertible.

21. Prove: If A is an $m \times n$ matrix and B is the $n \times 1$ matrix each of whose entries is $1/n$, then

$$AB = \begin{bmatrix} \bar{r}_1 \\ \bar{r}_2 \\ \vdots \\ \bar{r}_m \end{bmatrix}$$

where $\bar{r}_i$ is the average of the entries in the ith row of A.

22. (*Calculus required*) If the entries of the matrix

$$C = \begin{bmatrix} c_{11}(x) & c_{12}(x) & \cdots & c_{1n}(x) \\ c_{21}(x) & c_{22}(x) & \cdots & c_{2n}(x) \\ \vdots & \vdots & & \vdots \\ c_{m1}(x) & c_{m2}(x) & \cdots & c_{mn}(x) \end{bmatrix}$$

are differentiable functions of x, then we define

$$\frac{dC}{dx} = \begin{bmatrix} c'_{11}(x) & c'_{12}(x) & \cdots & c'_{1n}(x) \\ c'_{21}(x) & c'_{22}(x) & \cdots & c'_{2n}(x) \\ \vdots & \vdots & & \vdots \\ c'_{m1}(x) & c'_{m2}(x) & \cdots & c'_{mn}(x) \end{bmatrix}$$

Show that if the entries in A and B are differentiable functions of x and the sizes of the matrices are such that the stated operations can be performed, then

(a) $\dfrac{d}{dx}(kA) = k\dfrac{dA}{dx}$

(b) $\dfrac{d}{dx}(A + B) = \dfrac{dA}{dx} + \dfrac{dB}{dx}$

(c) $\dfrac{d}{dx}(AB) = \dfrac{dA}{dx}B + A\dfrac{dB}{dx}$

23. (*Calculus required*) Use part (c) of Exercise 22 to show that

$$\frac{dA^{-1}}{dx} = -A^{-1}\frac{dA}{dx}A^{-1}$$

State all the assumptions you make in obtaining this formula.

24. Assuming that the stated inverses exist, prove the following equalities.

(a) $(C^{-1} + D^{-1})^{-1} = C(C + D)^{-1}D$

(b) $(I + CD)^{-1}C = C(I + DC)^{-1}$

(c) $(C + DD^T)^{-1}D = C^{-1}D(I + D^TC^{-1}D)^{-1}$

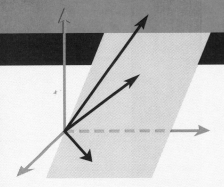

CHAPTER 2

Determinants

INTRODUCTION   In this chapter we will study "determinants" or, more precisely, "determinant functions." Unlike real-valued functions, such as $f(x) = x^2$, that assign a real number to a real variable x, determinant functions assign a real number $f(A)$ to a matrix variable A. Although determinants first arose in the context of solving systems of linear equations, they are no longer used for that purpose in real-world applications. Although they can be useful for solving very small linear systems (say two or three unknowns), our main interest in them stems from the fact that they link together various concepts in linear algebra and provide a useful formula for the inverse of a matrix.

2.1 Determinants by Cofactor Expansion

In this section we will define the notion of a "determinant." This will enable us to give a specific formula for the inverse of an invertible matrix, whereas up to now we have had only a computational procedure for finding it. This, in turn, will eventually provide us with a formula for solutions of certain kinds of linear systems.

Recall from Theorem 1.4.5 that the 2×2 matrix

$$A = \begin{bmatrix} a & b \\ c & d \end{bmatrix}$$

is invertible if and only if $ad - bc \neq 0$ and that the expression $ad - bc$ is called the **determinant** of the matrix A. Recall also that this determinant is denoted by writing

> **WARNING** It is important to keep in mind that $\det(A)$ is a *number*, whereas A is a *matrix*.

$$\det(A) = ad - bc \quad \text{or} \quad \begin{vmatrix} a & b \\ c & d \end{vmatrix} = ad - bc \tag{1}$$

and that the inverse of A can be expressed in terms of the determinant as

$$A^{-1} = \frac{1}{\det(A)} \begin{bmatrix} d & -b \\ -c & a \end{bmatrix} \tag{2}$$

Minors and Cofactors   One of our main goals in this chapter is to obtain an analog of Formula (2) that is applicable to square matrices of *all orders*. For this purpose we will find it convenient to use subscripted entries when writing matrices or determinants. Thus, if we denote a 2×2 matrix as

$$A = \begin{bmatrix} a_{11} & a_{12} \\ a_{21} & a_{22} \end{bmatrix}$$

We define the determinant of a 1×1 matrix $A = [a_{11}]$ as

$$\det[A] = \det[a_{11}] = a_{11}$$

then the two equations in (1) take the form

$$\det(A) = \begin{vmatrix} a_{11} & a_{12} \\ a_{21} & a_{22} \end{vmatrix} = a_{11}a_{22} - a_{12}a_{21} \tag{3}$$

The following definition will be key to our goal of extending the definition of a determinant to higher order matrices.

DEFINITION 1 If A is a square matrix, then the ***minor of entry*** a_{ij} is denoted by M_{ij} and is defined to be the determinant of the submatrix that remains after the ith row and jth column are deleted from A. The number $(-1)^{i+j}M_{ij}$ is denoted by C_{ij} and is called the ***cofactor of entry*** a_{ij}.

▶ **EXAMPLE 1 Finding Minors and Cofactors**

Let

$$A = \begin{bmatrix} 3 & 1 & -4 \\ 2 & 5 & 6 \\ 1 & 4 & 8 \end{bmatrix}$$

The minor of entry a_{11} is

WARNING We have followed the standard convention of using capital letters to denote minors and cofactors even though they are numbers, not matrices.

$$M_{11} = \begin{vmatrix} \cancel{3} & \cancel{1} & \cancel{-4} \\ 2 & 5 & 6 \\ 1 & 4 & 8 \end{vmatrix} = \begin{vmatrix} 5 & 6 \\ 4 & 8 \end{vmatrix} = 16$$

The cofactor of a_{11} is

$$C_{11} = (-1)^{1+1}M_{11} = M_{11} = 16$$

Similarly, the minor of entry a_{32} is

$$M_{32} = \begin{vmatrix} 3 & \cancel{1} & -4 \\ 2 & \cancel{5} & 6 \\ \cancel{1} & \cancel{4} & \cancel{8} \end{vmatrix} = \begin{vmatrix} 3 & -4 \\ 2 & 6 \end{vmatrix} = 26$$

The cofactor of a_{32} is

$$C_{32} = (-1)^{3+2}M_{32} = -M_{32} = -26 \quad ◀$$

Historical Note The term *determinant* was first introduced by the German mathematician Carl Friedrich Gauss in 1801 (see p. 15), who used them to "determine" properties of certain kinds of functions. Interestingly, the term *matrix* is derived from a Latin word for "womb" because it was viewed as a container of determinants.

Historical Note The term *minor* is apparently due to the English mathematician James Sylvester (see p. 34), who wrote the following in a paper published in 1850: "Now conceive any one line and any one column be struck out, we get...a square, one term less in breadth and depth than the original square; and by varying in every possible selection of the line and column excluded, we obtain, supposing the original square to consist of n lines and n columns, n^2 such minor squares, each of which will represent what I term a "First Minor Determinant" relative to the principal or complete determinant."

Remark Note that a minor M_{ij} and its corresponding cofactor C_{ij} are either the same or negatives of each other and that the relating sign $(-1)^{i+j}$ is either $+1$ or -1 in accordance with the pattern in the "checkerboard" array

$$\begin{bmatrix} + & - & + & - & + & \cdots \\ - & + & - & + & - & \cdots \\ + & - & + & - & + & \cdots \\ - & + & - & + & - & \cdots \\ \vdots & \vdots & \vdots & \vdots & \vdots & \end{bmatrix}$$

For example,

$$C_{11} = M_{11}, \quad C_{21} = -M_{21}, \quad C_{22} = M_{22}$$

and so forth. Thus, it is never really necessary to calculate $(-1)^{i+j}$ to calculate C_{ij}—you can simply compute the minor M_{ij} and then adjust the sign in accordance with the checkerboard pattern. Try this in Example 1.

▶ **EXAMPLE 2 Cofactor Expansions of a 2 × 2 Matrix**

The checkerboard pattern for a 2×2 matrix $A = [a_{ij}]$ is

$$\begin{bmatrix} + & - \\ - & + \end{bmatrix}$$

so that

$$C_{11} = M_{11} = a_{22} \qquad C_{12} = -M_{12} = -a_{21}$$
$$C_{21} = -M_{21} = -a_{12} \qquad C_{22} = M_{22} = a_{11}$$

We leave it for you to use Formula (3) to verify that $\det(A)$ can be expressed in terms of cofactors in the following four ways:

$$\begin{aligned} \det(A) = \begin{vmatrix} a_{11} & a_{12} \\ a_{21} & a_{22} \end{vmatrix} & \\ &= a_{11}C_{11} + a_{12}C_{12} \\ &= a_{21}C_{21} + a_{22}C_{22} \\ &= a_{11}C_{11} + a_{21}C_{21} \\ &= a_{12}C_{12} + a_{22}C_{22} \end{aligned} \qquad (4)$$

Each of last four equations is called a *cofactor expansion* of det[A]. In each cofactor expansion the entries and cofactors all come from the same row or same column of A. For example, in the first equation the entries and cofactors all come from the first row of A, in the second they all come from the second row of A, in the third they all come from the first column of A, and in the fourth they all come from the second column of A. ◀

Definition of a General Determinant

Formula (4) is a special case of the following general result, which we will state without proof.

> **THEOREM 2.1.1** *If A is an $n \times n$ matrix, then regardless of which row or column of A is chosen, the number obtained by multiplying the entries in that row or column by the corresponding cofactors and adding the resulting products is always the same.*

This result allows us to make the following definition.

DEFINITION 2 If A is an $n \times n$ matrix, then the number obtained by multiplying the entries in any row or column of A by the corresponding cofactors and adding the resulting products is called the ***determinant of*** A, and the sums themselves are called ***cofactor expansions of*** A. That is,

$$\det(A) = a_{1j}C_{1j} + a_{2j}C_{2j} + \cdots + a_{nj}C_{nj} \tag{5}$$

[cofactor expansion along the jth column]

and

$$\det(A) = a_{i1}C_{i1} + a_{i2}C_{i2} + \cdots + a_{in}C_{in} \tag{6}$$

[cofactor expansion along the ith row]

▶ **EXAMPLE 3** **Cofactor Expansion Along the First Row**

Find the determinant of the matrix

$$A = \begin{bmatrix} 3 & 1 & 0 \\ -2 & -4 & 3 \\ 5 & 4 & -2 \end{bmatrix}$$

by cofactor expansion along the first row.

Solution

$$\det(A) = \begin{vmatrix} 3 & 1 & 0 \\ -2 & -4 & 3 \\ 5 & 4 & -2 \end{vmatrix} = 3 \begin{vmatrix} -4 & 3 \\ 4 & -2 \end{vmatrix} - 1 \begin{vmatrix} -2 & 3 \\ 5 & -2 \end{vmatrix} + 0 \begin{vmatrix} -2 & -4 \\ 5 & 4 \end{vmatrix}$$

$$= 3(-4) - (1)(-11) + 0 = -1$$

▶ **EXAMPLE 4** **Cofactor Expansion Along the First Column**

Let A be the matrix in Example 3, and evaluate $\det(A)$ by cofactor expansion along the first column of A.

Solution

$$\det(A) = \begin{vmatrix} 3 & 1 & 0 \\ -2 & -4 & 3 \\ 5 & 4 & -2 \end{vmatrix} = 3 \begin{vmatrix} -4 & 3 \\ 4 & -2 \end{vmatrix} - (-2) \begin{vmatrix} 1 & 0 \\ 4 & -2 \end{vmatrix} + 5 \begin{vmatrix} 1 & 0 \\ -4 & 3 \end{vmatrix}$$

$$= 3(-4) - (-2)(-2) + 5(3) = -1$$

This agrees with the result obtained in Example 3.

Note that in Example 4 we had to compute three cofactors, whereas in Example 3 only two were needed because the third was multiplied by zero. As a rule, the best strategy for cofactor expansion is to expand along a row or column with the most zeros.

Charles Lutwidge Dodgson (Lewis Carroll) (1832–1898)

Historical Note Cofactor expansion is not the only method for expressing the determinant of a matrix in terms of determinants of lower order. For example, although it is not well known, the English mathematician Charles Dodgson, who was the author of *Alice's Adventures in Wonderland* and *Through the Looking Glass* under the pen name of Lewis Carroll, invented such a method, called "*condensation*." That method has recently been resurrected from obscurity because of its suitability for parallel processing on computers.

[*Image: Time & Life Pictures/Getty Images, Inc.*]

▶ EXAMPLE 5 Smart Choice of Row or Column

If A is the 4×4 matrix

$$A = \begin{bmatrix} 1 & 0 & 0 & -1 \\ 3 & 1 & 2 & 2 \\ 1 & 0 & -2 & 1 \\ 2 & 0 & 0 & 1 \end{bmatrix}$$

then to find $\det(A)$ it will be easiest to use cofactor expansion along the second column, since it has the most zeros:

$$\det(A) = 1 \cdot \begin{vmatrix} 1 & 0 & -1 \\ 1 & -2 & 1 \\ 2 & 0 & 1 \end{vmatrix}$$

For the 3×3 determinant, it will be easiest to use cofactor expansion along its second column, since it has the most zeros:

$$\det(A) = 1 \cdot -2 \cdot \begin{vmatrix} 1 & -1 \\ 2 & 1 \end{vmatrix}$$

$$= -2(1 + 2)$$

$$= -6$$

▶ EXAMPLE 6 Determinant of an Upper Triangular Matrix

The following computation shows that the determinant of a 4×4 upper triangular matrix is the product of its diagonal entries. Each part of the computation uses a cofactor expansion along the first row.

$$\begin{vmatrix} a_{11} & 0 & 0 & 0 \\ a_{21} & a_{22} & 0 & 0 \\ a_{31} & a_{32} & a_{33} & 0 \\ a_{41} & a_{42} & a_{43} & a_{44} \end{vmatrix} = a_{11} \begin{vmatrix} a_{22} & 0 & 0 \\ a_{32} & a_{33} & 0 \\ a_{42} & a_{43} & a_{44} \end{vmatrix}$$

$$= a_{11}a_{22} \begin{vmatrix} a_{33} & 0 \\ a_{43} & a_{44} \end{vmatrix}$$

$$= a_{11}a_{22}a_{33}|a_{44}| = a_{11}a_{22}a_{33}a_{44} \quad ◀$$

The method illustrated in Example 6 can be easily adapted to prove the following general result.

> **THEOREM 2.1.2** *If A is an $n \times n$ triangular matrix (upper triangular, lower triangular, or diagonal), then $\det(A)$ is the product of the entries on the main diagonal of the matrix; that is, $\det(A) = a_{11}a_{22} \cdots a_{nn}$.*

A Useful Technique for Evaluating 2×2 and 3×3 Determinants

Determinants of 2×2 and 3×3 matrices can be evaluated very efficiently using the pattern suggested in Figure 2.1.1.

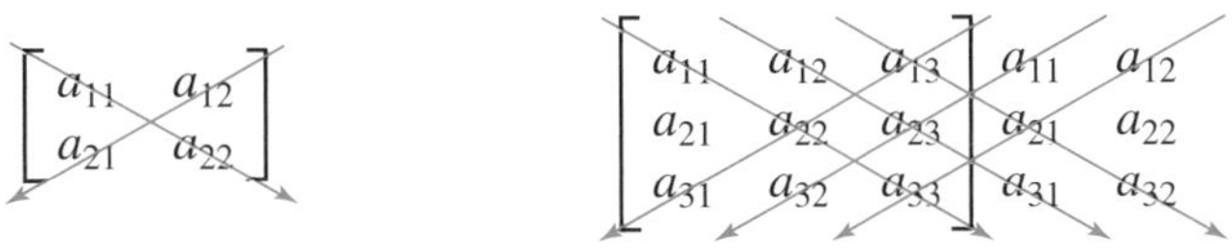

▶ Figure 2.1.1

In the 2×2 case, the determinant can be computed by forming the product of the entries on the rightward arrow and subtracting the product of the entries on the leftward arrow. In the 3×3 case we first recopy the first and second columns as shown in the figure, after

WARNING The arrow technique only works for determinants of 2×2 and 3×3 matrices.

which we can compute the determinant by summing the products of the entries on the rightward arrows and subtracting the products on the leftward arrows. These procedures execute the computations

$$\begin{vmatrix} a_{11} & a_{12} \\ a_{21} & a_{22} \end{vmatrix} = a_{11}a_{22} - a_{12}a_{21}$$

$$\begin{vmatrix} a_{11} & a_{12} & a_{13} \\ a_{21} & a_{22} & a_{23} \\ a_{31} & a_{32} & a_{33} \end{vmatrix} = a_{11} \begin{vmatrix} a_{22} & a_{23} \\ a_{32} & a_{33} \end{vmatrix} - a_{12} \begin{vmatrix} a_{21} & a_{23} \\ a_{31} & a_{33} \end{vmatrix} + a_{13} \begin{vmatrix} a_{21} & a_{22} \\ a_{31} & a_{32} \end{vmatrix}$$

$$= a_{11}(a_{22}a_{33} - a_{23}a_{32}) - a_{12}(a_{21}a_{33} - a_{23}a_{31}) + a_{13}(a_{21}a_{32} - a_{22}a_{31})$$

$$= a_{11}a_{22}a_{33} + a_{12}a_{23}a_{31} + a_{13}a_{21}a_{32} - a_{13}a_{22}a_{31} - a_{12}a_{21}a_{33} - a_{11}a_{23}a_{32}$$

which agrees with the cofactor expansions along the first row.

▶ **EXAMPLE 7** **A Technique for Evaluating 2 × 2 and 3 × 3 Determinants**

$$\begin{vmatrix} 3 & 1 \\ 4 & -2 \end{vmatrix} = (3)(-2) - (1)(4) = -10$$

$$\begin{vmatrix} 1 & 2 & 3 \\ -4 & 5 & 6 \\ 7 & -8 & 9 \end{vmatrix} =$$

$$= [45 + 84 + 96] - [105 - 48 - 72] = 240 \quad ◀$$

Concept Review

- Determinant
- Minor
- Cofactor
- Cofactor expansion

Skills

- Find the minors and cofactors of a square matrix.
- Use cofactor expansion to evaluate the determinant of a square matrix.
- Use the arrow technique to evaluate the determinant of a 2×2 or 3×3 matrix.
- Use the determinant of a 2×2 invertible matrix to find the inverse of that matrix.
- Find the determinant of an upper triangular, lower triangular, or diagonal matrix by inspection.

Exercise Set 2.1

▷ In Exercises 1–2, find all the minors and cofactors of the matrix A. ◀

1. $A = \begin{bmatrix} 1 & -2 & 3 \\ 6 & 7 & -1 \\ -3 & 1 & 4 \end{bmatrix}$

2. $A = \begin{bmatrix} 1 & 1 & 2 \\ 3 & 3 & 6 \\ 0 & 1 & 4 \end{bmatrix}$

3. Let

$$A = \begin{bmatrix} 4 & -1 & 1 & 6 \\ 0 & 0 & -3 & 3 \\ 4 & 1 & 0 & 14 \\ 4 & 1 & 3 & 2 \end{bmatrix}$$

Find

(a) M_{13} and C_{13}.

(b) M_{23} and C_{23}.

(c) M_{22} and C_{22}.

(d) M_{21} and C_{21}.

4. Let

$$A = \begin{bmatrix} 2 & 3 & -1 & 1 \\ -3 & 2 & 0 & 3 \\ 3 & -2 & 1 & 0 \\ 3 & -2 & 1 & 4 \end{bmatrix}$$

Find

(a) M_{32} and C_{32}.

(b) M_{44} and C_{44}.

(c) M_{41} and C_{41}.

(d) M_{24} and C_{24}.

▷ In Exercises 5–8, evaluate the determinant of the given matrix. If the matrix is invertible, use Equation (2) to find its inverse. ◁

5. $\begin{bmatrix} 3 & 5 \\ -2 & 4 \end{bmatrix}$

6. $\begin{bmatrix} 4 & 1 \\ 8 & 2 \end{bmatrix}$

7. $\begin{bmatrix} -5 & 7 \\ -7 & -2 \end{bmatrix}$

8. $\begin{bmatrix} \sqrt{2} & \sqrt{6} \\ 4 & \sqrt{3} \end{bmatrix}$

▷ In Exercises 9–14, use the arrow technique to evaluate the determinant of the given matrix. ◁

9. $\begin{bmatrix} a-3 & 5 \\ -3 & a-2 \end{bmatrix}$

10. $\begin{bmatrix} -2 & 7 & 6 \\ 5 & 1 & -2 \\ 3 & 8 & 4 \end{bmatrix}$

11. $\begin{bmatrix} -2 & 1 & 4 \\ 3 & 5 & -7 \\ 1 & 6 & 2 \end{bmatrix}$

12. $\begin{bmatrix} -1 & 1 & 2 \\ 3 & 0 & -5 \\ 1 & 7 & 2 \end{bmatrix}$

13. $\begin{bmatrix} 3 & 0 & 0 \\ 2 & -1 & 5 \\ 1 & 9 & -4 \end{bmatrix}$

14. $\begin{bmatrix} c & -4 & 3 \\ 2 & 1 & c^2 \\ 4 & c-1 & 2 \end{bmatrix}$

▷ In Exercises 15–18, find all values of λ for which $\det(A) = 0$. ◁

15. $A = \begin{bmatrix} \lambda-2 & 1 \\ -5 & \lambda+4 \end{bmatrix}$

16. $A = \begin{bmatrix} \lambda-4 & 0 & 0 \\ 0 & \lambda & 2 \\ 0 & 3 & \lambda-1 \end{bmatrix}$

17. $A = \begin{bmatrix} \lambda-1 & 0 \\ 2 & \lambda+1 \end{bmatrix}$

18. $A = \begin{bmatrix} \lambda-4 & 4 & 0 \\ -1 & \lambda & 0 \\ 0 & 0 & \lambda-5 \end{bmatrix}$

19. Evaluate the determinant of the matrix in Exercise 13 by a cofactor expansion along

(a) the first row.

(b) the first column.

(c) the second row.

(d) the second column.

(e) the third row.

(f) the third column.

20. Evaluate the determinant of the matrix in Exercise 12 by a cofactor expansion along

(a) the first row.

(b) the first column.

(c) the second row.

(d) the second column.

(e) the third row.

(f) the third column.

▷ In Exercises 21–26, evaluate $\det(A)$ by a cofactor expansion along a row or column of your choice. ◁

21. $A = \begin{bmatrix} -3 & 0 & 7 \\ 2 & 5 & 1 \\ -1 & 0 & 5 \end{bmatrix}$

22. $A = \begin{bmatrix} 3 & 3 & 1 \\ 1 & 0 & -4 \\ 1 & -3 & 5 \end{bmatrix}$

23. $A = \begin{bmatrix} 1 & k & k^2 \\ 1 & k & k^2 \\ 1 & k & k^2 \end{bmatrix}$

24. $A = \begin{bmatrix} k+1 & k-1 & 7 \\ 2 & k-3 & 4 \\ 5 & k+1 & k \end{bmatrix}$

25. $A = \begin{bmatrix} 3 & 3 & 0 & 5 \\ 2 & 2 & 0 & -2 \\ 4 & 1 & -3 & 0 \\ 2 & 10 & 3 & 2 \end{bmatrix}$

26. $A = \begin{bmatrix} 4 & 0 & 0 & 1 & 0 \\ 3 & 3 & 3 & -1 & 0 \\ 1 & 2 & 4 & 2 & 3 \\ 9 & 4 & 6 & 2 & 3 \\ 2 & 2 & 4 & 2 & 3 \end{bmatrix}$

▷ In Exercises 27–32, evaluate the determinant of the given matrix by inspection. ◁

27. $\begin{bmatrix} 1 & 0 & 0 \\ 0 & -1 & 0 \\ 0 & 0 & 1 \end{bmatrix}$

28. $\begin{bmatrix} 2 & 0 & 0 \\ 0 & 2 & 0 \\ 0 & 0 & 2 \end{bmatrix}$

29. $\begin{bmatrix} 0 & 0 & 0 & 0 \\ 1 & 2 & 0 & 0 \\ 0 & 4 & 3 & 0 \\ 1 & 2 & 3 & 8 \end{bmatrix}$

30. $\begin{bmatrix} 1 & 1 & 1 & 1 \\ 0 & 2 & 2 & 2 \\ 0 & 0 & 3 & 3 \\ 0 & 0 & 0 & 4 \end{bmatrix}$

31. $\begin{bmatrix} 1 & 2 & 7 & -3 \\ 0 & 1 & -4 & 1 \\ 0 & 0 & 2 & 7 \\ 0 & 0 & 0 & 3 \end{bmatrix}$

32. $\begin{bmatrix} -3 & 0 & 0 & 0 \\ 1 & 2 & 0 & 0 \\ 40 & 10 & -1 & 0 \\ 100 & 200 & -23 & 3 \end{bmatrix}$

33. Show that the value of the following determinant is independent of θ.

$$\begin{vmatrix} \sin(\theta) & \cos(\theta) & 0 \\ -\cos(\theta) & \sin(\theta) & 0 \\ \sin(\theta) - \cos(\theta) & \sin(\theta) + \cos(\theta) & 1 \end{vmatrix}$$

34. Show that the matrices

$$A = \begin{bmatrix} a & b \\ 0 & c \end{bmatrix} \quad \text{and} \quad B = \begin{bmatrix} d & e \\ 0 & f \end{bmatrix}$$

commute if and only if

$$\begin{vmatrix} b & a-c \\ e & d-f \end{vmatrix} = 0$$

35. By inspection, what is the relationship between the following determinants?

$$d_1 = \begin{vmatrix} a & b & c \\ d & 1 & f \\ g & 0 & 1 \end{vmatrix} \quad \text{and} \quad d_2 = \begin{vmatrix} a+\lambda & b & c \\ d & 1 & f \\ g & 0 & 1 \end{vmatrix}$$

36. Show that

$$\det(A) = \frac{1}{2} \begin{vmatrix} \text{tr}(A) & 1 \\ \text{tr}(A^2) & \text{tr}(A) \end{vmatrix}$$

for every 2×2 matrix A.

37. What can you say about an nth-order determinant all of whose entries are 1? Explain your reasoning.

38. What is the maximum number of zeros that a 3×3 matrix can have without having a zero determinant? Explain your reasoning.

39. What is the maximum number of zeros that a 4×4 matrix can have without having a zero determinant? Explain your reasoning.

40. Prove that (x_1, y_1), (x_2, y_2), and (x_3, y_3) are collinear points if and only if

$$\begin{vmatrix} x_1 & y_1 & 1 \\ x_2 & y_2 & 1 \\ x_3 & y_3 & 1 \end{vmatrix} = 0$$

41. Prove that the equation of the line through the distinct points (a_1, b_1) and (a_2, b_2) can be written as

$$\begin{vmatrix} x & y & 1 \\ a_1 & b_1 & 1 \\ a_2 & b_2 & 1 \end{vmatrix} = 0$$

42. Prove that if A is upper triangular and B_{ij} is the matrix that results when the ith row and jth column of A are deleted, then B_{ij} is upper triangular if $i < j$.

True-False Exercises

In parts (a)–(j) determine whether the statement is true or false, and justify your answer.

(a) The determinant of the 2×2 matrix $\begin{bmatrix} a & b \\ c & d \end{bmatrix}$ is $ad + bc$.

(b) Two square matrices A and B can have the same determinant only if they are the same size.

(c) The minor M_{ij} is the same as the cofactor C_{ij} if and only if $i + j$ is even.

(d) If A is a 3×3 symmetric matrix, then $C_{ij} = C_{ji}$ for all i and j.

(e) The value of a cofactor expansion of a matrix A is independent of the row or column chosen for the expansion.

(f) The determinant of a lower triangular matrix is the sum of the entries along its main diagonal.

(g) For every square matrix A and every scalar c, we have $\det(cA) = c \det(A)$.

(h) For all square matrices A and B, we have $\det(A + B) = \det(A) + \det(B)$.

(i) For every 2×2 matrix A, we have $\det(A^2) = (\det(A))^2$.

2.2 Evaluating Determinants by Row Reduction

In this section we will show how to evaluate a determinant by reducing the associated matrix to row echelon form. In general, this method requires less computation than cofactor expansion and hence is the method of choice for large matrices.

A Basic Theorem   We begin with a fundamental theorem that will lead us to an efficient procedure for evaluating the determinant of a square matrix of any size.

> **THEOREM 2.2.1** *Let A be a square matrix. If A has a row of zeros or a column of zeros, then $\det(A) = 0$.*

Proof   Since the determinant of A can be found by a cofactor expansion along any row or column, we can use the row or column of zeros. Thus, if we let $C_1, C_2, \ldots, C_n$ denote the cofactors of A along that row or column, then it follows from Formula (5) or (6) in Section 2.1 that

$$\det(A) = 0 \cdot C_1 + 0 \cdot C_2 + \cdots + 0 \cdot C_n = 0 \quad \blacktriangleleft$$

The following useful theorem relates the determinant of a matrix and the determinant of its transpose.

Because transposing a matrix changes its columns to rows and its rows to columns, almost every theorem about the rows of a determinant has a companion version about columns, and vice versa.

THEOREM 2.2.2 *Let A be a square matrix. Then* $\det(A) = \det(A^T)$.

Proof   Since transposing a matrix changes its columns to rows and its rows to columns, the cofactor expansion of A along any row is the same as the cofactor expansion of A^T along the corresponding column. Thus, both have the same determinant. ◀

Elementary Row Operations

The next theorem shows how an elementary row operation on a square matrix affects the value of its determinant. In place of a formal proof we have provided a table to illustrate the ideas in the 3×3 case (see Table 1).

THEOREM 2.2.3 *Let A be an $n \times n$ matrix.*

(a) *If B is the matrix that results when a single row or single column of A is multiplied by a scalar k, then* $\det(B) = k \det(A)$.

(b) *If B is the matrix that results when two rows or two columns of A are interchanged, then* $\det(B) = -\det(A)$.

(c) *If B is the matrix that results when a multiple of one row of A is added to another row or when a multiple of one column is added to another column, then* $\det(B) = \det(A)$.

The first panel of Table 1 shows that you can bring a common factor from any row (column) of a determinant through the determinant sign. This is a slightly different way of thinking about part (a) of Theorem 2.2.3.

Table 1

Relationship	Operation
$\begin{vmatrix} ka_{11} & ka_{12} & ka_{13} \\ a_{21} & a_{22} & a_{23} \\ a_{31} & a_{32} & a_{33} \end{vmatrix} = k \begin{vmatrix} a_{11} & a_{12} & a_{13} \\ a_{21} & a_{22} & a_{23} \\ a_{31} & a_{32} & a_{33} \end{vmatrix}$ $\det(B) = k\det(A)$	The first row of A is multiplied by k.
$\begin{vmatrix} a_{21} & a_{22} & a_{23} \\ a_{11} & a_{12} & a_{13} \\ a_{31} & a_{32} & a_{33} \end{vmatrix} = - \begin{vmatrix} a_{11} & a_{12} & a_{13} \\ a_{21} & a_{22} & a_{23} \\ a_{31} & a_{32} & a_{33} \end{vmatrix}$ $\det(B) = -\det(A)$	The first and second rows of A are interchanged.
$\begin{vmatrix} a_{11}+ka_{21} & a_{12}+ka_{22} & a_{13}+ka_{23} \\ a_{21} & a_{22} & a_{23} \\ a_{31} & a_{32} & a_{33} \end{vmatrix} = \begin{vmatrix} a_{11} & a_{12} & a_{13} \\ a_{21} & a_{22} & a_{23} \\ a_{31} & a_{32} & a_{33} \end{vmatrix}$ $\det(B) = \det(A)$	A multiple of the second row of A is added to the first row.

We will verify the first equation in Table 1 and leave the other two for you. To start, note that the determinants on the two sides of the equation differ only in the first row, so these determinants have the same cofactors, C_{11}, C_{12}, C_{13}, along that row (since those cofactors depend only on the entries in the *second* two rows). Thus, expanding the left side by cofactors along the first row yields

$$\begin{vmatrix} ka_{11} & ka_{12} & ka_{13} \\ a_{21} & a_{22} & a_{23} \\ a_{31} & a_{32} & a_{33} \end{vmatrix} = ka_{11}C_{11} + ka_{12}C_{12} + ka_{33}C_{13}$$

$$= k(a_{11}C_{11} + a_{12}C_{12} + a_{33}C_{13})$$

$$= k \begin{vmatrix} a_{11} & a_{12} & a_{13} \\ a_{21} & a_{22} & a_{23} \\ a_{31} & a_{32} & a_{33} \end{vmatrix}$$

Elementary Matrices   It will be useful to consider the special case of Theorem 2.2.3 in which $A = I_n$ is the $n \times n$ identity matrix and E (rather than B) denotes the elementary matrix that results when the row operation is performed on I_n. In this special case Theorem 2.2.3 implies the following result.

THEOREM 2.2.4 *Let E be an $n \times n$ elementary matrix.*
(a) If E results from multiplying a row of I_n by a nonzero number k, then $\det(E) = k$.
(b) If E results from interchanging two rows of I_n, then $\det(E) = -1$.
(c) If E results from adding a multiple of one row of I_n to another, then $\det(E) = 1$.

▶ **EXAMPLE 1 Determinants of Elementary Matrices**

The following determinants of elementary matrices, which are evaluated by inspection, illustrate Theorem 2.2.4.

Observe that the determinant of an elementary matrix cannot be zero.

$$\begin{vmatrix} 1 & 0 & 0 & 0 \\ 0 & 3 & 0 & 0 \\ 0 & 0 & 1 & 0 \\ 0 & 0 & 0 & 1 \end{vmatrix} = 3, \quad \begin{vmatrix} 0 & 0 & 0 & 1 \\ 0 & 1 & 0 & 0 \\ 0 & 0 & 1 & 0 \\ 1 & 0 & 0 & 0 \end{vmatrix} = -1, \quad \begin{vmatrix} 1 & 0 & 0 & 7 \\ 0 & 1 & 0 & 0 \\ 0 & 0 & 1 & 0 \\ 0 & 0 & 0 & 1 \end{vmatrix} = 1 \blacktriangleleft$$

The second row of I_4 was multiplied by 3.   **The first and last rows of I_4 were interchanged.**   **7 times the last row of I_4 was added to the first row.**

Matrices with Proportional Rows or Columns   If a square matrix A has two proportional rows, then a row of zeros can be introduced by adding a suitable multiple of one of the rows to the other. Similarly for columns. But adding a multiple of one row or column to another does not change the determinant, so from Theorem 2.2.1, we must have $\det(A) = 0$. This proves the following theorem.

THEOREM 2.2.5 *If A is a square matrix with two proportional rows or two proportional columns, then $\det(A) = 0$.*

▶ **EXAMPLE 2 Introducing Zero Rows**

The following computation shows how to introduce a row of zeros when there are two proportional rows.

$$\begin{vmatrix} 1 & 3 & -2 & 4 \\ 2 & 6 & -4 & 8 \\ 3 & 9 & 1 & 5 \\ 1 & 1 & 4 & 8 \end{vmatrix} = \begin{vmatrix} 1 & 3 & -2 & 4 \\ 0 & 0 & 0 & 0 \\ 3 & 9 & 1 & 5 \\ 1 & 1 & 4 & 8 \end{vmatrix} = 0$$

The second row is 2 times the first, so we added −2 times the first row to the second to introduce a row of zeros.

Each of the following matrices has two proportional rows or columns; thus, each has a determinant of zero.

$$\begin{bmatrix} -1 & 4 \\ -2 & 8 \end{bmatrix}, \qquad \begin{bmatrix} 1 & -2 & 7 \\ -4 & 8 & 5 \\ 2 & -4 & 3 \end{bmatrix}, \qquad \begin{bmatrix} 3 & -1 & 4 & -5 \\ 6 & -2 & 5 & 2 \\ 5 & 8 & 1 & 4 \\ -9 & 3 & -12 & 15 \end{bmatrix} \blacktriangleleft$$

Evaluating Determinants by Row Reduction

We will now give a method for evaluating determinants that involves substantially less computation than cofactor expansion. The idea of the method is to reduce the given matrix to upper triangular form by elementary row operations, then compute the determinant of the upper triangular matrix (an easy computation), and then relate that determinant to that of the original matrix. Here is an example.

▶ **EXAMPLE 3** **Using Row Reduction to Evaluate a Determinant**

Evaluate $\det(A)$ where

$$A = \begin{bmatrix} 0 & 1 & 5 \\ 3 & -6 & 9 \\ 2 & 6 & 1 \end{bmatrix}$$

Solution We will reduce A to row echelon form (which is upper triangular) and then apply Theorem 2.1.2.

$$\det(A) = \begin{vmatrix} 0 & 1 & 5 \\ 3 & -6 & 9 \\ 2 & 6 & 1 \end{vmatrix} = -\begin{vmatrix} 3 & -6 & 9 \\ 0 & 1 & 5 \\ 2 & 6 & 1 \end{vmatrix}$$
    ⟵ The first and second rows of *A* were interchanged.

$$= -3\begin{vmatrix} 1 & -2 & 3 \\ 0 & 1 & 5 \\ 2 & 6 & 1 \end{vmatrix}$$
    ⟵ A common factor of 3 from the first row was taken through the determinant sign.

$$= -3\begin{vmatrix} 1 & -2 & 3 \\ 0 & 1 & 5 \\ 0 & 10 & -5 \end{vmatrix}$$
    ⟵ −2 times the first row was added to the third row.

$$= -3\begin{vmatrix} 1 & -2 & 3 \\ 0 & 1 & 5 \\ 0 & 0 & -55 \end{vmatrix}$$
    ⟵ −10 times the second row was added to the third row.

$$= (-3)(-55)\begin{vmatrix} 1 & -2 & 3 \\ 0 & 1 & 5 \\ 0 & 0 & 1 \end{vmatrix}$$
    ⟵ A common factor of −55 from the last row was taken through the determinant sign.

$$= (-3)(-55)(1) = 165$$

Even with today's fastest computers it would take millions of years to calculate a 25 × 25 determinant by cofactor expansion, so methods based on row reduction are often used for large determinants. For determinants of small size (such as those in this text), cofactor expansion is often a reasonable choice.

▶ **EXAMPLE 4** **Using Column Operations to Evaluate a Determinant**

Compute the determinant of

$$A = \begin{bmatrix} 1 & 0 & 0 & 3 \\ 2 & 7 & 0 & 6 \\ 0 & 6 & 3 & 0 \\ 7 & 3 & 1 & -5 \end{bmatrix}$$

Solution This determinant could be computed as above by using elementary row operations to reduce A to row echelon form, but we can put A in lower triangular form in one step by adding -3 times the first column to the fourth to obtain

Example 4 points out that it is always wise to keep an eye open for column operations that can shorten computations.

$$\det(A) = \det \begin{bmatrix} 1 & 0 & 0 & 0 \\ 2 & 7 & 0 & 0 \\ 0 & 6 & 3 & 0 \\ 7 & 3 & 1 & -26 \end{bmatrix} = (1)(7)(3)(-26) = -546 \blacktriangleleft$$

Cofactor expansion and row or column operations can sometimes be used in combination to provide an effective method for evaluating determinants. The following example illustrates this idea.

▶ **EXAMPLE 5 Row Operations and Cofactor Expansion**

Evaluate $\det(A)$ where

$$A = \begin{bmatrix} 3 & 5 & -2 & 6 \\ 1 & 2 & -1 & 1 \\ 2 & 4 & 1 & 5 \\ 3 & 7 & 5 & 3 \end{bmatrix}$$

Solution By adding suitable multiples of the second row to the remaining rows, we obtain

$$\det(A) = \begin{vmatrix} 0 & -1 & 1 & 3 \\ 1 & 2 & -1 & 1 \\ 0 & 0 & 3 & 3 \\ 0 & 1 & 8 & 0 \end{vmatrix}$$

$$= - \begin{vmatrix} -1 & 1 & 3 \\ 0 & 3 & 3 \\ 1 & 8 & 0 \end{vmatrix} \quad \longleftarrow \quad \text{Cofactor expansion along the first column}$$

$$= - \begin{vmatrix} -1 & 1 & 3 \\ 0 & 3 & 3 \\ 0 & 9 & 3 \end{vmatrix} \quad \longleftarrow \quad \text{We added the first row to the third row.}$$

$$= -(-1) \begin{vmatrix} 3 & 3 \\ 9 & 3 \end{vmatrix} \quad \longleftarrow \quad \text{Cofactor expansion along the first column}$$

$$= -18 \blacktriangleleft$$

Skills

- Know the effect of elementary row operations on the value of a determinant.
- Know the determinants of the three types of elementary matrices.
- Know how to introduce zeros into the rows or columns of a matrix to facilitate the evaluation of its determinant.

- Use row reduction to evaluate the determinant of a matrix.
- Use column operations to evaluate the determinant of a matrix.
- Combine the use of row reduction and cofactor expansion to evaluate the determinant of a matrix.

Exercise Set 2.2

In Exercises 1–4, verify that $\det(A) = \det(A^T)$.

1. $A = \begin{bmatrix} -2 & 3 \\ 1 & 4 \end{bmatrix}$

2. $A = \begin{bmatrix} -6 & 1 \\ 2 & -2 \end{bmatrix}$

3. $A = \begin{bmatrix} 2 & -1 & 3 \\ 1 & 2 & 4 \\ 5 & -3 & 6 \end{bmatrix}$

4. $A = \begin{bmatrix} 4 & 2 & -1 \\ 0 & 2 & -3 \\ -1 & 1 & 5 \end{bmatrix}$

In Exercises 5–9, find the determinant of the given elementary matrix by inspection.

5. $\begin{bmatrix} 1 & 0 & 0 & 0 \\ 0 & 1 & 0 & 0 \\ 0 & 0 & -5 & 0 \\ 0 & 0 & 0 & 1 \end{bmatrix}$

6. $\begin{bmatrix} 1 & 0 & 0 \\ 0 & 1 & 0 \\ -5 & 0 & 1 \end{bmatrix}$

7. $\begin{bmatrix} 1 & 0 & 0 & 0 \\ 0 & 0 & 1 & 0 \\ 0 & 1 & 0 & 0 \\ 0 & 0 & 0 & 1 \end{bmatrix}$

8. $\begin{bmatrix} 1 & 0 & 0 & 0 \\ 0 & -\frac{1}{3} & 0 & 0 \\ 0 & 0 & 1 & 0 \\ 0 & 0 & 0 & 1 \end{bmatrix}$

9. $\begin{bmatrix} 1 & 0 & 0 & 0 \\ 0 & 1 & 0 & -9 \\ 0 & 0 & 1 & 0 \\ 0 & 0 & 0 & 1 \end{bmatrix}$

In Exercises 10–17, evaluate the determinant of the given matrix by reducing the matrix to row echelon form.

10. $\begin{bmatrix} 3 & 6 & -9 \\ 0 & 0 & -2 \\ -2 & 1 & 5 \end{bmatrix}$

11. $\begin{bmatrix} 0 & 3 & 1 \\ 1 & 1 & 2 \\ 3 & 2 & 4 \end{bmatrix}$

12. $\begin{bmatrix} 1 & -3 & 0 \\ -2 & 4 & 1 \\ 5 & -2 & 2 \end{bmatrix}$

13. $\begin{bmatrix} 3 & -6 & 9 \\ -2 & 7 & -2 \\ 0 & 1 & 5 \end{bmatrix}$

14. $\begin{bmatrix} 1 & -2 & 3 & 1 \\ 5 & -9 & 6 & 3 \\ -1 & 2 & -6 & -2 \\ 2 & 8 & 6 & 1 \end{bmatrix}$

15. $\begin{bmatrix} 2 & 1 & 3 & 1 \\ 1 & 0 & 1 & 1 \\ 0 & 2 & 1 & 0 \\ 0 & 1 & 2 & 3 \end{bmatrix}$

16. $\begin{bmatrix} 0 & 1 & 1 & 1 \\ \frac{1}{2} & \frac{1}{2} & 1 & \frac{1}{2} \\ \frac{2}{3} & \frac{1}{3} & \frac{1}{3} & 0 \\ -\frac{1}{3} & \frac{2}{3} & 0 & 0 \end{bmatrix}$

17. $\begin{bmatrix} 1 & 3 & 1 & 5 & 3 \\ -2 & -7 & 0 & -4 & 2 \\ 0 & 0 & 1 & 0 & 1 \\ 0 & 0 & 2 & 1 & 1 \\ 0 & 0 & 0 & 1 & 1 \end{bmatrix}$

18. Repeat Exercises 11–13 by using a combination of row operations and cofactor expansion.

19. Repeat Exercises 14–17 by using a combination of row operations and cofactor expansion.

In Exercises 20–27, evaluate the determinant, given that

$$\begin{vmatrix} a & b & c \\ d & e & f \\ g & h & i \end{vmatrix} = -6$$

20. $\begin{vmatrix} g & h & i \\ d & e & f \\ a & b & c \end{vmatrix}$

21. $\begin{vmatrix} d & e & f \\ g & h & i \\ a & b & c \end{vmatrix}$

22. $\begin{vmatrix} a & b & c \\ d & e & f \\ 2a & 2b & 2c \end{vmatrix}$

23. $\begin{vmatrix} 3a & 3b & 3c \\ -d & -e & -f \\ 4g & 4h & 4i \end{vmatrix}$

24. $\begin{vmatrix} a+d & b+e & c+f \\ -d & -e & -f \\ g & h & i \end{vmatrix}$

25. $\begin{vmatrix} a+g & b+h & c+i \\ d & e & f \\ g & h & i \end{vmatrix}$

26. $\begin{vmatrix} a & b & c \\ 2d & 2e & 2f \\ g+3a & h+3b & i+3c \end{vmatrix}$

27. $\begin{vmatrix} -3a & -3b & -3c \\ d & e & f \\ g-4d & h-4e & i-4f \end{vmatrix}$

28. Show that

(a) $\det \begin{bmatrix} 0 & 0 & a_{13} \\ 0 & a_{22} & a_{23} \\ a_{31} & a_{32} & a_{33} \end{bmatrix} = -a_{13}a_{22}a_{31}$

(b) $\det \begin{bmatrix} 0 & 0 & 0 & a_{14} \\ 0 & 0 & a_{23} & a_{24} \\ 0 & a_{32} & a_{33} & a_{34} \\ a_{41} & a_{42} & a_{43} & a_{44} \end{bmatrix} = a_{14}a_{23}a_{32}a_{41}$

29. Use row reduction to show that

$$\begin{vmatrix} 1 & 1 & 1 \\ a & b & c \\ a^2 & b^2 & c^2 \end{vmatrix} = (b-a)(c-a)(c-b)$$

In Exercises 30–33, confirm the identities without evaluating the determinants directly.

30. $\begin{vmatrix} a_1 + b_1 t & a_2 + b_2 t & a_3 + b_3 t \\ a_1 t + b_1 & a_2 t + b_2 & a_3 t + b_3 \\ c_1 & c_2 & c_3 \end{vmatrix} = (1 - t^2) \begin{vmatrix} a_1 & a_2 & a_3 \\ b_1 & b_2 & b_3 \\ c_1 & c_2 & c_3 \end{vmatrix}$

31. $\begin{vmatrix} a_1 & b_1 & a_1 + b_1 + c_1 \\ a_2 & b_2 & a_2 + b_2 + c_2 \\ a_3 & b_3 & a_3 + b_3 + c_3 \end{vmatrix} = \begin{vmatrix} a_1 & b_1 & c_1 \\ a_2 & b_2 & c_2 \\ a_3 & b_3 & c_3 \end{vmatrix}$

32. $\begin{vmatrix} a_1 & b_1 + ta_1 & c_1 + rb_1 + sa_1 \\ a_2 & b_2 + ta_2 & c_2 + rb_2 + sa_2 \\ a_3 & b_3 + ta_3 & c_3 + rb_3 + sa_3 \end{vmatrix} = \begin{vmatrix} a_1 & a_2 & a_3 \\ b_1 & b_2 & b_3 \\ c_1 & c_2 & c_3 \end{vmatrix}$

33. $\begin{vmatrix} a_1 + b_1 & a_1 - b_1 & c_1 \\ a_2 + b_2 & a_2 - b_2 & c_2 \\ a_3 + b_3 & a_3 - b_3 & c_3 \end{vmatrix} = -2\begin{vmatrix} a_1 & b_1 & c_1 \\ a_2 & b_2 & c_2 \\ a_3 & b_3 & c_3 \end{vmatrix}$

34. Find the determinant of the following matrix.

$$\begin{bmatrix} a & b & b & b \\ b & a & b & b \\ b & b & a & b \\ b & b & b & a \end{bmatrix}$$

▶ In Exercises 35–36, show that $\det(A) = 0$ without directly evaluating the determinant. ◀

35. $A = \begin{bmatrix} -2 & 8 & 1 & 4 \\ 3 & 2 & 5 & 1 \\ 1 & 10 & 6 & 5 \\ 4 & -6 & 4 & -3 \end{bmatrix}$

36. $A = \begin{bmatrix} -4 & 1 & 1 & 1 & 1 \\ 1 & -4 & 1 & 1 & 1 \\ 1 & 1 & -4 & 1 & 1 \\ 1 & 1 & 1 & -4 & 1 \\ 1 & 1 & 1 & 1 & -4 \end{bmatrix}$

True-False Exercises

In parts (a)–(f) determine whether the statement is true or false, and justify your answer.

(a) If A is a 4×4 matrix and B is obtained from A by interchanging the first two rows and then interchanging the last two rows, then $\det(B) = \det(A)$.

(b) If A is a 3×3 matrix and B is obtained from A by multiplying the first column by 4 and multiplying the third column by $\frac{3}{4}$, then $\det(B) = 3\det(A)$.

(c) If A is a 3×3 matrix and B is obtained from A by adding 5 times the first row to each of the second and third rows, then $\det(B) = 25\det(A)$.

(d) If A is an $n \times n$ matrix and B is obtained from A by multiplying each row of A by its row number, then

$$\det(B) = \frac{n(n+1)}{2}\det(A)$$

(e) If A is a square matrix with two identical columns, then $\det(A) = 0$.

(f) If the sum of the second and fourth row vectors of a 6×6 matrix A is equal to the last row vector, then $\det(A) = 0$.

2.3 Properties of Determinants; Cramer's Rule

In this section we will develop some fundamental properties of matrices, and we will use these results to derive a formula for the inverse of an invertible matrix and formulas for the solutions of certain kinds of linear systems.

Basic Properties of Determinants   Suppose that A and B are $n \times n$ matrices and k is any scalar. We begin by considering possible relationships between $\det(A)$, $\det(B)$, and

$$\det(kA), \quad \det(A + B), \quad \text{and} \quad \det(AB)$$

Since a common factor of any row of a matrix can be moved through the determinant sign, and since each of the n rows in kA has a common factor of k, it follows that

$$\det(kA) = k^n \det(A) \tag{1}$$

For example,

$$\begin{vmatrix} ka_{11} & ka_{12} & ka_{13} \\ ka_{21} & ka_{22} & ka_{23} \\ ka_{31} & ka_{32} & ka_{33} \end{vmatrix} = k^3 \begin{vmatrix} a_{11} & a_{12} & a_{13} \\ a_{21} & a_{22} & a_{23} \\ a_{31} & a_{32} & a_{33} \end{vmatrix}$$

Unfortunately, no simple relationship exists among $\det(A)$, $\det(B)$, and $\det(A + B)$. In particular, we emphasize that $\det(A + B)$ will usually *not* be equal to $\det(A) + \det(B)$. The following example illustrates this fact.

▶ EXAMPLE 1 $\det(A + B) \neq \det(A) + \det(B)$

Consider

$$A = \begin{bmatrix} 1 & 2 \\ 2 & 5 \end{bmatrix}, \quad B = \begin{bmatrix} 3 & 1 \\ 1 & 3 \end{bmatrix}, \quad A + B = \begin{bmatrix} 4 & 3 \\ 3 & 8 \end{bmatrix}$$

We have $\det(A) = 1$, $\det(B) = 8$, and $\det(A + B) = 23$; thus

$$\det(A + B) \neq \det(A) + \det(B) \quad ◀$$

In spite of the previous example, there is a useful relationship concerning sums of determinants that is applicable when the matrices involved are the same except for *one* row (column). For example, consider the following two matrices that differ only in the second row:

$$A = \begin{bmatrix} a_{11} & a_{12} \\ a_{21} & a_{22} \end{bmatrix} \quad \text{and} \quad B = \begin{bmatrix} a_{11} & a_{12} \\ b_{21} & b_{22} \end{bmatrix}$$

Calculating the determinants of A and B we obtain

$$\det(A) + \det(B) = (a_{11}a_{22} - a_{12}a_{21}) + (a_{11}b_{22} - a_{12}b_{21})$$
$$= a_{11}(a_{22} + b_{22}) - a_{12}(a_{21} + b_{21})$$
$$= \det \begin{bmatrix} a_{11} & a_{12} \\ a_{21} + b_{21} & a_{22} + b_{22} \end{bmatrix}$$

Thus

$$\det \begin{bmatrix} a_{11} & a_{12} \\ a_{21} & a_{22} \end{bmatrix} + \det \begin{bmatrix} a_{11} & a_{12} \\ b_{21} & b_{22} \end{bmatrix} = \det \begin{bmatrix} a_{11} & a_{12} \\ a_{21} + b_{21} & a_{22} + b_{22} \end{bmatrix}$$

This is a special case of the following general result.

THEOREM 2.3.1 *Let A, B, and C be $n \times n$ matrices that differ only in a single row, say the rth, and assume that the rth row of C can be obtained by adding corresponding entries in the rth rows of A and B. Then*

$$\det(C) = \det(A) + \det(B)$$

The same result holds for columns.

▶ EXAMPLE 2 **Sums of Determinants**

We leave it to you to confirm the following equality by evaluating the determinants.

$$\det \begin{bmatrix} 1 & 7 & 5 \\ 2 & 0 & 3 \\ 1+0 & 4+1 & 7+(-1) \end{bmatrix} = \det \begin{bmatrix} 1 & 7 & 5 \\ 2 & 0 & 3 \\ 1 & 4 & 7 \end{bmatrix} + \det \begin{bmatrix} 1 & 7 & 5 \\ 2 & 0 & 3 \\ 0 & 1 & -1 \end{bmatrix} \quad ◀$$

Determinant of a Matrix Product

Considering the complexity of the formulas for determinants and matrix multiplication, it would seem unlikely that a simple relationship should exist between them. This is what makes the simplicity of our next result so surprising. We will show that if A and B are square matrices of the same size, then

$$\det(AB) = \det(A)\det(B) \tag{2}$$

The proof of this theorem is fairly intricate, so we will have to develop some preliminary results first. We begin with the special case of (2) in which A is an elementary matrix. Because this special case is only a prelude to (2), we call it a lemma.

> **LEMMA 2.3.2** *If B is an $n \times n$ matrix and E is an $n \times n$ elementary matrix, then*
>
> $$\det(EB) = \det(E)\det(B)$$

Proof We will consider three cases, each in accordance with the row operation that produces the matrix E.

Case 1 If E results from multiplying a row of I_n by k, then by Theorem 1.5.1, EB results from B by multiplying the corresponding row by k; so from Theorem 2.2.3(a) we have

$$\det(EB) = k\det(B)$$

But from Theorem 2.2.4(a) we have $\det(E) = k$, so

$$\det(EB) = \det(E)\det(B)$$

Cases 2 and 3 The proofs of the cases where E results from interchanging two rows of I_n or from adding a multiple of one row to another follow the same pattern as Case 1 and are left as exercises. ◄

Remark It follows by repeated applications of Lemma 2.3.2 that if B is an $n \times n$ matrix and $E_1, E_2, \ldots, E_r$ are $n \times n$ elementary matrices, then

$$\det(E_1 E_2 \cdots E_r B) = \det(E_1)\det(E_2)\cdots\det(E_r)\det(B) \tag{3}$$

Determinant Test for Invertibility

Our next theorem provides an important criterion for determining whether a matrix is invertible. It also takes us a step closer to establishing Formula (2).

> **THEOREM 2.3.3** *A square matrix A is invertible if and only if $\det(A) \neq 0$.*

Proof Let R be the reduced row echelon form of A. As a preliminary step, we will show that $\det(A)$ and $\det(R)$ are both zero or both nonzero: Let $E_1, E_2, \ldots, E_r$ be the elementary matrices that correspond to the elementary row operations that produce R from A. Thus

$$R = E_r \cdots E_2 E_1 A$$

and from (3),

$$\det(R) = \det(E_r) \cdots \det(E_2)\det(E_1)\det(A) \tag{4}$$

We pointed out in the margin note that accompanies Theorem 2.2.4 that the determinant of an elementary matrix is nonzero. Thus, it follows from Formula (4) that $\det(A)$ and $\det(R)$ are either both zero or both nonzero, which sets the stage for the main part of the proof. If we assume first that A is invertible, then it follows from Theorem 1.6.4 that $R = I$ and hence that $\det(R) = 1\ (\neq 0)$. This, in turn, implies that $\det(A) \neq 0$, which is what we wanted to show.

It follows from Theorems 2.3.3 and 2.2.5 that a square matrix with two proportional rows or two proportional columns is not invertible.

Conversely, assume that $\det(A) \neq 0$. It follows from this that $\det(R) \neq 0$, which tells us that R cannot have a row of zeros. Thus, it follows from Theorem 1.4.3 that $R = I$ and hence that A is invertible by Theorem 1.6.4. ◄

▶ EXAMPLE 3 **Determinant Test for Invertibility**

Since the first and third rows of

$$A = \begin{bmatrix} 1 & 2 & 3 \\ 1 & 0 & 1 \\ 2 & 4 & 6 \end{bmatrix}$$

are proportional, $\det(A) = 0$. Thus A is not invertible. ◀

We are now ready for the main result concerning products of matrices.

> **THEOREM 2.3.4** *If A and B are square matrices of the same size, then*
>
> $$\det(AB) = \det(A)\det(B)$$

Proof We divide the proof into two cases that depend on whether or not A is invertible. If the matrix A is not invertible, then by Theorem 1.6.5 neither is the product AB. Thus, from Theorem 2.3.3, we have $\det(AB) = 0$ and $\det(A) = 0$, so it follows that $\det(AB) = \det(A)\det(B)$.

Now

assume that A is invertible. By Theorem 1.6.4, the matrix A is expressible as a product of elementary matrices, say

$$A = E_1 E_2 \cdots E_r \tag{5}$$

so

$$AB = E_1 E_2 \cdots E_r B$$

Applying (3) to this equation yields

$$\det(AB) = \det(E_1)\det(E_2)\cdots\det(E_r)\det(B)$$

and applying (3) again yields

$$\det(AB) = \det(E_1 E_2 \cdots E_r)\det(B)$$

which, from (5), can be written as $\det(AB) = \det(A)\det(B)$. ◀

**Augustin Louis Cauchy
(1789–1857)**

Historical Note In 1815 the great French mathematician Augustin Cauchy published a landmark paper in which he gave the first systematic and modern treatment of determinants. It was in that paper that Theorem 2.3.4 was stated and proved in full generality for the first time. Special cases of the theorem had been stated and proved earlier, but it was Cauchy who made the final jump.

[*Image: The Granger Collection, New York*]

▶ EXAMPLE 4 **Verifying That $\det(AB) = \det(A)\det(B)$**

Consider the matrices

$$A = \begin{bmatrix} 3 & 1 \\ 2 & 1 \end{bmatrix}, \quad B = \begin{bmatrix} -1 & 3 \\ 5 & 8 \end{bmatrix}, \quad AB = \begin{bmatrix} 2 & 17 \\ 3 & 14 \end{bmatrix}$$

We leave it for you to verify that

$$\det(A) = 1, \quad \det(B) = -23, \quad \text{and} \quad \det(AB) = -23$$

Thus $\det(AB) = \det(A)\det(B)$, as guaranteed by Theorem 2.3.4. ◀

The following theorem gives a useful relationship between the determinant of an invertible matrix and the determinant of its inverse.

> **THEOREM 2.3.5** *If A is invertible, then*
> $$\det(A^{-1}) = \frac{1}{\det(A)}$$

Proof Since $A^{-1}A = I$, it follows that $\det(A^{-1}A) = \det(I)$. Therefore, we must have $\det(A^{-1})\det(A) = 1$. Since $\det(A) \neq 0$, the proof can be completed by dividing through by $\det(A)$. ◄

Adjoint of a Matrix

It follows from Theorems 2.3.5 and 2.1.2 that

$$\det(A^{-1}) = \frac{1}{a_{11}}\frac{1}{a_{22}}\cdots\frac{1}{a_{nn}}$$

Moreover, by using the adjoint formula it is possible to show that

$$\frac{1}{a_{11}}, \ \frac{1}{a_{22}}, \dots, \ \frac{1}{a_{nn}}$$

are actually the successive diagonal entries of A^{-1} (compare A and A^{-1} in Example 3 of Section 1.7).

In a cofactor expansion we compute $\det(A)$ by multiplying the entries in a row or column by their cofactors and adding the resulting products. It turns out that if one multiplies the entries in any row by the corresponding cofactors from a *different* row, the sum of these products is always zero. (This result also holds for columns.) Although we omit the general proof, the next example illustrates the idea of the proof in a special case.

▶ **EXAMPLE 5** Entries and Cofactors from Different Rows

Let

$$A = \begin{bmatrix} a_{11} & a_{12} & a_{13} \\ a_{21} & a_{22} & a_{23} \\ a_{31} & a_{32} & a_{33} \end{bmatrix}$$

Consider the quantity

$$a_{11}C_{31} + a_{12}C_{32} + a_{13}C_{33}$$

that is formed by multiplying the entries in the first row by the cofactors of the corresponding entries in the third row and adding the resulting products. We can show that this quantity is equal to zero by the following trick: Construct a new matrix A' by replacing the third row of A with another copy of the first row. That is,

$$A' = \begin{bmatrix} a_{11} & a_{12} & a_{13} \\ a_{21} & a_{22} & a_{23} \\ a_{11} & a_{12} & a_{13} \end{bmatrix}$$

Let C'_{31}, C'_{32}, C'_{33} be the cofactors of the entries in the third row of A'. Since the first two rows of A and A' are the same, and since the computations of $C_{31}, C_{32}, C_{33}, C'_{31}, C'_{32}$, and C'_{33} involve only entries from the first two rows of A and A', it follows that

$$C_{31} = C'_{31}, \quad C_{32} = C'_{32}, \quad C_{33} = C'_{33}$$

Since A' has two identical rows, it follows from (3) that

$$\det(A') = 0 \tag{6}$$

On the other hand, evaluating $\det(A')$ by cofactor expansion along the third row gives

$$\det(A') = a_{11}C'_{31} + a_{12}C'_{32} + a_{13}C'_{33} = a_{11}C_{31} + a_{12}C_{32} + a_{13}C_{33} \tag{7}$$

From (6) and (7) we obtain

$$a_{11}C_{31} + a_{12}C_{32} + a_{13}C_{33} = 0 \quad ◄$$

DEFINITION 1   If A is any $n \times n$ matrix and C_{ij} is the cofactor of a_{ij}, then the matrix

$$\begin{bmatrix} C_{11} & C_{12} & \cdots & C_{1n} \\ C_{21} & C_{22} & \cdots & C_{2n} \\ \vdots & \vdots & & \vdots \\ C_{n1} & C_{n2} & \cdots & C_{nn} \end{bmatrix}$$

is called the ***matrix of cofactors from A***. The transpose of this matrix is called the ***adjoint of A*** and is denoted by $\mathrm{adj}(A)$.

Leonard Eugene Dickson (1874–1954)

Historical Note   The use of the term *adjoint* for the transpose of the matrix of cofactors appears to have been introduced by the American mathematician L. E. Dickson in a research paper that he published in 1902.
[*Image: Courtesy of the American Mathematical Society*]

▶ **EXAMPLE 6   Adjoint of a 3 × 3 Matrix**

Let

$$A = \begin{bmatrix} 3 & 2 & -1 \\ 1 & 6 & 3 \\ 2 & -4 & 0 \end{bmatrix}$$

The cofactors of A are

$$\begin{array}{ccc} C_{11} = 12 & C_{12} = 6 & C_{13} = -16 \\ C_{21} = 4 & C_{22} = 2 & C_{23} = 16 \\ C_{31} = 12 & C_{32} = -10 & C_{33} = 16 \end{array}$$

so the matrix of cofactors is

$$\begin{bmatrix} 12 & 6 & -16 \\ 4 & 2 & 16 \\ 12 & -10 & 16 \end{bmatrix}$$

and the adjoint of A is

$$\mathrm{adj}(A) = \begin{bmatrix} 12 & 4 & 12 \\ 6 & 2 & -10 \\ -16 & 16 & 16 \end{bmatrix} \quad \blacktriangleleft$$

In Theorem 1.4.5 we gave a formula for the inverse of a 2×2 invertible matrix. Our next theorem extends that result to $n \times n$ invertible matrices.

THEOREM 2.3.6   Inverse of a Matrix Using Its Adjoint

If A is an invertible matrix, then

$$A^{-1} = \frac{1}{\det(A)} \mathrm{adj}(A) \tag{8}$$

Proof   We show first that

$$A \, \mathrm{adj}(A) = \det(A) I$$

Consider the product

$$A \, \mathrm{adj}(A) = \begin{bmatrix} a_{11} & a_{12} & \cdots & a_{1n} \\ a_{21} & a_{22} & \cdots & a_{2n} \\ \vdots & \vdots & & \vdots \\ a_{i1} & a_{i2} & \cdots & a_{in} \\ \vdots & \vdots & & \vdots \\ a_{n1} & a_{n2} & \cdots & a_{nn} \end{bmatrix} \begin{bmatrix} C_{11} & C_{21} & \cdots & C_{j1} & \cdots & C_{n1} \\ C_{12} & C_{22} & \cdots & C_{j2} & \cdots & C_{n2} \\ \vdots & \vdots & & \vdots & & \vdots \\ C_{1n} & C_{2n} & \cdots & C_{jn} & \cdots & C_{nn} \end{bmatrix}$$

The entry in the ith row and jth column of the product $A \text{ adj}(A)$ is

$$a_{i1}C_{j1} + a_{i2}C_{j2} + \cdots + a_{in}C_{jn} \tag{9}$$

(see the shaded lines above).

If $i = j$, then (9) is the cofactor expansion of $\det(A)$ along the ith row of A (Theorem 2.1.1), and if $i \neq j$, then the a's and the cofactors come from different rows of A, so the value of (9) is zero. Therefore,

$$A \text{ adj}(A) = \begin{bmatrix} \det(A) & 0 & \cdots & 0 \\ 0 & \det(A) & \cdots & 0 \\ \vdots & \vdots & & \vdots \\ 0 & 0 & \cdots & \det(A) \end{bmatrix} = \det(A)I \tag{10}$$

Since A is invertible, $\det(A) \neq 0$. Therefore, Equation (10) can be rewritten as

$$\frac{1}{\det(A)}[A \text{ adj}(A)] = I \quad \text{or} \quad A\left[\frac{1}{\det(A)}\text{adj}(A)\right] = I$$

Multiplying both sides on the left by A^{-1} yields

$$A^{-1} = \frac{1}{\det(A)}\text{adj}(A) \quad \blacktriangleleft$$

▶ **EXAMPLE 7 Using the Adjoint to Find an Inverse Matrix**

Use (8) to find the inverse of the matrix A in Example 6.

Solution We leave it for you to check that $\det(A) = 64$. Thus

$$A^{-1} = \frac{1}{\det(A)}\text{adj}(A) = \frac{1}{64}\begin{bmatrix} 12 & 4 & 12 \\ 6 & 2 & -10 \\ -16 & 16 & 16 \end{bmatrix} = \begin{bmatrix} \frac{12}{64} & \frac{4}{64} & \frac{12}{64} \\ \frac{6}{64} & \frac{2}{64} & -\frac{10}{64} \\ -\frac{16}{64} & \frac{16}{64} & \frac{16}{64} \end{bmatrix} \quad \blacktriangleleft$$

Cramer's Rule   Our next theorem uses the formula for the inverse of an invertible matrix to produce a formula, called ***Cramer's rule***, for the solution of a linear system $A\mathbf{x} = \mathbf{b}$ of n equations in n unknowns in the case where the coefficient matrix A is invertible (or, equivalently, when $\det(A) \neq 0$).

THEOREM 2.3.7 Cramer's Rule

If $A\mathbf{x} = \mathbf{b}$ is a system of n linear equations in n unknowns such that $\det(A) \neq 0$, then the system has a unique solution. This solution is

$$x_1 = \frac{\det(A_1)}{\det(A)}, \quad x_2 = \frac{\det(A_2)}{\det(A)}, \ldots, \quad x_n = \frac{\det(A_n)}{\det(A)}$$

where A_j is the matrix obtained by replacing the entries in the jth column of A by the entries in the matrix

$$\mathbf{b} = \begin{bmatrix} b_1 \\ b_2 \\ \vdots \\ b_n \end{bmatrix}$$

Proof If $\det(A) \neq 0$, then A is invertible, and by Theorem 1.6.2, $\mathbf{x} = A^{-1}\mathbf{b}$ is the unique solution of $A\mathbf{x} = \mathbf{b}$. Therefore, by Theorem 2.3.6 we have

$$\mathbf{x} = A^{-1}\mathbf{b} = \frac{1}{\det(A)}\text{adj}(A)\mathbf{b} = \frac{1}{\det(A)}\begin{bmatrix} C_{11} & C_{21} & \cdots & C_{n1} \\ C_{12} & C_{22} & \cdots & C_{n2} \\ \vdots & \vdots & & \vdots \\ C_{1n} & C_{2n} & \cdots & C_{nn} \end{bmatrix}\begin{bmatrix} b_1 \\ b_2 \\ \vdots \\ b_n \end{bmatrix}$$

Multiplying the matrices out gives

$$\mathbf{x} = \frac{1}{\det(A)}\begin{bmatrix} b_1 C_{11} + b_2 C_{21} + \cdots + b_n C_{n1} \\ b_1 C_{12} + b_2 C_{22} + \cdots + b_n C_{n2} \\ \vdots & \vdots & \vdots \\ b_1 C_{1n} + b_2 C_{2n} + \cdots + b_n C_{nn} \end{bmatrix}$$

The entry in the jth row of $\mathbf{x}$ is therefore

$$x_j = \frac{b_1 C_{1j} + b_2 C_{2j} + \cdots + b_n C_{nj}}{\det(A)} \tag{11}$$

Now let

$$A_j = \begin{bmatrix} a_{11} & a_{12} & \cdots & a_{1j-1} & b_1 & a_{1j+1} & \cdots & a_{1n} \\ a_{21} & a_{22} & \cdots & a_{2j-1} & b_2 & a_{2j+1} & \cdots & a_{2n} \\ \vdots & \vdots & & \vdots & \vdots & \vdots & & \vdots \\ a_{n1} & a_{n2} & \cdots & a_{nj-1} & b_n & a_{nj+1} & \cdots & a_{nn} \end{bmatrix}$$

Since A_j differs from A only in the jth column, it follows that the cofactors of entries $b_1, b_2, \ldots, b_n$ in A_j are the same as the cofactors of the corresponding entries in the jth column of A. The cofactor expansion of $\det(A_j)$ along the jth column is therefore

$$\det(A_j) = b_1 C_{1j} + b_2 C_{2j} + \cdots + b_n C_{nj}$$

Substituting this result in (11) gives

$$x_j = \frac{\det(A_j)}{\det(A)} \blacktriangleleft$$

Gabriel Cramer
(1704–1752)

▶ **EXAMPLE 8 Using Cramer's Rule to Solve a Linear System**

Use Cramer's rule to solve

$$\begin{aligned} x_1 + \quad\quad + 2x_3 &= 6 \\ -3x_1 + 4x_2 + 6x_3 &= 30 \\ -x_1 - 2x_2 + 3x_3 &= 8 \end{aligned}$$

Solution

$$A = \begin{bmatrix} 1 & 0 & 2 \\ -3 & 4 & 6 \\ -1 & -2 & 3 \end{bmatrix}, \quad A_1 = \begin{bmatrix} 6 & 0 & 2 \\ 30 & 4 & 6 \\ 8 & -2 & 3 \end{bmatrix},$$

$$A_2 = \begin{bmatrix} 1 & 6 & 2 \\ -3 & 30 & 6 \\ -1 & 8 & 3 \end{bmatrix}, \quad A_3 = \begin{bmatrix} 1 & 0 & 6 \\ -3 & 4 & 30 \\ -1 & -2 & 8 \end{bmatrix}$$

For $n > 3$, it is usually more efficient to solve a linear system with n equations in n unknowns by Gauss–Jordan elimination than by Cramer's rule. Its main use is for obtaining properties of solutions of a linear system without actually solving the system.

Therefore,

$$x_1 = \frac{\det(A_1)}{\det(A)} = \frac{-40}{44} = \frac{-10}{11}, \quad x_2 = \frac{\det(A_2)}{\det(A)} = \frac{72}{44} = \frac{18}{11},$$

$$x_3 = \frac{\det(A_3)}{\det(A)} = \frac{152}{44} = \frac{38}{11} \blacktriangleleft$$

Equivalence Theorem   In Theorem 1.6.4 we listed five results that are equivalent to the invertibility of a matrix A. We conclude this section by merging Theorem 2.3.3 with that list to produce the following theorem that relates all of the major topics we have studied thus far.

THEOREM 2.3.8 **Equivalent Statements**

If A is an $n \times n$ matrix, then the following statements are equivalent.

(a)   *A is invertible.*

(b)   *$A\mathbf{x} = \mathbf{0}$ has only the trivial solution.*

(c)   *The reduced row echelon form of A is I_n.*

(d)   *A can be expressed as a product of elementary matrices.*

(e)   *$A\mathbf{x} = \mathbf{b}$ is consistent for every $n \times 1$ matrix $\mathbf{b}$.*

(f)   *$A\mathbf{x} = \mathbf{b}$ has exactly one solution for every $n \times 1$ matrix $\mathbf{b}$.*

(g)   *$\det(A) \neq 0$.*

OPTIONAL   We now have all of the machinery necessary to prove the following two results, which we stated without proof in Theorem 1.7.1:

- **Theorem 1.7.1(c)** A triangular matrix is invertible if and only if its diagonal entries are all nonzero.

- **Theorem 1.7.1(d)** The inverse of an invertible lower triangular matrix is lower triangular, and the inverse of an invertible upper triangular matrix is upper triangular.

Proof of Theorem 1.7.1(c)   Let $A = [a_{ij}]$ be a triangular matrix, so that its diagonal entries are

$$a_{11}, a_{22}, \ldots, a_{nn}$$

From Theorem 2.1.2, the matrix A is invertible if and only if

$$\det(A) = a_{11}a_{22}\cdots a_{nn}$$

is nonzero, which is true if and only if the diagonal entries are all nonzero.

Proof of Theorem 1.7.1(d)   We will prove the result for upper triangular matrices and leave the lower triangular case for you. Assume that A is upper triangular and invertible. Since

$$A^{-1} = \frac{1}{\det(A)}\operatorname{adj}(A)$$

we can prove that A^{-1} is upper triangular by showing that $\operatorname{adj}(A)$ is upper triangular or, equivalently, that the matrix of cofactors is lower triangular. We can do this by showing that every cofactor C_{ij} with $i < j$ (i.e., above the main diagonal) is zero. Since

$$C_{ij} = (-1)^{i+j} M_{ij}$$

it suffices to show that each minor M_{ij} with $i < j$ is zero. For this purpose, let B_{ij} be the matrix that results when the ith row and jth column of A are deleted, so

$$M_{ij} = \det(B_{ij}) \tag{12}$$

From the assumption that $i < j$, it follows that B_{ij} is upper triangular (see Figure 1.7.1). Since A is upper triangular, its $(i + 1)$-st row begins with at least i zeros. But the ith row of B_{ij} is the $(i + 1)$-st row of A with the entry in the jth column removed. Since $i < j$, none of the first i zeros is removed by deleting the jth column; thus the ith row of B_{ij} starts with at least i zeros, which implies that this row has a zero on the main diagonal. It now follows from Theorem 2.1.2 that $\det(B_{ij}) = 0$ and from (12) that $M_{ij} = 0$. ◄

Concept Review

- Determinant test for invertibility
- Matrix of cofactors
- Adjoint of a matrix
- Cramer's rule
- Equivalent statements about an invertible matrix

Skills

- Know how determinants behave with respect to basic arithmetic operations, as given in Equation (1), Theorem 2.3.1, Lemma 2.3.2, and Theorem 2.3.4.
- Use the determinant to test a matrix for invertibility.
- Know how $\det(A)$ and $\det(A^{-1})$ are related.
- Compute the matrix of cofactors for a square matrix A.
- Compute $\text{adj}(A)$ for a square matrix A.
- Use the adjoint of an invertible matrix to find its inverse.
- Use Cramer's rule to solve linear systems of equations.
- Know the equivalent characterizations of an invertible matrix given in Theorem 2.3.8.

Exercise Set 2.3

In Exercises 1–4, verify that $\det(kA) = k^n \det(A)$.

1. $A = \begin{bmatrix} -1 & 2 \\ 3 & 4 \end{bmatrix}$; $k = 2$ **2.** $A = \begin{bmatrix} 2 & 2 \\ 5 & -2 \end{bmatrix}$; $k = -4$

3. $A = \begin{bmatrix} 2 & -1 & 3 \\ 3 & 2 & 1 \\ 1 & 4 & 5 \end{bmatrix}$; $k = -2$

4. $A = \begin{bmatrix} 1 & 1 & 1 \\ 0 & 2 & 3 \\ 0 & 1 & -2 \end{bmatrix}$; $k = 3$

In Exercises 5–6, verify that $\det(AB) = \det(BA)$ and determine whether the equality $\det(A + B) = \det(A) + \det(B)$ holds.

5. $A = \begin{bmatrix} 2 & 1 & 0 \\ 3 & 4 & 0 \\ 0 & 0 & 2 \end{bmatrix}$ and $B = \begin{bmatrix} 1 & -1 & 3 \\ 7 & 1 & 2 \\ 5 & 0 & 1 \end{bmatrix}$

6. $A = \begin{bmatrix} -1 & 8 & 2 \\ 1 & 0 & -1 \\ -2 & 2 & 2 \end{bmatrix}$ and $B = \begin{bmatrix} 2 & -1 & -4 \\ 1 & 1 & 3 \\ 0 & 3 & -1 \end{bmatrix}$

In Exercises 7–14, use determinants to decide whether the given matrix is invertible.

7. $A = \begin{bmatrix} 2 & 5 & 5 \\ -1 & -1 & 0 \\ 2 & 4 & 3 \end{bmatrix}$ **8.** $A = \begin{bmatrix} 2 & 0 & 3 \\ 0 & 3 & 2 \\ -2 & 0 & -4 \end{bmatrix}$

9. $A = \begin{bmatrix} 2 & -3 & 5 \\ 0 & 1 & -3 \\ 0 & 0 & 2 \end{bmatrix}$ **10.** $A = \begin{bmatrix} -3 & 0 & 1 \\ 5 & 0 & 6 \\ 8 & 0 & 3 \end{bmatrix}$

11. $A = \begin{bmatrix} 4 & 2 & 8 \\ -2 & 1 & -4 \\ 3 & 1 & 6 \end{bmatrix}$ **12.** $A = \begin{bmatrix} 1 & 0 & -1 \\ 9 & -1 & 4 \\ 8 & 9 & -1 \end{bmatrix}$

13. $A = \begin{bmatrix} 2 & 0 & 0 \\ 8 & 1 & 0 \\ -5 & 3 & 6 \end{bmatrix}$ **14.** $A = \begin{bmatrix} \sqrt{2} & -\sqrt{7} & 0 \\ 3\sqrt{2} & -3\sqrt{7} & 0 \\ 5 & -9 & 0 \end{bmatrix}$

In Exercises 15–18, find the values of k for which A is invertible.

15. $A = \begin{bmatrix} k - 3 & -2 \\ -2 & k - 2 \end{bmatrix}$ **16.** $A = \begin{bmatrix} k & 2 \\ 2 & k \end{bmatrix}$

17. $A = \begin{bmatrix} 1 & 2 & 4 \\ 3 & 1 & 6 \\ k & 3 & 2 \end{bmatrix}$ **18.** $A = \begin{bmatrix} 1 & 2 & 0 \\ k & 1 & k \\ 0 & 2 & 1 \end{bmatrix}$

In Exercises 19–23, decide whether the given matrix is invertible, and if so, use the adjoint method to find its inverse.

19. $A = \begin{bmatrix} 2 & 5 & 5 \\ -1 & -1 & 0 \\ 2 & 4 & 3 \end{bmatrix}$ **20.** $A = \begin{bmatrix} 2 & 0 & 3 \\ 0 & 3 & 2 \\ -2 & 0 & -4 \end{bmatrix}$

21. $A = \begin{bmatrix} 2 & -3 & 5 \\ 0 & 1 & -3 \\ 0 & 0 & 2 \end{bmatrix}$ **22.** $A = \begin{bmatrix} 2 & 0 & 0 \\ 8 & 1 & 0 \\ -5 & 3 & 6 \end{bmatrix}$

23. $A = \begin{bmatrix} 1 & 3 & 1 & 1 \\ 2 & 5 & 2 & 2 \\ 1 & 3 & 8 & 9 \\ 1 & 3 & 2 & 2 \end{bmatrix}$

In Exercises 24–29, solve by Cramer's rule, where it applies.

24. $\begin{aligned} 7x_1 - 2x_2 &= 3 \\ 3x_1 + x_2 &= 5 \end{aligned}$ **25.** $\begin{aligned} 4x + 5y &= 2 \\ 11x + y + 2z &= 3 \\ x + 5y + 2z &= 1 \end{aligned}$

26. $\begin{aligned} x - 4y + z &= 6 \\ 4x - y + 2z &= -1 \\ 2x + 2y - 3z &= -20 \end{aligned}$ **27.** $\begin{aligned} x_1 - 3x_2 + x_3 &= 4 \\ 2x_1 - x_2 &= -2 \\ 4x_1 - 3x_3 &= 0 \end{aligned}$

28.
$$\begin{aligned} -x_1 - 4x_2 + 2x_3 + x_4 &= -32 \\ 2x_1 - x_2 + 7x_3 + 9x_4 &= 14 \\ -x_1 + x_2 + 3x_3 + x_4 &= 11 \\ x_1 - 2x_2 + x_3 - 4x_4 &= -4 \end{aligned}$$

29.
$$\begin{aligned} 3x_1 - x_2 + x_3 &= 4 \\ -x_1 + 7x_2 - 2x_3 &= 1 \\ 2x_1 + 6x_2 - x_3 &= 5 \end{aligned}$$

30. Show that the matrix

$$A = \begin{bmatrix} \cos\theta & \sin\theta & 0 \\ -\sin\theta & \cos\theta & 0 \\ 0 & 0 & 1 \end{bmatrix}$$

is invertible for all values of θ; then find A^{-1} using Theorem 2.3.6.

31. Use Cramer's rule to solve for y without solving for the unknowns x, z, and w.

$$\begin{aligned} 4x + y + z + w &= 6 \\ 3x + 7y - z + w &= 1 \\ 7x + 3y - 5z + 8w &= -3 \\ x + y + z + 2w &= 3 \end{aligned}$$

32. Let $A\mathbf{x} = \mathbf{b}$ be the system in Exercise 31.

(a) Solve by Cramer's rule.

(b) Solve by Gauss–Jordan elimination.

(c) Which method involves fewer computations?

33. Prove that if $\det(A) = 1$ and all the entries in A are integers, then all the entries in A^{-1} are integers.

34. Let $A\mathbf{x} = \mathbf{b}$ be a system of n linear equations in n unknowns with integer coefficients and integer constants. Prove that if $\det(A) = 1$, the solution $\mathbf{x}$ has integer entries.

35. Let

$$A = \begin{bmatrix} a & b & c \\ d & e & f \\ g & h & i \end{bmatrix}$$

Assuming that $\det(A) = -7$, find

(a) $\det(3A)$ (b) $\det(A^{-1})$ (c) $\det(2A^{-1})$

(d) $\det((2A)^{-1})$ (e) $\det \begin{bmatrix} a & g & d \\ b & h & e \\ c & i & f \end{bmatrix}$

36. In each part, find the determinant given that A is a 4×4 matrix for which $\det(A) = -2$.

(a) $\det(-A)$ (b) $\det(A^{-1})$ (c) $\det(2A^T)$ (d) $\det(A^3)$

37. In each part, find the determinant given that A is a 3×3 matrix for which $\det(A) = 7$.

(a) $\det(3A)$ (b) $\det(A^{-1})$

(c) $\det(2A^{-1})$ (d) $\det((2A)^{-1})$

38. Prove that a square matrix A is invertible if and only if A^TA is invertible.

39. Show that if A is a square matrix, then $\det(A^TA) = \det(AA^T)$.

True-False Exercises

In parts (a)–(l) determine whether the statement is true or false, and justify your answer.

(a) If A is a 3×3 matrix, then $\det(2A) = 2\det(A)$.

(b) If A and B are square matrices of the same size such that $\det(A) = \det(B)$, then $\det(A + B) = 2\det(A)$.

(c) If A and B are square matrices of the same size and A is invertible, then

$$\det(A^{-1}BA) = \det(B)$$

(d) A square matrix A is invertible if and only if $\det(A) = 0$.

(e) The matrix of cofactors of A is precisely $[\mathrm{adj}(A)]^T$.

(f) For every $n \times n$ matrix A, we have

$$A \cdot \mathrm{adj}(A) = (\det(A))I_n$$

(g) If A is a square matrix and the linear system $A\mathbf{x} = \mathbf{0}$ has multiple solutions for $\mathbf{x}$, then $\det(A) = 0$.

(h) If A is an $n \times n$ matrix and there exists an $n \times 1$ matrix $\mathbf{b}$ such that the linear system $A\mathbf{x} = \mathbf{b}$ has no solutions, then the reduced row echelon form of A cannot be I_n.

(i) If E is an elementary matrix, then $E\mathbf{x} = \mathbf{0}$ has only the trivial solution.

(j) If A is an invertible matrix, then the linear system $A\mathbf{x} = \mathbf{0}$ has only the trivial solution if and only if the linear system $A^{-1}\mathbf{x} = \mathbf{0}$ has only the trivial solution.

(k) If A is invertible, then $\mathrm{adj}(A)$ must also be invertible.

(l) If A has a row of zeros, then so does $\mathrm{adj}(A)$.

Chapter 2 Supplementary Exercises

In Exercises 1–8, evaluate the determinant of the given matrix by (a) cofactor expansion and (b) using elementary row operations to introduce zeros into the matrix.

1. $\begin{bmatrix} -4 & 2 \\ 3 & 3 \end{bmatrix}$

2. $\begin{bmatrix} 7 & -1 \\ -2 & -6 \end{bmatrix}$

3. $\begin{bmatrix} -1 & 5 & 2 \\ 0 & 2 & -1 \\ -3 & 1 & 1 \end{bmatrix}$

4. $\begin{bmatrix} -1 & -2 & -3 \\ -4 & -5 & -6 \\ -7 & -8 & -9 \end{bmatrix}$

5. $\begin{bmatrix} 3 & 0 & -1 \\ 1 & 1 & 1 \\ 0 & 4 & 2 \end{bmatrix}$

6. $\begin{bmatrix} -5 & 1 & 4 \\ 3 & 0 & 2 \\ 1 & -2 & 2 \end{bmatrix}$

7. $\begin{bmatrix} 3 & 6 & 0 & 1 \\ -2 & 3 & 1 & 4 \\ 1 & 0 & -1 & 1 \\ -9 & 2 & -2 & 2 \end{bmatrix}$

8. $\begin{bmatrix} -1 & -2 & -3 & -4 \\ 4 & 3 & 2 & 1 \\ 1 & 2 & 3 & 4 \\ -4 & -3 & -2 & -1 \end{bmatrix}$

9. Evaluate the determinants in Exercises 3–6 by using the arrow technique (see Example 7 in Section 2.1).

10. (a) Construct a 4×4 matrix whose determinant is easy to compute using cofactor expansion but hard to evaluate using elementary row operations.

(b) Construct a 4×4 matrix whose determinant is easy to compute using elementary row operations but hard to evaluate using cofactor expansion.

11. Use the determinant to decide whether the matrices in Exercises 1–4 are invertible.

12. Use the determinant to decide whether the matrices in Exercises 5–8 are invertible.

In Exercises 13–15, find the determinant of the given matrix by any method.

13. $\begin{vmatrix} 5 & b-3 \\ b-2 & -3 \end{vmatrix}$

14. $\begin{vmatrix} 3 & -4 & a \\ a^2 & 1 & 2 \\ 2 & a-1 & 4 \end{vmatrix}$

15. $\begin{vmatrix} 0 & 0 & 0 & 0 & -3 \\ 0 & 0 & 0 & -4 & 0 \\ 0 & 0 & -1 & 0 & 0 \\ 0 & 2 & 0 & 0 & 0 \\ 5 & 0 & 0 & 0 & 0 \end{vmatrix}$

16. Solve for x.

$$\begin{vmatrix} x & -1 \\ 3 & 1-x \end{vmatrix} = \begin{vmatrix} 1 & 0 & -3 \\ 2 & x & -6 \\ 1 & 3 & x-5 \end{vmatrix}$$

In Exercises 17–24, use the adjoint method (Theorem 2.3.6) to find the inverse of the given matrix, if it exists.

17. The matrix in Exercise 1.   18. The matrix in Exercise 2.

19. The matrix in Exercise 3.   20. The matrix in Exercise 4.

21. The matrix in Exercise 5.   22. The matrix in Exercise 6.

23. The matrix in Exercise 7.   24. The matrix in Exercise 8.

25. Use Cramer's rule to solve for x' and y' in terms of x and y.

$$x = \tfrac{3}{5}x' - \tfrac{4}{5}y'$$
$$y = \tfrac{4}{5}x' + \tfrac{3}{5}y'$$

26. Use Cramer's rule to solve for x' and y' in terms of x and y.

$$x = x'\cos\theta - y'\sin\theta$$
$$y = x'\sin\theta + y'\cos\theta$$

27. By examining the determinant of the coefficient matrix, show that the following system has a nontrivial solution if and only if $\alpha = \beta$.

$$\begin{aligned} x + y + \alpha z &= 0 \\ x + y + \beta z &= 0 \\ \alpha x + \beta y + z &= 0 \end{aligned}$$

28. Let A be a 3×3 matrix, each of whose entries is 1 or 0. What is the largest possible value for $\det(A)$?

29. (a) For the triangle in the accompanying figure, use trigonometry to show that

$$b\cos\gamma + c\cos\beta = a$$
$$c\cos\alpha + a\cos\gamma = b$$
$$a\cos\beta + b\cos\alpha = c$$

and then apply Cramer's rule to show that

$$\cos\alpha = \frac{b^2 + c^2 - a^2}{2bc}$$

(b) Use Cramer's rule to obtain similar formulas for $\cos\beta$ and $\cos\gamma$.

◀ Figure Ex-29

30. Use determinants to show that for all real values of λ, the only solution of

$$x - 2y = \lambda x$$
$$x - y = \lambda y$$

is $x = 0$, $y = 0$.

31. Prove: If A is invertible, then $\mathrm{adj}(A)$ is invertible and

$$[\mathrm{adj}(A)]^{-1} = \frac{1}{\det(A)}A = \mathrm{adj}(A^{-1})$$

32. Prove: If A is an $n \times n$ matrix, then

$$\det[\operatorname{adj}(A)] = [\det(A)]^{n-1}$$

33. Prove: If the entries in each row of an $n \times n$ matrix A add up to zero, then the determinant of A is zero. [*Hint:* Consider the product $A\mathbf{x}$, where $\mathbf{x}$ is the $n \times 1$ matrix, each of whose entries is one.]

34. (a) In the accompanying figure, the area of the triangle ABC can be expressed as

area ABC = area $ADEC$ + area $CEFB$ − area $ADFB$

Use this and the fact that the area of a trapezoid equals $\frac{1}{2}$ the altitude times the sum of the parallel sides to show that

$$\text{area } ABC = \frac{1}{2} \begin{vmatrix} x_1 & y_1 & 1 \\ x_2 & y_2 & 1 \\ x_3 & y_3 & 1 \end{vmatrix}$$

[*Note:* In the derivation of this formula, the vertices are labeled such that the triangle is traced counterclockwise proceeding from (x_1, y_1) to (x_2, y_2) to (x_3, y_3). For a clockwise orientation, the determinant above yields the *negative* of the area.]

(b) Use the result in (a) to find the area of the triangle with vertices $(3, 3)$, $(4, 0)$, $(-2, -1)$.

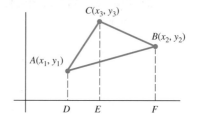

◀ Figure Ex-34

35. Use the fact that 21,375, 38,798, 34,162, 40,223, and 79,154 are all divisible by 19 to show that

$$\begin{vmatrix} 2 & 1 & 3 & 7 & 5 \\ 3 & 8 & 7 & 9 & 8 \\ 3 & 4 & 1 & 6 & 2 \\ 4 & 0 & 2 & 2 & 3 \\ 7 & 9 & 1 & 5 & 4 \end{vmatrix}$$

is divisible by 19 without directly evaluating the determinant.

36. Without directly evaluating the determinant, show that

$$\begin{vmatrix} \sin\alpha & \cos\alpha & \sin(\alpha+\delta) \\ \sin\beta & \cos\beta & \sin(\beta+\delta) \\ \sin\gamma & \cos\gamma & \sin(\gamma+\delta) \end{vmatrix} = 0$$

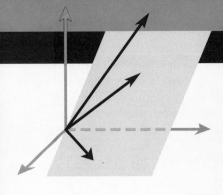

CHAPTER 3

Euclidean Vector Spaces

INTRODUCTION Engineers and physicists distinguish between two types of physical quantities— *scalars*, which are quantities that can be described by a numerical value alone, and *vectors*, which are quantities that require both a number and a direction for their complete physical description. For example, temperature, length, and speed are scalars because they can be fully described by a number that tells "how much"—a temperature of 20°C, a length of 5 cm, or a speed of 75 km/h. In contrast, velocity and force are vectors because they require a number that tells "how much" and a direction that tells "which way"—say, a boat moving at 10 knots in a direction 45° northeast, or a force of 100 lb acting vertically. Although the notions of vectors and scalars that we will study in this text have their origins in physics and engineering, we will be more concerned with using them to build mathematical structures and then applying those structures to such diverse fields as genetics, computer science, economics, telecommunications, and environmental science.

3.1 Vectors in 2-Space, 3-Space, and n-Space

Linear algebra is concerned with two kinds of mathematical objects, "matrices" and "vectors." We are already familiar with the basic ideas about matrices, so in this section we will introduce some of the basic ideas about vectors. As we progress through this text we will see that vectors and matrices are closely related and that much of linear algebra is concerned with that relationship.

Geometric Vectors

Engineers and physicists represent vectors in two dimensions (also called **2-*space***) or in three dimensions (also called **3-*space***) by arrows. The direction of the arrowhead specifies the **direction** of the vector and the **length** of the arrow specifies the magnitude. Mathematicians call these **geometric vectors**. The tail of the arrow is called the **initial point** of the vector and the tip the **terminal point** (Figure 3.1.1).

In this text we will denote vectors in boldface type such as $\mathbf{a}$, $\mathbf{b}$, $\mathbf{v}$, $\mathbf{w}$, and $\mathbf{x}$, and we will denote scalars in lowercase italic type such as a, k, v, w, and x. When we want to indicate that a vector $\mathbf{v}$ has initial point A and terminal point B, then, as shown in Figure 3.1.2, we will write

$$\mathbf{v} = \overrightarrow{AB}$$

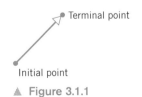

Terminal point

Initial point

▲ Figure 3.1.1

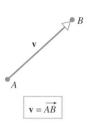

$$\boxed{\mathbf{v} = \overrightarrow{AB}}$$

▲ Figure 3.1.2

Vectors with the same length and direction, such as those in Figure 3.1.3, are said to be *equivalent*. Since we want a vector to be determined solely by its length and direction, equivalent vectors are regarded to be the same vector even though they may be in different positions. Equivalent vectors are also said to be *equal*, which we indicate by writing

$$\mathbf{v} = \mathbf{w}$$

The vector whose initial and terminal points coincide has length zero, so we call this the *zero vector* and denote it by **0**. The zero vector has no natural direction, so we will agree that it can be assigned any direction that is convenient for the problem at hand.

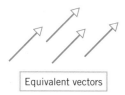

Equivalent vectors

▲ Figure 3.1.3

Vector Addition

There are a number of important algebraic operations on vectors, all of which have their origin in laws of physics.

> **Parallelogram Rule for Vector Addition** If **v** and **w** are vectors in 2-space or 3-space that are positioned so their initial points coincide, then the two vectors form adjacent sides of a parallelogram, and the *sum* **v** + **w** is the vector represented by the arrow from the common initial point of **v** and **w** to the opposite vertex of the parallelogram (Figure 3.1.4*a*).

Here is another way to form the sum of two vectors.

> **Triangle Rule for Vector Addition** If **v** and **w** are vectors in 2-space or 3-space that are positioned so the initial point of **w** is at the terminal point of **v**, then the *sum* **v** + **w** is represented by the arrow from the initial point of **v** to the terminal point of **w** (Figure 3.1.4*b*).

In Figure 3.1.4*c* we have constructed the sums **v** + **w** and **w** + **v** by the triangle rule. This construction makes it evident that

$$\mathbf{v} + \mathbf{w} = \mathbf{w} + \mathbf{v} \tag{1}$$

and that the sum obtained by the triangle rule is the same as the sum obtained by the parallelogram rule.

▶ Figure 3.1.4

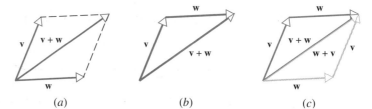

(a)　　　　　(b)　　　　　(c)

Vector addition can also be viewed as a process of translating points.

> **Vector Addition Viewed as Translation** If **v**, **w**, and **v** + **w** are positioned so their initial points coincide, then the terminal point of **v** + **w** can be viewed in two ways:
>
> 1. The terminal point of **v** + **w** is the point that results when the terminal point of **v** is translated in the direction of **w** by a distance equal to the length of **w** (Figure 3.1.5*a*).
> 2. The terminal point of **v** + **w** is the point that results when the terminal point of **w** is translated in the direction of **v** by a distance equal to the length of **v** (Figure 3.1.5*b*).
>
> Accordingly, we say that **v** + **w** is the *translation of* **v** *by* **w** or, alternatively, the *translation of* **w** *by* **v**.

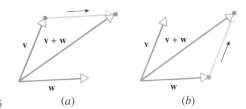

► Figure 3.1.5 (*a*) (*b*)

Vector Subtraction In ordinary arithmetic we can write $a - b = a + (-b)$, which expresses subtraction in terms of addition. There is an analogous idea in vector arithmetic.

> **Vector Subtraction** The ***negative*** of a vector **v**, denoted by $-\mathbf{v}$, is the vector that has the same length as **v** but is oppositely directed (Figure 3.1.6*a*), and the ***difference*** of **v** from **w**, denoted by $\mathbf{w} - \mathbf{v}$, is taken to be the sum
>
> $$\mathbf{w} - \mathbf{v} = \mathbf{w} + (-\mathbf{v}) \tag{2}$$

The difference of **v** from **w** can be obtained geometrically by the parallelogram method shown in Figure 3.1.6*b*, or more directly by positioning **w** and **v** so their initial points coincide and drawing the vector from the terminal point of **v** to the terminal point of **w** (Figure 3.1.6*c*).

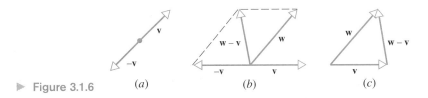

► Figure 3.1.6 (*a*) (*b*) (*c*)

Scalar Multiplication Sometimes there is a need to change the length of a vector or change its length and reverse its direction. This is accomplished by a type of multiplication in which vectors are multiplied by scalars. As an example, the product 2**v** denotes the vector that has the same direction as **v** but twice the length, and the product $-2\mathbf{v}$ denotes the vector that is oppositely directed to **v** and has twice the length. Here is the general result.

> **Scalar Multiplication** If **v** is a nonzero vector in 2-space or 3-space, and if k is a nonzero scalar, then we define the ***scalar product of* v *by* k** to be the vector whose length is $|k|$ times the length of **v** and whose direction is the same as that of **v** if k is positive and opposite to that of **v** if k is negative. If $k = 0$ or $\mathbf{v} = \mathbf{0}$, then we define $k\mathbf{v}$ to be **0**.

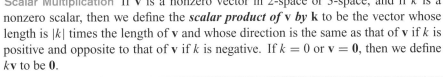

▲ Figure 3.1.7

Figure 3.1.7 shows the geometric relationship between a vector **v** and some of its scalar multiples. In particular, observe that $(-1)\mathbf{v}$ has the same length as **v** but is oppositely directed; therefore,

$$(-1)\mathbf{v} = -\mathbf{v} \tag{3}$$

Parallel and Collinear Vectors Suppose that **v** and **w** are vectors in 2-space or 3-space with a common initial point. If one of the vectors is a scalar multiple of the other, then the vectors lie on a common line, so it is reasonable to say that they are *collinear* (Figure 3.1.8*a*). However, if we translate one of the vectors, as indicated in Figure 3.1.8*b*, then the vectors are *parallel* but no longer collinear. This creates a linguistic problem because translating a vector does not change it. The only way to resolve this problem is to agree that the terms *parallel* and

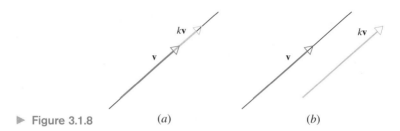

► Figure 3.1.8   (a)   (b)

collinear mean the same thing when applied to vectors. Although the vector **0** has no clearly defined direction, we will regard it to be parallel to all vectors when convenient.

Sums of Three or More Vectors

Vector addition satisfies the ***associative law for addition***, meaning that when we add three vectors, say **u**, **v**, and **w**, it does not matter which two we add first; that is,

$$\mathbf{u} + (\mathbf{v} + \mathbf{w}) = (\mathbf{u} + \mathbf{v}) + \mathbf{w}$$

It follows from this that there is no ambiguity in the expression $\mathbf{u} + \mathbf{v} + \mathbf{w}$ because the same result is obtained no matter how the vectors are grouped.

A simple way to construct $\mathbf{u} + \mathbf{v} + \mathbf{w}$ is to place the vectors "tip to tail" in succession and then draw the vector from the initial point of **u** to the terminal point of **w** (Figure 3.1.9a). The tip-to-tail method also works for four or more vectors (Figure 3.1.9b). The tip-to-tail method also makes it evident that if **u**, **v**, and **w** are vectors in 3-space with a *common initial point*, then $\mathbf{u} + \mathbf{v} + \mathbf{w}$ is the diagonal of the parallelepiped that has the three vectors as adjacent sides (Figure 3.1.9c).

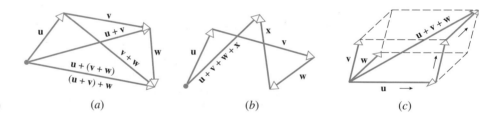

► Figure 3.1.9   (a)   (b)   (c)

Vectors in Coordinate Systems

Up until now we have discussed vectors without reference to a coordinate system. However, as we will soon see, computations with vectors are much simpler to perform if a coordinate system is present to work with.

If a vector **v** in 2-space or 3-space is positioned with its initial point at the origin of a rectangular coordinate system, then the vector is completely determined by the coordinates of its terminal point (Figure 3.1.10). We call these coordinates the ***components*** of **v** relative to the coordinate system. We will write $\mathbf{v} = (v_1, v_2)$ to denote a vector **v** in 2-space with components (v_1, v_2), and $\mathbf{v} = (v_1, v_2, v_3)$ to denote a vector **v** in 3-space with components (v_1, v_2, v_3).

The component forms of the zero vector are $\mathbf{0} = (0, 0)$ in 2-space and $\mathbf{0} = (0, 0, 0)$ in 3-space.

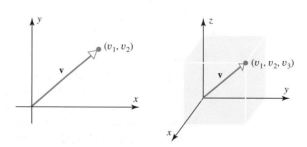

► Figure 3.1.10

It should be evident geometrically that two vectors in 2-space or 3-space are equivalent if and only if they have the same terminal point when their initial points are at the origin. Algebraically, this means that two vectors are equivalent if and only if their corresponding components are equal. Thus, for example, the vectors

$$\mathbf{v} = (v_1, v_2, v_3) \quad \text{and} \quad \mathbf{w} = (w_1, w_2, w_3)$$

in 3-space are equivalent if and only if

$$v_1 = w_1, \quad v_2 = w_2, \quad v_3 = w_3$$

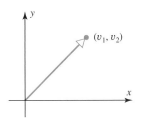

▲ Figure 3.1.11 The ordered pair (v_1, v_2) can represent a point or a vector.

Remark   It may have occurred to you that an ordered pair (v_1, v_2) can represent either a vector with *components* v_1 and v_2 or a point with *components* v_1 and v_2 (and similarly for ordered triples). Both are valid geometric interpretations, so the appropriate choice will depend on the geometric viewpoint that we want to emphasize (Figure 3.1.11).

Vectors Whose Initial Point
Is Not at the Origin

It is sometimes necessary to consider vectors whose initial points are not at the origin. If $\overrightarrow{P_1 P_2}$ denotes the vector with initial point $P_1(x_1, y_1)$ and terminal point $P_2(x_2, y_2)$, then the components of this vector are given by the formula

$$\overrightarrow{P_1 P_2} = (x_2 - x_1, y_2 - y_1) \tag{4}$$

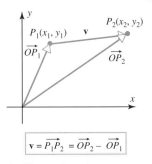

▲ Figure 3.1.12

That is, the components of $\overrightarrow{P_1 P_2}$ are obtained by subtracting the coordinates of the initial point from the coordinates of the terminal point. For example, in Figure 3.1.12 the vector $\overrightarrow{P_1 P_2}$ is the difference of vectors $\overrightarrow{OP_2}$ and $\overrightarrow{OP_1}$, so

$$\overrightarrow{P_1 P_2} = \overrightarrow{OP_2} - \overrightarrow{OP_1} = (x_2, y_2) - (x_1, y_1) = (x_2 - x_1, y_2 - y_1)$$

As you might expect, the components of a vector in 3-space that has initial point $P_1(x_1, y_1, z_1)$ and terminal point $P_2(x_2, y_2, z_2)$ are given by

$$\overrightarrow{P_1 P_2} = (x_2 - x_1, y_2 - y_1, z_2 - z_1) \tag{5}$$

▶ **EXAMPLE 1   Finding the Components of a Vector**

The components of the vector $\mathbf{v} = \overrightarrow{P_1 P_2}$ with initial point $P_1(2, -1, 4)$ and terminal point $P_2(7, 5, -8)$ are

$$\mathbf{v} = (7 - 2, 5 - (-1), (-8) - 4) = (5, 6, -12) \blacktriangleleft$$

n-Space

The idea of using ordered pairs and triples of real numbers to represent points in two-dimensional space and three-dimensional space was well known in the eighteenth and nineteenth centuries. By the dawn of the twentieth century, mathematicians and physicists were exploring the use of "higher-dimensional" spaces in mathematics and physics. Today, even the layman is familiar with the notion of time as a fourth dimension, an idea used by Albert Einstein in developing the general theory of relativity. Today, physicists working in the field of "string theory" commonly use 11-dimensional space in their quest for a unified theory that will explain how the fundamental forces of nature work. Much of the remaining work in this section is concerned with extending the notion of space to n-dimensions.

To explore these ideas further, we start with some terminology and notation. The set of all real numbers can be viewed geometrically as a line. It is called the ***real line*** and is denoted by R or R^1. The superscript reinforces the intuitive idea that a line is one-dimensional. The set of all ordered pairs of real numbers (called **2-*tuples***) and the set of all ordered triples of real numbers (called **3-*tuples***) are denoted by R^2 and

R^3, respectively. The superscript reinforces the idea that the ordered pairs correspond to points in the plane (two-dimensional) and ordered triples to points in space (three-dimensional). The following definition extends this idea.

DEFINITION 1 If n is a positive integer, then an **ordered n-tuple** is a sequence of n real numbers $(v_1, v_2, \ldots, v_n)$. The set of all ordered n-tuples is called **n-space** and is denoted by R^n.

Remark You can think of the numbers in an n-tuple $(v_1, v_2, \ldots, v_n)$ as either the coordinates of a *generalized point* or the components of a *generalized vector*, depending on the geometric image you want to bring to mind—the choice makes no difference mathematically, since it is the algebraic properties of n-tuples that are of concern.

Here are some typical applications that lead to n-tuples.

- **Experimental Data**—A scientist performs an experiment and makes n numerical measurements each time the experiment is performed. The result of each experiment can be regarded as a vector $\mathbf{y} = (y_1, y_2, \ldots, y_n)$ in R^n in which $y_1, y_2, \ldots, y_n$ are the measured values.

- **Storage and Warehousing**—A national trucking company has 15 depots for storing and servicing its trucks. At each point in time the distribution of trucks in the service depots can be described by a 15-tuple $\mathbf{x} = (x_1, x_2, \ldots, x_{15})$ in which x_1 is the number of trucks in the first depot, x_2 is the number in the second depot, and so forth.

- **Electrical Circuits**—A certain kind of processing chip is designed to receive four input voltages and produces three output voltages in response. The input voltages can be regarded as vectors in R^4 and the output voltages as vectors in R^3. Thus, the chip can be viewed as a device that transforms an input vector $\mathbf{v} = (v_1, v_2, v_3, v_4)$ in R^4 into an output vector $\mathbf{w} = (w_1, w_2, w_3)$ in R^3.

- **Graphical Images**—One way in which color images are created on computer screens is by assigning each pixel (an addressable point on the screen) three numbers that describe the *hue*, *saturation*, and *brightness* of the pixel. Thus, a complete color image can be viewed as a set of 5-tuples of the form $\mathbf{v} = (x, y, h, s, b)$ in which x and y are the screen coordinates of a pixel and h, s, and b are its hue, saturation, and brightness.

- **Economics**—One approach to economic analysis is to divide an economy into sectors (manufacturing, services, utilities, and so forth) and measure the output of each sector by a dollar value. Thus, in an economy with 10 sectors the economic output of the entire economy can be represented by a 10-tuple $\mathbf{s} = (s_1, s_2, \ldots, s_{10})$ in which the numbers $s_1, s_2, \ldots, s_{10}$ are the outputs of the individual sectors.

Albert Einstein
(1879–1955)

Historical Note The German-born physicist Albert Einstein immigrated to the United States in 1935, where he settled at Princeton University. Einstein spent the last three decades of his life working unsuccessfully at producing a *unified field theory* that would establish an underlying link between the forces of gravity and electromagnetism. Recently, physicists have made progress on the problem using a framework known as *string theory*. In this theory the smallest, indivisible components of the Universe are not particles but loops that behave like vibrating strings. Whereas Einstein's space-time universe was four-dimensional, strings reside in an 11-dimensional world that is the focus of current research.

[*Image: ©Bettmann/©Corbis*]

• **Mechanical Systems**—Suppose that six particles move along the same coordinate line so that at time t their coordinates are $x_1, x_2, \ldots, x_6$ and their velocities are $v_1, v_2, \ldots, v_6$, respectively. This information can be represented by the vector

$$\mathbf{v} = (x_1, x_2, x_3, x_4, x_5, x_6, v_1, v_2, v_3, v_4, v_5, v_6, t)$$

in R^{13}. This vector is called the *state* of the particle system at time t.

Operations on Vectors in R^n

Our next goal is to define useful operations on vectors in R^n. These operations will all be natural extensions of the familiar operations on vectors in R^2 and R^3. We will denote a vector $\mathbf{v}$ in R^n using the notation

$$\mathbf{v} = (v_1, v_2, \ldots, v_n)$$

and we will call $\mathbf{0} = (0, 0, \ldots, 0)$ the *zero vector*.

We noted earlier that in R^2 and R^3 two vectors are equivalent (equal) if and only if their corresponding components are the same. Thus, we make the following definition.

DEFINITION 2 Vectors $\mathbf{v} = (v_1, v_2, \ldots, v_n)$ and $\mathbf{w} = (w_1, w_2, \ldots, w_n)$ in R^n are said to be *equivalent* (also called *equal*) if

$$v_1 = w_1, \quad v_2 = w_2, \ldots, \quad v_n = w_n$$

We indicate this by writing $\mathbf{v} = \mathbf{w}$.

▶ **EXAMPLE 2** **Equality of Vectors**

$$(a, b, c, d) = (1, -4, 2, 7)$$

if and only if $a = 1, b = -4, c = 2$, and $d = 7$. ◀

Our next objective is to define the operations of addition, subtraction, and scalar multiplication for vectors in R^n. To motivate these ideas, we will consider how these operations can be performed on vectors in R^2 using components. By studying Figure 3.1.13 you should be able to deduce that if $\mathbf{v} = (v_1, v_2)$ and $\mathbf{w} = (w_1, w_2)$, then

$$\mathbf{v} + \mathbf{w} = (v_1 + w_1, v_2 + w_2) \tag{6}$$

$$k\mathbf{v} = (kv_1, kv_2) \tag{7}$$

In particular, it follows from (7) that

$$-\mathbf{v} = (-1)\mathbf{v} = (-v_1, -v_2) \tag{8}$$

and hence that

$$\mathbf{w} - \mathbf{v} = \mathbf{w} + (-\mathbf{v}) = (w_1 - v_1, w_2 - v_2) \tag{9}$$

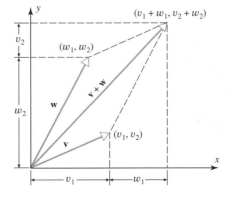
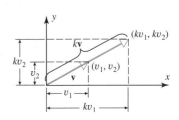

▶ Figure 3.1.13

Motivated by Formulas (6)–(9), we make the following definition.

DEFINITION 3 If $\mathbf{v} = (v_1, v_2, \ldots, v_n)$ and $\mathbf{w} = (w_1, w_2, \ldots, w_n)$ are vectors in R^n, and if k is any scalar, then we define

$$\mathbf{v} + \mathbf{w} = (v_1 + w_1, v_2 + w_2, \ldots, v_n + w_n) \tag{10}$$

$$k\mathbf{v} = (kv_1, kv_2, \ldots, kv_n) \tag{11}$$

$$-\mathbf{v} = (-v_1, -v_2, \ldots, -v_n) \tag{12}$$

$$\mathbf{w} - \mathbf{v} = \mathbf{w} + (-\mathbf{v}) = (w_1 - v_1, w_2 - v_2, \ldots, w_n - v_n) \tag{13}$$

In words, vectors are added (or subtracted) by adding (or subtracting) their corresponding components, and a vector is multiplied by a scalar by multiplying each component by that scalar.

▶ **EXAMPLE 3 Algebraic Operations Using Components**

If $\mathbf{v} = (1, -3, 2)$ and $\mathbf{w} = (4, 2, 1)$, then

$$\mathbf{v} + \mathbf{w} = (5, -1, 3), \qquad 2\mathbf{v} = (2, -6, 4)$$
$$-\mathbf{w} = (-4, -2, -1), \qquad \mathbf{v} - \mathbf{w} = \mathbf{v} + (-\mathbf{w}) = (-3, -5, 1) \quad ◀$$

The following theorem summarizes the most important properties of vector operations.

THEOREM 3.1.1 *If* $\mathbf{u}$, $\mathbf{v}$, *and* $\mathbf{w}$ *are vectors in* R^n, *and if* k *and* m *are scalars, then*:

(a) $\mathbf{u} + \mathbf{v} = \mathbf{v} + \mathbf{u}$

(b) $(\mathbf{u} + \mathbf{v}) + \mathbf{w} = \mathbf{u} + (\mathbf{v} + \mathbf{w})$

(c) $\mathbf{u} + \mathbf{0} = \mathbf{0} + \mathbf{u} = \mathbf{u}$

(d) $\mathbf{u} + (-\mathbf{u}) = \mathbf{0}$

(e) $k(\mathbf{u} + \mathbf{v}) = k\mathbf{u} + k\mathbf{v}$

(f) $(k + m)\mathbf{u} = k\mathbf{u} + m\mathbf{u}$

(g) $k(m\mathbf{u}) = (km)\mathbf{u}$

(h) $1\mathbf{u} = \mathbf{u}$

We will prove part (b) and leave some of the other proofs as exercises.

Proof (b) Let $\mathbf{u} = (u_1, u_2, \ldots, u_n)$, $\mathbf{v} = (v_1, v_2, \ldots, v_n)$, and $\mathbf{w} = (w_1, w_2, \ldots, w_n)$. Then

$$\begin{aligned}
(\mathbf{u} + \mathbf{v}) + \mathbf{w} &= \big((u_1, u_2, \ldots, u_n) + (v_1, v_2, \ldots, v_n)\big) + (w_1, w_2, \ldots, w_n) \\
&= (u_1 + v_1, u_2 + v_2, \ldots, u_n + v_n) + (w_1, w_2, \ldots, w_n) & \text{[Vector addition]} \\
&= \big((u_1 + v_1) + w_1, (u_2 + v_2) + w_2, \ldots, (u_n + v_n) + w_n\big) & \text{[Vector addition]} \\
&= \big(u_1 + (v_1 + w_1), u_2 + (v_2 + w_2), \ldots, u_n + (v_n + w_n)\big) & \text{[Regroup]} \\
&= (u_1, u_2, \ldots, u_n) + (v_1 + w_1, v_2 + w_2, \ldots, v_n + w_n) & \text{[Vector addition]} \\
&= \mathbf{u} + (\mathbf{v} + \mathbf{w}) \quad ◀
\end{aligned}$$

The following additional properties of vectors in R^n can be deduced easily by expressing the vectors in terms of components (verify).

THEOREM 3.1.2 *If* $\mathbf{v}$ *is a vector in* R^n *and* k *is a scalar, then*:

(a) $0\mathbf{v} = \mathbf{0}$

(b) $k\mathbf{0} = \mathbf{0}$

(c) $(-1)\mathbf{v} = -\mathbf{v}$

Calculating Without
Components

One of the powerful consequences of Theorems 3.1.1 and 3.1.2 is that they allow calculations to be performed without expressing the vectors in terms of components. For example, suppose that $\mathbf{x}$, $\mathbf{a}$, and $\mathbf{b}$ are vectors in R^n, and we want to solve the vector equation $\mathbf{x} + \mathbf{a} = \mathbf{b}$ for the vector $\mathbf{x}$ without using components. We could proceed as follows:

$\mathbf{x} + \mathbf{a} = \mathbf{b}$	[Given]
$(\mathbf{x} + \mathbf{a}) + (-\mathbf{a}) = \mathbf{b} + (-\mathbf{a})$	[Add the negative of a to both sides]
$\mathbf{x} + (\mathbf{a} + (-\mathbf{a})) = \mathbf{b} - \mathbf{a}$	[Part (b) of Theorem 3.1.1]
$\mathbf{x} + \mathbf{0} = \mathbf{b} - \mathbf{a}$	[Part (d) of Theorem 3.1.1]
$\mathbf{x} = \mathbf{b} - \mathbf{a}$	[Part (c) of Theorem 3.1.1]

While this method is obviously more cumbersome than computing with components in R^n, it will become important later in the text where we will encounter more general kinds of vectors.

Linear Combinations

Addition, subtraction, and scalar multiplication are frequently used in combination to form new vectors. For example, if $\mathbf{v}_1$, $\mathbf{v}_2$, and $\mathbf{v}_3$ are vectors in R^n, then the vectors

$$\mathbf{u} = 2\mathbf{v}_1 + 3\mathbf{v}_2 + \mathbf{v}_3 \quad \text{and} \quad \mathbf{w} = 7\mathbf{v}_1 - 6\mathbf{v}_2 + 8\mathbf{v}_3$$

are formed in this way. In general, we make the following definition.

DEFINITION 4 If $\mathbf{w}$ is a vector in R^n, then $\mathbf{w}$ is said to be a ***linear combination*** of the vectors $\mathbf{v}_1, \mathbf{v}_2, \ldots, \mathbf{v}_r$ in R^n if it can be expressed in the form

$$\mathbf{w} = k_1\mathbf{v}_1 + k_2\mathbf{v}_2 + \cdots + k_r\mathbf{v}_r \tag{14}$$

where $k_1, k_2, \ldots, k_r$ are scalars. These scalars are called the ***coefficients*** of the linear combination. In the case where $r = 1$, Formula (14) becomes $\mathbf{w} = k_1\mathbf{v}_1$, so that a linear combination of a single vector is just a scalar muliple of that vector.

Note that this definition of a linear combination is consistent with that given in the context of matrices (see Definition 6 in Section 1.3).

Application of Linear Combinations to Color Models

Colors on computer monitors are commonly based on what is called the **RGB** *color model*. Colors in this system are created by adding together percentages of the primary colors red (R), green (G), and blue (B). One way to do this is to identify the primary colors with the vectors

$\mathbf{r} = (1, 0, 0)$ (pure red),
$\mathbf{g} = (0, 1, 0)$ (pure green),
$\mathbf{b} = (0, 0, 1)$ (pure blue)

in R^3 and to create all other colors by forming linear combinations of $\mathbf{r}$, $\mathbf{g}$, and $\mathbf{b}$ using coefficients between 0 and 1, inclusive; these coefficients represent the percentage of each pure color in the mix.

The set of all such color vectors is called **RGB** *space* or the **RGB** *color cube* (Figure 3.1.14). Thus, each color vector $\mathbf{c}$ in this cube is expressible as a linear combination of the form

$$\begin{aligned} \mathbf{c} &= k_1\mathbf{r} + k_2\mathbf{g} + k_3\mathbf{b} \\ &= k_1(1, 0, 0) + k_2(0, 1, 0) + k_3(0, 0, 1) \\ &= (k_1, k_2, k_3) \end{aligned}$$

where $0 \leq k_i \leq 1$. As indicated in the figure, the corners of the cube represent the pure primary colors together with the colors black, white, magenta, cyan, and yellow. The vectors along the diagonal running from black to white correspond to shades of gray.

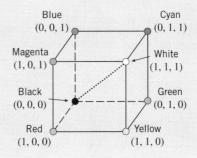

▶ **Figure 3.1.14**

Alternative Notations for Vectors

Up to now we have been writing vectors in R^n using the notation

$$\mathbf{v} = (v_1, v_2, \ldots, v_n) \tag{15}$$

We call this the **comma-delimited** form. However, since a vector in R^n is just a list of its n components in a specific order, any notation that displays those components in the correct order is a valid way of representing the vector. For example, the vector in (15) can be written as

$$\mathbf{v} = [v_1 \quad v_2 \quad \cdots \quad v_n] \tag{16}$$

which is called **row-matrix** form, or as

$$\mathbf{v} = \begin{bmatrix} v_1 \\ v_2 \\ \vdots \\ v_n \end{bmatrix} \tag{17}$$

which is called **column-matrix** form. The choice of notation is often a matter of taste or convenience, but sometimes the nature of a problem will suggest a preferred notation. Notations (15), (16), and (17) will all be used at various places in this text.

Concept Review

- Geometric vector
- Direction
- Length
- Initial point
- Terminal point
- Equivalent vectors
- Zero vector

- Vector addition: parallelogram rule and triangle rule
- Vector subtraction
- Negative of a vector
- Scalar multiplication
- Collinear (i.e., parallel) vectors
- Components of a vector

- Coordinates of a point
- n-tuple
- n-space
- Vector operations in n-space: addition, subtraction, scalar multiplication
- Linear combination of vectors

Skills

- Perform geometric operations on vectors: addition, subtraction, and scalar multiplication.
- Perform algebraic operations on vectors: addition, subtraction, and scalar multiplication.
- Determine whether two vectors are equivalent.
- Determine whether two vectors are collinear.

- Sketch vectors whose initial and terminal points are given.
- Find components of a vector whose initial and terminal points are given.
- Prove basic algebraic properties of vectors (Theorems 3.1.1 and 3.1.2).

Exercise Set 3.1

▶ In Exercises 1–2, draw a coordinate system (as in Figure 3.1.10) and locate the points whose coordinates are given.

1. (a) $(3, 4, 5)$ (b) $(-3, 4, 5)$ (c) $(3, -4, 5)$
   (d) $(3, 4, -5)$ (e) $(-3, -4, 5)$ (f) $(-3, 4, -5)$

2. (a) $(0, 3, -3)$ (b) $(3, -3, 0)$ (c) $(-3, 0, 0)$
   (d) $(3, 0, 3)$ (e) $(0, 0, -3)$ (f) $(0, 3, 0)$

▶ In Exercises 3–4, sketch the following vectors with the initial points located at the origin.

3. (a) $\mathbf{v}_1 = (3, 6)$ (b) $\mathbf{v}_2 = (-4, -8)$
   (c) $\mathbf{v}_3 = (-4, -3)$ (d) $\mathbf{v}_4 = (3, 4, 5)$
   (e) $\mathbf{v}_5 = (3, 3, 0)$ (f) $\mathbf{v}_6 = (-1, 0, 2)$

4. (a) $\mathbf{v}_1 = (5, -4)$ (b) $\mathbf{v}_2 = (3, 0)$
   (c) $\mathbf{v}_3 = (0, -7)$ (d) $\mathbf{v}_4 = (0, 0, -3)$
   (e) $\mathbf{v}_5 = (0, 4, -1)$ (f) $\mathbf{v}_6 = (2, 2, 2)$

▷ In Exercises 5–6, sketch the vector $\overrightarrow{P_1 P_2}$ with the initial point located at the origin. ◁

5. (a) $P_1(4, 8)$, $P_2(3, 7)$

   (b) $P_1(3, -5)$, $P_2(-4, -7)$

   (c) $P_1(3, -7, 2)$, $P_2(-2, 5, -4)$

6. (a) $P_1(-5, 0)$, $P_2(-3, 1)$

   (b) $P_1(0, 0)$, $P_2(3, 4)$

   (c) $P_1(-1, 0, 2)$, $P_2(0, -1, 0)$

   (d) $P_1(2, 2, 2)$, $P_2(0, 0, 0)$

▷ In Exercises 7–8, find the components of the vector $\overrightarrow{P_1 P_2}$. ◁

7. (a) $P_1(3, 5)$, $P_2(2, 8)$

   (b) $P_1(5, -2, 1)$, $P_2(2, 4, 2)$

8. (a) $P_1(-6, 2)$, $P_2(-4, -1)$

   (b) $P_1(0, 0, 0)$, $P_2(-1, 6, 1)$

9. (a) Find the terminal point of the vector that is equivalent to $\mathbf{u} = (1, 2)$ and whose initial point is $A(1, 1)$.

   (b) Find the initial point of the vector that is equivalent to $\mathbf{u} = (1, 1, 3)$ and whose terminal point is $B(-1, -1, 2)$.

10. (a) Find the initial point of the vector that is equivalent to $\mathbf{u} = (1, 2)$ and whose terminal point is $B(2, 0)$.

    (b) Find the terminal point of the vector that is equivalent to $\mathbf{u} = (1, 1, 3)$ and whose initial point is $A(0, 2, 0)$.

11. Find a nonzero vector $\mathbf{u}$ with terminal point $Q(3, 0, -5)$ such that

    (a) $\mathbf{u}$ has the same direction as $\mathbf{v} = (4, -2, -1)$.

    (b) $\mathbf{u}$ is oppositely directed to $\mathbf{v} = (4, -2, -1)$.

12. Find a nonzero vector $\mathbf{u}$ with initial point $P(-1, 3, -5)$ such that

    (a) $\mathbf{u}$ has the same direction as $\mathbf{v} = (6, 7, -3)$.

    (b) $\mathbf{u}$ is oppositely directed to $\mathbf{v} = (6, 7, -3)$.

13. Let $\mathbf{u} = (4, -1)$, $\mathbf{v} = (0, 5)$, and $\mathbf{w} = (-3, -3)$. Find the components of

    (a) $\mathbf{u} + \mathbf{w}$ (b) $\mathbf{v} - 3\mathbf{u}$

    (c) $2(\mathbf{u} - 5\mathbf{w})$ (d) $3\mathbf{v} - 2(\mathbf{u} + 2\mathbf{w})$

    (e) $-3(\mathbf{w} - 2\mathbf{u} + \mathbf{v})$ (f) $(-2\mathbf{u} - \mathbf{v}) - 5(\mathbf{v} + 3\mathbf{w})$

14. Let $\mathbf{u} = (-3, 1, 2)$, $\mathbf{v} = (4, 0, -8)$, and $\mathbf{w} = (6, -1, -4)$. Find the components of

    (a) $\mathbf{v} - \mathbf{w}$ (b) $6\mathbf{u} + 2\mathbf{v}$

    (c) $-\mathbf{v} + \mathbf{u}$ (d) $5(\mathbf{v} - 4\mathbf{u})$

    (e) $-3(\mathbf{v} - 8\mathbf{w})$ (f) $(2\mathbf{u} - 7\mathbf{w}) - (8\mathbf{v} + \mathbf{u})$

15. Let $\mathbf{u} = (-3, 2, 1, 0)$, $\mathbf{v} = (4, 7, -3, 2)$, and $\mathbf{w} = (5, -2, 8, 1)$. Find the components of

    (a) $\mathbf{v} - \mathbf{w}$ (b) $2\mathbf{u} + 7\mathbf{v}$

    (c) $-\mathbf{u} + (\mathbf{v} - 4\mathbf{w})$ (d) $6(\mathbf{u} - 3\mathbf{v})$

    (e) $-\mathbf{v} - \mathbf{w}$ (f) $(6\mathbf{v} - \mathbf{w}) - (4\mathbf{u} + \mathbf{v})$

16. Let $\mathbf{u}$, $\mathbf{v}$, and $\mathbf{w}$ be the vectors in Exercise 15. Find the vector $\mathbf{x}$ that satisfies $5\mathbf{x} - 2\mathbf{v} = 2(\mathbf{w} - 5\mathbf{x})$.

17. Let $\mathbf{u} = (5, -1, 0, 3, -3)$, $\mathbf{v} = (-1, -1, 7, 2, 0)$, and $\mathbf{w} = (-4, 2, -3, -5, 2)$. Find the components of

    (a) $\mathbf{w} - \mathbf{u}$ (b) $2\mathbf{v} + 3\mathbf{u}$

    (c) $-\mathbf{w} + 3(\mathbf{v} - \mathbf{u})$ (d) $5(-\mathbf{v} + 4\mathbf{u} - \mathbf{w})$

    (e) $-2(3\mathbf{w} + \mathbf{v}) + (2\mathbf{u} + \mathbf{w})$ (f) $\frac{1}{2}(\mathbf{w} - 5\mathbf{v} + 2\mathbf{u}) + \mathbf{v}$

18. Let $\mathbf{u} = (1, 2, -3, 5, 0)$, $\mathbf{v} = (0, 4, -1, 1, 2)$, and $\mathbf{w} = (7, 1, -4, -2, 3)$. Find the components of

    (a) $\mathbf{v} + \mathbf{w}$ (b) $3(2\mathbf{u} - \mathbf{v})$

    (c) $(3\mathbf{u} - \mathbf{v}) - (2\mathbf{u} + 4\mathbf{w})$

19. Let $\mathbf{u} = (-3, 1, 2, 4, 4)$, $\mathbf{v} = (4, 0, -8, 1, 2)$, and $\mathbf{w} = (6, -1, -4, 3, -5)$. Find the components of

    (a) $\mathbf{v} - \mathbf{w}$ (b) $6\mathbf{u} + 2\mathbf{v}$

    (c) $(2\mathbf{u} - 7\mathbf{w}) - (8\mathbf{v} + \mathbf{u})$

20. Let $\mathbf{u}$, $\mathbf{v}$, and $\mathbf{w}$ be the vectors in Exercise 18. Find the components of the vector $\mathbf{x}$ that satisfies the equation $3\mathbf{u} + \mathbf{v} - 2\mathbf{w} = 3\mathbf{x} + 2\mathbf{w}$.

21. Let $\mathbf{u}$, $\mathbf{v}$, and $\mathbf{w}$ be the vectors in Exercise 19. Find the components of the vector $\mathbf{x}$ that satisfies the equation $2\mathbf{u} - \mathbf{v} + \mathbf{x} = 7\mathbf{x} + \mathbf{w}$.

22. For what value(s) of t, if any, is the given vector parallel to $\mathbf{u} = (4, -1)$?

    (a) $(8t, -2)$ (b) $(8t, 2t)$ (c) $(1, t^2)$

23. Which of the following vectors in R^6 are parallel to $\mathbf{u} = (-2, 1, 0, 3, 5, 1)$?

    (a) $(4, 2, 0, 6, 10, 2)$

    (b) $(4, -2, 0, -6, -10, -2)$

    (c) $(0, 0, 0, 0, 0, 0)$

24. Let $\mathbf{u} = (2, 1, 0, 1, -1)$ and $\mathbf{v} = (-2, 3, 1, 0, 2)$. Find scalars a and b so that $a\mathbf{u} + b\mathbf{v} = (-8, 8, 3, -1, 7)$.

25. Let $\mathbf{u} = (1, -1, 3, 5)$ and $\mathbf{v} = (2, 1, 0, -3)$. Find scalars a and b so that $a\mathbf{u} + b\mathbf{v} = (1, -4, 9, 18)$.

26. Find all scalars c_1, c_2, and c_3 such that

    $c_1(1, 2, 0) + c_2(2, 1, 1) + c_3(0, 3, 1) = (0, 0, 0)$

27. Find all scalars c_1, c_2, and c_3 such that

    $c_1(1, -1, 0) + c_2(3, 2, 1) + c_3(0, 1, 4) = (-1, 1, 19)$

28. Find all scalars c_1, c_2, and c_3 such that

    $c_1(-1, 0, 2) + c_2(2, 2, -2) + c_3(1, -2, 1) = (-6, 12, 4)$

29. Let $\mathbf{u}_1 = (-1, 3, 2, 0)$, $\mathbf{u}_2 = (2, 0, 4, -1)$, $\mathbf{u}_3 = (7, 1, 1, 4)$, and $\mathbf{u}_4 = (6, 3, 1, 2)$. Find scalars c_1, c_2, c_3, and c_4 such that $c_1\mathbf{u}_1 + c_2\mathbf{u}_2 + c_3\mathbf{u}_3 + c_4\mathbf{u}_4 = (0, 5, 6, -3)$.

30. Show that there do not exist scalars c_1, c_2, and c_3 such that $c_1(1, 0, 1, 0) + c_2(1, 0, -2, 1) + c_3(2, 0, 1, 2) = (1, -2, 2, 3)$

31. Show that there do not exist scalars c_1, c_2, and c_3 such that $c_1(-2, 9, 6) + c_2(-3, 2, 1) + c_3(1, 7, 5) = (0, 5, 4)$

32. Consider Figure 3.1.12. Discuss a geometric interpretation of the vector
$$\mathbf{u} = \overrightarrow{OP_1} + \tfrac{1}{2}(\overrightarrow{OP_2} - \overrightarrow{OP_1})$$

33. Let P be the point $(2, 3, -2)$ and Q the point $(7, -4, 1)$.
    (a) Find the midpoint of the line segment connecting P and Q.
    (b) Find the point on the line segment connecting P and Q that is $\tfrac{3}{4}$ of the way from P to Q.

34. Let P be the point $(1, 3, 7)$. If the point $(4, 0, -6)$ is the midpoint of the line segment connecting P and Q, what is Q?

35. Prove parts (a), (c), and (d) of Theorem 3.1.1.

36. Prove parts (e)–(h) of Theorem 3.1.1.

37. Prove parts (a)–(c) of Theorem 3.1.2.

True-False Exercises

In parts (a)–(k) determine whether the statement is true or false, and justify your answer.

(a) Two equivalent vectors must have the same initial point.

(b) The vectors (a, b) and $(a, b, 0)$ are equivalent.

(c) If k is a scalar and $\mathbf{v}$ is a vector, then $\mathbf{v}$ and $k\mathbf{v}$ are parallel if and only if $k \geq 0$.

(d) The vectors $\mathbf{v} + (\mathbf{u} + \mathbf{w})$ and $(\mathbf{w} + \mathbf{v}) + \mathbf{u}$ are the same.

(e) If $\mathbf{u} + \mathbf{v} = \mathbf{u} + \mathbf{w}$, then $\mathbf{v} = \mathbf{w}$.

(f) If a and b are scalars such that $a\mathbf{u} + b\mathbf{v} = \mathbf{0}$, then $\mathbf{u}$ and $\mathbf{v}$ are parallel vectors.

(g) Collinear vectors with the same length are equal.

(h) If $(a, b, c) + (x, y, z) = (x, y, z)$, then (a, b, c) must be the zero vector.

(i) If k and m are scalars and $\mathbf{u}$ and $\mathbf{v}$ are vectors, then
$$(k + m)(\mathbf{u} + \mathbf{v}) = k\mathbf{u} + m\mathbf{v}$$

(j) If the vectors $\mathbf{v}$ and $\mathbf{w}$ are given, then the vector equation
$$3(2\mathbf{v} - \mathbf{x}) = 5\mathbf{x} - 4\mathbf{w} + \mathbf{v}$$
can be solved for $\mathbf{x}$.

(k) The linear combinations $a_1\mathbf{v}_1 + a_2\mathbf{v}_2$ and $b_1\mathbf{v}_1 + b_2\mathbf{v}_2$ can only be equal if $a_1 = b_1$ and $a_2 = b_2$.

3.2 Norm, Dot Product, and Distance in R^n

In this section we will be concerned with the notions of length and distance as they relate to vectors. We will first discuss these ideas in R^2 and R^3 and then extend them algebraically to R^n.

Norm of a Vector   In this text we will denote the length of a vector $\mathbf{v}$ by the symbol $\|\mathbf{v}\|$, which is read as the *norm* of $\mathbf{v}$, the *length* of $\mathbf{v}$, or the *magnitude* of $\mathbf{v}$ (the term "norm" being a common mathematical synonym for length). As suggested in Figure 3.2.1a, it follows from the Theorem of Pythagoras that the norm of a vector (v_1, v_2) in R^2 is
$$\|\mathbf{v}\| = \sqrt{v_1^2 + v_2^2} \tag{1}$$

Similarly, for a vector (v_1, v_2, v_3) in R^3, it follows from Figure 3.2.1b and two applications of the Theorem of Pythagoras that
$$\|\mathbf{v}\|^2 = (OR)^2 + (RP)^2 = (OQ)^2 + (QR)^2 + (RP)^2 = v_1^2 + v_2^2 + v_3^2$$

and hence that
$$\|\mathbf{v}\| = \sqrt{v_1^2 + v_2^2 + v_3^2} \tag{2}$$

Motivated by the pattern of Formulas (1) and (2) we make the following definition.

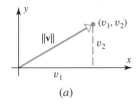

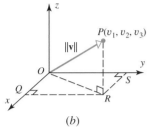

▲ Figure 3.2.1

> **DEFINITION 1** If $\mathbf{v} = (v_1, v_2, \ldots, v_n)$ is a vector in R^n, then the **norm** of $\mathbf{v}$ (also called the **length** of $\mathbf{v}$ or the **magnitude** of $\mathbf{v}$) is denoted by $\|\mathbf{v}\|$, and is defined by the formula
>
> $$\|\mathbf{v}\| = \sqrt{v_1^2 + v_2^2 + v_3^2 + \cdots + v_n^2} \tag{3}$$

▶ **EXAMPLE 1 Calculating Norms**

It follows from Formula (2) that the norm of the vector $\mathbf{v} = (-3, 2, 1)$ in R^3 is

$$\|\mathbf{v}\| = \sqrt{(-3)^2 + 2^2 + 1^2} = \sqrt{14}$$

and it follows from Formula (3) that the norm of the vector $\mathbf{v} = (2, -1, 3, -5)$ in R^4 is

$$\|\mathbf{v}\| = \sqrt{2^2 + (-1)^2 + 3^2 + (-5)^2} = \sqrt{39} \blacktriangleleft$$

Our first theorem in this section will generalize to R^n the following three familiar facts about vectors in R^2 and R^3:

- Distances are nonnegative.
- The zero vector is the only vector of length zero.
- Multiplying a vector by a scalar multiplies its length by the absolute value of that scalar.

It is important to recognize that just because these results hold in R^2 and R^3 does not guarantee that they hold in R^n—their validity in R^n must be *proved* using algebraic properties of n-tuples.

> **THEOREM 3.2.1** *If $\mathbf{v}$ is a vector in R^n, and if k is any scalar, then*:
> (a) $\|\mathbf{v}\| \geq 0$
> (b) $\|\mathbf{v}\| = 0$ *if and only if* $\mathbf{v} = \mathbf{0}$
> (c) $\|k\mathbf{v}\| = |k| \|\mathbf{v}\|$

We will prove part (c) and leave (a) and (b) as exercises.

Proof (c)   If $\mathbf{v} = (v_1, v_2, \ldots, v_n)$, then $k\mathbf{v} = (kv_1, kv_2, \ldots, kv_n)$, so

$$\|k\mathbf{v}\| = \sqrt{(kv_1)^2 + (kv_2)^2 + \cdots + (kv_n)^2}$$

$$= \sqrt{(k^2)(v_1^2 + v_2^2 + \cdots + v_n^2)}$$

$$= |k|\sqrt{v_1^2 + v_2^2 + \cdots + v_n^2}$$

$$= |k| \|\mathbf{v}\| \quad \triangleleft$$

Unit Vectors   A vector of norm 1 is called a **unit vector**. Such vectors are useful for specifying a direction when length is not relevant to the problem at hand. You can obtain a unit vector in a desired direction by choosing any *nonzero* vector $\mathbf{v}$ in that direction and multiplying $\mathbf{v}$ by the reciprocal of its length. For example, if $\mathbf{v}$ is a vector of length 2 in R^2 or R^3, then $\frac{1}{2}\mathbf{v}$ is a unit vector in the same direction as $\mathbf{v}$. More generally, if $\mathbf{v}$ is any nonzero vector in R^n, then

$$\mathbf{u} = \frac{1}{\|\mathbf{v}\|}\mathbf{v} \tag{4}$$

WARNING Sometimes you will see Formula (4) expressed as

$$\mathbf{u} = \frac{\mathbf{v}}{\|\mathbf{v}\|}$$

This is just a more compact way of writing that formula and is *not* intended to convey that **v** is being divided by $\|\mathbf{v}\|$.

defines a unit vector that is in the same direction as **v**. We can confirm that (4) is a unit vector by applying part (*c*) of Theorem 3.2.1 with $k = 1/\|\mathbf{v}\|$ to obtain

$$\|\mathbf{u}\| = \|k\mathbf{v}\| = |k|\|\mathbf{v}\| = k\|\mathbf{v}\| = \frac{1}{\|\mathbf{v}\|}\|\mathbf{v}\| = 1$$

The process of multiplying a nonzero vector by the reciprocal of its length to obtain a unit vector is called **normalizing v**.

▶ **EXAMPLE 2 Normalizing a Vector**

Find the unit vector **u** that has the same direction as $\mathbf{v} = (2, 2, -1)$.

Solution The vector **v** has length

$$\|\mathbf{v}\| = \sqrt{2^2 + 2^2 + (-1)^2} = 3$$

Thus, from (4)

$$\mathbf{u} = \tfrac{1}{3}(2, 2, -1) = \left(\tfrac{2}{3}, \tfrac{2}{3}, -\tfrac{1}{3}\right)$$

As a check, you may want to confirm that $\|\mathbf{u}\| = 1$. ◀

The Standard Unit Vectors

When a rectangular coordinate system is introduced in R^2 or R^3, the unit vectors in the positive directions of the coordinate axes are called the **standard unit vectors**. In R^2 these vectors are denoted by

$$\mathbf{i} = (1, 0) \quad \text{and} \quad \mathbf{j} = (0, 1)$$

and in R^3 by

$$\mathbf{i} = (1, 0, 0), \quad \mathbf{j} = (0, 1, 0), \quad \text{and} \quad \mathbf{k} = (0, 0, 1)$$

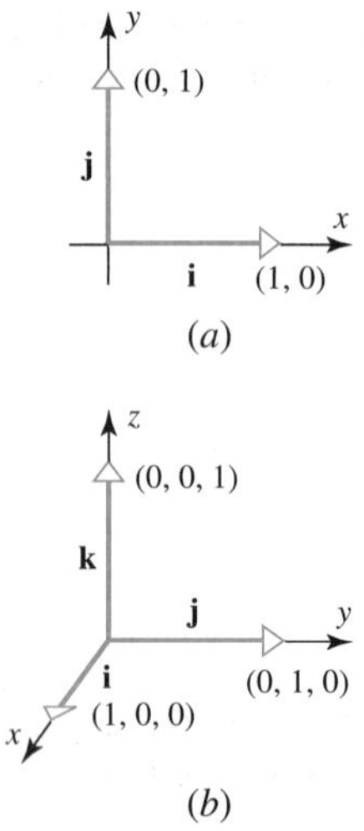

(Figure 3.2.2). Every vector $\mathbf{v} = (v_1, v_2)$ in R^2 and every vector $\mathbf{v} = (v_1, v_2, v_3)$ in R^3 can be expressed as a linear combination of standard unit vectors by writing

$$\mathbf{v} = (v_1, v_2) = v_1(1, 0) + v_2(0, 1) = v_1\mathbf{i} + v_2\mathbf{j} \tag{5}$$

$$\mathbf{v} = (v_1, v_2, v_3) = v_1(1, 0, 0) + v_2(0, 1, 0) + v_3(0, 0, 1) = v_1\mathbf{i} + v_2\mathbf{j} + v_3\mathbf{k} \tag{6}$$

Moreover, we can generalize these formulas to R^n by defining the **standard unit vectors in R^n** to be

$$\mathbf{e}_1 = (1, 0, 0, \ldots, 0), \quad \mathbf{e}_2 = (0, 1, 0, \ldots, 0), \ldots, \quad \mathbf{e}_n = (0, 0, 0, \ldots, 1) \tag{7}$$

in which case every vector $\mathbf{v} = (v_1, v_2, \ldots, v_n)$ in R^n can be expressed as

$$\mathbf{v} = (v_1, v_2, \ldots, v_n) = v_1\mathbf{e}_1 + v_2\mathbf{e}_2 + \cdots + v_n\mathbf{e}_n \tag{8}$$

(a)

(b)

▲ Figure 3.2.2

▶ **EXAMPLE 3 Linear Combinations of Standard Unit Vectors**

$$(2, -3, 4) = 2\mathbf{i} - 3\mathbf{j} + 4\mathbf{k}$$
$$(7, 3, -4, 5) = 7\mathbf{e}_1 + 3\mathbf{e}_2 - 4\mathbf{e}_3 + 5\mathbf{e}_4 ◀$$

Distance in R^n

If P_1 and P_2 are points in R^2 or R^3, then the length of the vector $\overrightarrow{P_1P_2}$ is equal to the distance d between the two points (Figure 3.2.3). Specifically, if $P_1(x_1, y_1)$ and $P_2(x_2, y_2)$ are points in R^2, then Formula (4) of Section 3.1 implies that

$$d = \|\overrightarrow{P_1P_2}\| = \sqrt{(x_2 - x_1)^2 + (y_2 - y_1)^2} \tag{9}$$

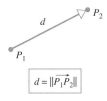

We noted in the previous section that n-tuples can be viewed either as vectors or points in R^n. In Definition 2 we chose to describe them as points, as that seemed the more natural interpretation.

This is the familiar distance formula from analytic geometry. Similarly, the distance between the points $P_1(x_1, y_1, z_1)$ and $P_2(x_2, y_2, z_2)$ in 3-space is

$$d(\mathbf{u}, \mathbf{v}) = \|\overrightarrow{P_1 P_2}\| = \sqrt{(x_2 - x_1)^2 + (y_2 - y_1)^2 + (z_2 - z_1)^2} \qquad (10)$$

Motivated by Formulas (9) and (10), we make the following definition.

DEFINITION 2 If $\mathbf{u} = (u_1, u_2, \ldots, u_n)$ and $\mathbf{v} = (v_1, v_2, \ldots, v_n)$ are points in R^n, then we denote the **distance** between $\mathbf{u}$ and $\mathbf{v}$ by $d(\mathbf{u}, \mathbf{v})$ and define it to be

$$d(\mathbf{u}, \mathbf{v}) = \|\mathbf{u} - \mathbf{v}\| = \sqrt{(u_1 - v_1)^2 + (u_2 - v_2)^2 + \cdots + (u_n - v_n)^2} \qquad (11)$$

▶ **EXAMPLE 4 Calculating Distance in R^n**

If

$$\mathbf{u} = (1, 3, -2, 7) \quad \text{and} \quad \mathbf{v} = (0, 7, 2, 2)$$

then the distance between $\mathbf{u}$ and $\mathbf{v}$ is

$$d(\mathbf{u}, \mathbf{v}) = \sqrt{(1 - 0)^2 + (3 - 7)^2 + (-2 - 2)^2 + (7 - 2)^2} = \sqrt{58} \quad ◀$$

Dot Product   Our next objective is to define a useful multiplication operation on vectors in R^2 and R^3 and then extend that operation to R^n. To do this we will first need to define exactly what we mean by the "angle" between two vectors in R^2 or R^3. For this purpose, let $\mathbf{u}$ and $\mathbf{v}$ be nonzero vectors in R^2 or R^3 that have been positioned so that their initial points coincide. We define the **angle between $\mathbf{u}$ and $\mathbf{v}$** to be the angle θ determined by $\mathbf{u}$ and $\mathbf{v}$ that satisfies the inequalities $0 \leq \theta \leq \pi$ (Figure 3.2.4).

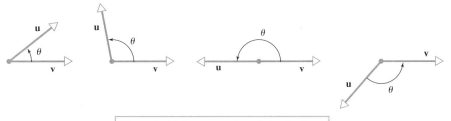

The angle θ between $\mathbf{u}$ and $\mathbf{v}$ satisfies $0 \leq \theta \leq \pi$.

▶ Figure 3.2.4

DEFINITION 3 If $\mathbf{u}$ and $\mathbf{v}$ are nonzero vectors in R^2 or R^3, and if θ is the angle between $\mathbf{u}$ and $\mathbf{v}$, then the **dot product** (also called the **Euclidean inner product**) of $\mathbf{u}$ and $\mathbf{v}$ is denoted by $\mathbf{u} \cdot \mathbf{v}$ and is defined as

$$\mathbf{u} \cdot \mathbf{v} = \|\mathbf{u}\| \|\mathbf{v}\| \cos \theta \qquad (12)$$

If $\mathbf{u} = \mathbf{0}$ or $\mathbf{v} = \mathbf{0}$, then we define $\mathbf{u} \cdot \mathbf{v}$ to be 0.

The sign of the dot product reveals information about the angle θ that we can obtain by rewriting Formula (12) as

$$\cos \theta = \frac{\mathbf{u} \cdot \mathbf{v}}{\|\mathbf{u}\| \|\mathbf{v}\|} \qquad (13)$$

Since $0 \leq \theta \leq \pi$, it follows from Formula (13) and properties of the cosine function studied in trigonometry that

- θ is acute if $\mathbf{u} \cdot \mathbf{v} > 0$.
- θ is obtuse if $\mathbf{u} \cdot \mathbf{v} < 0$.
- $\theta = \pi/2$ if $\mathbf{u} \cdot \mathbf{v} = 0$.

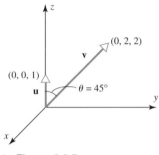

▲ Figure 3.2.5

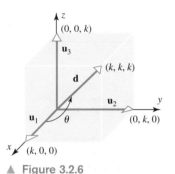

▲ Figure 3.2.6

Note that the angle θ obtained in Example 6 does not involve k. Why was this to be expected?

Component Form of the Dot Product

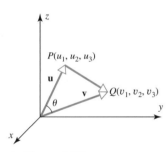

▲ Figure 3.2.7

▶ **EXAMPLE 5 Dot Product**

Find the dot product of the vectors shown in Figure 3.2.5.

Solution The lengths of the vectors are

$$\|\mathbf{u}\| = 1 \quad \text{and} \quad \|\mathbf{v}\| = \sqrt{8} = 2\sqrt{2}$$

and the cosine of the angle θ between them is

$$\cos(45°) = 1/\sqrt{2}$$

Thus, it follows from Formula (12) that

$$\mathbf{u} \cdot \mathbf{v} = \|\mathbf{u}\|\|\mathbf{v}\| \cos\theta = (1)(2\sqrt{2})(1/\sqrt{2}) = 2$$

▶ **EXAMPLE 6 A Geometry Problem Solved Using Dot Product**

Find the angle between a diagonal of a cube and one of its edges.

Solution Let k be the length of an edge and introduce a coordinate system as shown in Figure 3.2.6. If we let $\mathbf{u}_1 = (k, 0, 0)$, $\mathbf{u}_2 = (0, k, 0)$, and $\mathbf{u}_3 = (0, 0, k)$, then the vector

$$\mathbf{d} = (k, k, k) = \mathbf{u}_1 + \mathbf{u}_2 + \mathbf{u}_3$$

is a diagonal of the cube. It follows from Formula (13) that the angle θ between $\mathbf{d}$ and the edge $\mathbf{u}_1$ satisfies

$$\cos\theta = \frac{\mathbf{u}_1 \cdot \mathbf{d}}{\|\mathbf{u}_1\|\|\mathbf{d}\|} = \frac{k^2}{(k)(\sqrt{3k^2})} = \frac{1}{\sqrt{3}}$$

With the help of a calculator we obtain

$$\theta = \cos^{-1}\left(\frac{1}{\sqrt{3}}\right) \approx 54.74° \quad ◀$$

For computational purposes it is desirable to have a formula that expresses the dot product of two vectors in terms of components. We will derive such a formula for vectors in 3-space; the derivation for vectors in 2-space is similar.

Let $\mathbf{u} = (u_1, u_2, u_3)$ and $\mathbf{v} = (v_1, v_2, v_3)$ be two nonzero vectors. If, as shown in Figure 3.2.7, θ is the angle between $\mathbf{u}$ and $\mathbf{v}$, then the law of cosines yields

$$\|\overrightarrow{PQ}\|^2 = \|\mathbf{u}\|^2 + \|\mathbf{v}\|^2 - 2\|\mathbf{u}\|\|\mathbf{v}\| \cos\theta \tag{14}$$

Historical Note The dot product notation was first introduced by the American physicist and mathematician J. Willard Gibbs in a pamphlet distributed to his students at Yale University in the 1880s. The product was originally written on the baseline, rather than centered as today, and was referred to as the *direct product*. Gibbs's pamphlet was eventually incorporated into a book entitled *Vector Analysis* that was published in 1901 and coauthored with one of his students. Gibbs made major contributions to the fields of thermodynamics and electromagnetic theory and is generally regarded as the greatest American physicist of the nineteenth century.

[*Image: The Granger Collection, New York*]

Josiah Willard Gibbs
(1839–1903)

Since $\overrightarrow{PQ} = \mathbf{v} - \mathbf{u}$, we can rewrite (14) as

$$\|\mathbf{u}\|\,\|\mathbf{v}\| \cos\theta = \tfrac{1}{2}(\|\mathbf{u}\|^2 + \|\mathbf{v}\|^2 - \|\mathbf{v} - \mathbf{u}\|^2)$$

or

$$\mathbf{u} \cdot \mathbf{v} = \tfrac{1}{2}(\|\mathbf{u}\|^2 + \|\mathbf{v}\|^2 - \|\mathbf{v} - \mathbf{u}\|^2)$$

Substituting

$$\|\mathbf{u}\|^2 = u_1^2 + u_2^2 + u_3^2, \qquad \|\mathbf{v}\|^2 = v_1^2 + v_2^2 + v_3^2$$

and

$$\|\mathbf{v} - \mathbf{u}\|^2 = (v_1 - u_1)^2 + (v_2 - u_2)^2 + (v_3 - u_3)^2$$

we obtain, after simplifying,

$$\mathbf{u} \cdot \mathbf{v} = u_1 v_1 + u_2 v_2 + u_3 v_3 \tag{15}$$

The companion formula for vectors in 2-space is

$$\mathbf{u} \cdot \mathbf{v} = u_1 v_1 + u_2 v_2 \tag{16}$$

Motivated by the pattern in Formulas (15) and (16), we make the following definition.

> Although we derived Formula (15) and its 2-space companion under the assumption that $\mathbf{u}$ and $\mathbf{v}$ are nonzero, it turned out that these formulas are also applicable if $\mathbf{u} = \mathbf{0}$ or $\mathbf{v} = \mathbf{0}$ (verify).

> In words, *to calculate the dot product (Euclidean inner product) multiply corresponding components and add the resulting products.*

DEFINITION 4 If $\mathbf{u} = (u_1, u_2, \ldots, u_n)$ and $\mathbf{v} = (v_1, v_2, \ldots, v_n)$ are vectors in R^n, then the **dot product** (also called the **Euclidean inner product**) of $\mathbf{u}$ and $\mathbf{v}$ is denoted by $\mathbf{u} \cdot \mathbf{v}$ and is defined by

$$\mathbf{u} \cdot \mathbf{v} = u_1 v_1 + u_2 v_2 + \cdots + u_n v_n \tag{17}$$

▶ **EXAMPLE 7 Calculating Dot Products Using Components**

(a) Use Formula (15) to compute the dot product of the vectors $\mathbf{u}$ and $\mathbf{v}$ in Example 5.

(b) Calculate $\mathbf{u} \cdot \mathbf{v}$ for the following vectors in R^4:

$$\mathbf{u} = (-1, 3, 5, 7), \quad \mathbf{v} = (-3, -4, 1, 0)$$

Solution (*a*) The component forms of the vectors are $\mathbf{u} = (0, 0, 1)$ and $\mathbf{v} = (0, 2, 2)$. Thus,

$$\mathbf{u} \cdot \mathbf{v} = (0)(0) + (0)(2) + (1)(2) = 2$$

which agrees with the result obtained geometrically in Example 5.

Solution (*b*)

$$\mathbf{u} \cdot \mathbf{v} = (-1)(-3) + (3)(-4) + (5)(1) + (7)(0) = -4 \quad ◀$$

Algebraic Properties of the Dot Product

In the special case where $\mathbf{u} = \mathbf{v}$ in Definition 4, we obtain the relationship

$$\mathbf{v} \cdot \mathbf{v} = v_1^2 + v_2^2 + \cdots + v_n^2 = \|\mathbf{v}\|^2 \tag{18}$$

This yields the following formula for expressing the length of a vector in terms of a dot product:

$$\|\mathbf{v}\| = \sqrt{\mathbf{v} \cdot \mathbf{v}} \tag{19}$$

Dot products have many of the same algebraic properties as products of real numbers.

THEOREM 3.2.2 *If* $\mathbf{u}$, $\mathbf{v}$, *and* $\mathbf{w}$ *are vectors in* R^n, *and if* k *is a scalar, then:*

(*a*) $\mathbf{u} \cdot \mathbf{v} = \mathbf{v} \cdot \mathbf{u}$ **[Symmetry property]**

(*b*) $\mathbf{u} \cdot (\mathbf{v} + \mathbf{w}) = \mathbf{u} \cdot \mathbf{v} + \mathbf{u} \cdot \mathbf{w}$ **[Distributive property]**

(*c*) $k(\mathbf{u} \cdot \mathbf{v}) = (k\mathbf{u}) \cdot \mathbf{v}$ **[Homogeneity property]**

(*d*) $\mathbf{v} \cdot \mathbf{v} \geq 0$ *and* $\mathbf{v} \cdot \mathbf{v} = 0$ *if and only if* $\mathbf{v} = 0$ **[Positivity property]**

We will prove parts (*c*) and (*d*) and leave the other proofs as exercises.

Proof (*c*)   Let $\mathbf{u} = (u_1, u_2, \ldots, u_n)$ and $\mathbf{v} = (v_1, v_2, \ldots, v_n)$. Then

$$k(\mathbf{u} \cdot \mathbf{v}) = k(u_1 v_1 + u_2 v_2 + \cdots + u_n v_n)$$
$$= (ku_1)v_1 + (ku_2)v_2 + \cdots + (ku_n)v_n = (k\mathbf{u}) \cdot \mathbf{v}$$

Proof (*d*)   The result follows from parts (*a*) and (*b*) of Theorem 3.2.1 and the fact that

$$\mathbf{v} \cdot \mathbf{v} = v_1 v_1 + v_2 v_2 + \cdots + v_n v_n = v_1^2 + v_2^2 + \cdots + v_n^2 = \|\mathbf{v}\|^2 \quad \blacktriangleleft$$

The next theorem gives additional properties of dot products. The proofs can be obtained either by expressing the vectors in terms of components or by using the algebraic properties established in Theorem 3.2.2.

THEOREM 3.2.3 *If* $\mathbf{u}$, $\mathbf{v}$, *and* $\mathbf{w}$ *are vectors in* R^n, *and if* k *is a scalar, then:*

(*a*) $\mathbf{0} \cdot \mathbf{v} = \mathbf{v} \cdot \mathbf{0} = 0$

(*b*) $(\mathbf{u} + \mathbf{v}) \cdot \mathbf{w} = \mathbf{u} \cdot \mathbf{w} + \mathbf{v} \cdot \mathbf{w}$

(*c*) $\mathbf{u} \cdot (\mathbf{v} - \mathbf{w}) = \mathbf{u} \cdot \mathbf{v} - \mathbf{u} \cdot \mathbf{w}$

(*d*) $(\mathbf{u} - \mathbf{v}) \cdot \mathbf{w} = \mathbf{u} \cdot \mathbf{w} - \mathbf{v} \cdot \mathbf{w}$

(*e*) $k(\mathbf{u} \cdot \mathbf{v}) = \mathbf{u} \cdot (k\mathbf{v})$

We will show how Theorem 3.2.2 can be used to prove part (*b*) without breaking the vectors into components. The other proofs are left as exercises.

Proof (*b*)

$$(\mathbf{u} + \mathbf{v}) \cdot \mathbf{w} = \mathbf{w} \cdot (\mathbf{u} + \mathbf{v}) \quad \text{[By symmetry]}$$
$$= \mathbf{w} \cdot \mathbf{u} + \mathbf{w} \cdot \mathbf{v} \quad \text{[By distributivity]}$$
$$= \mathbf{u} \cdot \mathbf{w} + \mathbf{v} \cdot \mathbf{w} \quad \text{[By symmetry]} \quad \blacktriangleleft$$

Formulas (18) and (19) together with Theorems 3.2.2 and 3.2.3 make it possible to manipulate expressions involving dot products using familiar algebraic techniques.

▶ **EXAMPLE 8 Calculating with Dot Products**

$$(\mathbf{u} - 2\mathbf{v}) \cdot (3\mathbf{u} + 4\mathbf{v}) = \mathbf{u} \cdot (3\mathbf{u} + 4\mathbf{v}) - 2\mathbf{v} \cdot (3\mathbf{u} + 4\mathbf{v})$$
$$= 3(\mathbf{u} \cdot \mathbf{u}) + 4(\mathbf{u} \cdot \mathbf{v}) - 6(\mathbf{v} \cdot \mathbf{u}) - 8(\mathbf{v} \cdot \mathbf{v})$$
$$= 3\|\mathbf{u}\|^2 - 2(\mathbf{u} \cdot \mathbf{v}) - 8\|\mathbf{v}\|^2 \quad \blacktriangleleft$$

Cauchy–Schwarz Inequality and Angles in R^n

Our next objective is to extend to R^n the notion of "angle" between nonzero vectors $\mathbf{u}$ and $\mathbf{v}$. We will do this by starting with the formula

$$\theta = \cos^{-1}\left(\frac{\mathbf{u}\cdot\mathbf{v}}{\|\mathbf{u}\|\|\mathbf{v}\|}\right) \tag{20}$$

which we previously derived for nonzero vectors in R^2 and R^3. Since dot products and norms have been defined for vectors in R^n, it would seem that this formula has all the ingredients to serve as a *definition* of the angle θ between two vectors, $\mathbf{u}$ and $\mathbf{v}$, in R^n. However, there is a fly in the ointment, the problem being that the inverse cosine in Formula (20) is not defined unless its argument satisfies the inequalities

$$-1 \le \frac{\mathbf{u}\cdot\mathbf{v}}{\|\mathbf{u}\|\|\mathbf{v}\|} \le 1 \tag{21}$$

Fortunately, these inequalities *do* hold for all nonzero vectors in R^n as a result of the following fundamental result known as the ***Cauchy–Schwarz inequality***.

THEOREM 3.2.4 Cauchy–Schwarz Inequality

If $\mathbf{u} = (u_1, u_2, \ldots, u_n)$ *and* $\mathbf{v} = (v_1, v_2, \ldots, v_n)$ *are vectors in* R^n, *then*

$$|\mathbf{u}\cdot\mathbf{v}| \le \|\mathbf{u}\|\|\mathbf{v}\| \tag{22}$$

or in terms of components

$$|u_1 v_1 + u_2 v_2 + \cdots + u_n v_n| \le (u_1^2 + u_2^2 + \cdots + u_n^2)^{1/2}(v_1^2 + v_2^2 + \cdots + v_n^2)^{1/2} \tag{23}$$

We will omit the proof of this theorem because later in the text we will prove a more general version of which this will be a special case. Our goal for now will be to use this theorem to prove that the inequalities in (21) hold for all nonzero vectors in R^n. Once that is done we will have established all the results required to use Formula (20) as our *definition* of the angle between nonzero vectors $\mathbf{u}$ and $\mathbf{v}$ in R^n.

To prove that the inequalities in (21) hold for all nonzero vectors in R^n, divide both sides of Formula (22) by the product $\|\mathbf{u}\|\|\mathbf{v}\|$ to obtain

$$\frac{|\mathbf{u}\cdot\mathbf{v}|}{\|\mathbf{u}\|\|\mathbf{v}\|} \le 1 \quad \text{or equivalently} \quad \left|\frac{\mathbf{u}\cdot\mathbf{v}}{\|\mathbf{u}\|\|\mathbf{v}\|}\right| \le 1$$

from which (21) follows.

Hermann Amandus Schwarz (1843–1921)

Viktor Yakovlevich Bunyakovsky (1804–1889)

Historical Note The Cauchy–Schwarz inequality is named in honor of the French mathematician Augustin Cauchy (see p. 109) and the German mathematician Hermann Schwarz. Variations of this inequality occur in many different settings and under various names. Depending on the context in which the inequality occurs, you may find it called Cauchy's inequality, the Schwarz inequality, or sometimes even the Bunyakovsky inequality, in recognition of the Russian mathematician who published his version of the inequality in 1859, about 25 years before Schwarz.
[*Images: wikipedia (Schwarz); wikipedia (Bunyakovsky)*]

Geometry in R^n

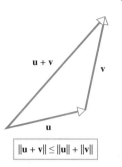

$$\|\mathbf{u} + \mathbf{v}\| \leq \|\mathbf{u}\| + \|\mathbf{v}\|$$

▲ Figure 3.2.8

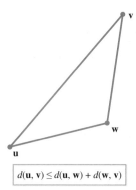

$$d(\mathbf{u}, \mathbf{v}) \leq d(\mathbf{u}, \mathbf{w}) + d(\mathbf{w}, \mathbf{v})$$

▲ Figure 3.2.9

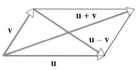

▲ Figure 3.2.10

Earlier in this section we extended various concepts to R^n with the idea that familiar results that we can visualize in R^2 and R^3 might be valid in R^n as well. Here are two fundamental theorems from plane geometry whose validity extends to R^n:

- The sum of the lengths of two side of a triangle is at least as large as the third (Figure 3.2.8).
- The shortest distance between two points is a straight line (Figure 3.2.9).

The following theorem generalizes these theorems to R^n.

THEOREM 3.2.5 *If $\mathbf{u}$, $\mathbf{v}$, and $\mathbf{w}$ are vectors in R^n, and if k is any scalar, then:*

(a) $\|\mathbf{u} + \mathbf{v}\| \leq \|\mathbf{u}\| + \|\mathbf{v}\|$ [Triangle inequality for vectors]

(b) $d(\mathbf{u}, \mathbf{v}) \leq d(\mathbf{u}, \mathbf{w}) + d(\mathbf{w}, \mathbf{v})$ [Triangle inequality for distances]

Proof (a)

$$
\begin{aligned}
\|\mathbf{u} + \mathbf{v}\|^2 &= (\mathbf{u} + \mathbf{v}) \cdot (\mathbf{u} + \mathbf{v}) = (\mathbf{u} \cdot \mathbf{u}) + 2(\mathbf{u} \cdot \mathbf{v}) + (\mathbf{v} \cdot \mathbf{v}) \\
&= \|\mathbf{u}\|^2 + 2(\mathbf{u} \cdot \mathbf{v}) + \|\mathbf{v}\|^2 \\
&\leq \|\mathbf{u}\|^2 + 2|\mathbf{u} \cdot \mathbf{v}| + \|\mathbf{v}\|^2 \qquad \longleftarrow \text{Property of absolute value} \\
&\leq \|\mathbf{u}\|^2 + 2\|\mathbf{u}\|\,\|\mathbf{v}\| + \|\mathbf{v}\|^2 \qquad \longleftarrow \text{Cauchy–Schwarz inequality} \\
&= (\|\mathbf{u}\| + \|\mathbf{v}\|)^2
\end{aligned}
$$

Proof (b)   It follows from part (a) and Formula (11) that

$$
\begin{aligned}
d(\mathbf{u}, \mathbf{v}) &= \|\mathbf{u} - \mathbf{v}\| = \|(\mathbf{u} - \mathbf{w}) + (\mathbf{w} - \mathbf{v})\| \\
&\leq \|\mathbf{u} - \mathbf{w}\| + \|\mathbf{w} - \mathbf{v}\| = d(\mathbf{u}, \mathbf{w}) + d(\mathbf{w}, \mathbf{v}) \quad \blacktriangleleft
\end{aligned}
$$

It is proved in plane geometry that for any parallelogram the sum of the squares of the diagonals is equal to the sum of the squares of the four sides (Figure 3.2.10). The following theorem generalizes that result to R^n.

THEOREM 3.2.6 **Parallelogram Equation for Vectors**

If $\mathbf{u}$ and $\mathbf{v}$ are vectors in R^n, then

$$\|\mathbf{u} + \mathbf{v}\|^2 + \|\mathbf{u} - \mathbf{v}\|^2 = 2\left(\|\mathbf{u}\|^2 + \|\mathbf{v}\|^2\right) \tag{24}$$

Proof

$$
\begin{aligned}
\|\mathbf{u} + \mathbf{v}\|^2 + \|\mathbf{u} - \mathbf{v}\|^2 &= (\mathbf{u} + \mathbf{v}) \cdot (\mathbf{u} + \mathbf{v}) + (\mathbf{u} - \mathbf{v}) \cdot (\mathbf{u} - \mathbf{v}) \\
&= 2(\mathbf{u} \cdot \mathbf{u}) + 2(\mathbf{v} \cdot \mathbf{v}) \\
&= 2\left(\|\mathbf{u}\|^2 + \|\mathbf{v}\|^2\right) \quad \blacktriangleleft
\end{aligned}
$$

We could state and prove many more theorems from plane geometry that generalize to R^n, but the ones already given should suffice to convince you that R^n is not so different from R^2 and R^3 even though we cannot visualize it directly. The next theorem establishes a fundamental relationship between the dot product and norm in R^n.

> **THEOREM 3.2.7** *If* **u** *and* **v** *are vectors in* R^n *with the Euclidean inner product, then*
> $$\mathbf{u} \cdot \mathbf{v} = \tfrac{1}{4}\|\mathbf{u} + \mathbf{v}\|^2 - \tfrac{1}{4}\|\mathbf{u} - \mathbf{v}\|^2 \tag{25}$$

Note that Formula (25) expresses the dot product in terms of norms.

Proof
$$\|\mathbf{u} + \mathbf{v}\|^2 = (\mathbf{u} + \mathbf{v}) \cdot (\mathbf{u} + \mathbf{v}) = \|\mathbf{u}\|^2 + 2(\mathbf{u} \cdot \mathbf{v}) + \|\mathbf{v}\|^2$$
$$\|\mathbf{u} - \mathbf{v}\|^2 = (\mathbf{u} - \mathbf{v}) \cdot (\mathbf{u} - \mathbf{v}) = \|\mathbf{u}\|^2 - 2(\mathbf{u} \cdot \mathbf{v}) + \|\mathbf{v}\|^2$$
from which (25) follows by simple algebra. ◀

Dot Products as Matrix Multiplication

There are various ways to express the dot product of vectors using matrix notation. The formulas depend on whether the vectors are expressed as row matrices or column matrices. Here are the possibilities.

Table 1

Form	Dot Product	Example
u a column matrix and **v** a column matrix	$\mathbf{u} \cdot \mathbf{v} = \mathbf{u}^T\mathbf{v} = \mathbf{v}^T\mathbf{u}$	$\mathbf{u} = \begin{bmatrix} 1 \\ -3 \\ 5 \end{bmatrix}$ $\mathbf{v} = \begin{bmatrix} 5 \\ 4 \\ 0 \end{bmatrix}$ $\mathbf{u}^T\mathbf{v} = \begin{bmatrix} 1 & -3 & 5 \end{bmatrix}\begin{bmatrix} 5 \\ 4 \\ 0 \end{bmatrix} = -7$ $\mathbf{v}^T\mathbf{u} = \begin{bmatrix} 5 & 4 & 0 \end{bmatrix}\begin{bmatrix} 1 \\ -3 \\ 5 \end{bmatrix} = -7$
u a row matrix and **v** a column matrix	$\mathbf{u} \cdot \mathbf{v} = \mathbf{u}\mathbf{v} = \mathbf{v}^T\mathbf{u}^T$	$\mathbf{u} = \begin{bmatrix} 1 & -3 & 5 \end{bmatrix}$ $\mathbf{v} = \begin{bmatrix} 5 \\ 4 \\ 0 \end{bmatrix}$ $\mathbf{u}\mathbf{v} = \begin{bmatrix} 1 & -3 & 5 \end{bmatrix}\begin{bmatrix} 5 \\ 4 \\ 0 \end{bmatrix} = -7$ $\mathbf{v}^T\mathbf{u}^T = \begin{bmatrix} 5 & 4 & 0 \end{bmatrix}\begin{bmatrix} 1 \\ -3 \\ 5 \end{bmatrix} = -7$
u a column matrix and **v** a row matrix	$\mathbf{u} \cdot \mathbf{v} = \mathbf{v}\mathbf{u} = \mathbf{u}^T\mathbf{v}^T$	$\mathbf{u} = \begin{bmatrix} 1 \\ -3 \\ 5 \end{bmatrix}$ $\mathbf{v} = \begin{bmatrix} 5 & 4 & 0 \end{bmatrix}$ $\mathbf{v}\mathbf{u} = \begin{bmatrix} 5 & 4 & 0 \end{bmatrix}\begin{bmatrix} 1 \\ -3 \\ 5 \end{bmatrix} = -7$ $\mathbf{u}^T\mathbf{v}^T = \begin{bmatrix} 1 & -3 & 5 \end{bmatrix}\begin{bmatrix} 5 \\ 4 \\ 0 \end{bmatrix} = -7$
u a row matrix and **v** a row matrix	$\mathbf{u} \cdot \mathbf{v} = \mathbf{u}\mathbf{v}^T = \mathbf{v}\mathbf{u}^T$	$\mathbf{u} = \begin{bmatrix} 1 & -3 & 5 \end{bmatrix}$ $\mathbf{v} = \begin{bmatrix} 5 & 4 & 0 \end{bmatrix}$ $\mathbf{u}\mathbf{v}^T = \begin{bmatrix} 1 & -3 & 5 \end{bmatrix}\begin{bmatrix} 5 \\ 4 \\ 0 \end{bmatrix} = -7$ $\mathbf{v}\mathbf{u}^T = \begin{bmatrix} 5 & 4 & 0 \end{bmatrix}\begin{bmatrix} 1 \\ -3 \\ 5 \end{bmatrix} = -7$

If A is an $n \times n$ matrix and **u** and **v** are $n \times 1$ matrices, then it follows from the first row in Table 1 and properties of the transpose that
$$A\mathbf{u} \cdot \mathbf{v} = \mathbf{v}^T(A\mathbf{u}) = (\mathbf{v}^T A)\mathbf{u} = (A^T\mathbf{v})^T\mathbf{u} = \mathbf{u} \cdot A^T\mathbf{v}$$
$$\mathbf{u} \cdot A\mathbf{v} = (A\mathbf{v})^T\mathbf{u} = (\mathbf{v}^T A^T)\mathbf{u} = \mathbf{v}^T(A^T\mathbf{u}) = A^T\mathbf{u} \cdot \mathbf{v}$$

The resulting formulas

$$A\mathbf{u} \cdot \mathbf{v} = \mathbf{u} \cdot A^T \mathbf{v} \tag{26}$$

$$\mathbf{u} \cdot A\mathbf{v} = A^T \mathbf{u} \cdot \mathbf{v} \tag{27}$$

provide an important link between multiplication by an $n \times n$ matrix A and multiplication by A^T.

▶ **EXAMPLE 9 Verifying That $A\mathbf{u} \cdot \mathbf{v} = \mathbf{u} \cdot A^T\mathbf{v}$**

Suppose that

$$A = \begin{bmatrix} 1 & -2 & 3 \\ 2 & 4 & 1 \\ -1 & 0 & 1 \end{bmatrix}, \quad \mathbf{u} = \begin{bmatrix} -1 \\ 2 \\ 4 \end{bmatrix}, \quad \mathbf{v} = \begin{bmatrix} -2 \\ 0 \\ 5 \end{bmatrix}$$

Then

$$A\mathbf{u} = \begin{bmatrix} 1 & -2 & 3 \\ 2 & 4 & 1 \\ -1 & 0 & 1 \end{bmatrix} \begin{bmatrix} -1 \\ 2 \\ 4 \end{bmatrix} = \begin{bmatrix} 7 \\ 10 \\ 5 \end{bmatrix}$$

$$A^T\mathbf{v} = \begin{bmatrix} 1 & 2 & -1 \\ -2 & 4 & 0 \\ 3 & 1 & 1 \end{bmatrix} \begin{bmatrix} -2 \\ 0 \\ 5 \end{bmatrix} = \begin{bmatrix} -7 \\ 4 \\ -1 \end{bmatrix}$$

from which we obtain

$$A\mathbf{u} \cdot \mathbf{v} = 7(-2) + 10(0) + 5(5) = 11$$
$$\mathbf{u} \cdot A^T\mathbf{v} = (-1)(-7) + 2(4) + 4(-1) = 11$$

Thus, $A\mathbf{u} \cdot \mathbf{v} = \mathbf{u} \cdot A^T\mathbf{v}$ as guaranteed by Formula (8). We leave it for you to verify that Formula (27) also holds. ◀

A Dot Product View of Matrix Multiplication

Dot products provide another way of thinking about matrix multiplication. Recall that if $A = [a_{ij}]$ is an $m \times r$ matrix and $B = [b_{ij}]$ is an $r \times n$ matrix, then the ijth entry of AB is

$$a_{i1}b_{1j} + a_{i2}b_{2j} + \cdots + a_{ir}b_{rj}$$

which is the dot product of the ith row vector of A

$$[a_{i1} \quad a_{i2} \quad \cdots \quad a_{ir}]$$

and the jth column vector of B

$$\begin{bmatrix} b_{1j} \\ b_{2j} \\ \vdots \\ b_{rj} \end{bmatrix}$$

Thus, if the row vectors of A are $\mathbf{r}_1, \mathbf{r}_2, \ldots, \mathbf{r}_m$ and the column vectors of B are $\mathbf{c}_1, \mathbf{c}_2, \ldots, \mathbf{c}_n$, then the matrix product AB can be expressed as

$$AB = \begin{bmatrix} \mathbf{r}_1 \cdot \mathbf{c}_1 & \mathbf{r}_1 \cdot \mathbf{c}_2 & \cdots & \mathbf{r}_1 \cdot \mathbf{c}_n \\ \mathbf{r}_2 \cdot \mathbf{c}_1 & \mathbf{r}_2 \cdot \mathbf{c}_2 & \cdots & \mathbf{r}_2 \cdot \mathbf{c}_n \\ \vdots & \vdots & & \vdots \\ \mathbf{r}_m \cdot \mathbf{c}_1 & \mathbf{r}_m \cdot \mathbf{c}_2 & \cdots & \mathbf{r}_m \cdot \mathbf{c}_n \end{bmatrix} \tag{28}$$

Application of Dot Products to ISBN Numbers

Although the system has recently changed, most books published in the last 25 years have been assigned a unique 10-digit number called an *International Standard Book Number* or ISBN. The first nine digits of this number are split into three groups—the first group representing the country or group of countries in which the book originates, the second identifying the publisher, and the third assigned to the book title itself. The tenth and final digit, called a *check digit*, is computed from the first nine digits and is used to ensure that an electronic transmission of the ISBN, say over the Internet, occurs without error.

To explain how this is done, regard the first nine digits of the ISBN as a vector **b** in R^9, and let **a** be the vector

$$\mathbf{a} = (1, 2, 3, 4, 5, 6, 7, 8, 9)$$

Then the check digit c is computed using the following procedure:

1. Form the dot product $\mathbf{a} \cdot \mathbf{b}$.
2. Divide $\mathbf{a} \cdot \mathbf{b}$ by 11, thereby producing a remainder c that is an integer between 0 and 10, inclusive. The check digit is taken to be c, with the proviso that $c = 10$ is written as X to avoid double digits.

For example, the ISBN of the brief edition of *Calculus*, sixth edition, by Howard Anton is

$$0\text{-}471\text{-}15307\text{-}9$$

which has a check digit of 9. This is consistent with the first nine digits of the ISBN, since

$$\mathbf{a} \cdot \mathbf{b} = (1, 2, 3, 4, 5, 6, 7, 8, 9) \cdot (0, 4, 7, 1, 1, 5, 3, 0, 7) = 152$$

Dividing 152 by 11 produces a quotient of 13 and a remainder of 9, so the check digit is $c = 9$. If an electronic order is placed for a book with a certain ISBN, then the warehouse can use the above procedure to verify that the check digit is consistent with the first nine digits, thereby reducing the possibility of a costly shipping error.

Concept Review

- Norm (or length or magnitude) of a vector
- Unit vector
- Normalized vector
- Standard unit vectors
- Distance between points in R^n
- Angle between two vectors in R^n
- Dot product (or Euclidean inner product) of two vectors in R^n
- Cauchy–Schwarz inequality
- Triangle inequality
- Parallelogram equation for vectors

Skills

- Compute the norm of a vector in R^n.
- Determine whether a given vector in R^n is a unit vector.
- Normalize a nonzero vector in R^n.
- Determine the distance between two vectors in R^n.
- Compute the dot product of two vectors in R^n.
- Compute the angle between two nonzero vectors in R^n.
- Prove basic properties pertaining to norms and dot products (Theorems 3.2.1–3.2.3 and 3.2.5–3.2.7).

Exercise Set 3.2

In Exercises 1–2, find the norm of **v**, a unit vector that has the same direction as **v**, and a unit vector that is oppositely directed to **v**.

1. (a) $\mathbf{v} = (4, -3)$ (b) $\mathbf{v} = (2, 2, 2)$
   (c) $\mathbf{v} = (1, 0, 2, 1, 3)$

2. (a) $\mathbf{v} = (-5, 12)$ (b) $\mathbf{v} = (1, -1, 2)$
   (c) $\mathbf{v} = (-2, 3, 3, -1)$

In Exercises 3–4, evaluate the given expression with $\mathbf{u} = (2, -2, 3)$, $\mathbf{v} = (1, -3, 4)$, and $\mathbf{w} = (3, 6, -4)$.

3. (a) $\|\mathbf{u} + \mathbf{v}\|$ (b) $\|\mathbf{u}\| + \|\mathbf{v}\|$
   (c) $\|-2\mathbf{u} + 2\mathbf{v}\|$ (d) $\|3\mathbf{u} - 5\mathbf{v} + \mathbf{w}\|$

4. (a) $\|\mathbf{u} + \mathbf{v} + \mathbf{w}\|$ (b) $\|\mathbf{u} - \mathbf{v}\|$
   (c) $\|3\mathbf{v}\| - 3\|\mathbf{v}\|$ (d) $\|\mathbf{u}\| - \|\mathbf{v}\|$

In Exercises 5–6, evaluate the given expression with $\mathbf{u} = (-2, -1, 4, 5)$, $\mathbf{v} = (3, 1, -5, 7)$, and $\mathbf{w} = (-6, 2, 1, 1)$.

5. (a) $\|3\mathbf{u} - 5\mathbf{v} + \mathbf{w}\|$ (b) $\|3\mathbf{u}\| - 5\|\mathbf{v}\| + \|\mathbf{w}\|$
   (c) $\|-\|\mathbf{u}\|\mathbf{v}\|$

6. (a) $\|\mathbf{u}\| - 2\|\mathbf{v}\| - 3\|\mathbf{w}\|$ (b) $\|\mathbf{u}\| + \|-2\mathbf{v}\| + \|-3\mathbf{w}\|$
   (c) $\|\|\mathbf{u} - \mathbf{v}\|\mathbf{w}\|$

7. Let $\mathbf{v} = (-2, 3, 0, 6)$. Find all scalars k such that $\|k\mathbf{v}\| = 5$.

8. Let $\mathbf{v} = (1, 1, 2, -3, 1)$. Find all scalars k such that $\|k\mathbf{v}\| = 4$.

In Exercises 9–10, find $\mathbf{u} \cdot \mathbf{v}$, $\mathbf{u} \cdot \mathbf{u}$, and $\mathbf{v} \cdot \mathbf{v}$.

9. (a) $\mathbf{u} = (3, 1, 4)$, $\mathbf{v} = (2, 2, -4)$

(b) $\mathbf{u} = (1, 1, 4, 6)$, $\mathbf{v} = (2, -2, 3, -2)$

10. (a) $\mathbf{u} = (1, 1, -2, 3)$, $\mathbf{v} = (-1, 0, 5, 1)$

(b) $\mathbf{u} = (2, -1, 1, 0, -2)$, $\mathbf{v} = (1, 2, 2, 2, 1)$

In Exercises 11–12, find the Euclidean distance between $\mathbf{u}$ and $\mathbf{v}$.

11. (a) $\mathbf{u} = (3, 3, 3)$, $\mathbf{v} = (1, 0, 4)$

(b) $\mathbf{u} = (0, -2, -1, 1)$, $\mathbf{v} = (-3, 2, 4, 4)$

(c) $\mathbf{u} = (3, -3, -2, 0, -3, 13, 5)$,
$\mathbf{v} = (-4, 1, -1, 5, 0, -11, 4)$

12. (a) $\mathbf{u} = (1, 2, -3, 0)$, $\mathbf{v} = (5, 1, 2, -2)$

(b) $\mathbf{u} = (2, -1, -4, 1, 0, 6, -3, 1)$,
$\mathbf{v} = (-2, -1, 0, 3, 7, 2, -5, 1)$

(c) $\mathbf{u} = (0, 1, 1, 1, 2)$, $\mathbf{v} = (2, 1, 0, -1, 3)$

13. Find the cosine of the angle between the vectors in each part of Exercise 11, and then state whether the angle is acute, obtuse, or 90°.

14. Find the cosine of the angle between the vectors in each part of Exercise 12, and then state whether the angle is acute, obtuse, or 90°.

15. Suppose that a vector $\mathbf{a}$ in the xy-plane has a length of 9 units and points in a direction that is 120° counterclockwise from the positive x-axis, and a vector $\mathbf{b}$ in that plane has a length of 5 units and points in the positive y-direction. Find $\mathbf{a} \cdot \mathbf{b}$.

16. Suppose that a vector $\mathbf{a}$ in the xy-plane points in a direction that is 47° counterclockwise from the positive x-axis, and a vector $\mathbf{b}$ in that plane points in a direction that is 43° clockwise from the positive x-axis. What can you say about the value of $\mathbf{a} \cdot \mathbf{b}$?

In Exercises 17–18, determine whether the expression makes sense mathematically. If not, explain why.

17. (a) $\mathbf{u} \cdot (\mathbf{v} \cdot \mathbf{w})$ (b) $\mathbf{u} \cdot (\mathbf{v} + \mathbf{w})$

(c) $\|\mathbf{u} \cdot \mathbf{v}\|$ (d) $(\mathbf{u} \cdot \mathbf{v}) - \|\mathbf{u}\|$

18. (a) $\|\mathbf{u}\| \cdot \|\mathbf{v}\|$ (b) $(\mathbf{u} \cdot \mathbf{v}) - \mathbf{w}$

(c) $(\mathbf{u} \cdot \mathbf{v}) - k$ (d) $k \cdot \mathbf{u}$

19. Find a unit vector that has the same direction as the given vector.

(a) $(-4, -3)$ (b) $(1, 7)$

(c) $(-3, 2, \sqrt{3})$ (d) $(1, 2, 3, 4, 5)$

20. Find a unit vector that is oppositely directed to the given vector.

(a) $(-12, -5)$ (b) $(3, -3, -3)$

(c) $(-6, 8)$ (d) $(-3, 1, \sqrt{6}, 3)$

21. State a procedure for finding a vector of a specified length m that points in the same direction as a given vector $\mathbf{v}$.

22. If $\|\mathbf{v}\| = 2$ and $\|\mathbf{w}\| = 3$, what are the largest and smallest values possible for $\|\mathbf{v} - \mathbf{w}\|$? Give a geometric explanation of your results.

23. Find the cosine of the angle θ between $\mathbf{u}$ and $\mathbf{v}$.

(a) $\mathbf{u} = (2, 3)$, $\mathbf{v} = (5, -7)$

(b) $\mathbf{u} = (-6, -2)$, $\mathbf{v} = (4, 0)$

(c) $\mathbf{u} = (1, -5, 4)$, $\mathbf{v} = (3, 3, 3)$

(d) $\mathbf{u} = (-2, 2, 3)$, $\mathbf{v} = (1, 7, -4)$

24. Find the radian measure of the angle θ (with $0 \leq \theta \leq \pi$) between $\mathbf{u}$ and $\mathbf{v}$.

(a) $(1, -7)$ and $(21, 3)$ (b) $(0, 2)$ and $(3, -3)$

(c) $(-1, 1, 0)$ and $(0, -1, 1)$ (d) $(1, -1, 0)$ and $(1, 0, 0)$

In Exercises 25–26, verify that the Cauchy–Schwarz inequality holds.

25. (a) $\mathbf{u} = (3, 2)$, $\mathbf{v} = (4, -1)$

(b) $\mathbf{u} = (-3, 1, 0)$, $\mathbf{v} = (2, -1, 3)$

(c) $\mathbf{u} = (0, 2, 2, 1)$, $\mathbf{v} = (1, 1, 1, 1)$

26. (a) $\mathbf{u} = (4, 1, 1)$, $\mathbf{v} = (1, 2, 3)$

(b) $\mathbf{u} = (1, 2, 1, 2, 3)$, $\mathbf{v} = (0, 1, 1, 5, -2)$

(c) $\mathbf{u} = (1, 3, 5, 2, 0, 1)$, $\mathbf{v} = (0, 2, 4, 1, 3, 5)$

27. Let $\mathbf{p}_0 = (x_0, y_0, z_0)$ and $\mathbf{p} = (x, y, z)$. Describe the set of all points (x, y, z) for which $\|\mathbf{p} - \mathbf{p}_0\| = 1$.

28. (a) Show that the components of the vector $\mathbf{v} = (v_1, v_2)$ in Figure Ex-28a are $v_1 = \|\mathbf{v}\| \cos \theta$ and $v_2 = \|\mathbf{v}\| \sin \theta$.

(b) Let $\mathbf{u}$ and $\mathbf{v}$ be the vectors in Figure Ex-28b. Use the result in part (a) to find the components of $4\mathbf{u} - 5\mathbf{v}$.

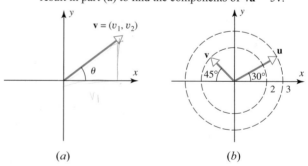

(a)   (b)

▲ Figure Ex-28

29. Prove parts (a) and (b) of Theorem 3.2.1.

30. Prove parts (a) and (c) of Theorem 3.2.3.

31. Prove parts (d) and (e) of Theorem 3.2.3.

32. Under what conditions will the triangle inequality (Theorem 3.2.5a) be an equality? Explain your answer geometrically.

33. What can you say about two nonzero vectors, $\mathbf{u}$ and $\mathbf{v}$, that satisfy the equation $\|\mathbf{u} + \mathbf{v}\| = \|\mathbf{u}\| + \|\mathbf{v}\|$?

34. (a) What relationship must hold for the point $\mathbf{p} = (a, b, c)$ to be equidistant from the origin and the xz-plane? Make sure that the relationship you state is valid for positive and negative values of a, b, and c.

   (b) What relationship must hold for the point $\mathbf{p} = (a, b, c)$ to be farther from the origin than from the xz-plane? Make sure that the relationship you state is valid for positive and negative values of a, b, and c

True-False Exercises

In parts (a)–(j) determine whether the statement is true or false, and justify your answer.

(a) If each component of a vector in R^3 is doubled, the norm of that vector is doubled.

(b) In R^2, the vectors of norm 5 whose initial points are at the origin have terminal points lying on a circle of radius 5 centered at the origin.

(c) Every vector in R^n has a positive norm.

(d) If $\mathbf{v}$ is a nonzero vector in R^n, there are exactly two unit vectors that are parallel to $\mathbf{v}$.

(e) If $\|\mathbf{u}\| = 2$, $\|\mathbf{v}\| = 1$, and $\mathbf{u} \cdot \mathbf{v} = 1$, then the angle between $\mathbf{u}$ and $\mathbf{v}$ is $\pi/3$ radians.

(f) The expressions $(\mathbf{u} \cdot \mathbf{v}) + \mathbf{w}$ and $\mathbf{u} \cdot (\mathbf{v} + \mathbf{w})$ are both meaningful and equal to each other.

(g) If $\mathbf{u} \cdot \mathbf{v} = \mathbf{u} \cdot \mathbf{w}$, then $\mathbf{v} = \mathbf{w}$.

(h) If $\mathbf{u} \cdot \mathbf{v} = 0$, then either $\mathbf{u} = \mathbf{0}$ or $\mathbf{v} = \mathbf{0}$.

(i) In R^2, if $\mathbf{u}$ lies in the first quadrant and $\mathbf{v}$ lies in the third quadrant, then $\mathbf{u} \cdot \mathbf{v}$ cannot be positive.

(j) For all vectors $\mathbf{u}$, $\mathbf{v}$, and $\mathbf{w}$ in R^n, we have

$$\|\mathbf{u} + \mathbf{v} + \mathbf{w}\| \le \|\mathbf{u}\| + \|\mathbf{v}\| + \|\mathbf{w}\|$$

3.3 Orthogonality

In the last section we defined the notion of "angle" between vectors in R^n. In this section we will focus on the notion of "perpendicularity." Perpendicular vectors in R^n play an important role in a wide variety of applications.

Orthogonal Vectors   Recall from Formula (20) in the previous section that the angle θ between two *nonzero* vectors $\mathbf{u}$ and $\mathbf{v}$ in R^n is defned by the formula

$$\theta = \cos^{-1}\left(\frac{\mathbf{u} \cdot \mathbf{v}}{\|\mathbf{u}\|\|\mathbf{v}\|}\right)$$

It follows from this that $\theta = \pi/2$ if and only if $\mathbf{u} \cdot \mathbf{v} = 0$. Thus, we make the following definition.

DEFINITION 1 Two nonzero vectors $\mathbf{u}$ and $\mathbf{v}$ in R^n are said to be *orthogonal* (or *perpendicular*) if $\mathbf{u} \cdot \mathbf{v} = 0$. We will also agree that the zero vector in R^n is orthogonal to *every* vector in R^n. A nonempty set of vectors in R^n is called an *orthogonal set* if all pairs of distinct vectors in the set are orthogonal. An orthogonal set of unit vectors is called an *orthonormal set*.

▶ **EXAMPLE 1 Orthogonal Vectors**

(a)   Show that $\mathbf{u} = (-2, 3, 1, 4)$ and $\mathbf{v} = (1, 2, 0, -1)$ are orthogonal vectors in R^4.

(b)   Show that the set $S = \{\mathbf{i}, \mathbf{j}, \mathbf{k}\}$ of standard unit vectors is an orthogonal set in R^3.

Solution (a) The vectors are orthogonal since

$$\mathbf{u} \cdot \mathbf{v} = (-2)(1) + (3)(2) + (1)(0) + (4)(-1) = 0$$

In Example 1 there is no need to check that

$$\mathbf{j} \cdot \mathbf{i} = \mathbf{k} \cdot \mathbf{i} = \mathbf{k} \cdot \mathbf{j} = 0$$

since this follows from computations in the example and the symmetry property of the dot product.

Solution (b) We must show that all pairs of distinct vectors are orthogonal, that is,

$$\mathbf{i} \cdot \mathbf{j} = \mathbf{i} \cdot \mathbf{k} = \mathbf{j} \cdot \mathbf{k} = 0$$

This is evident geometrically (Figure 3.2.2), but it can be seen as well from the computations

$$\mathbf{i} \cdot \mathbf{j} = (1, 0, 0) \cdot (0, 1, 0) = 0$$
$$\mathbf{i} \cdot \mathbf{k} = (1, 0, 0) \cdot (0, 0, 1) = 0$$
$$\mathbf{j} \cdot \mathbf{k} = (0, 1, 0) \cdot (0, 0, 1) = 0 \quad \blacktriangleleft$$

Lines and Planes Determined by Points and Normals

One learns in analytic geometry that a line in R^2 is determined uniquely by its slope and one of its points, and that a plane in R^3 is determined uniquely by its "inclination" and one of its points. One way of specifying slope and inclination is to use a *nonzero* vector **n**, called a ***normal***, that is orthogonal to the line or plane in question. For example, Figure 3.3.1 shows the line through the point $P_0(x_0, y_0)$ that has normal $\mathbf{n} = (a, b)$ and the plane through the point $P_0(x_0, y_0, z_0)$ that has normal $\mathbf{n} = (a, b, c)$. Both the line and the plane are represented by the vector equation

$$\mathbf{n} \cdot \overrightarrow{P_0 P} = 0 \tag{1}$$

where P is either an arbitrary point (x, y) on the line or an arbitrary point (x, y, z) in the plane. The vector $\overrightarrow{P_0 P}$ can be expressed in terms of components as

$$\overrightarrow{P_0 P} = (x - x_0, y - y_0) \quad \text{[line]}$$
$$\overrightarrow{P_0 P} = (x - x_0, y - y_0, z - z_0) \quad \text{[plane]}$$

Thus, Equation (1) can be written as

$$a(x - x_0) + b(y - y_0) = 0 \quad \text{[line]} \tag{2}$$

$$a(x - x_0) + b(y - y_0) + c(z - z_0) = 0 \quad \text{[plane]} \tag{3}$$

These are called the ***point-normal*** equations of the line and plane.

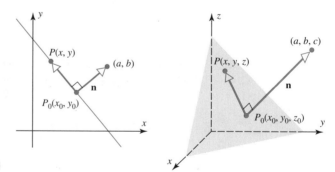

▶ Figure 3.3.1

▶ **EXAMPLE 2 Point-Normal Equations**

It follows from (2) that in R^2 the equation

$$6(x - 3) + (y + 7) = 0$$

represents the line through the point $(3, -7)$ with normal $\mathbf{n} = (6, 1)$; and it follows from (3) that in R^3 the equation

$$4(x - 3) + 2y - 5(z - 7) = 0$$

represents the plane through the point $(3, 0, 7)$ with normal $\mathbf{n} = (4, 2, -5)$. ◄

When convenient, the terms in Equations (2) and (3) can be multiplied out and the constants combined. This leads to the following theorem.

THEOREM 3.3.1

(a) *If a and b are constants that are not both zero, then an equation of the form*

$$ax + by + c = 0 \tag{4}$$

represents a line in R^2 with normal $\mathbf{n} = (a, b)$.

(b) *If a, b, and c are constants that are not all zero, then an equation of the form*

$$ax + by + cz + d = 0 \tag{5}$$

represents a plane in R^3 with normal $\mathbf{n} = (a, b, c)$.

► **EXAMPLE 3 Vectors Orthogonal to Lines and Planes Through the Origin**

(a) The equation $ax + by = 0$ represents a line through the origin in R^2. Show that the vector $\mathbf{n}_1 = (a, b)$ formed from the coefficients of the equation is orthogonal to the line, that is, orthogonal to every vector along the line.

(b) The equation $ax + by + cz = 0$ represents a plane through the origin in R^3. Show that the vector $\mathbf{n}_2 = (a, b, c)$ formed from the coefficients of the equation is orthogonal to the plane, that is, orthogonal to every vector that lies in the plane.

Solution We will solve both problems together. The two equations can be written as

$$(a, b) \cdot (x, y) = 0 \quad \text{and} \quad (a, b, c) \cdot (x, y, z) = 0$$

or, alternatively, as

$$\mathbf{n}_1 \cdot (x, y) = 0 \quad \text{and} \quad \mathbf{n}_2 \cdot (x, y, z) = 0$$

These equations show that $\mathbf{n}_1$ is orthogonal to every vector (x, y) on the line and that $\mathbf{n}_2$ is orthogonal to every vector (x, y, z) in the plane (Figure 3.3.1). ◄

Recall that

$$ax + by = 0 \quad \text{and} \quad ax + by + cz = 0$$

are called *homogeneous equations*. Example 3 illustrates that homogeneous equations in two or three unknowns can be written in the vector form

Referring to Table 1 of Section 3.2, in what other ways can you write (6) if $\mathbf{n}$ and $\mathbf{x}$ are expressed in matrix form?

$$\mathbf{n} \cdot \mathbf{x} = 0 \tag{6}$$

where $\mathbf{n}$ is the vector of coefficients and $\mathbf{x}$ is the vector of unknowns. In R^2 this is called the ***vector form of a line*** through the origin, and in R^3 it is called the ***vector form of a plane*** through the origin.

Orthogonal Projections In many applications it is necessary to "decompose" a vector $\mathbf{u}$ into a sum of two terms, one term being a scalar multiple of a specified nonzero vector $\mathbf{a}$ and the other term being orthogonal to $\mathbf{a}$. For example, if $\mathbf{u}$ and $\mathbf{a}$ are vectors in R^2 that are positioned so their initial points coincide at a point Q, then we can create such a decomposition as follows (Figure 3.3.2):

- Drop a perpendicular from the tip of $\mathbf{u}$ to the line through $\mathbf{a}$.
- Construct the vector $\mathbf{w}_1$ from Q to the foot of the perpendicular.
- Construct the vector $\mathbf{w}_2 = \mathbf{u} - \mathbf{w}_1$.

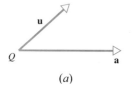 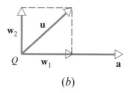 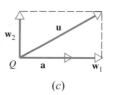 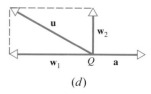

(a)   (b)   (c)   (d)

▲ Figure 3.3.2 In parts (b) through (d), $\mathbf{u} = \mathbf{w}_1 + \mathbf{w}_2$, where $\mathbf{w}_1$ is parallel to $\mathbf{a}$ and $\mathbf{w}_2$ is orthogonal to $\mathbf{a}$.

Since

$$\mathbf{w}_1 + \mathbf{w}_2 = \mathbf{w}_1 + (\mathbf{u} - \mathbf{w}_1) = \mathbf{u}$$

we have decomposed $\mathbf{u}$ into a sum of two orthogonal vectors, the first term being a scalar multiple of $\mathbf{a}$ and the second being orthogonal to $\mathbf{a}$.

The following theorem shows that the foregoing results, which we illustrated using vectors in R^2, apply as well in R^n.

THEOREM 3.3.2 Projection Theorem

If $\mathbf{u}$ and $\mathbf{a}$ are vectors in R^n, and if $\mathbf{a} \neq 0$, then $\mathbf{u}$ can be expressed in exactly one way in the form $\mathbf{u} = \mathbf{w}_1 + \mathbf{w}_2$, where $\mathbf{w}_1$ is a scalar multiple of $\mathbf{a}$ and $\mathbf{w}_2$ is orthogonal to $\mathbf{a}$.

Proof Since the vector $\mathbf{w}_1$ is to be a scalar multiple of $\mathbf{a}$, it must have the form

$$\mathbf{w}_1 = k\mathbf{a} \tag{7}$$

Our goal is to find a value of the scalar k and a vector $\mathbf{w}_2$ that is orthogonal to $\mathbf{a}$ such that

$$\mathbf{u} = \mathbf{w}_1 + \mathbf{w}_2 \tag{8}$$

We can determine k by using (7) to rewrite (8) as

$$\mathbf{u} = \mathbf{w}_1 + \mathbf{w}_2 = k\mathbf{a} + \mathbf{w}_2$$

and then applying Theorems 3.2.2 and 3.2.3 to obtain

$$\mathbf{u} \cdot \mathbf{a} = (k\mathbf{a} + \mathbf{w}_2) \cdot \mathbf{a} = k\|\mathbf{a}\|^2 + (\mathbf{w}_2 \cdot \mathbf{a}) \tag{9}$$

Since $\mathbf{w}_2$ is to be orthogonal to $\mathbf{a}$, the last term in (9) must be 0, and hence k must satisfy the equation

$$\mathbf{u} \cdot \mathbf{a} = k\|\mathbf{a}\|^2$$

from which we obtain

$$k = \frac{\mathbf{u} \cdot \mathbf{a}}{\|\mathbf{a}\|^2}$$

as the only possible value for k. The proof can be completed by rewriting (8) as

$$\mathbf{w}_2 = \mathbf{u} - \mathbf{w}_1 = \mathbf{u} - k\mathbf{a} = \mathbf{u} - \frac{\mathbf{u} \cdot \mathbf{a}}{\|\mathbf{a}\|^2}\mathbf{a}$$

and then confirming that $\mathbf{w}_2$ is orthogonal to $\mathbf{a}$ by showing that $\mathbf{w}_2 \cdot \mathbf{a} = 0$ (we leave the details for you). ◄

The vectors $\mathbf{w}_1$ and $\mathbf{w}_2$ in the Projection Theorem have associated names—the vector $\mathbf{w}_1$ is called the ***orthogonal projection of*** $\mathbf{u}$ ***on*** $\mathbf{a}$ or sometimes ***the vector component of*** $\mathbf{u}$ ***along*** $\mathbf{a}$, and the vector $\mathbf{w}_2$ is called the vector ***component of*** $\mathbf{u}$ ***orthogonal to*** $\mathbf{a}$. The vector $\mathbf{w}_1$ is commonly denoted by the symbol $\text{proj}_\mathbf{a}\mathbf{u}$, in which case it follows from (8) that $\mathbf{w}_2 = \mathbf{u} - \text{proj}_\mathbf{a}\mathbf{u}$. In summary,

$$\text{proj}_\mathbf{a}\mathbf{u} = \frac{\mathbf{u} \cdot \mathbf{a}}{\|\mathbf{a}\|^2}\mathbf{a} \quad \textit{(vector component of } \mathbf{u} \textit{ along } \mathbf{a}) \tag{10}$$

$$\mathbf{u} - \text{proj}_\mathbf{a}\mathbf{u} = \mathbf{u} - \frac{\mathbf{u} \cdot \mathbf{a}}{\|\mathbf{a}\|^2}\mathbf{a} \quad \textit{(vector component of } \mathbf{u} \textit{ orthogonal to } \mathbf{a}) \tag{11}$$

► **EXAMPLE 4** **Orthogonal Projection on a Line**

Find the orthogonal projections of the vectors $\mathbf{e}_1 = (1, 0)$ and $\mathbf{e}_2 = (0, 1)$ on the line L that makes an angle θ with the positive x-axis in R^2.

Solution As illustrated in Figure 3.3.3, $\mathbf{a} = (\cos\theta, \sin\theta)$ is a unit vector along the line L, so our first problem is to find the orthogonal projection of $\mathbf{e}_1$ along $\mathbf{a}$. Since

$$\|\mathbf{a}\| = \sqrt{\sin^2\theta + \cos^2\theta} = 1 \quad \text{and} \quad \mathbf{e}_1 \cdot \mathbf{a} = (1, 0) \cdot (\cos\theta, \sin\theta) = \cos\theta$$

it follows from Formula (10) that this projection is

$$\text{proj}_\mathbf{a}\mathbf{e}_1 = \frac{\mathbf{e}_1 \cdot \mathbf{a}}{\|\mathbf{a}\|^2}\mathbf{a} = (\cos\theta)(\cos\theta, \sin\theta) = (\cos^2\theta, \sin\theta\cos\theta)$$

Similarly, since $\mathbf{e}_2 \cdot \mathbf{a} = (0, 1) \cdot (\cos\theta, \sin\theta) = \sin\theta$, it follows from Formula (10) that

$$\text{proj}_\mathbf{a}\mathbf{e}_2 = \frac{\mathbf{e}_2 \cdot \mathbf{a}}{\|\mathbf{a}\|^2}\mathbf{a} = (\sin\theta)(\cos\theta, \sin\theta) = (\sin\theta\cos\theta, \sin^2\theta)$$

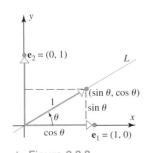

▲ Figure 3.3.3

► **EXAMPLE 5** **Vector Component of u Along a**

Let $\mathbf{u} = (2, -1, 3)$ and $\mathbf{a} = (4, -1, 2)$. Find the vector component of $\mathbf{u}$ along $\mathbf{a}$ and the vector component of $\mathbf{u}$ orthogonal to $\mathbf{a}$.

Solution

$$\mathbf{u} \cdot \mathbf{a} = (2)(4) + (-1)(-1) + (3)(2) = 15$$
$$\|\mathbf{a}\|^2 = 4^2 + (-1)^2 + 2^2 = 21$$

Thus the vector component of $\mathbf{u}$ along $\mathbf{a}$ is

$$\text{proj}_\mathbf{a}\mathbf{u} = \frac{\mathbf{u} \cdot \mathbf{a}}{\|\mathbf{a}\|^2}\mathbf{a} = \tfrac{15}{21}(4, -1, 2) = \left(\tfrac{20}{7}, -\tfrac{5}{7}, \tfrac{10}{7}\right)$$

and the vector component of $\mathbf{u}$ orthogonal to $\mathbf{a}$ is

$$\mathbf{u} - \text{proj}_\mathbf{a}\mathbf{u} = (2, -1, 3) - \left(\tfrac{20}{7}, -\tfrac{5}{7}, \tfrac{10}{7}\right) = \left(-\tfrac{6}{7}, -\tfrac{2}{7}, \tfrac{11}{7}\right)$$

As a check, you may wish to verify that the vectors $\mathbf{u} - \text{proj}_\mathbf{a}\mathbf{u}$ and $\mathbf{a}$ are perpendicular by showing that their dot product is zero. ◄

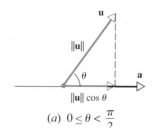

(a) $0 \leq \theta < \dfrac{\pi}{2}$

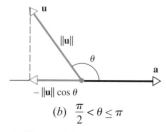

(b) $\dfrac{\pi}{2} < \theta \leq \pi$

▲ Figure 3.3.4

Sometimes we will be more interested in the *norm* of the vector component of **u** along **a** than in the vector component itself. A formula for this norm can be derived as follows:

$$\|\text{proj}_a \mathbf{u}\| = \left\| \frac{\mathbf{u} \cdot \mathbf{a}}{\|\mathbf{a}\|^2} \mathbf{a} \right\| = \left| \frac{\mathbf{u} \cdot \mathbf{a}}{\|\mathbf{a}\|^2} \right| \|\mathbf{a}\| = \frac{|\mathbf{u} \cdot \mathbf{a}|}{\|\mathbf{a}\|^2} \|\mathbf{a}\|$$

where the second equality follows from part (c) of Theorem 3.2.1 and the third from the fact that $\|\mathbf{a}\|^2 > 0$. Thus,

$$\|\text{proj}_a \mathbf{u}\| = \frac{|\mathbf{u} \cdot \mathbf{a}|}{\|\mathbf{a}\|} \tag{12}$$

If θ denotes the angle between **u** and **a**, then $\mathbf{u} \cdot \mathbf{a} = \|\mathbf{u}\| \|\mathbf{a}\| \cos\theta$, so (12) can also be written as

$$\|\text{proj}_a \mathbf{u}\| = \|\mathbf{u}\| |\cos\theta| \tag{13}$$

(Verify.) A geometric interpretation of this result is given in Figure 3.3.4.

The Theorem of Pythagoras   In Section 3.2 we found that many theorems about vectors in R^2 and R^3 also hold in R^n. Another example of this is the following generalization of the Theorem of Pythagoras (Figure 3.3.5).

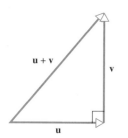

▲ Figure 3.3.5

THEOREM 3.3.3 Theorem of Pythagoras in R^n

If **u** *and* **v** *are orthogonal vectors in* R^n *with the Euclidean inner product, then*

$$\|\mathbf{u} + \mathbf{v}\|^2 = \|\mathbf{u}\|^2 + \|\mathbf{v}\|^2 \tag{14}$$

Proof   Since **u** and **v** are orthogonal, we have $\mathbf{u} \cdot \mathbf{v} = 0$, from which it follows that

$$\|\mathbf{u} + \mathbf{v}\|^2 = (\mathbf{u} + \mathbf{v}) \cdot (\mathbf{u} + \mathbf{v}) = \|\mathbf{u}\|^2 + 2(\mathbf{u} \cdot \mathbf{v}) + \|\mathbf{v}\|^2 = \|\mathbf{u}\|^2 + \|\mathbf{v}\|^2 \quad ◄$$

▶ **EXAMPLE 6** Theorem of Pythagoras in R^4

We showed in Example 1 that the vectors

$$\mathbf{u} = (-2, 3, 1, 4) \quad \text{and} \quad \mathbf{v} = (1, 2, 0, -1)$$

are orthogonal. Verify the Theorem of Pythagoras for these vectors.

Solution   We leave it for you to confirm that

$$\mathbf{u} + \mathbf{v} = (-1, 5, 1, 3)$$
$$\|\mathbf{u} + \mathbf{v}\|^2 = 36$$
$$\|\mathbf{u}\|^2 + \|\mathbf{v}\|^2 = 30 + 6$$

Thus, $\|\mathbf{u} + \mathbf{v}\|^2 = \|\mathbf{u}\|^2 + \|\mathbf{v}\|^2$ ◄

OPTIONAL
Distance Problems

We will now show how orthogonal projections can be used to solve the following three distance problems:

Problem 1. Find the distance between a point and a line in R^2.

Problem 2. Find the distance between a point and a plane in R^3.

Problem 3. Find the distance between two parallel planes in R^3.

A method for solving the first two problems is provided by the next theorem. Since the proofs of the two parts are similar, we will prove part (b) and leave part (a) as an exercise.

THEOREM 3.3.4

(a) *In R^2 the distance D between the point $P_0(x_0, y_0)$ and the line $ax + by + c = 0$ is*

$$D = \frac{|ax_0 + by_0 + c|}{\sqrt{a^2 + b^2}} \qquad (15)$$

(b) *In R^3 the distance D between the point $P_0(x_0, y_0, z_0)$ and the plane $ax + by + cz + d = 0$ is*

$$D = \frac{|ax_0 + by_0 + cz_0 + d|}{\sqrt{a^2 + b^2 + c^2}} \qquad (16)$$

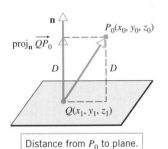

Distance from P_0 to plane.

▲ Figure 3.3.6

Proof (b)   Let $Q(x_1, y_1, z_1)$ be any point in the plane. Position the normal $\mathbf{n} = (a, b, c)$ so that its initial point is at Q. As illustrated in Figure 3.3.6, the distance D is equal to the length of the orthogonal projection of $\overrightarrow{QP_0}$ on $\mathbf{n}$. Thus, it follows from Formula (16) that

$$D = \|\operatorname{proj}_{\mathbf{n}} \overrightarrow{QP_0}\| = \frac{|\overrightarrow{QP_0} \cdot \mathbf{n}|}{\|\mathbf{n}\|}$$

But

$$\overrightarrow{QP_0} = (x_0 - x_1, y_0 - y_1, z_0 - z_1)$$

$$\overrightarrow{QP_0} \cdot \mathbf{n} = a(x_0 - x_1) + b(y_0 - y_1) + c(z_0 - z_1)$$

$$\|\mathbf{n}\| = \sqrt{a^2 + b^2 + c^2}$$

Thus

$$D = \frac{|a(x_0 - x_1) + b(y_0 - y_1) + c(z_0 - z_1)|}{\sqrt{a^2 + b^2 + c^2}} \qquad (17)$$

Since the point $Q(x_1, y_1, z_1)$ lies in the given plane, its coordinates satisfy the equation of that plane; thus

$$ax_1 + by_1 + cz_1 + d = 0$$

or

$$d = -ax_1 - by_1 - cz_1$$

Substituting this expression in (17) yields (16). ◀

▶ **EXAMPLE 7** **Distance Between a Point and a Plane**

Find the distance D between the point $(1, -4, -3)$ and the plane $2x - 3y + 6z = -1$.

Solution Since the distance formulas in Theorem 3.3.4 require that the equations of the line and plane be written with zero on the right side, we first need to rewrite the equation of the plane as

$$2x - 3y + 6z + 1 = 0$$

from which we obtain

$$D = \frac{|2(1) + (-3)(-4) + 6(-3) + 1|}{\sqrt{2^2 + (-3)^2 + 6^2}} = \frac{|-3|}{7} = \frac{3}{7} \quad ◀$$

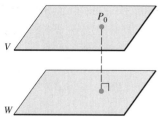

Figure 3.3.7 The distance between the parallel planes V and W is equal to the distance between P_0 and W.

The third distance problem posed above is to find the distance between two parallel planes in R^3. As suggested in Figure 3.3.7, the distance between a plane V and a plane W can be obtained by finding any point P_0 in one of the planes, and computing the distance between that point and the other plane. Here is an example.

▶ **EXAMPLE 8 Distance Between Parallel Planes**

The planes

$$x + 2y - 2z = 3 \quad \text{and} \quad 2x + 4y - 4z = 7$$

are parallel since their normals, $(1, 2, -2)$ and $(2, 4, -4)$, are parallel vectors. Find the distance between these planes.

Solution To find the distance D between the planes, we can select an arbitrary point in one of the planes and compute its distance to the other plane. By setting $y = z = 0$ in the equation $x + 2y - 2z = 3$, we obtain the point $P_0(3, 0, 0)$ in this plane. From (16), the distance between P_0 and the plane $2x + 4y - 4z = 7$ is

$$D = \frac{|2(3) + 4(0) + (-4)(0) - 7|}{\sqrt{2^2 + 4^2 + (-4)^2}} = \frac{1}{6} \quad \blacktriangleleft$$

Concept Review

- Orthogonal (perpendicular) vectors
- Orthogonal set of vectors
- Normal to a line
- Normal to a plane

- Point-normal equations
- Vector form of a line
- Vector form of a plane
- Orthogonal projection of **u** on **a**
- Vector component of **u** along **a**

- Vector component of **u** orthogonal to **a**
- Theorem of Pythagoras

Skills

- Determine whether two vectors are orthogonal.
- Determine whether a given set of vectors forms an orthogonal set.
- Find equations for lines (or planes) by using a normal vector and a point on the line (or plane).
- Find the vector form of a line or plane through the origin.

- Compute the vector component of **u** along **a** and orthogonal to **a**.
- Find the distance between a point and a line in R^2 or R^3.
- Find the distance between two parallel planes in R^3.
- Find the distance between a point and a plane.

Exercise Set 3.3

▷ In Exercises 1–2, determine whether **u** and **v** are orthogonal vectors. ◁

1. (a) $\mathbf{u} = (6, 1, 4)$, $\mathbf{v} = (2, 0, -3)$

(b) $\mathbf{u} = (0, 0, -1)$, $\mathbf{v} = (1, 1, 1)$

(c) $\mathbf{u} = (-6, 0, 4)$, $\mathbf{v} = (3, 1, 6)$

(d) $\mathbf{u} = (2, 4, -8)$, $\mathbf{v} = (5, 3, 7)$

2. (a) $\mathbf{u} = (2, 3)$, $\mathbf{v} = (5, -7)$

(b) $\mathbf{u} = (-6, -2)$, $\mathbf{v} = (4, 0)$

(c) $\mathbf{u} = (1, -5, 4)$, $\mathbf{v} = (3, 3, 3)$

(d) $\mathbf{u} = (-2, 2, 3)$, $\mathbf{v} = (1, 7, -4)$

▷ In Exercises 3–4, determine whether the vectors form an orthogonal set. ◁

3. (a) $\mathbf{v}_1 = (2, 3)$, $\mathbf{v}_2 = (3, 2)$

(b) $\mathbf{v}_1 = (-1, 1)$, $\mathbf{v}_2 = (1, 1)$

(c) $\mathbf{v}_1 = (-2, 1, 1)$, $\mathbf{v}_2 = (1, 0, 2)$, $\mathbf{v}_3 = (-2, -5, 1)$

(d) $\mathbf{v}_1 = (-3, 4, -1)$, $\mathbf{v}_2 = (1, 2, 5)$, $\mathbf{v}_3 = (4, -3, 0)$

4. (a) $\mathbf{v}_1 = (2, 3)$, $\mathbf{v}_2 = (-3, 2)$

(b) $\mathbf{v}_1 = (1, -2)$, $\mathbf{v}_2 = (-2, 1)$

(c) $\mathbf{v}_1 = (1, 0, 1)$, $\mathbf{v}_2 = (1, 1, 1)$, $\mathbf{v}_3 = (-1, 0, 1)$

(d) $\mathbf{v}_1 = (2, -2, 1)$, $\mathbf{v}_2 = (2, 1, -2)$, $\mathbf{v}_3 = (1, 2, 2)$

5. Find a unit vector that is orthogonal to both $\mathbf{u} = (1, 0, 1)$ and $\mathbf{v} = (0, 1, 1)$.

6. (a) Show that $\mathbf{v} = (a, b)$ and $\mathbf{w} = (-b, a)$ are orthogonal vectors.

(b) Use the result in part (a) to find two vectors that are orthogonal to $\mathbf{v} = (2, -3)$.

(c) Find two unit vectors that are orthogonal to $(-3, 4)$.

7. Do the points $A(1, 1, 1)$, $B(-2, 0, 3)$, and $C(-3, -1, 1)$ form the vertices of a right triangle? Explain your answer.

8. Repeat Exercise 7 for the points $A(3, 0, 2)$, $B(4, 3, 0)$, and $C(8, 1, -1)$.

In Exercises 9–12, find a point-normal form of the equation of the plane passing through P and having $\mathbf{n}$ as a normal.

9. $P(-1, 3, -2)$; $\mathbf{n} = (-2, 1, -1)$

10. $P(1, 1, 4)$; $\mathbf{n} = (1, 9, 8)$ **11.** $P(2, 0, 0)$; $\mathbf{n} = (0, 0, 2)$

12. $P(0, 0, 0)$; $\mathbf{n} = (1, 2, 3)$

In Exercises 13–16, determine whether the given planes are parallel.

13. $4x - y + 2z = 5$ and $7x - 3y + 4z = 8$

14. $x - 4y - 3z - 2 = 0$ and $3x - 12y - 9z - 7 = 0$

15. $2y = 8x - 4z + 5$ and $x = \frac{1}{2}z + \frac{1}{4}y$

16. $(-4, 1, 2) \cdot (x, y, z) = 0$ and $(8, -2, -4) \cdot (x, y, z) = 0$

In Exercises 17–18, determine whether the given planes are perpendicular.

17. $3x - y + z - 4 = 0$, $x + 2z = -1$

18. $x - 2y + 3z = 4$, $-2x + 5y + 4z = -1$

In Exercises 19–20, find $\|\text{proj}_{\mathbf{a}}\mathbf{u}\|$.

19. (a) $\mathbf{u} = (1, -2)$, $\mathbf{a} = (-4, -3)$

(b) $\mathbf{u} = (3, 0, 4)$, $\mathbf{a} = (2, 3, 3)$

20. (a) $\mathbf{u} = (5, 6)$, $\mathbf{a} = (2, -1)$

(b) $\mathbf{u} = (3, -2, 6)$, $\mathbf{a} = (1, 2, -7)$

In Exercises 21–28, find the vector component of $\mathbf{u}$ along $\mathbf{a}$ and the vector component of $\mathbf{u}$ orthogonal to $\mathbf{a}$.

21. $\mathbf{u} = (6, 2)$, $\mathbf{a} = (3, -9)$ **22.** $\mathbf{u} = (-1, -2)$, $\mathbf{a} = (-2, 3)$

23. $\mathbf{u} = (3, 1, -7)$, $\mathbf{a} = (1, 0, 5)$

24. $\mathbf{u} = (1, 0, 0)$, $\mathbf{a} = (4, 3, 8)$

25. $\mathbf{u} = (1, 1, 1)$, $\mathbf{a} = (0, 2, -1)$

26. $\mathbf{u} = (2, 0, 1)$, $\mathbf{a} = (1, 2, 3)$

27. $\mathbf{u} = (2, 1, 1, 2)$, $\mathbf{a} = (4, -4, 2, -2)$

28. $\mathbf{u} = (5, 0, -3, 7)$, $\mathbf{a} = (2, 1, -1, -1)$

In Exercises 29–32, find the distance between the point and the line.

29. $(-3, 1)$; $4x + 3y + 4 = 0$

30. $(-1, 4)$; $x - 3y + 2 = 0$

31. $(2, -5)$; $y = -4x + 2$

32. $(1, 8)$; $3x + y = 5$

In Exercises 33–36, find the distance between the point and the plane.

33. $(3, 1, -2)$; $x + 2y - 2z = 4$

34. $(-1, -1, 2)$; $2x + 5y - 6z = 4$

35. $(-1, 2, 1)$; $2x + 3y - 4z = 1$

36. $(0, 3, -2)$; $x - y - z = 3$

In Exercises 37–40, find the distance between the given parallel planes.

37. $2x - y - z = 5$ and $-4x + 2y + 2z = 12$

38. $3x - 4y + z = 1$ and $6x - 8y + 2z = 3$

39. $-4x + y - 3z = 0$ and $8x - 2y + 6z = 0$

40. $2x - y + z = 1$ and $2x - y + z = -1$

41. Let $\mathbf{i}, \mathbf{j}$, and $\mathbf{k}$ be unit vectors along the positive x, y, and z axes of a rectangular coordinate system in 3-space. If $\mathbf{v} = (a, b, c)$ is a nonzero vector, then the angles α, β, and γ between $\mathbf{v}$ and the vectors $\mathbf{i}$, $\mathbf{j}$, and $\mathbf{k}$, respectively, are called the **direction angles** of $\mathbf{v}$ (Figure Ex-41), and the numbers $\cos \alpha$, $\cos \beta$, and $\cos \gamma$ are called the **direction cosines** of $\mathbf{v}$.

(a) Show that $\cos \alpha = a/\|\mathbf{v}\|$.

(b) Find $\cos \beta$ and $\cos \gamma$.

(c) Show that $\mathbf{v}/\|\mathbf{v}\| = (\cos \alpha, \cos \beta, \cos \gamma)$.

(d) Show that $\cos^2 \alpha + \cos^2 \beta + \cos^2 \gamma = 1$.

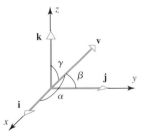

◀ Figure Ex-41

42. Use the result in Exercise 41 to estimate, to the nearest degree, the angles that a diagonal of a box with dimensions 10 cm × 15 cm × 25 cm makes with the edges of the box.

43. Show that if $\mathbf{v}$ is orthogonal to both $\mathbf{w}_1$ and $\mathbf{w}_2$, then $\mathbf{v}$ is orthogonal to $k_1\mathbf{w}_1 + k_2\mathbf{w}_2$ for all scalars k_1 and k_2.

44. Let $\mathbf{u}$ and $\mathbf{v}$ be nonzero vectors in 2- or 3-space, and let $k = \|\mathbf{u}\|$ and $l = \|\mathbf{v}\|$. Show that the vector $\mathbf{w} = l\mathbf{u} + k\mathbf{v}$ bisects the angle between $\mathbf{u}$ and $\mathbf{v}$.

45. Prove part (*a*) of Theorem 3.3.4.

46. Is it possible to have

$$\text{proj}_a \mathbf{u} = \text{proj}_u \mathbf{a}?$$

Explain your reasoning.

True-False Exercises

In parts (a)–(g) determine whether the statement is true or false, and justify your answer.

(a) The vectors $(3, -1, 2)$ and $(0, 0, 0)$ are orthogonal.

(b) If $\mathbf{u}$ and $\mathbf{v}$ are orthogonal vectors, then for all nonzero scalars k and m, $k\mathbf{u}$ and $m\mathbf{v}$ are orthogonal vectors.

(c) The orthogonal projection of $\mathbf{u}$ on $\mathbf{a}$ is perpendicular to the vector component of $\mathbf{u}$ orthogonal to $\mathbf{a}$.

(d) If $\mathbf{a}$ and $\mathbf{b}$ are orthogonal vectors, then for every nonzero vector $\mathbf{u}$, we have

$$\text{proj}_a (\text{proj}_b (\mathbf{u})) = \mathbf{0}$$

(e) If $\mathbf{a}$ and $\mathbf{u}$ are nonzero vectors, then

$$\text{proj}_a (\text{proj}_a (\mathbf{u})) = \text{proj}_a (\mathbf{u})$$

(f) If the relationship

$$\text{proj}_a \mathbf{u} = \text{proj}_a \mathbf{v}$$

holds for some nonzero vector $\mathbf{a}$, then $\mathbf{u} = \mathbf{v}$.

(g) For all vectors $\mathbf{u}$ and $\mathbf{v}$, it is true that

$$\|\mathbf{u} + \mathbf{v}\| = \|\mathbf{u}\| + \|\mathbf{v}\|$$

3.4 The Geometry of Linear Systems

In this section we will use parametric and vector methods to study general systems of linear equations. This work will enable us to interpret solution sets of linear systems with n unknowns as geometric objects in R^n just as we interpreted solution sets of linear systems with two and three unknowns as points, lines, and planes in R^2 and R^3.

Vector and Parametric Equations of Lines in R^2 and R^3

In the last section we derived equations of lines and planes that are determined by a point and a normal vector. However, there are other useful ways of specifying lines and planes. For example, a unique line in R^2 or R^3 is determined by a point $\mathbf{x}_0$ on the line and a nonzero vector $\mathbf{v}$ parallel to the line, and a unique plane in R^3 is determined by a point $\mathbf{x}_0$ in the plane and two noncollinear vectors $\mathbf{v}_1$ and $\mathbf{v}_2$ parallel to the plane. The best way to visualize this is to translate the vectors so their initial points are at $\mathbf{x}_0$ (Figure 3.4.1).

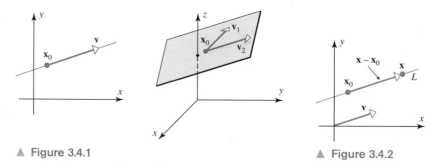

▲ Figure 3.4.1   ▲ Figure 3.4.2

Let us begin by deriving an equation for the line L that contains the point $\mathbf{x}_0$ and is parallel to $\mathbf{v}$. If $\mathbf{x}$ is a general point on such a line, then, as illustrated in Figure 3.4.2, the vector $\mathbf{x} - \mathbf{x}_0$ will be some scalar multiple of $\mathbf{v}$, say

$$\mathbf{x} - \mathbf{x}_0 = t\mathbf{v} \quad \text{or equivalently} \quad \mathbf{x} = \mathbf{x}_0 + t\mathbf{v}$$

As the variable t (called a ***parameter***) varies from $-\infty$ to ∞, the point $\mathbf{x}$ traces out the line L. Accordingly, we have the following result.

Although it is not stated explicitly, it is understood in Formulas (1) and (2) that the parameter t varies from $-\infty$ to ∞. This applies to all vector and parametric equations in this text except where stated otherwise.

THEOREM 3.4.1 *Let L be the line in R^2 or R^3 that contains the point $\mathbf{x}_0$ and is parallel to the nonzero vector $\mathbf{v}$. Then the equation of the line through $\mathbf{x}_0$ that is parallel to $\mathbf{v}$ is*

$$\mathbf{x} = \mathbf{x}_0 + t\mathbf{v} \tag{1}$$

If $\mathbf{x}_0 = \mathbf{0}$, then the line passes through the origin and the equation has the form

$$\mathbf{x} = t\mathbf{v} \tag{2}$$

Vector and Parametric Equations of Planes in R^3

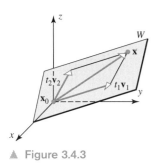

▲ Figure 3.4.3

Next we will derive an equation for the plane W that contains the point $\mathbf{x}_0$ and is parallel to the noncollinear vectors $\mathbf{v}_1$ and $\mathbf{v}_2$. As shown in Figure 3.4.3, if $\mathbf{x}$ is any point in the plane, then by forming suitable scalar multiples of $\mathbf{v}_1$ and $\mathbf{v}_2$, say $t_1\mathbf{v}_1$ and $t_2\mathbf{v}_2$, we can create a parallelogram with diagonal $\mathbf{x} - \mathbf{x}_0$ and adjacent sides $t_1\mathbf{v}_1$ and $t_2\mathbf{v}_2$. Thus, we have

$$\mathbf{x} - \mathbf{x}_0 = t_1\mathbf{v}_1 + t_2\mathbf{v}_2 \quad \text{or equivalently} \quad \mathbf{x} = \mathbf{x}_0 + t_1\mathbf{v}_1 + t_2\mathbf{v}_2$$

As the variables t_1 and t_2 (called *parameters*) vary independently from $-\infty$ to ∞, the point $\mathbf{x}$ varies over the entire plane W. Accordingly, we make the following definition.

THEOREM 3.4.2 *Let W be the plane in R^3 that contains the point $\mathbf{x}_0$ and is parallel to the noncollinear vectors $\mathbf{v}_1$ and $\mathbf{v}_2$. Then an equation of the plane through $\mathbf{x}_0$ that is parallel to $\mathbf{v}_1$ and $\mathbf{v}_2$ is given by*

$$\mathbf{x} = \mathbf{x}_0 + t_1\mathbf{v}_1 + t_2\mathbf{v}_2 \tag{3}$$

If $\mathbf{x}_0 = \mathbf{0}$, then the plane passes through the origin and the equation has the form

$$\mathbf{x} = t_1\mathbf{v}_1 + t_2\mathbf{v}_2 \tag{4}$$

Remark   Observe that the line through $\mathbf{x}_0$ represented by Equation (1) is the translation by $\mathbf{x}_0$ of the line through the origin represented by Equation (2) and that the plane through $\mathbf{x}_0$ represented by Equation (3) is the translation by $\mathbf{x}_0$ of the plane through the origin represented by Equation (4) (Figure 3.4.4).

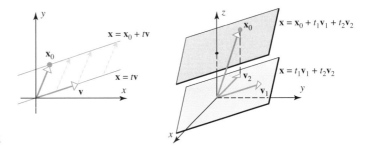

▶ Figure 3.4.4

Motivated by the forms of Formulas (1) to (4), we can extend the notions of line and plane to R^n by making the following definitions.

DEFINITION 1   If $\mathbf{x}_0$ and $\mathbf{v}$ are vectors in R^n, and if $\mathbf{v}$ is nonzero, then the equation

$$\mathbf{x} = \mathbf{x}_0 + t\mathbf{v} \tag{5}$$

defines the ***line through $\mathbf{x}_0$ that is parallel to $\mathbf{v}$***. In the special case where $\mathbf{x}_0 = \mathbf{0}$, the line is said to ***pass through the origin***.

> **DEFINITION 2** If x_0, v_1, and v_2 are vectors in R^n, and if v_1 and v_2 are not collinear, then the equation
>
> $$\mathbf{x} = \mathbf{x}_0 + t_1\mathbf{v}_1 + t_2\mathbf{v}_2 \qquad (6)$$
>
> defines the **plane through x_0 that is parallel to v_1 and v_2**. In the special case where $x_0 = 0$, the plane is said to **pass through the origin**.

Equations (5) and (6) are called **vector forms** of a line and plane in R^n. If the vectors in these equations are expressed in terms of their components and the corresponding components on each side are equated, then the resulting equations are called **parametric equations** of the line and plane. Here are some examples.

▶ **EXAMPLE 1 Vector and Parametric Equations of Lines in R^2 and R^3**

(a) Find a vector equation and parametric equations of the line in R^2 that passes through the origin and is parallel to the vector $\mathbf{v} = (-2, 3)$.

(b) Find a vector equation and parametric equations of the line in R^3 that passes through the point $P_0(1, 2, -3)$ and is parallel to the vector $\mathbf{v} = (4, -5, 1)$.

(c) Use the vector equation obtained in part (b) to find two points on the line that are different from P_0.

Solution (a) It follows from (5) with $\mathbf{x}_0 = \mathbf{0}$ that a vector equation of the line is $\mathbf{x} = t\mathbf{v}$. If we let $\mathbf{x} = (x, y)$, then this equation can be expressed in vector form as

$$(x, y) = t(-2, 3)$$

Equating corresponding components on the two sides of this equation yields the parametric equations

$$x = -2t, \quad y = 3t$$

Solution (b) It follows from (5) that a vector equation of the line is $\mathbf{x} = \mathbf{x}_0 + t\mathbf{v}$. If we let $\mathbf{x} = (x, y, z)$, and if we take $\mathbf{x}_0 = (1, 2, -3)$, then this equation can be expressed in vector form as

$$(x, y, z) = (1, 2, -3) + t(4, -5, 1) \qquad (7)$$

Equating corresponding components on the two sides of this equation yields the parametric equations

$$x = 1 + 4t, \quad y = 2 - 5t, \quad z = -3 + t$$

Solution (c) A point on the line represented by Equation (7) can be obtained by substituting a specific numerical value for the parameter t. However, since $t = 0$ produces $(x, y, z) = (1, 2, -3)$, which is the point P_0, this value of t does not serve our purpose. Taking $t = 1$ produces the point $(5, -3, -2)$ and taking $t = -1$ produces the point $(-3, 7, -4)$. Any other distinct values for t (except $t = 0$) would work just as well. ◀

▶ **EXAMPLE 2 Vector and Parametric Equations of a Plane in R^3**

Find vector and parametric equations of the plane $x - y + 2z = 5$.

Solution We will find the parametric equations first. We can do this by solving the equation for any one of the variables in terms of the other two and then using those two variables as parameters. For example, solving for x in terms of y and z yields

$$x = 5 + y - 2z \qquad (8)$$

We would have obtained different parametric and vector equations in Example 2 had we solved (8) for y or z rather than x. However, one can show the same plane results in all three cases as the parameters vary from $-\infty$ to ∞.

and then using y and z as parameters t_1 and t_2, respectively, yields the parametric equations

$$x = 5 + t_1 - 2t_2, \quad y = t_1, \quad z = t_2$$

To obtain a vector equation of the plane we rewrite these parametric equations as

$$(x, y, z) = (5 + t_1 - 2t_2, t_1, t_2)$$

or, equivalently, as

$$(x, y, z) = (5, 0, 0) + t_1(1, 1, 0) + t_2(-2, 0, 1)$$

▶ **EXAMPLE 3 Vector and Parametric Equations of Lines and Planes in R^4**

(a) Find vector and parametric equations of the line through the origin of R^4 that is parallel to the vector $\mathbf{v} = (5, -3, 6, 1)$.

(b) Find vector and parametric equations of the plane in R^4 that passes through the point $\mathbf{x}_0 = (2, -1, 0, 3)$ and is parallel to both $\mathbf{v}_1 = (1, 5, 2, -4)$ and $\mathbf{v}_2 = (0, 7, -8, 6)$.

Solution (a) If we let $\mathbf{x} = (x_1, x_2, x_3, x_4)$, then the vector equation $\mathbf{x} = t\mathbf{v}$ can be expressed as

$$(x_1, x_2, x_3, x_4) = t(5, -3, 6, 1)$$

Equating corresponding components yields the parametric equations

$$x_1 = 5t, \quad x_2 = -3t, \quad x_3 = 6t, \quad x_4 = t$$

Solution (b) The vector equation $\mathbf{x} = \mathbf{x}_0 + t_1\mathbf{v}_1 + t_2\mathbf{v}_2$ can be expressed as

$$(x_1, x_2, x_3, x_4) = (2, -1, 0, 3) + t_1(1, 5, 2, -4) + t_2(0, 7, -8, 6)$$

which yields the parametric equations

$$x_1 = 2 + t_1$$
$$x_2 = -1 + 5t_1 + 7t_2$$
$$x_3 = 2t_1 - 8t_2$$
$$x_4 = 3 - 4t_1 + 6t_2 \quad ◀$$

Lines Through Two Points in R^n

If $\mathbf{x}_0$ and $\mathbf{x}_1$ are distinct points in R^n, then the line determined by these points is parallel to the vector $\mathbf{v} = \mathbf{x}_1 - \mathbf{x}_0$ (Figure 3.4.5), so it follows from (5) that the line can be expressed in vector form as

$$\mathbf{x} = \mathbf{x}_0 + t(\mathbf{x}_1 - \mathbf{x}_0) \tag{9}$$

or, equivalently, as

$$\mathbf{x} = (1 - t)\mathbf{x}_0 + t\mathbf{x}_1 \tag{10}$$

These are called the *two-point vector equations* of a line in R^n.

▲ Figure 3.4.5

▶ **EXAMPLE 4 A Line Through Two Points in R^2**

Find vector and parametric equations for the line in R^2 that passes through the points $P(0, 7)$ and $Q(5, 0)$.

Solution We will see below that it does not matter which point we take to be $\mathbf{x}_0$ and which we take to be $\mathbf{x}_1$, so let us choose $\mathbf{x}_0 = (0, 7)$ and $\mathbf{x}_1 = (5, 0)$. It follows that $\mathbf{x}_1 - \mathbf{x}_0 = (5, -7)$ and hence that

$$(x, y) = (0, 7) + t(5, -7) \tag{11}$$

which we can rewrite in parametric form as

$$x = 5t, \quad y = 7 - 7t$$

Had we reversed our choices and taken $\mathbf{x}_0 = (5, 0)$ and $\mathbf{x}_1 = (0, 7)$, then the resulting vector equation would have been

$$(x, y) = (5, 0) + t(-5, 7) \tag{12}$$

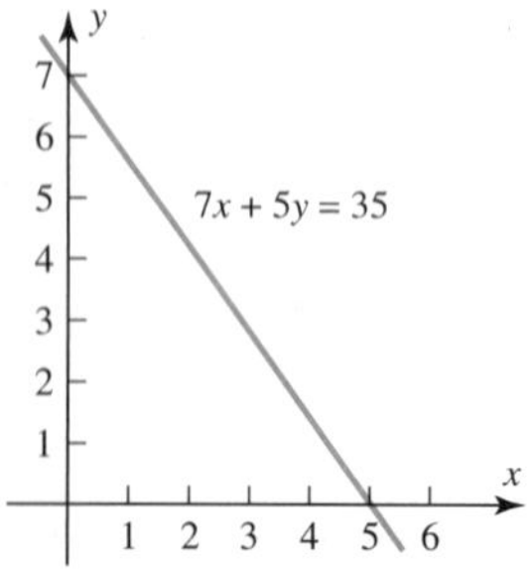

and the parametric equations would have been

$$x = 5 - 5t, \quad y = 7t$$

(verify). Although (11) and (12) look different, they both represent the line whose equation in rectangular coordinates is

$$7x + 5y = 35$$

(Figure 3.4.6). This can be seen by eliminating the parameter t from the parametric equations (verify). ◀

▲ Figure 3.4.6

The point $\mathbf{x} = (x, y)$ in Equations (10) and (11) traces an entire line in R^2 as the parameter t varies over the interval $(-\infty, \infty)$. If, however, we restrict the parameter to vary from $t = 0$ to $t = 1$, then $\mathbf{x}$ will not trace the entire line but rather just the *line segment* joining the points $\mathbf{x}_0$ and $\mathbf{x}_1$. The point $\mathbf{x}$ will start at $\mathbf{x}_0$ when $t = 0$ and end at $\mathbf{x}_1$ when $t = 1$. Accordingly, we make the following definition.

DEFINITION 3   If $\mathbf{x}_0$ and $\mathbf{x}_1$ are vectors in R^n, then the equation

$$\mathbf{x} = \mathbf{x}_0 + t(\mathbf{x}_1 - \mathbf{x}_0) \quad (0 \le t \le 1) \tag{13}$$

defines the **line segment from $\mathbf{x}_0$ to $\mathbf{x}_1$**. When convenient, Equation (13) can be written as

$$\mathbf{x} = (1 - t)\mathbf{x}_0 + t\mathbf{x}_1 \quad (0 \le t \le 1) \tag{14}$$

▶ **EXAMPLE 5   A Line Segment from One Point to Another in R^2**

It follows from (13) and (14) that the line segment in R^2 from $\mathbf{x}_0 = (1, -3)$ to $\mathbf{x}_1 = (5, 6)$ can be represented either by the equation

$$\mathbf{x} = (1, -3) + t(4, 9) \quad (0 \le t \le 1)$$

or by

$$\mathbf{x} = (1 - t)(1, -3) + t(5, 6) \quad (0 \le t \le 1) \; ◀$$

Dot Product Form of a Linear System

Our next objective is to show how to express linear equations and linear systems in dot product notation. This will lead us to some important results about orthogonality and linear systems.

Recall that a *linear equation* in the variables $x_1, x_2, \ldots, x_n$ has the form

$$a_1x_1 + a_2x_2 + \cdots + a_nx_n = b \quad (a_1, a_2, \ldots, a_n \text{ not all zero}) \tag{15}$$

and that the corresponding *homogeneous* equation is

$$a_1x_1 + a_2x_2 + \cdots + a_nx_n = 0 \quad (a_1, a_2, \ldots, a_n \text{ not all zero}) \tag{16}$$

These equations can be rewritten in vector form by letting

$$\mathbf{a} = (a_1, a_2, \ldots, a_n) \quad \text{and} \quad \mathbf{x} = (x_1, x_2, \ldots, x_n)$$

in which case Formula (15) can be written as

$$\mathbf{a} \cdot \mathbf{x} = b \tag{17}$$

and Formula (16) as

$$\mathbf{a} \cdot \mathbf{x} = 0 \tag{18}$$

Except for a notational change from $\mathbf{n}$ to $\mathbf{a}$, Formula (18) is the extension to R^n of Formula (6) in Section 3.3. This equation reveals that *each solution vector* $\mathbf{x}$ *of a homogeneous equation is orthogonal to the coefficient vector* $\mathbf{a}$. To take this geometric observation a step further, consider the homogeneous system

$$\begin{aligned}
a_{11}x_1 + a_{12}x_2 + \cdots + a_{1n}x_n &= 0 \\
a_{21}x_1 + a_{22}x_2 + \cdots + a_{2n}x_n &= 0 \\
\vdots \qquad \vdots \qquad\qquad \vdots \qquad \vdots \\
a_{m1}x_1 + a_{m2}x_2 + \cdots + a_{mn}x_n &= 0
\end{aligned}$$

If we denote the successive row vectors of the coefficient matrix by $\mathbf{r}_1, \mathbf{r}_2, \ldots, \mathbf{r}_m$, then we can rewrite this system in dot product form as

$$\begin{aligned}
\mathbf{r}_1 \cdot \mathbf{x} &= 0 \\
\mathbf{r}_2 \cdot \mathbf{x} &= 0 \\
\vdots \qquad \vdots \\
\mathbf{r}_m \cdot \mathbf{x} &= 0
\end{aligned} \tag{19}$$

from which we see that every solution vector $\mathbf{x}$ is orthogonal to every row vector of the coefficient matrix. In summary, we have the following result.

THEOREM 3.4.3 *If A is an $m \times n$ matrix, then the solution set of the homogeneous linear system $A\mathbf{x} = \mathbf{0}$ consists of all vectors in R^n that are orthogonal to every row vector of A.*

▶ **EXAMPLE 6 Orthogonality of Row Vectors and Solution Vectors**

We showed in Example 6 of Section 1.2 that the general solution of the homogeneous linear system

$$\begin{bmatrix} 1 & 3 & -2 & 0 & 2 & 0 \\ 2 & 6 & -5 & -2 & 4 & -3 \\ 0 & 0 & 5 & 10 & 0 & 15 \\ 2 & 6 & 0 & 8 & 4 & 18 \end{bmatrix} \begin{bmatrix} x_1 \\ x_2 \\ x_3 \\ x_4 \\ x_5 \\ x_6 \end{bmatrix} = \begin{bmatrix} 0 \\ 0 \\ 0 \\ 0 \end{bmatrix}$$

is

$$x_1 = -3r - 4s - 2t, \quad x_2 = r, \quad x_3 = -2s, \quad x_4 = s, \quad x_5 = t, \quad x_6 = 0$$

which we can rewrite in vector form as

$$\mathbf{x} = (-3r - 4s - 2t, r, -2s, s, t, 0)$$

According to Theorem 3.4.3, the vector $\mathbf{x}$ must be orthogonal to each of the row vectors

$$\begin{aligned}
\mathbf{r}_1 &= (1, 3, -2, 0, 2, 0) \\
\mathbf{r}_2 &= (2, 6, -5, -2, 4, -3) \\
\mathbf{r}_3 &= (0, 0, 5, 10, 0, 15) \\
\mathbf{r}_4 &= (2, 6, 0, 8, 4, 18)
\end{aligned}$$

We will confirm that $\mathbf{x}$ is orthogonal to $\mathbf{r}_1$, and leave it for you to verify that $\mathbf{x}$ is orthogonal to the other three row vectors as well. The dot product of $\mathbf{r}_1$ and $\mathbf{x}$ is

$$\mathbf{r}_1 \cdot \mathbf{x} = 1(-3r - 4s - 2t) + 3(r) + (-2)(-2s) + 0(s) + 2(t) + 0(0) = 0$$

which establishes the orthogonality. ◄

The Relationship Between $A\mathbf{x} = \mathbf{0}$ *and* $A\mathbf{x} = \mathbf{b}$

We will conclude this section by exploring the relationship between the solutions of a homogeneous linear system $A\mathbf{x} = \mathbf{0}$ and the solutions (if any) of a nonhomogeneous linear system $A\mathbf{x} = \mathbf{b}$ that has the same coefficient matrix. These are called *corresponding linear systems*.

To motivate the result we are seeking, let us compare the solutions of the corresponding linear systems

$$\begin{bmatrix} 1 & 3 & -2 & 0 & 2 & 0 \\ 2 & 6 & -5 & -2 & 4 & -3 \\ 0 & 0 & 5 & 10 & 0 & 15 \\ 2 & 6 & 0 & 8 & 4 & 18 \end{bmatrix} \begin{bmatrix} x_1 \\ x_2 \\ x_3 \\ x_4 \\ x_5 \\ x_6 \end{bmatrix} = \begin{bmatrix} 0 \\ 0 \\ 0 \\ 0 \end{bmatrix} \quad \text{and} \quad \begin{bmatrix} 1 & 3 & -2 & 0 & 2 & 0 \\ 2 & 6 & -5 & -2 & 4 & -3 \\ 0 & 0 & 5 & 10 & 0 & 15 \\ 2 & 6 & 0 & 8 & 4 & 18 \end{bmatrix} \begin{bmatrix} x_1 \\ x_2 \\ x_3 \\ x_4 \\ x_5 \\ x_6 \end{bmatrix} = \begin{bmatrix} 0 \\ -1 \\ 5 \\ 6 \end{bmatrix}$$

We showed in Examples 5 and 6 of Section 1.2 that the general solutions of these linear systems can be written in parametric form as

homogeneous $\longrightarrow$ $x_1 = -3r - 4s - 2t$, $x_2 = r$, $x_3 = -2s$, $x_4 = s$, $x_5 = t$, $x_6 = 0$

nonhomogeneous $\longrightarrow$ $x_1 = -3r - 4s - 2t$, $x_2 = r$, $x_3 = -2s$, $x_4 = s$, $x_5 = t$, $x_6 = \frac{1}{3}$

which we can then rewrite in vector form as

homogeneous $\longrightarrow$ $(x_1, x_2, x_3, x_4, x_5) = (-3r - 4s - 2t, r, -2s, s, t, 0)$

nonhomogeneous $\longrightarrow$ $(x_1, x_2, x_3, x_4, x_5) = \left(-3r - 4s - 2t, r, -2s, s, t, \frac{1}{3} \right)$

By splitting the vectors on the right apart and collecting terms with like parameters, we can rewrite these equations as

homogeneous $\longrightarrow$ $(x_1, x_2, x_3, x_4, x_5) = r(-3, 1, 0, 0, 0) + s(-4, 0, -2, 1, 0, 0) + t(-2, 0, 0, 0, 1, 0)$ (20)

nonhomogeneous $\longrightarrow$ $(x_1, x_2, x_3, x_4, x_5) = r(-3, 1, 0, 0, 0) + s(-4, 0, -2, 1, 0, 0)$
$$+ t(-2, 0, 0, 0, 1, 0) + \left(0, 0, 0, 0, 0, \frac{1}{3} \right) \quad (21)$$

Formulas (20) and (21) reveal that each solution of the nonhomogeneous system can be obtained by adding the fixed vector $\left(0, 0, 0, 0, 0, \frac{1}{3} \right)$ to the corresponding solution of the homogeneous system. This is a special case of the following general result.

> **THEOREM 3.4.4** *The general solution of a consistent linear system* $A\mathbf{x} = \mathbf{b}$ *can be obtained by adding any specific solution of* $A\mathbf{x} = \mathbf{b}$ *to the general solution of* $A\mathbf{x} = \mathbf{0}$.

Proof Let $\mathbf{x}_0$ be any specific solution of $A\mathbf{x} = \mathbf{b}$, let W denote the solution set of $A\mathbf{x} = \mathbf{0}$, and let $\mathbf{x}_0 + W$ denote the set of all vectors that result by adding $\mathbf{x}_0$ to each vector in W. We must show that if $\mathbf{x}$ is a vector in $\mathbf{x}_0 + W$, then $\mathbf{x}$ is a solution of $A\mathbf{x} = \mathbf{b}$, and conversely, that every solution of $A\mathbf{x} = \mathbf{b}$ is in the set $\mathbf{x}_0 + W$.

Assume first that $\mathbf{x}$ is a vector in $\mathbf{x}_0 + W$. This implies that $\mathbf{x}$ is expressible in the form $\mathbf{x} = \mathbf{x}_0 + \mathbf{w}$, where $A\mathbf{x}_0 = \mathbf{b}$ and $A\mathbf{w} = \mathbf{0}$. Thus,

$$A\mathbf{x} = A(\mathbf{x}_0 + \mathbf{w}) = A\mathbf{x}_0 + A\mathbf{w} = \mathbf{b} + \mathbf{0} = \mathbf{b}$$

which shows that $\mathbf{x}$ is a solution of $A\mathbf{x} = \mathbf{b}$.

Conversely, let $\mathbf{x}$ be any solution of $A\mathbf{x} = \mathbf{b}$. To show that $\mathbf{x}$ is in the set $\mathbf{x}_0 + W$ we must show that $\mathbf{x}$ is expressible in the form

$$\mathbf{x} = \mathbf{x}_0 + \mathbf{w} \qquad (22)$$

where $\mathbf{w}$ is in W (i.e., $A\mathbf{w} = \mathbf{0}$). We can do this by taking $\mathbf{w} = \mathbf{x} - \mathbf{x}_0$. This vector obviously satisfies (22), and it is in W since

$$A\mathbf{w} = A(\mathbf{x} - \mathbf{x}_0) = A\mathbf{x} - A\mathbf{x}_0 = \mathbf{b} - \mathbf{b} = \mathbf{0} \quad \blacktriangleleft$$

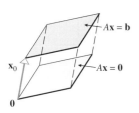

▲ Figure 3.4.7 The solution set of $A\mathbf{x} = \mathbf{b}$ is a translation of the solution space of $A\mathbf{x} = \mathbf{0}$.

Remark   Theorem 3.4.4 has a useful geometric interpretation that is illustrated in Figure 3.4.7. If, as discussed in Section 3.1, we interpret vector addition as translation, then the theorem states that if $\mathbf{x}_0$ is *any* specific solution of $A\mathbf{x} = \mathbf{b}$, then the *entire* solution set of $A\mathbf{x} = \mathbf{b}$ can be obtained by translating the solution set of $A\mathbf{x} = \mathbf{0}$ by the vector $\mathbf{x}_0$.

Concept Review

- Parameters
- Parametric equations of lines
- Parametric equations of planes
- Two-point vector equations of a line
- Vector equation of a line
- Vector equation of a plane

Skills

- Express the equations of lines in R^2 and R^3 using either vector or parametric equations.
- Express the equations of planes in R^n using either vector or parametric equations.
- Express the equation of a line containing two given points in R^2 or R^3 using either vector or parametric equations.
- Find equations of a line and a line segment.

- Verify the orthogonality of the row vectors of a linear system of equations and a solution vector.
- Use a specific solution to the nonhomogeneous linear system $A\mathbf{x} = \mathbf{b}$ and the general solution of the corresponding linear system $A\mathbf{x} = \mathbf{0}$ to obtain the general solution to $A\mathbf{x} = \mathbf{b}$.

Exercise Set 3.4

In Exercises 1–4, find vector and parametric equations of the line containing the point and parallel to the vector. ◀

1. Point: $(-4, 1)$; vector: $\mathbf{v} = (0, -8)$

2. Point: $(2, -1)$; vector: $\mathbf{v} = (-4, -2)$

3. Point: $(0, 0, 0)$; vector: $\mathbf{v} = (-3, 0, 1)$

4. Point: $(-9, 3, 4)$; vector: $\mathbf{v} = (-1, 6, 0)$

In Exercises 5–8, use the given equation of a line to find a point on the line and a vector parallel to the line. ◀

5. $\mathbf{x} = (3 - 5t, -6 - t)$ **6.** $(x, y, z) = (4t, 7, 4 + 3t)$

7. $\mathbf{x} = (1 - t)(4, 6) + t(-2, 0)$

8. $\mathbf{x} = (1 - t)(0, -5, 1)$

In Exercises 9–12, find vector and parametric equations of the plane containing the given point and parallel vectors. ◀

9. Point: $(-3, 1, 0)$; vectors: $\mathbf{v}_1 = (0, -3, 6)$ and $\mathbf{v}_2 = (-5, 1, 2)$

10. Point: $(0, 6, -2)$; vectors: $\mathbf{v}_1 = (0, 9, -1)$ and $\mathbf{v}_2 = (0, -3, 0)$

11. Point: $(-1, 1, 4)$; vectors: $\mathbf{v}_1 = (6, -1, 0)$ and $\mathbf{v}_2 = (-1, 3, 1)$

12. Point: $(0, 5, -4)$; vectors: $\mathbf{v}_1 = (0, 0, -5)$ and $\mathbf{v}_2 = (1, -3, -2)$

In Exercises 13–14, find vector and parametric equations of the line in R^2 that passes through the origin and is orthogonal to $\mathbf{v}$.

13. $\mathbf{v} = (-2, 3)$ **14.** $\mathbf{v} = (1, -4)$

In Exercises 15–16, find vector and parametric equations of the plane in R^3 that passes through the origin and is orthogonal to $\mathbf{v}$. ◀

15. $\mathbf{v} = (4, 0, -5)$ [*Hint:* Construct two nonparallel vectors orthogonal to $\mathbf{v}$ in R^3].

16. $\mathbf{v} = (3, 1, -6)$

In Exercises 17–20, find the general solution to the linear system and confirm that the row vectors of the coefficient matrix are orthogonal to the solution vectors. ◀

17. $\begin{aligned} x_1 + x_2 + x_3 &= 0 \\ 2x_1 + 2x_2 + 2x_3 &= 0 \\ 3x_1 + 3x_2 + 3x_3 &= 0 \end{aligned}$

18. $\begin{aligned} x_1 + 3x_2 - 4x_3 &= 0 \\ 2x_1 + 6x_2 - 8x_3 &= 0 \end{aligned}$

19. $\begin{aligned} x_1 + 5x_2 + x_3 + 2x_4 - x_5 &= 0 \\ x_1 - 2x_2 - x_3 + 3x_4 + 2x_5 &= 0 \end{aligned}$

20. $\begin{aligned} x_1 + 3x_2 - 4x_3 &= 0 \\ x_1 + 2x_2 + 3x_3 &= 0 \end{aligned}$

21. (a) The equation $x + y + z = 1$ can be viewed as a linear system of one equation in three unknowns. Express a general solution of this equation as a particular solution plus a general solution of the associated homogeneous system.

(b) Give a geometric interpretation of the result in part (a).

22. (a) The equation $x + y = 1$ can be viewed as a linear system of one equation in two unknowns. Express a general solution of this equation as a particular solution plus a general solution of the associated homogeneous system.

(b) Give a geometric interpretation of the result in part (a).

23. (a) Find a homogeneous linear system of two equations in three unknowns whose solution space consists of those vectors in R^3 that are orthogonal to $\mathbf{a} = (1, 1, 1)$ and $\mathbf{b} = (-2, 3, 0)$.

(b) What kind of geometric object is the solution space?

(c) Find a general solution of the system obtained in part (a), and confirm that Theorem 3.4.3 holds.

24. (a) Find a homogeneous linear system of two equations in three unknowns whose solution space consists of those vectors in R^3 that are orthogonal to $\mathbf{a} = (-3, 2, -1)$ and $\mathbf{b} = (0, -2, -2)$.

(b) What kind of geometric object is the solution space?

(c) Find a general solution of the system obtained in part (a), and confirm that Theorem 3.4.3 holds.

25. Consider the linear systems

$$\begin{bmatrix} 3 & 2 & -1 \\ 6 & 4 & -2 \\ -3 & -2 & 1 \end{bmatrix} \begin{bmatrix} x_1 \\ x_2 \\ x_3 \end{bmatrix} = \begin{bmatrix} 0 \\ 0 \\ 0 \end{bmatrix}$$

and

$$\begin{bmatrix} 3 & 2 & -1 \\ 6 & 4 & -2 \\ -3 & -2 & 1 \end{bmatrix} \begin{bmatrix} x_1 \\ x_2 \\ x_3 \end{bmatrix} = \begin{bmatrix} 2 \\ 4 \\ -2 \end{bmatrix}$$

(a) Find a general solution of the homogeneous system.

(b) Confirm that $x_1 = 1, x_2 = 0, x_3 = 1$ is a solution of the nonhomogeneous system.

(c) Use the results in parts (a) and (b) to find a general solution of the nonhomogeneous system.

(d) Check your result in part (c) by solving the nonhomogeneous system directly.

26. Consider the linear systems

$$\begin{bmatrix} 1 & -2 & 3 \\ 2 & 1 & 4 \\ 1 & -7 & 5 \end{bmatrix} \begin{bmatrix} x_1 \\ x_2 \\ x_3 \end{bmatrix} = \begin{bmatrix} 0 \\ 0 \\ 0 \end{bmatrix}$$

and

$$\begin{bmatrix} 1 & -2 & 3 \\ 2 & 1 & 4 \\ 1 & -7 & 5 \end{bmatrix} \begin{bmatrix} x_1 \\ x_2 \\ x_3 \end{bmatrix} = \begin{bmatrix} 2 \\ 7 \\ -1 \end{bmatrix}$$

(a) Find a general solution of the homogeneous system.

(b) Confirm that $x_1 = 1, x_2 = 1, x_3 = 1$ is a solution of the nonhomogeneous system.

(c) Use the results in parts (a) and (b) to find a general solution of the nonhomogeneous system.

(d) Check your result in part (c) by solving the nonhomogeneous system directly.

In Exercises 27–28, find a general solution of the system, and use that solution to find a general solution of the associated homogeneous system and a particular solution of the given system.

27. $\begin{bmatrix} 3 & 4 & 1 & 2 \\ 6 & 8 & 2 & 5 \\ 9 & 12 & 3 & 10 \end{bmatrix} \begin{bmatrix} x_1 \\ x_2 \\ x_3 \\ x_4 \end{bmatrix} = \begin{bmatrix} 3 \\ 7 \\ 13 \end{bmatrix}$

28. $\begin{bmatrix} 9 & -3 & 5 & 6 \\ 6 & -2 & 3 & 1 \\ 3 & -1 & 3 & 14 \end{bmatrix} \begin{bmatrix} x_1 \\ x_2 \\ x_3 \\ x_4 \end{bmatrix} = \begin{bmatrix} 4 \\ 5 \\ -8 \end{bmatrix}$

True-False Exercises

In parts (a)–(f) determine whether the statement is true or false, and justify your answer.

(a) The vector equation of a line can be determined from any point lying on the line and a nonzero vector parallel to the line.

(b) The vector equation of a plane can be determined from any point lying in the plane and a nonzero vector parallel to the plane.

(c) The points lying on a line through the origin in R^2 or R^3 are all scalar multiples of any nonzero vector on the line.

(d) All solution vectors of the linear system $A\mathbf{x} = \mathbf{b}$ are orthogonal to the row vectors of the matrix A if and only if $\mathbf{b} = \mathbf{0}$.

(e) The general solution of the nonhomogeneous linear system $A\mathbf{x} = \mathbf{b}$ can be obtained by adding $\mathbf{b}$ to the general solution of the homogeneous linear system $A\mathbf{x} = \mathbf{0}$.

(f) If $\mathbf{x}_1$ and $\mathbf{x}_2$ are two solutions of the nonhomogeneous linear system $A\mathbf{x} = \mathbf{b}$, then $\mathbf{x}_1 - \mathbf{x}_2$ is a solution of the corresponding homogeneous linear system.

3.5 Cross Product

This optional section is concerned with properties of vectors in 3-space that are important to physicists and engineers. It can be omitted, if desired, since subsequent sections do not depend on its content. Among other things, we define an operation that provides a way of constructing a vector in 3-space that is perpendicular to two given vectors, and we give a geometric interpretation of 3×3 determinants.

Cross Product of Vectors   In Section 3.2 we defined the dot product of two vectors $\mathbf{u}$ and $\mathbf{v}$ in n-space. That operation produced a *scalar* as its result. We will now define a type of vector multiplication that produces a *vector* as the result but which is applicable only to vectors in 3-space.

DEFINITION 1 If $\mathbf{u} = (u_1, u_2, u_3)$ and $\mathbf{v} = (v_1, v_2, v_3)$ are vectors in 3-space, then the **cross product** $\mathbf{u} \times \mathbf{v}$ is the vector defined by

$$\mathbf{u} \times \mathbf{v} = (u_2 v_3 - u_3 v_2, u_3 v_1 - u_1 v_3, u_1 v_2 - u_2 v_1)$$

or, in determinant notation,

$$\mathbf{u} \times \mathbf{v} = \left(\begin{vmatrix} u_2 & u_3 \\ v_2 & v_3 \end{vmatrix}, -\begin{vmatrix} u_1 & u_3 \\ v_1 & v_3 \end{vmatrix}, \begin{vmatrix} u_1 & u_2 \\ v_1 & v_2 \end{vmatrix} \right) \tag{1}$$

Remark   Instead of memorizing (1), you can obtain the components of $\mathbf{u} \times \mathbf{v}$ as follows:

- Form the 2×3 matrix $\begin{bmatrix} u_1 & u_2 & u_3 \\ v_1 & v_2 & v_3 \end{bmatrix}$ whose first row contains the components of $\mathbf{u}$ and whose second row contains the components of $\mathbf{v}$.

- To find the first component of $\mathbf{u} \times \mathbf{v}$, delete the first column and take the determinant; to find the second component, delete the second column and take the negative of the determinant; and to find the third component, delete the third column and take the determinant.

▶ **EXAMPLE 1 Calculating a Cross Product**

Find $\mathbf{u} \times \mathbf{v}$, where $\mathbf{u} = (1, 2, -2)$ and $\mathbf{v} = (3, 0, 1)$.

Solution   From either (1) or the mnemonic in the preceding remark, we have

$$\mathbf{u} \times \mathbf{v} = \left(\begin{vmatrix} 2 & -2 \\ 0 & 1 \end{vmatrix}, -\begin{vmatrix} 1 & -2 \\ 3 & 1 \end{vmatrix}, \begin{vmatrix} 1 & 2 \\ 3 & 0 \end{vmatrix} \right)$$
$$= (2, -7, -6) \ ◀$$

The following theorem gives some important relationships between the dot product and cross product and also shows that $\mathbf{u} \times \mathbf{v}$ is orthogonal to both $\mathbf{u}$ and $\mathbf{v}$.

Historical Note The cross product notation $A \times B$ was introduced by the American physicist and mathematician J. Willard Gibbs, (see p. 134) in a series of unpublished lecture notes for his students at Yale University. It appeared in a published work for the first time in the second edition of the book *Vector Analysis*, by Edwin Wilson (1879–1964), a student of Gibbs. Gibbs originally referred to $A \times B$ as the "skew product."

THEOREM 3.5.1 **Relationships Involving Cross Product and Dot Product**

If **u**, **v**, *and* **w** *are vectors in 3-space, then*

(a) $\mathbf{u} \cdot (\mathbf{u} \times \mathbf{v}) = 0$ (**u** × **v** *is orthogonal to* **u**)

(b) $\mathbf{v} \cdot (\mathbf{u} \times \mathbf{v}) = 0$ (**u** × **v** *is orthogonal to* **v**)

(c) $\|\mathbf{u} \times \mathbf{v}\|^2 = \|\mathbf{u}\|^2 \|\mathbf{v}\|^2 - (\mathbf{u} \cdot \mathbf{v})^2$ (*Lagrange's identity*)

(d) $\mathbf{u} \times (\mathbf{v} \times \mathbf{w}) = (\mathbf{u} \cdot \mathbf{w})\mathbf{v} - (\mathbf{u} \cdot \mathbf{v})\mathbf{w}$ (*relationship between cross and dot products*)

(e) $(\mathbf{u} \times \mathbf{v}) \times \mathbf{w} = (\mathbf{u} \cdot \mathbf{w})\mathbf{v} - (\mathbf{v} \cdot \mathbf{w})\mathbf{u}$ (*relationship between cross and dot products*)

Proof (a)   Let $\mathbf{u} = (u_1, u_2, u_3)$ and $\mathbf{v} = (v_1, v_2, v_3)$. Then

$$\mathbf{u} \cdot (\mathbf{u} \times \mathbf{v}) = (u_1, u_2, u_3) \cdot (u_2 v_3 - u_3 v_2, u_3 v_1 - u_1 v_3, u_1 v_2 - u_2 v_1)$$
$$= u_1(u_2 v_3 - u_3 v_2) + u_2(u_3 v_1 - u_1 v_3) + u_3(u_1 v_2 - u_2 v_1) = 0$$

Proof (b)   Similar to (*a*).

Proof (c)   Since

$$\|\mathbf{u} \times \mathbf{v}\|^2 = (u_2 v_3 - u_3 v_2)^2 + (u_3 v_1 - u_1 v_3)^2 + (u_1 v_2 - u_2 v_1)^2 \tag{2}$$

and

$$\|\mathbf{u}\|^2 \|\mathbf{v}\|^2 - (\mathbf{u} \cdot \mathbf{v})^2 = (u_1^2 + u_2^2 + u_3^2)(v_1^2 + v_2^2 + v_3^2) - (u_1 v_1 + u_2 v_2 + u_3 v_3)^2 \tag{3}$$

the proof can be completed by "multiplying out" the right sides of (2) and (3) and verifying their equality.

Proof (d) and (e)   See Exercises 38 and 39. ◄

▶ **EXAMPLE 2** **u × v Is Perpendicular to u and to v**

Consider the vectors

$$\mathbf{u} = (1, 2, -2) \quad \text{and} \quad \mathbf{v} = (3, 0, 1)$$

In Example 1 we showed that

$$\mathbf{u} \times \mathbf{v} = (2, -7, -6)$$

Since

$$\mathbf{u} \cdot (\mathbf{u} \times \mathbf{v}) = (1)(2) + (2)(-7) + (-2)(-6) = 0$$

and

$$\mathbf{v} \cdot (\mathbf{u} \times \mathbf{v}) = (3)(2) + (0)(-7) + (1)(-6) = 0$$

u × **v** is orthogonal to both **u** and **v**, as guaranteed by Theorem 3.5.1. ◄

Historical Note   Joseph Louis Lagrange was a French-Italian mathematician and astronomer. Although his father wanted him to become a lawyer, Lagrange was attracted to mathematics and astronomy after reading a memoir by the astronomer Halley. At age 16 he began to study mathematics on his own and by age 19 was appointed to a professorship at the Royal Artillery School in Turin. The following year he solved some famous problems using new methods that eventually blossomed into a branch of mathematics called the *calculus of variations*. These methods and Lagrange's applications of them to problems in celestial mechanics were so monumental that by age 25 he was regarded by many of his contemporaries as the greatest living mathematician. One of Lagrange's most famous works is a memoir, *Mécanique Analytique*, in which he reduced the theory of mechanics to a few general formulas from which all other necessary equations could be derived. Napoleon was a great admirer of Lagrange and showered him with many honors. In spite of his fame, Lagrange was a shy and modest man. On his death, he was buried with honor in the Pantheon.

[*Image:* ©SSPL/The Image Works]

Joseph Louis Lagrange
(1736–1813)

The main arithmetic properties of the cross product are listed in the next theorem.

THEOREM 3.5.2 Properties of Cross Product

If **u**, **v**, *and* **w** *are any vectors in 3-space and k is any scalar, then*:

(*a*) $\mathbf{u} \times \mathbf{v} = -(\mathbf{v} \times \mathbf{u})$

(*b*) $\mathbf{u} \times (\mathbf{v} + \mathbf{w}) = (\mathbf{u} \times \mathbf{v}) + (\mathbf{u} \times \mathbf{w})$

(*c*) $(\mathbf{u} + \mathbf{v}) \times \mathbf{w} = (\mathbf{u} \times \mathbf{w}) + (\mathbf{v} \times \mathbf{w})$

(*d*) $k(\mathbf{u} \times \mathbf{v}) = (k\mathbf{u}) \times \mathbf{v} = \mathbf{u} \times (k\mathbf{v})$

(*e*) $\mathbf{u} \times \mathbf{0} = \mathbf{0} \times \mathbf{u} = \mathbf{0}$

(*f*) $\mathbf{u} \times \mathbf{u} = \mathbf{0}$

The proofs follow immediately from Formula (1) and properties of determinants; for example, part (*a*) can be proved as follows.

Proof (a)   Interchanging **u** and **v** in (1) interchanges the rows of the three determinants on the right side of (1) and hence changes the sign of each component in the cross product. Thus $\mathbf{u} \times \mathbf{v} = -(\mathbf{v} \times \mathbf{u})$. ◄

The proofs of the remaining parts are left as exercises.

▶ **EXAMPLE 3 Standard Unit Vectors**

Consider the vectors

$$\mathbf{i} = (1, 0, 0), \quad \mathbf{j} = (0, 1, 0), \quad \mathbf{k} = (0, 0, 1)$$

These vectors each have length 1 and lie along the coordinate axes (Figure 3.5.1). They are called the ***standard unit vectors*** in 3-space. Every vector $\mathbf{v} = (v_1, v_2, v_3)$ in 3-space is expressible in terms of **i**, **j**, and **k** since we can write

$$\mathbf{v} = (v_1, v_2, v_3) = v_1(1, 0, 0) + v_2(0, 1, 0) + v_3(0, 0, 1) = v_1\mathbf{i} + v_2\mathbf{j} + v_3\mathbf{k}$$

For example,

$$(2, -3, 4) = 2\mathbf{i} - 3\mathbf{j} + 4\mathbf{k}$$

From (1) we obtain

$$\mathbf{i} \times \mathbf{j} = \left(\begin{vmatrix} 0 & 0 \\ 1 & 0 \end{vmatrix}, -\begin{vmatrix} 1 & 0 \\ 0 & 0 \end{vmatrix}, \begin{vmatrix} 1 & 0 \\ 0 & 1 \end{vmatrix} \right) = (0, 0, 1) = \mathbf{k} \quad ◄$$

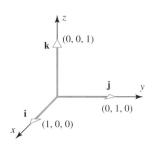

▲ Figure 3.5.1 The standard unit vectors.

You should have no trouble obtaining the following results:

$$\begin{array}{ccc} \mathbf{i} \times \mathbf{i} = \mathbf{0} & \mathbf{j} \times \mathbf{j} = \mathbf{0} & \mathbf{k} \times \mathbf{k} = \mathbf{0} \\ \mathbf{i} \times \mathbf{j} = \mathbf{k} & \mathbf{j} \times \mathbf{k} = \mathbf{i} & \mathbf{k} \times \mathbf{i} = \mathbf{j} \\ \mathbf{j} \times \mathbf{i} = -\mathbf{k} & \mathbf{k} \times \mathbf{j} = -\mathbf{i} & \mathbf{i} \times \mathbf{k} = -\mathbf{j} \end{array}$$

▲ Figure 3.5.2

Figure 3.5.2 is helpful for remembering these results. Referring to this diagram, the cross product of two consecutive vectors going clockwise is the next vector around, and the cross product of two consecutive vectors going counterclockwise is the negative of the next vector around.

Determinant Form of Cross Product

It is also worth noting that a cross product can be represented symbolically in the form

$$\mathbf{u} \times \mathbf{v} = \begin{vmatrix} \mathbf{i} & \mathbf{j} & \mathbf{k} \\ u_1 & u_2 & u_3 \\ v_1 & v_2 & v_3 \end{vmatrix} = \begin{vmatrix} u_2 & u_3 \\ v_2 & v_3 \end{vmatrix} \mathbf{i} - \begin{vmatrix} u_1 & u_3 \\ v_1 & v_3 \end{vmatrix} \mathbf{j} + \begin{vmatrix} u_1 & u_2 \\ v_1 & v_2 \end{vmatrix} \mathbf{k} \qquad (4)$$

For example, if $\mathbf{u} = (1, 2, -2)$ and $\mathbf{v} = (3, 0, 1)$, then

$$\mathbf{u} \times \mathbf{v} = \begin{vmatrix} \mathbf{i} & \mathbf{j} & \mathbf{k} \\ 1 & 2 & -2 \\ 3 & 0 & 1 \end{vmatrix} = 2\mathbf{i} - 7\mathbf{j} - 6\mathbf{k}$$

which agrees with the result obtained in Example 1.

WARNING It is not true in general that $\mathbf{u} \times (\mathbf{v} \times \mathbf{w}) = (\mathbf{u} \times \mathbf{v}) \times \mathbf{w}$. For example,

$$\mathbf{i} \times (\mathbf{j} \times \mathbf{j}) = \mathbf{i} \times \mathbf{0} = \mathbf{0}$$

and

$$(\mathbf{i} \times \mathbf{j}) \times \mathbf{j} = \mathbf{k} \times \mathbf{j} = -\mathbf{i}$$

so

$$\mathbf{i} \times (\mathbf{j} \times \mathbf{j}) \neq (\mathbf{i} \times \mathbf{j}) \times \mathbf{j}$$

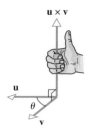

$\mathbf{u} \times \mathbf{v}$

$\mathbf{u}$

θ

$\mathbf{v}$

▲ Figure 3.5.3

We know from Theorem 3.5.1 that $\mathbf{u} \times \mathbf{v}$ is orthogonal to both $\mathbf{u}$ and $\mathbf{v}$. If $\mathbf{u}$ and $\mathbf{v}$ are nonzero vectors, it can be shown that the direction of $\mathbf{u} \times \mathbf{v}$ can be determined using the following "right-hand rule" (Figure 3.5.3): Let θ be the angle between $\mathbf{u}$ and $\mathbf{v}$, and suppose $\mathbf{u}$ is rotated through the angle θ until it coincides with $\mathbf{v}$. If the fingers of the right hand are cupped so that they point in the direction of rotation, then the thumb indicates (roughly) the direction of $\mathbf{u} \times \mathbf{v}$.

You may find it instructive to practice this rule with the products

$$\mathbf{i} \times \mathbf{j} = \mathbf{k}, \quad \mathbf{j} \times \mathbf{k} = \mathbf{i}, \quad \mathbf{k} \times \mathbf{i} = \mathbf{j}$$

Geometric Interpretation of Cross Product

If $\mathbf{u}$ and $\mathbf{v}$ are vectors in 3-space, then the norm of $\mathbf{u} \times \mathbf{v}$ has a useful geometric interpretation. Lagrange's identity, given in Theorem 3.5.1, states that

$$\|\mathbf{u} \times \mathbf{v}\|^2 = \|\mathbf{u}\|^2 \|\mathbf{v}\|^2 - (\mathbf{u} \cdot \mathbf{v})^2 \qquad (5)$$

If θ denotes the angle between $\mathbf{u}$ and $\mathbf{v}$, then $\mathbf{u} \cdot \mathbf{v} = \|\mathbf{u}\| \|\mathbf{v}\| \cos\theta$, so (5) can be rewritten as

$$\begin{aligned}\|\mathbf{u} \times \mathbf{v}\|^2 &= \|\mathbf{u}\|^2 \|\mathbf{v}\|^2 - \|\mathbf{u}\|^2 \|\mathbf{v}\|^2 \cos^2\theta \\ &= \|\mathbf{u}\|^2 \|\mathbf{v}\|^2 (1 - \cos^2\theta) \\ &= \|\mathbf{u}\|^2 \|\mathbf{v}\|^2 \sin^2\theta \end{aligned}$$

Since $0 \leq \theta \leq \pi$, it follows that $\sin\theta \geq 0$, so this can be rewritten as

$$\|\mathbf{u} \times \mathbf{v}\| = \|\mathbf{u}\| \|\mathbf{v}\| \sin\theta \qquad (6)$$

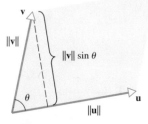

$\mathbf{v}$

$\|\mathbf{v}\|$

$\|\mathbf{v}\| \sin\theta$

θ

$\mathbf{u}$

$\|\mathbf{u}\|$

▲ Figure 3.5.4

But $\|\mathbf{v}\| \sin\theta$ is the altitude of the parallelogram determined by $\mathbf{u}$ and $\mathbf{v}$ (Figure 3.5.4). Thus, from (6), the area A of this parallelogram is given by

$$A = (\text{base})(\text{altitude}) = \|\mathbf{u}\| \|\mathbf{v}\| \sin\theta = \|\mathbf{u} \times \mathbf{v}\|$$

This result is even correct if $\mathbf{u}$ and $\mathbf{v}$ are collinear, since the parallelogram determined by $\mathbf{u}$ and $\mathbf{v}$ has zero area and from (6) we have $\mathbf{u} \times \mathbf{v} = \mathbf{0}$ because $\theta = 0$ in this case. Thus we have the following theorem.

> **THEOREM 3.5.3** **Area of a Parallelogram**
>
> *If $\mathbf{u}$ and $\mathbf{v}$ are vectors in 3-space, then $\|\mathbf{u} \times \mathbf{v}\|$ is equal to the area of the parallelogram determined by $\mathbf{u}$ and $\mathbf{v}$.*

▶ **EXAMPLE 4** **Area of a Triangle**

Find the area of the triangle determined by the points $P_1(2, 2, 0)$, $P_2(-1, 0, 2)$, and $P_3(0, 4, 3)$.

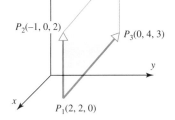

▲ Figure 3.5.5

Solution The area A of the triangle is $\frac{1}{2}$ the area of the parallelogram determined by the vectors $\overrightarrow{P_1 P_2}$ and $\overrightarrow{P_1 P_3}$ (Figure 3.5.5). Using the method discussed in Example 1 of Section 3.1, $\overrightarrow{P_1 P_2} = (-3, -2, 2)$ and $\overrightarrow{P_1 P_3} = (-2, 2, 3)$. It follows that

$$\overrightarrow{P_1 P_2} \times \overrightarrow{P_1 P_3} = (-10, 5, -10)$$

(verify) and consequently that

$$A = \tfrac{1}{2}\|\overrightarrow{P_1 P_2} \times \overrightarrow{P_1 P_3}\| = \tfrac{1}{2}(15) = \tfrac{15}{2} \quad ◀$$

> **DEFINITION 2** If $\mathbf{u}$, $\mathbf{v}$, and $\mathbf{w}$ are vectors in 3-space, then
>
> $$\mathbf{u} \cdot (\mathbf{v} \times \mathbf{w})$$
>
> is called the ***scalar triple product*** of $\mathbf{u}$, $\mathbf{v}$, and $\mathbf{w}$.

The scalar triple product of $\mathbf{u} = (u_1, u_2, u_3)$, $\mathbf{v} = (v_1, v_2, v_3)$, and $\mathbf{w} = (w_1, w_2, w_3)$ can be calculated from the formula

$$\mathbf{u} \cdot (\mathbf{v} \times \mathbf{w}) = \begin{vmatrix} u_1 & u_2 & u_3 \\ v_1 & v_2 & v_3 \\ w_1 & w_2 & w_3 \end{vmatrix} \tag{7}$$

This follows from Formula (4) since

$$\mathbf{u} \cdot (\mathbf{v} \times \mathbf{w}) = \mathbf{u} \cdot \left(\begin{vmatrix} v_2 & v_3 \\ w_2 & w_3 \end{vmatrix} \mathbf{i} - \begin{vmatrix} v_1 & v_3 \\ w_1 & w_3 \end{vmatrix} \mathbf{j} + \begin{vmatrix} v_1 & v_2 \\ w_1 & w_2 \end{vmatrix} \mathbf{k} \right)$$

$$= \begin{vmatrix} v_2 & v_3 \\ w_2 & w_3 \end{vmatrix} u_1 - \begin{vmatrix} v_1 & v_3 \\ w_1 & w_3 \end{vmatrix} u_2 + \begin{vmatrix} v_1 & v_2 \\ w_1 & w_2 \end{vmatrix} u_3$$

$$= \begin{vmatrix} u_1 & u_2 & u_3 \\ v_1 & v_2 & v_3 \\ w_1 & w_2 & w_3 \end{vmatrix}$$

▶ **EXAMPLE 5** **Calculating a Scalar Triple Product**

Calculate the scalar triple product $\mathbf{u} \cdot (\mathbf{v} \times \mathbf{w})$ of the vectors

$$\mathbf{u} = 3\mathbf{i} - 2\mathbf{j} - 5\mathbf{k}, \quad \mathbf{v} = \mathbf{i} + 4\mathbf{j} - 4\mathbf{k}, \quad \mathbf{w} = 3\mathbf{j} + 2\mathbf{k}$$

Solution From (7),

$$\mathbf{u} \cdot (\mathbf{v} \times \mathbf{w}) = \begin{vmatrix} 3 & -2 & -5 \\ 1 & 4 & -4 \\ 0 & 3 & 2 \end{vmatrix}$$

$$= 3 \begin{vmatrix} 4 & -4 \\ 3 & 2 \end{vmatrix} - (-2) \begin{vmatrix} 1 & -4 \\ 0 & 2 \end{vmatrix} + (-5) \begin{vmatrix} 1 & 4 \\ 0 & 3 \end{vmatrix}$$

$$= 60 + 4 - 15 = 49 \quad \blacktriangleleft$$

Remark   The symbol $(\mathbf{u} \cdot \mathbf{v}) \times \mathbf{w}$ makes no sense because we cannot form the cross product of a scalar and a vector. Thus, no ambiguity arises if we write $\mathbf{u} \cdot \mathbf{v} \times \mathbf{w}$ rather than $\mathbf{u} \cdot (\mathbf{v} \times \mathbf{w})$. However, for clarity we will usually keep the parentheses.

▲ Figure 3.5.6

It follows from (7) that

$$\mathbf{u} \cdot (\mathbf{v} \times \mathbf{w}) = \mathbf{w} \cdot (\mathbf{u} \times \mathbf{v}) = \mathbf{v} \cdot (\mathbf{w} \times \mathbf{u})$$

since the 3×3 determinants that represent these products can be obtained from one another by *two* row interchanges. (Verify.) These relationships can be remembered by moving the vectors $\mathbf{u}$, $\mathbf{v}$, and $\mathbf{w}$ clockwise around the vertices of the triangle in Figure 3.5.6.

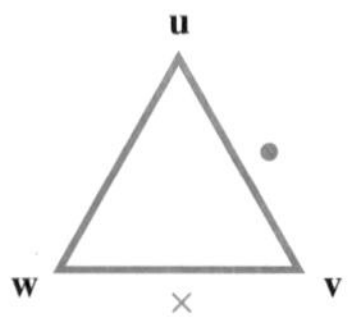

Geometric Interpretation of Determinants   The next theorem provides a useful geometric interpretation of 2×2 and 3×3 determinants.

THEOREM 3.5.4

(a)   *The absolute value of the determinant*

$$\det \begin{bmatrix} u_1 & u_2 \\ v_1 & v_2 \end{bmatrix}$$

is equal to the area of the parallelogram in 2-space determined by the vectors $\mathbf{u} = (u_1, u_2)$ *and* $\mathbf{v} = (v_1, v_2)$. *(See Figure 3.5.7a.)*

(b)   *The absolute value of the determinant*

$$\det \begin{bmatrix} u_1 & u_2 & u_3 \\ v_1 & v_2 & v_3 \\ w_1 & w_2 & w_3 \end{bmatrix}$$

is equal to the volume of the parallelepiped in 3-space determined by the vectors $\mathbf{u} = (u_1, u_2, u_3)$, $\mathbf{v} = (v_1, v_2, v_3)$, *and* $\mathbf{w} = (w_1, w_2, w_3)$. *(See Figure 3.5.7b.)*

Proof (a)   The key to the proof is to use Theorem 3.5.3. However, that theorem applies to vectors in 3-space, whereas $\mathbf{u} = (u_1, u_2)$ and $\mathbf{v} = (v_1, v_2)$ are vectors in 2-space. To circumvent this "dimension problem," we will view $\mathbf{u}$ and $\mathbf{v}$ as vectors in the xy-plane of an xyz-coordinate system (Figure 3.5.7c), in which case these vectors are expressed as $\mathbf{u} = (u_1, u_2, 0)$ and $\mathbf{v} = (v_1, v_2, 0)$. Thus

$$\mathbf{u} \times \mathbf{v} = \begin{vmatrix} \mathbf{i} & \mathbf{j} & \mathbf{k} \\ u_1 & u_2 & 0 \\ v_1 & v_2 & 0 \end{vmatrix} = \begin{vmatrix} u_1 & u_2 \\ v_1 & v_2 \end{vmatrix} \mathbf{k} = \det \begin{bmatrix} u_1 & u_2 \\ v_1 & v_2 \end{bmatrix} \mathbf{k}$$

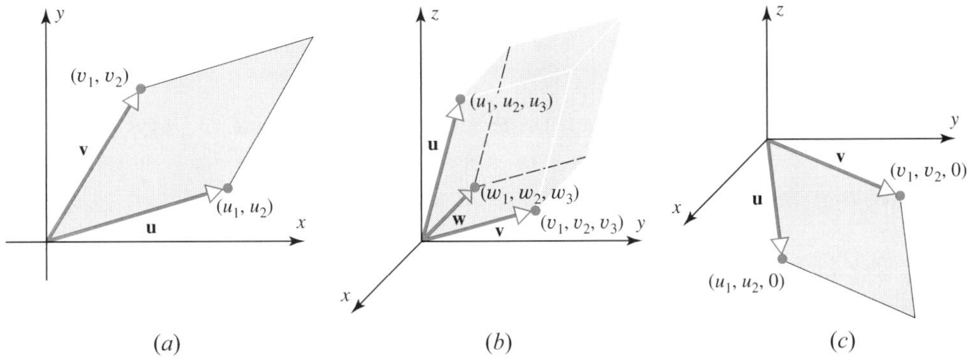

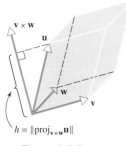

▲ Figure 3.5.7

It now follows from Theorem 3.5.3 and the fact that $\|\mathbf{k}\| = 1$ that the area A of the parallelogram determined by $\mathbf{u}$ and $\mathbf{v}$ is

$$A = \|\mathbf{u} \times \mathbf{v}\| = \left\| \det \begin{bmatrix} u_1 & u_2 \\ v_1 & v_2 \end{bmatrix} \mathbf{k} \right\| = \left| \det \begin{bmatrix} u_1 & u_2 \\ v_1 & v_2 \end{bmatrix} \right| \|\mathbf{k}\| = \left| \det \begin{bmatrix} u_1 & u_2 \\ v_1 & v_2 \end{bmatrix} \right|$$

which completes the proof.

Proof (b) As shown in Figure 3.5.8, take the base of the parallelepiped determined by $\mathbf{u}$, $\mathbf{v}$, and $\mathbf{w}$ to be the parallelogram determined by $\mathbf{v}$ and $\mathbf{w}$. It follows from Theorem 3.5.3 that the area of the base is $\|\mathbf{v} \times \mathbf{w}\|$ and, as illustrated in Figure 3.5.8, the height h of the parallelepiped is the length of the orthogonal projection of $\mathbf{u}$ on $\mathbf{v} \times \mathbf{w}$. Therefore, by Formula (12) of Section 3.3,

$$h = \|\text{proj}_{\mathbf{v} \times \mathbf{w}} \mathbf{u}\| = \frac{|\mathbf{u} \cdot (\mathbf{v} \times \mathbf{w})|}{\|\mathbf{v} \times \mathbf{w}\|}$$

It follows that the volume V of the parallelepiped is

$$V = (\text{area of base}) \cdot \text{height} = \|\mathbf{v} \times \mathbf{w}\| \frac{|\mathbf{u} \cdot (\mathbf{v} \times \mathbf{w})|}{\|\mathbf{v} \times \mathbf{w}\|} = |\mathbf{u} \cdot (\mathbf{v} \times \mathbf{w})|$$

so from (7),

$$V = \left| \det \begin{bmatrix} u_1 & u_2 & u_3 \\ v_1 & v_2 & v_3 \\ w_1 & w_2 & w_3 \end{bmatrix} \right| \tag{8}$$

which completes the proof. ◀

Remark If V denotes the volume of the parallelepiped determined by vectors $\mathbf{u}$, $\mathbf{v}$, and $\mathbf{w}$, then it follows from Formulas (7) and (8) that

$$V = \begin{bmatrix} \text{volume of parallelepiped} \\ \text{determined by } \mathbf{u}, \mathbf{v}, \text{ and } \mathbf{w} \end{bmatrix} = |\mathbf{u} \cdot (\mathbf{v} \times \mathbf{w})| \tag{9}$$

From this result and the discussion immediately following Definition 3 of Section 3.2, we can conclude that

$$\mathbf{u} \cdot (\mathbf{v} \times \mathbf{w}) = \pm V$$

where the $+$ or $-$ results depending on whether $\mathbf{u}$ makes an acute or an obtuse angle with $\mathbf{v} \times \mathbf{w}$.

Formula (9) leads to a useful test for ascertaining whether three given vectors lie in the same plane. Since three vectors not in the same plane determine a parallelepiped of

▲ Figure 3.5.8

$h = \|\text{proj}_{\mathbf{v} \times \mathbf{w}} \mathbf{u}\|$

positive volume, it follows from (9) that $|\mathbf{u} \cdot (\mathbf{v} \times \mathbf{w})| = 0$ if and only if the vectors $\mathbf{u}$, $\mathbf{v}$, and $\mathbf{w}$ lie in the same plane. Thus we have the following result.

> **THEOREM 3.5.5** *If the vectors* $\mathbf{u} = (u_1, u_2, u_3)$, $\mathbf{v} = (v_1, v_2, v_3)$, *and* $\mathbf{w} = (w_1, w_2, w_3)$ *have the same initial point, then they lie in the same plane if and only if*
> $$\mathbf{u} \cdot (\mathbf{v} \times \mathbf{w}) = \begin{vmatrix} u_1 & u_2 & u_3 \\ v_1 & v_2 & v_3 \\ w_1 & w_2 & w_3 \end{vmatrix} = 0$$

Concept Review

- Cross product of two vectors
- Determinant form of cross product
- Scalar triple product

Skills

- Compute the cross product of two vectors $\mathbf{u}$ and $\mathbf{v}$ in R^3.
- Know the geometric relationship between $\mathbf{u} \times \mathbf{v}$ to $\mathbf{u}$ and $\mathbf{v}$.
- Know the properties of the cross product (listed in Theorem 3.5.2).
- Compute the scalar triple product of three vectors in 3-space.
- Know the geometric interpretation of the scalar triple product.
- Compute the areas of triangles and parallelograms determined by two vectors or three points in 2-space or 3-space.
- Use the scalar triple product to determine whether three given vectors in 3-space are collinear.

Exercise Set 3.5

In Exercises 1–2, let $\mathbf{u} = (3, 2, -1)$, $\mathbf{v} = (0, 2, -3)$, and $\mathbf{w} = (2, 6, 7)$. Compute the indicated vectors.

1. (a) $\mathbf{v} \times \mathbf{w}$ (b) $\mathbf{u} \times (\mathbf{v} \times \mathbf{w})$ (c) $(\mathbf{u} \times \mathbf{v}) \times \mathbf{w}$

2. (a) $(\mathbf{u} \times \mathbf{v}) \times (\mathbf{v} \times \mathbf{w})$ (b) $\mathbf{u} \times (\mathbf{v} - 2\mathbf{w})$

   (c) $(\mathbf{u} \times \mathbf{v}) - 2\mathbf{w}$

In Exercises 3–6, use the cross product to find a vector that is orthogonal to both $\mathbf{u}$ and $\mathbf{v}$.

3. $\mathbf{u} = (-6, 4, 2)$, $\mathbf{v} = (3, 1, 5)$

4. $\mathbf{u} = (1, 1, -2)$, $\mathbf{v} = (2, -1, 2)$

5. $\mathbf{u} = (-2, 1, 5)$, $\mathbf{v} = (3, 0, -3)$

6. $\mathbf{u} = (3, 3, 1)$, $\mathbf{v} = (0, 4, 2)$

In Exercises 7–10, find the area of the parallelogram determined by the given vectors $\mathbf{u}$ and $\mathbf{v}$.

7. $\mathbf{u} = (1, -1, 2)$, $\mathbf{v} = (0, 3, 1)$

8. $\mathbf{u} = (3, -1, 4)$, $\mathbf{v} = (6, -2, 8)$

9. $\mathbf{u} = (2, 3, 0)$, $\mathbf{v} = (-1, 2, -2)$

10. $\mathbf{u} = (1, 1, 1)$, $\mathbf{v} = (3, 2, -5)$

In Exercises 11–12, find the area of the parallelogram with the given vertices.

11. $P_1(1, 2)$, $P_2(4, 4)$, $P_3(7, 5)$, $P_4(4, 3)$

12. $P_1(3, 2)$, $P_2(5, 4)$, $P_3(9, 4)$, $P_4(7, 2)$

In Exercises 13–14, find the area of the triangle with the given vertices.

13. $A(2, 0)$, $B(3, 4)$, $C(-1, 2)$

14. $A(1, 1)$, $B(2, 2)$, $C(3, -3)$

In Exercises 15–16, find the area of the triangle in 3-space that has the given vertices.

15. $P_1(2, 6, -1)$, $P_2(1, 1, 1)$, $P_3(4, 6, 2)$

16. $P(1, -1, 2)$, $Q(0, 3, 4)$, $R(6, 1, 8)$

In Exercises 17–18, find the volume of the parallelepiped with sides $\mathbf{u}$, $\mathbf{v}$, and $\mathbf{w}$.

17. $\mathbf{u} = (2, -6, 2)$, $\mathbf{v} = (0, 4, -2)$, $\mathbf{w} = (2, 2, -4)$

18. $\mathbf{u} = (3, 1, 2)$, $\mathbf{v} = (4, 5, 1)$, $\mathbf{w} = (1, 2, 4)$

In Exercises 19–20, determine whether $\mathbf{u}$, $\mathbf{v}$, and $\mathbf{w}$ lie in the same plane when positioned so that their initial points coincide.

19. $u = (-1, -2, 1)$, $v = (3, 0, -2)$, $w = (5, -4, 0)$

20. $u = (5, -2, 1)$, $v = (4, -1, 1)$, $w = (1, -1, 0)$

In Exercises 21–24, compute the scalar triple product $u \cdot (v \times w)$.

21. $u = (-2, 0, 6)$, $v = (1, -3, 1)$, $w = (-5, -1, 1)$

22. $u = (-1, 2, 4)$, $v = (3, 4, -2)$, $w = (-1, 2, 5)$

23. $u = (a, 0, 0)$, $v = (0, b, 0)$, $w = (0, 0, c)$

24. $u = (3, -1, 6)$, $v = (2, 4, 3)$, $w = (5, -1, 2)$

In Exercises 25–26, suppose that $u \cdot (v \times w) = 3$. Find

25. (a) $u \cdot (w \times v)$ (b) $(v \times w) \cdot u$ (c) $w \cdot (u \times v)$

26. (a) $v \cdot (u \times w)$ (b) $(u \times w) \cdot v$ (c) $v \cdot (w \times w)$

27. (a) Find the area of the triangle having vertices $A(1, 0, 1)$, $B(0, 2, 3)$, and $C(2, 1, 0)$.

    (b) Use the result of part (a) to find the length of the altitude from vertex C to side AB.

28. Use the cross product to find the sine of the angle between the vectors $u = (2, 3, -6)$ and $v = (2, 3, 6)$.

29. Simplify $(u + v) \times (u - v)$.

30. Let $a = (a_1, a_2, a_3)$, $b = (b_1, b_2, b_3)$, $c = (c_1, c_2, c_3)$, and $d = (d_1, d_2, d_3)$. Show that

$$(a + d) \cdot (b \times c) = a \cdot (b \times c) + d \cdot (b \times c)$$

31. Let u, v, and w be nonzero vectors in 3-space with the same initial point, but such that no two of them are collinear. Show that

    (a) $u \times (v \times w)$ lies in the plane determined by v and w.

    (b) $(u \times v) \times w$ lies in the plane determined by u and v.

32. Prove the following identities.

    (a) $(u + kv) \times v = u \times v$

    (b) $u \cdot (v \times z) = -(u \times z) \cdot v$

33. Prove: If a, b, c, and d lie in the same plane, then $(a \times b) \times (c \times d) = 0$.

34. Prove: If θ is the angle between u and v and $u \cdot v \neq 0$, then $\tan \theta = \|u \times v\|/(u \cdot v)$.

35. Show that if u, v, and w are vectors in R^3, no two of which are collinear, then $u \times (v \times w)$ lies in the plane determined by v and w.

36. It is a theorem of solid geometry that the volume of a tetrahedron is $\frac{1}{3}$(area of base) $\cdot$ (height). Use this result to prove

that the volume of a tetrahedron whose sides are the vectors a, b, and c is $\frac{1}{6}|a \cdot (b \times c)|$ (see the accompanying figure).

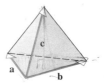

◀ Figure Ex-36

37. Use the result of Exercise 36 to find the volume of the tetrahedron with vertices P, Q, R, S.

    (a) $P(-1, 2, 0)$, $Q(2, 1, -3)$, $R(1, 1, 1)$, $S(3, -2, 3)$

    (b) $P(0, 0, 0)$, $Q(1, 2, -1)$, $R(3, 4, 0)$, $S(-1, -3, 4)$

38. Prove part (d) of Theorem 3.5.1. [*Hint:* First prove the result in the case where $w = i = (1, 0, 0)$, then when $w = j = (0, 1, 0)$, and then when $w = k = (0, 0, 1)$. Finally, prove it for an arbitrary vector $w = (w_1, w_2, w_3)$ by writing $w = w_1 i + w_2 j + w_3 k$.]

39. Prove part (e) of Theorem 3.5.1. [*Hint:* Apply part (a) of Theorem 3.5.2 to the result in part (d) of Theorem 3.5.1.]

40. Prove:

    (a) Prove (b) of Theorem 3.5.2.

    (b) Prove (c) of Theorem 3.5.2.

    (c) Prove (d) of Theorem 3.5.2.

    (d) Prove (e) of Theorem 3.5.2.

    (e) Prove (f) of Theorem 3.5.2.

True-False Exercises

In parts (a)–(f) determine whether the statement is true or false, and justify your answer.

    (a) The cross product of two nonzero vectors u and v is a nonzero vector if and only if u and v are not parallel.

    (b) A normal vector to a plane can be obtained by taking the cross product of two nonzero and noncollinear vectors lying in the plane.

    (c) The scalar triple product of u, v, and w determines a vector whose length is equal to the volume of the parallelepiped determined by u, v, and w.

    (d) If u and v are vectors in 3-space, then $\|v \times u\|$ is equal to the area of the parallelogram determined by u and v.

    (e) For all vectors u, v, and w in 3-space, the vectors $(u \times v) \times w$ and $u \times (v \times w)$ are the same.

    (f) If u, v, and w are vectors in R^3, where u is nonzero and $u \times v = u \times w$, then $v = w$.

Chapter 3 Supplementary Exercises

1. Let $\mathbf{u} = (-2, 0, 4)$, $\mathbf{v} = (3, -1, 6)$, and $\mathbf{w} = (2, -5, -5)$. Compute

   (a) $3\mathbf{v} - 2\mathbf{u}$ (b) $\|\mathbf{u} + \mathbf{v} + \mathbf{w}\|$

   (c) the distance between $-3\mathbf{u}$ and $\mathbf{v} + 5\mathbf{w}$

   (d) $\text{proj}_{\mathbf{w}}\mathbf{u}$ (e) $\mathbf{u} \cdot (\mathbf{v} \times \mathbf{w})$

   (f) $(-5\mathbf{v} + \mathbf{w}) \times ((\mathbf{u} \cdot \mathbf{v})\mathbf{w})$

2. Repeat Exercise 1 for the vectors $\mathbf{u} = 3\mathbf{i} - 5\mathbf{j} + \mathbf{k}$, $\mathbf{v} = -2\mathbf{i} + 2\mathbf{k}$, and $\mathbf{w} = -\mathbf{j} + 4\mathbf{k}$.

3. Repeat parts (a)–(d) of Exercise 1 for the vectors $\mathbf{u} = (-2, 6, 2, 1)$, $\mathbf{v} = (-3, 0, 8, 0)$, and $\mathbf{w} = (9, 1, -6, -6)$.

4. Repeat parts (a)–(d) of Exercise 1 for the vectors $\mathbf{u} = (0, 5, 0, -1, -2)$, $\mathbf{v} = (1, -1, 6, -2, 0)$, and $\mathbf{w} = (-4, -1, 4, 0, 2)$.

▷ In Exercises 5–6, determine whether the given set of vectors forms an orthogonal set. If so, normalize each vector to form an orthonormal set. ◁

5. $(-32, -1, 19)$, $(3, -1, 5)$, $(1, 6, 2)$

6. $(-2, 0, 1)$, $(1, 1, 2)$, $(1, -5, 2)$

7. (a) The set of all vectors in R^2 that are orthogonal to a nonzero vector is what kind of geometric object?

   (b) The set of all vectors in R^3 that are orthogonal to a nonzero vector is what kind of geometric object?

   (c) The set of all vectors in R^2 that are orthogonal to two noncollinear vectors is what kind of geometric object?

   (d) The set of all vectors in R^3 that are orthogonal to two noncollinear vectors is what kind of geometric object?

8. Show that $\mathbf{v}_1 = \left(\frac{2}{3}, \frac{1}{3}, \frac{2}{3}\right)$ and $\mathbf{v}_2 = \left(\frac{1}{3}, \frac{2}{3}, -\frac{2}{3}\right)$ are orthonormal vectors, and find a third vector $\mathbf{v}_3$ for which $\{\mathbf{v}_1, \mathbf{v}_2, \mathbf{v}_3\}$ is an orthonormal set.

9. *True or False:* If $\mathbf{u}$ and $\mathbf{v}$ are nonzero vectors such that $\|\mathbf{u} + \mathbf{v}\|^2 = \|\mathbf{u}\|^2 + \|\mathbf{v}\|^2$, then $\mathbf{u}$ and $\mathbf{v}$ are orthogonal.

10. *True or False:* If $\mathbf{u}$ is orthogonal to $\mathbf{v} + \mathbf{w}$, then $\mathbf{u}$ is orthogonal to $\mathbf{v}$ and $\mathbf{w}$.

11. Consider the points $P(3, -1, 4)$, $Q(6, 0, 2)$, and $R(5, 1, 1)$. Find the point S in R^3 whose first component is -1 and such that $\overrightarrow{PQ}$ is parallel to $\overrightarrow{RS}$.

12. Consider the points $P(-3, 1, 0, 6)$, $Q(0, 5, 1, -2)$, and $R(-4, 1, 4, 0)$. Find the point S in R^4 whose third component is 6 and such that $\overrightarrow{PQ}$ is parallel to $\overrightarrow{RS}$.

13. Using the points in Exercise 11, find the cosine of the angle between the vectors $\overrightarrow{PQ}$ and $\overrightarrow{PR}$.

14. Using the points in Exercise 12, find the cosine of the angle between the vectors $\overrightarrow{PQ}$ and $\overrightarrow{PR}$.

15. Find the distance between the point $P(-3, 1, 3)$ and the plane $5x + z = 3y - 4$.

16. Show that the planes $3x - y + 6z = 7$ and $-6x + 2y - 12z = 1$ are parallel, and find the distance between the planes.

▷ In Exercises 17–22, find vector and parametric equations for the line or plane in question. ◁

17. The plane in R^3 that contains the points $P(-2, 1, 3)$, $Q(-1, -1, 1)$, and $R(3, 0, -2)$.

18. The line in R^3 that contains the point $P(-1, 6, 0)$ and is orthogonal to the plane $4x - z = 5$.

19. The line in R^2 that is parallel to the vector $\mathbf{v} = (8, -1)$ and contains the point $P(0, -3)$.

20. The plane in R^3 that contains the point $P(-2, 1, 0)$ and is parallel to the plane $-8x + 6y - z = 4$.

21. The line in R^2 with equation $y = 3x - 5$.

22. The plane in R^3 with equation $2x - 6y + 3z = 5$.

▷ In Exercises 23–25, find a point-normal equation for the given plane. ◁

23. The plane that is represented by the vector equation $(x, y, z) = (-1, 5, 6) + t_1(0, -1, 3) + t_2(2, -1, 0)$.

24. The plane that contains the point $P(-5, 1, 0)$ and is orthogonal to the line with parametric equations $x = 3 - 5t$, $y = 2t$, and $z = 7$.

25. The plane that passes through the points $P(9, 0, 4)$, $Q(-1, 4, 3)$, and $R(0, 6, -2)$.

26. Suppose that $\{\mathbf{v}_1, \mathbf{v}_2, \mathbf{v}_3\}$ and $\{\mathbf{w}_1, \mathbf{w}_2\}$ are two sets of vectors such that $\mathbf{v}_i$ and $\mathbf{w}_j$ are orthogonal for all i and j. Prove that if a_1, a_2, a_3, b_1, b_2 are any scalars, then the vectors $\mathbf{v} = a_1\mathbf{v}_1 + a_2\mathbf{v}_2 + a_3\mathbf{v}_3$ and $\mathbf{w} = b_1\mathbf{w}_1 + b_2\mathbf{w}_2$ are orthogonal.

27. Prove that if two vectors $\mathbf{u}$ and $\mathbf{v}$ in R^2 are orthogonal to a nonzero vector $\mathbf{w}$ in R^2, then $\mathbf{u}$ and $\mathbf{v}$ are scalar multiples of each other.

28. Prove that $\|\mathbf{u} + \mathbf{v}\| = \|\mathbf{u}\| + \|\mathbf{v}\|$ if and only if $\mathbf{u}$ and $\mathbf{v}$ are parallel vectors.

29. The equation $Ax + By = 0$ represents a line through the origin in R^2 if A and B are not both zero. What does this equation represent in R^3 if you think of it as $Ax + By + 0z = 0$? Explain.

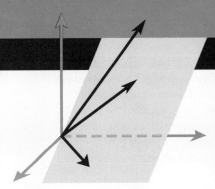

General Vector Spaces

INTRODUCTION   Recall that we began our study of vectors by viewing them as directed line segments (arrows). We then extended this idea by introducing rectangular coordinate systems, which enabled us to view vectors as ordered pairs and ordered triples of real numbers. As we developed properties of these vectors we noticed patterns in various formulas that enabled us to extend the notion of a vector to an n-tuple of real numbers. Although n-tuples took us outside the realm of our "visual experience," it gave us a valuable tool for understanding and studying systems of linear equations. In this chapter we will extend the concept of a vector yet again by using the most important algebraic properties of vectors in R^n as axioms. These axioms, if satisfied by a set of objects, will enable us to think of those objects as vectors.

4.1 Real Vector Spaces

In this section we will extend the concept of a vector by using the basic properties of vectors in R^n as axioms, which if satisfied by a set of objects, guarantee that those objects behave like familiar vectors.

Vector Space Axioms   The following definition consists of ten axioms, eight of which are properties of vectors in R^n that were stated in Theorem 3.1.1. It is important to keep in mind that one does not *prove* axioms; rather, they are assumptions that serve as the starting point for proving theorems.

DEFINITION 1 Let V be an arbitrary nonempty set of objects on which two operations are defined: addition, and multiplication by scalars. By **addition** we mean a rule for associating with each pair of objects **u** and **v** in V an object $\mathbf{u} + \mathbf{v}$, called the **sum** of **u** and **v**; by **scalar multiplication** we mean a rule for associating with each scalar k and each object **u** in V an object $k\mathbf{u}$, called the **scalar multiple** of **u** by k. If the following axioms are satisfied by all objects **u**, **v**, **w** in V and all scalars k and m, then we call V a **vector space** and we call the objects in V **vectors**.

1. If **u** and **v** are objects in V, then $\mathbf{u} + \mathbf{v}$ is in V.
2. $\mathbf{u} + \mathbf{v} = \mathbf{v} + \mathbf{u}$
3. $\mathbf{u} + (\mathbf{v} + \mathbf{w}) = (\mathbf{u} + \mathbf{v}) + \mathbf{w}$
4. There is an object **0** in V, called a **zero vector** for V, such that $\mathbf{0} + \mathbf{u} = \mathbf{u} + \mathbf{0} = \mathbf{u}$ for all **u** in V.
5. For each **u** in V, there is an object $-\mathbf{u}$ in V, called a **negative** of **u**, such that $\mathbf{u} + (-\mathbf{u}) = (-\mathbf{u}) + \mathbf{u} = \mathbf{0}$.
6. If k is any scalar and **u** is any object in V, then $k\mathbf{u}$ is in V.
7. $k(\mathbf{u} + \mathbf{v}) = k\mathbf{u} + k\mathbf{v}$
8. $(k + m)\mathbf{u} = k\mathbf{u} + m\mathbf{u}$
9. $k(m\mathbf{u}) = (km)(\mathbf{u})$
10. $1\mathbf{u} = \mathbf{u}$

Vector space scalars can be real numbers or complex numbers. Vector spaces with real scalars are called **real vector spaces** and those with complex scalars are called **complex vector spaces**. For now we will be concerned exclusively with real vector spaces. We will consider complex vector spaces later.

Observe that the definition of a vector space does not specify the nature of the vectors or the operations. Any kind of object can be a vector, and the operations of addition and scalar multiplication need not have any relationship to those on R^n. The only requirement is that the ten vector space axioms be satisfied. In the examples that follow we will use four basic steps to show that a set with two operations is a vector space.

To Show that a Set with Two Operations is a Vector Space

Step 1. Identify the set V of objects that will become vectors.

Step 2. Identify the addition and scalar multiplication operations on V.

Step 3. Verify Axioms 1 and 6; that is, adding two vectors in V produces a vector in V, and multiplying a vector in V by a scalar also produces a vector in V. Axiom 1 is called **closure under addition**, and Axiom 6 is called **closure under scalar multiplication**.

Step 4. Confirm that Axioms 2, 3, 4, 5, 7, 8, 9, and 10 hold.

Hermann Günther Grassmann (1809–1877)

Historical Note The notion of an "abstract vector space" evolved over many years and had many contributors. The idea crystallized with the work of the German mathematician H. G. Grassmann, who published a paper in 1862 in which he considered abstract systems of unspecified elements on which he defined formal operations of addition and scalar multiplication. Grassmann's work was controversial, and others, including Augustin Cauchy (p. 137), laid reasonable claim to the idea.

[*Image*: ©Sueddeutsche Zeitung Photo/The Image Works]

Our first example is the simplest of all vector spaces in that it contains only one object. Since Axiom 4 requires that every vector space contain a zero vector, the object will have to be that vector.

▶ **EXAMPLE 1 The Zero Vector Space**

Let V consist of a single object, which we denote by $\mathbf{0}$, and define

$$\mathbf{0} + \mathbf{0} = \mathbf{0} \quad \text{and} \quad k\mathbf{0} = \mathbf{0}$$

for all scalars k. It is easy to check that all the vector space axioms are satisfied. We call this the *zero vector space*. ◀

Our second example is one of the most important of all vector spaces—the familiar space R^n. It should not be surprising that the operations on R^n satisfy the vector space axioms because those axioms were based on known properties of operations on R^n.

▶ **EXAMPLE 2** R^n **Is a Vector Space**

Let $V = R^n$, and define the vector space operations on V to be the usual operations of addition and scalar multiplication of n-tuples; that is,

$$\mathbf{u} + \mathbf{v} = (u_1, u_2, \ldots, u_n) + (v_1, v_2, \ldots, v_n) = (u_1 + v_1, u_2 + v_2, \ldots, u_n + v_n)$$
$$k\mathbf{u} = (ku_1, ku_2, \ldots, ku_n)$$

The set $V = R^n$ is closed under addition and scalar multiplication because the foregoing operations produce n-tuples as their end result, and these operations satisfy Axioms 2, 3, 4, 5, 7, 8, 9, and 10 by virtue of Theorem 3.1.1. ◀

Our next example is a generalization of R^n in which we allow vectors to have infinitely many components.

▶ **EXAMPLE 3 The Vector Space of Infinite Sequences of Real Numbers**

Let V consist of objects of the form

$$\mathbf{u} = (u_1, u_2, \ldots, u_n, \ldots)$$

in which $u_1, u_2, \ldots, u_n, \ldots$ is an infinite sequence of real numbers. We define two infinite sequences to be *equal* if their corresponding components are equal, and we define addition and scalar multiplication componentwise by

$$\mathbf{u} + \mathbf{v} = (u_1, u_2, \ldots, u_n, \ldots) + (v_1, v_2, \ldots, v_n, \ldots)$$
$$= (u_1 + v_1, u_2 + v_2, \ldots, u_n + v_n, \ldots)$$
$$k\mathbf{u} = (ku_1, ku_2, \ldots, ku_n, \ldots)$$

We leave it as an exercise to confirm that V with these operations is a vector space. We will denote this vector space by the symbol R^∞. ◀

In the next example our vectors will be matrices. This may be a little confusing at first because matrices are composed of rows and columns, which are themselves vectors (row vectors and column vectors). However, here we will not be concerned with the individual rows and columns but rather with the properties of the matrix operations as they relate to the matrix as a whole.

▶ **EXAMPLE 4   A Vector Space of 2 × 2 Matrices**

Let V be the set of 2×2 matrices with real entries, and take the vector space operations on V to be the usual operations of matrix addition and scalar multiplication; that is,

Note that Equation (1) involves *three* different addition operations: the addition operation on vectors, the addition operation on matrices, and the addition operation on real numbers.

$$\mathbf{u} + \mathbf{v} = \begin{bmatrix} u_{11} & u_{12} \\ u_{21} & u_{22} \end{bmatrix} + \begin{bmatrix} v_{11} & v_{12} \\ v_{21} & v_{22} \end{bmatrix} = \begin{bmatrix} u_{11} + v_{11} & u_{12} + v_{12} \\ u_{21} + v_{21} & u_{22} + v_{22} \end{bmatrix} \tag{1}$$

$$k\mathbf{u} = k \begin{bmatrix} u_{11} & u_{12} \\ u_{21} & u_{22} \end{bmatrix} = \begin{bmatrix} ku_{11} & ku_{12} \\ ku_{21} & ku_{22} \end{bmatrix}$$

The set V is closed under addition and scalar multiplication because the foregoing operations produce 2×2 matrices as the end result. Thus, it remains to confirm that Axioms 2, 3, 4, 5, 7, 8, 9, and 10 hold. Some of these are standard properties of matrix operations. For example, Axiom 2 follows from Theorem 1.4.1*a* since

$$\mathbf{u} + \mathbf{v} = \begin{bmatrix} u_{11} & u_{12} \\ u_{21} & u_{22} \end{bmatrix} + \begin{bmatrix} v_{11} & v_{12} \\ v_{21} & v_{22} \end{bmatrix} = \begin{bmatrix} v_{11} & v_{12} \\ v_{21} & v_{22} \end{bmatrix} + \begin{bmatrix} u_{11} & u_{12} \\ u_{21} & u_{22} \end{bmatrix} = \mathbf{v} + \mathbf{u}$$

Similarly, Axioms 3, 7, 8, and 9 follow from parts (b), (h), (j), and (e), respectively, of that theorem (verify). This leaves Axioms 4, 5, and 10 that remain to be verified.

To confirm that Axiom 4 is satisfied, we must find a 2×2 matrix $\mathbf{0}$ in V for which $\mathbf{u} + \mathbf{0} = \mathbf{0} + \mathbf{u}$ for all 2×2 matrices in V. We can do this by taking

$$\mathbf{0} = \begin{bmatrix} 0 & 0 \\ 0 & 0 \end{bmatrix}$$

With this definition,

$$\mathbf{0} + \mathbf{u} = \begin{bmatrix} 0 & 0 \\ 0 & 0 \end{bmatrix} + \begin{bmatrix} u_{11} & u_{12} \\ u_{21} & u_{22} \end{bmatrix} = \begin{bmatrix} u_{11} & u_{12} \\ u_{21} & u_{22} \end{bmatrix} = \mathbf{u}$$

and similarly $\mathbf{u} + \mathbf{0} = \mathbf{u}$. To verify that Axiom 5 holds we must show that each object $\mathbf{u}$ in V has a negative $-\mathbf{u}$ in V such that $\mathbf{u} + (-\mathbf{u}) = \mathbf{0}$ and $(-\mathbf{u}) + \mathbf{u} = \mathbf{0}$. This can be done by defining the negative of $\mathbf{u}$ to be

$$-\mathbf{u} = \begin{bmatrix} -u_{11} & -u_{12} \\ -u_{21} & -u_{22} \end{bmatrix}$$

With this definition,

$$\mathbf{u} + (-\mathbf{u}) = \begin{bmatrix} u_{11} & u_{12} \\ u_{21} & u_{22} \end{bmatrix} + \begin{bmatrix} -u_{11} & -u_{12} \\ -u_{21} & -u_{22} \end{bmatrix} = \begin{bmatrix} 0 & 0 \\ 0 & 0 \end{bmatrix} = \mathbf{0}$$

and similarly $(-\mathbf{u}) + \mathbf{u} = \mathbf{0}$. Finally, Axiom 10 holds because

$$1\mathbf{u} = 1 \begin{bmatrix} u_{11} & u_{12} \\ u_{21} & u_{22} \end{bmatrix} = \begin{bmatrix} u_{11} & u_{12} \\ u_{21} & u_{22} \end{bmatrix} = \mathbf{u}$$

▶ **EXAMPLE 5   The Vector Space of *m* × *n* Matrices**

Example 4 is a special case of a more general class of vector spaces. You should have no trouble adapting the argument used in that example to show that the set V of all $m \times n$ matrices with the usual matrix operations of addition and scalar multiplication is a vector space. We will denote this vector space by the symbol M_{mn}. Thus, for example, the vector space in Example 4 is denoted as M_{22}.

▶ **EXAMPLE 6** **The Vector Space of Real-Valued Functions**

Let V be the set of real-valued functions that are defined at each x in the interval $(-\infty, \infty)$. If $\mathbf{f} = f(x)$ and $\mathbf{g} = g(x)$ are two functions in V and if k is any scalar, then define the operations of addition and scalar multiplication by

$$(\mathbf{f} + \mathbf{g})(x) = f(x) + g(x) \tag{2}$$

$$(k\mathbf{f})(x) = kf(x) \tag{3}$$

One way to think about these operations is to view the numbers $f(x)$ and $g(x)$ as "components" of $\mathbf{f}$ and $\mathbf{g}$ at the point x, in which case Equations (2) and (3) state that two functions are added by adding corresponding components, and a function is multiplied by a scalar by multiplying each component by that scalar—exactly as in R^n and R^∞. This idea is illustrated in parts (a) and (b) of Figure 4.1.1. The set V with these operations is denoted by the symbol $F(-\infty, \infty)$. We can prove that this is a vector space as follows:

Axioms 1 and 6: These closure axioms require that if we add two functions that are defined at each x in the interval $(-\infty, \infty)$, then sums and scalar multiples of those functions are also defined at each x in the interval $(-\infty, \infty)$. This follows from Formulas (2) and (3).

Axiom 4: This axiom requires that there exists a function $\mathbf{0}$ in $F(-\infty, \infty)$, which when added to any other function $\mathbf{f}$ in $F(-\infty, \infty)$ produces $\mathbf{f}$ back again as the result. The function, whose value at every point x in the interval $(-\infty, \infty)$ is zero, has this property. Geometrically, the graph of the function $\mathbf{0}$ is the line that coincides with the x-axis.

Axiom 5: This axiom requires that for each function $\mathbf{f}$ in $F(-\infty, \infty)$ there exists a function $-\mathbf{f}$ in $F(-\infty, \infty)$, which when added to $\mathbf{f}$ produces the function $\mathbf{0}$. The function defined by $-\mathbf{f}(x) = -f(x)$ has this property. The graph of $-\mathbf{f}$ can be obtained by reflecting the graph of $\mathbf{f}$ about the x-axis (Figure 4.1.1c).

Axioms 2, 3, 7, 8, 9, 10: The validity of each of these axioms follows from properties of real numbers. For example, if $\mathbf{f}$ and $\mathbf{g}$ are functions in $F(-\infty, \infty)$, then Axiom 2 requires that $\mathbf{f} + \mathbf{g} = \mathbf{g} + \mathbf{f}$. This follows from the computation

$$(\mathbf{f} + \mathbf{g})(x) = \mathbf{f}(x) + \mathbf{g}(x) = \mathbf{g}(x) + \mathbf{f}(x) = (\mathbf{g} + \mathbf{f})(x)$$

in which the first and last equalities follow from (2), and the middle equality is a property of real numbers. We will leave the proofs of the remaining parts as exercises. ◀

In Example 6 the functions were defined on the entire interval $(-\infty, \infty)$. However, the arguments used in that example apply as well on all subintervals of $(-\infty, \infty)$, such as a closed interval $[a, b]$ or an open interval (a, b). We will denote the vector spaces of functions on these intervals by $F[a, b]$ and $F(a, b)$, respectively.

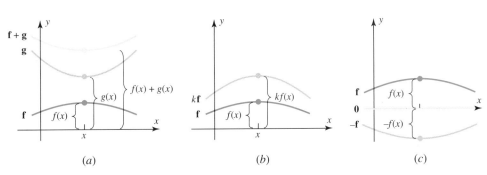

(a)   (b)   (c)

▲ Figure 4.1.1

It is important to recognize that you cannot impose any two operations on any set V and expect the vector space axioms to hold. For example, if V is the set of n-tuples with *positive* components, and if the standard operations from R^n are used, then V is not closed under scalar multiplication, because if $\mathbf{u}$ is a nonzero n-tuple in V, then $(-1)\mathbf{u}$ has at least one negative component and hence is not in V. The following is a less obvious example in which only one of the ten vector space axioms fails to hold.

▶ **EXAMPLE 7 A Set That Is Not a Vector Space**

Let $V = R^2$ and define addition and scalar multiplication operations as follows: If $\mathbf{u} = (u_1, u_2)$ and $\mathbf{v} = (v_1, v_2)$, then define

$$\mathbf{u} + \mathbf{v} = (u_1 + v_1, u_2 + v_2)$$

and if k is any real number, then define

$$k\mathbf{u} = (ku_1, 0)$$

For example, if $\mathbf{u} = (2, 4)$, $\mathbf{v} = (-3, 5)$, and $k = 7$, then

$$\mathbf{u} + \mathbf{v} = (2 + (-3), 4 + 5) = (-1, 9)$$
$$k\mathbf{u} = 7\mathbf{u} = (7 \cdot 2, 0) = (14, 0)$$

The addition operation is the standard one from R^2, but the scalar multiplication is not. In the exercises we will ask you to show that the first nine vector space axioms are satisfied. However, Axiom 10 fails to hold for certain vectors. For example, if $\mathbf{u} = (u_1, u_2)$ is such that $u_2 \neq 0$, then

$$1\mathbf{u} = 1(u_1, u_2) = (1 \cdot u_1, 0) = (u_1, 0) \neq \mathbf{u}$$

Thus, V is not a vector space with the stated operations. ◀

Our final example will be an unusual vector space that we have included to illustrate how varied vector spaces can be. Since the objects in this space will be real numbers, it will be important for you to keep track of which operations are intended as vector operations and which ones as ordinary operations on real numbers.

▶ **EXAMPLE 8 An Unusual Vector Space**

Let V be the set of positive real numbers, and define the operations on V to be

$$u + v = uv \quad \text{[Vector addition is numerical multiplication.]}$$
$$ku = u^k \quad \text{[Scalar multiplication is numerical exponentiation.]}$$

Thus, for example, $1 + 1 = 1$ and $(2)(1) = 1^2 = 1$—strange indeed, but nevertheless the set V with these operations satisfies the 10 vector space axioms and hence is a vector space. We will confirm Axioms 4, 5, and 7, and leave the others as exercises.

- Axiom 4—The zero vector in this space is the number 1 (i.e., $\mathbf{0} = 1$) since

$$u + 1 = u \cdot 1 = u$$

- Axiom 5—The negative of a vector u is its reciprocal (i.e., $-u = 1/u$) since

$$u + \frac{1}{u} = u\left(\frac{1}{u}\right) = 1 (= \mathbf{0})$$

- Axiom 7—$k(u + v) = (uv)^k = u^k v^k = (ku) + (kv)$. ◀

Some Properties of Vectors The following is our first theorem about general vector spaces. As you will see, its proof is very formal with each step being justified by a vector space axiom or a known property of real numbers. There will not be many rigidly formal proofs of this type in the text, but we have included these to reinforce the idea that the familiar properties of vectors can all be derived from the vector space axioms.

> **THEOREM 4.1.1** *Let V be a vector space, **u** a vector in V, and k a scalar; then:*
>
> (a) $0\mathbf{u} = \mathbf{0}$
>
> (b) $k\mathbf{0} = \mathbf{0}$
>
> (c) $(-1)\mathbf{u} = -\mathbf{u}$
>
> (d) *If* $k\mathbf{u} = \mathbf{0}$, *then* $k = 0$ *or* $\mathbf{u} = \mathbf{0}$.

We will prove parts (*a*) and (*c*) and leave proofs of the remaining parts as exercises.

Proof (a)   We can write

$$0\mathbf{u} + 0\mathbf{u} = (0+0)\mathbf{u} \quad \text{[Axiom 8]}$$
$$= 0\mathbf{u} \quad \text{[Property of the number 0]}$$

By Axiom 5 the vector $0\mathbf{u}$ has a negative, $-0\mathbf{u}$. Adding this negative to both sides above yields

$$[0\mathbf{u} + 0\mathbf{u}] + (-0\mathbf{u}) = 0\mathbf{u} + (-0\mathbf{u})$$

or

$$0\mathbf{u} + [0\mathbf{u} + (-0\mathbf{u})] = 0\mathbf{u} + (-0\mathbf{u}) \quad \text{[Axiom 3]}$$
$$0\mathbf{u} + \mathbf{0} = \mathbf{0} \quad \text{[Axiom 5]}$$
$$0\mathbf{u} = \mathbf{0} \quad \text{[Axiom 4]}$$

Proof (c)   To prove that $(-1)\mathbf{u} = -\mathbf{u}$, we must show that $\mathbf{u} + (-1)\mathbf{u} = \mathbf{0}$. The proof is as follows:

$$\mathbf{u} + (-1)\mathbf{u} = 1\mathbf{u} + (-1)\mathbf{u} \quad \text{[Axiom 10]}$$
$$= (1 + (-1))\mathbf{u} \quad \text{[Axiom 8]}$$
$$= 0\mathbf{u} \quad \text{[Property of numbers]}$$
$$= \mathbf{0} \quad \text{[Part (a) of this theorem]}$$

A Closing Observation   This section of the text is very important to the overall plan of linear algebra in that it establishes a common thread between such diverse mathematical objects as geometric vectors, vectors in R^n, infinite sequences, matrices, and real-valued functions, to name a few. As a result, whenever we discover a new theorem about general vector spaces, we will at the same time be discovering a theorem about geometric vectors, vectors in R^n, sequences, matrices, real-valued functions, and about any new kinds of vectors that we might discover.

To illustrate this idea, consider what the rather innocent-looking result in part (*a*) of Theorem 4.1.1 says about the vector space in Example 8. Keeping in mind that the vectors in that space are positive real numbers, that scalar multiplication means numerical exponentiation, and that the zero vector is the number 1, the equation

$$0\mathbf{u} = \mathbf{0}$$

is a statement of the fact that if u is a positive real number, then

$$u^0 = 1$$

Concept Review

- Vector space
- Closure under addition
- Closure under scalar multiplication
- Examples of vector spaces

Skills

- Determine whether a given set with two operations is a vector space.
- Show that a set with two operations is not a vector space by demonstrating that at least one of the vector space axioms fails.

Exercise Set 4.1

1. Let V be the set of all ordered pairs of real numbers, and consider the following addition and scalar multiplication operations on $\mathbf{u} = (u_1, u_2)$ and $\mathbf{v} = (v_1, v_2)$:

   $$\mathbf{u} + \mathbf{v} = (u_1 + v_1, u_2 + v_2), \quad k\mathbf{u} = (0, ku_2)$$

   (a) Compute $\mathbf{u} + \mathbf{v}$ and $k\mathbf{u}$ for $\mathbf{u} = (-1, 2)$, $\mathbf{v} = (3, 4)$, and $k = 3$.

   (b) In words, explain why V is closed under addition and scalar multiplication.

   (c) Since addition on V is the standard addition operation on R^2, certain vector space axioms hold for V because they are known to hold for R^2. Which axioms are they?

   (d) Show that Axioms 7, 8, and 9 hold.

   (e) Show that Axiom 10 fails and hence that V is not a vector space under the given operations.

2. Let V be the set of all ordered pairs of real numbers, and consider the following addition and scalar multiplication operations on $\mathbf{u} = (u_1, u_2)$ and $\mathbf{v} = (v_1, v_2)$:

   $$\mathbf{u} + \mathbf{v} = (u_1 + v_1 + 1, u_2 + v_2 + 1), \quad k\mathbf{u} = (ku_1, ku_2)$$

   (a) Compute $\mathbf{u} + \mathbf{v}$ and $k\mathbf{u}$ for $\mathbf{u} = (0, 4)$, $\mathbf{v} = (1, -3)$, and $k = 2$.

   (b) Show that $(0, 0) \neq \mathbf{0}$.

   (c) Show that $(-1, -1) = \mathbf{0}$.

   (d) Show that Axiom 5 holds by producing an ordered pair $-\mathbf{u}$ such that $\mathbf{u} + (-\mathbf{u}) = \mathbf{0}$ for $\mathbf{u} = (u_1, u_2)$.

   (e) Find two vector space axioms that fail to hold.

   In Exercises 3–12, determine whether each set equipped with the given operations is a vector space. For those that are not vector spaces identify the vector space axioms that fail. ◄

3. The set of all real numbers with the standard operations of addition and multiplication.

4. The set of all pairs of real numbers of the form $(x, 0)$ with the standard operations on R^2.

5. The set of all pairs of real numbers of the form (x, y), where $x \geq 0$, with the standard operations on R^2.

6. The set of all n-tuples of real numbers that have the form $(x, x, \ldots, x)$ with the standard operations on R^n.

7. The set of all triples of real numbers with the standard vector addition but with scalar multiplication defined by

   $$k(x, y, z) = (k^2 x, k^2 y, k^2 z)$$

8. The set of all 2×2 invertible matrices with the standard matrix addition and scalar multiplication.

9. The set of all 2×2 matrices of the form

   $$\begin{bmatrix} a & 0 \\ 0 & b \end{bmatrix}$$

   with the standard matrix addition and scalar multiplication.

10. The set of all real-valued functions f defined everywhere on the real line and such that $f(1) = 0$ with the operations used in Example 6.

11. The set of all pairs of real numbers of the form $(1, x)$ with the operations

    $$(1, y) + (1, y') = (1, y + y') \quad \text{and} \quad k(1, y) = (1, ky)$$

12. The set of polynomials of the form $a_0 + a_1 x$ with the operations

    $$(a_0 + a_1 x) + (b_0 + b_1 x) = (a_0 + b_0) + (a_1 + b_1)x$$

    and

    $$k(a_0 + a_1 x) = (ka_0) + (ka_1)x$$

13. Verify Axioms 3, 7, 8, and 9 for the vector space given in Example 4.

14. Verify Axioms 1, 2, 3, 7, 8, 9, and 10 for the vector space given in Example 6.

15. With the addition and scalar multiplication operations defined in Example 7, show that $V = R^2$ satisfies Axioms 1–9.

16. Verify Axioms 1, 2, 3, 6, 8, 9, and 10 for the vector space given in Example 8.

17. Show that the set of all points in R^2 lying on a line is a vector space with respect to the standard operations of vector addition and scalar multiplication if and only if the line passes through the origin.

18. Show that the set of all points in R^3 lying in a plane is a vector space with respect to the standard operations of vector addition and scalar multiplication if and only if the plane passes through the origin.

In Exercises 19–21, prove that the given set with the stated operations is a vector space.

19. The set $V = \{\mathbf{0}\}$ with the operations of addition and scalar multiplication given in Example 1.

20. The set R^∞ of all infinite sequences of real numbers with the operations of addition and scalar multiplication given in Example 3.

21. The set M_{mn} of all $m \times n$ matrices with the usual operations of addition and scalar multiplication.

22. Prove part (d) of Theorem 4.1.1.

23. The argument that follows proves that if $\mathbf{u}$, $\mathbf{v}$, and $\mathbf{w}$ are vectors in a vector space V such that $\mathbf{u} + \mathbf{w} = \mathbf{v} + \mathbf{w}$, then $\mathbf{u} = \mathbf{v}$ (the *cancellation law* for vector addition). As illustrated, justify the steps by filling in the blanks.

$\mathbf{u} + \mathbf{w} = \mathbf{v} + \mathbf{w}$	Hypothesis
$(\mathbf{u} + \mathbf{w}) + (-\mathbf{w}) = (\mathbf{v} + \mathbf{w}) + (-\mathbf{w})$	Add $-\mathbf{w}$ to both sides.
$\mathbf{u} + [\mathbf{w} + (-\mathbf{w})] = \mathbf{v} + [\mathbf{w} + (-\mathbf{w})]$	_____
$\mathbf{u} + \mathbf{0} = \mathbf{v} + \mathbf{0}$	_____
$\mathbf{u} = \mathbf{v}$	_____

24. Let $\mathbf{v}$ be any vector in a vector space V. Prove that $0\mathbf{v} = \mathbf{0}$.

25. Below is a seven-step proof of part (b) of Theorem 4.1.1. Justify each step either by stating that it is true by *hypothesis* or by specifying which of the ten vector space axioms applies.

Hypothesis: Let $\mathbf{u}$ be any vector in a vector space V, let $\mathbf{0}$ be the zero vector in V, and let k be a scalar.

Conclusion: Then $k\mathbf{0} = \mathbf{0}$.

Proof: (1) $k\mathbf{0} + k\mathbf{u} = k(\mathbf{0} + \mathbf{u})$
(2) $\qquad\quad = k\mathbf{u}$
(3) Since $k\mathbf{u}$ is in V, $-k\mathbf{u}$ is in V.
(4) Therefore, $(k\mathbf{0} + k\mathbf{u}) + (-k\mathbf{u}) = k\mathbf{u} + (-k\mathbf{u})$.
(5) $\qquad k\mathbf{0} + (k\mathbf{u} + (-k\mathbf{u})) = k\mathbf{u} + (-k\mathbf{u})$
(6) $\qquad\qquad\qquad k\mathbf{0} + \mathbf{0} = \mathbf{0}$
(7) $\qquad\qquad\qquad\qquad k\mathbf{0} = \mathbf{0}$

26. Let $\mathbf{v}$ be any vector in a vector space V. Prove that $-\mathbf{v} = (-1)\mathbf{v}$.

27. Prove: If $\mathbf{u}$ is a vector in a vector space V and k a scalar such that $k\mathbf{u} = \mathbf{0}$, then either $k = 0$ or $\mathbf{u} = \mathbf{0}$. [*Suggestion:* Show that if $k\mathbf{u} = \mathbf{0}$ and $k \neq 0$, then $\mathbf{u} = \mathbf{0}$. The result then follows as a logical consequence of this.]

True-False Exercises

In parts (a)–(e) determine whether the statement is true or false, and justify your answer.

(a) A vector is a directed line segment (an arrow).

(b) A vector is an n-tuple of real numbers.

(c) A vector is any element of a vector space.

(d) There is a vector space consisting of exactly two distinct vectors.

(e) The set of polynomials with degree exactly 1 is a vector space under the operations defined in Exercise 12.

4.2 Subspaces

It is possible for one vector space to be contained within another. We will explore this idea in this section, we will discuss how to recognize such vector spaces, and we will give a variety of examples that will be used in our later work.

We will begin with some terminology.

> **DEFINITION 1** A subset W of a vector space V is called a ***subspace*** of V if W is itself a vector space under the addition and scalar multiplication defined on V.

In general, to show that a nonempty set W with two operations is a vector space one must verify the ten vector space axioms. However, if W is a subspace of a known vector space V, then certain axioms need not be verified because they are "inherited" from V. For example, it is *not* necessary to verify that $\mathbf{u} + \mathbf{v} = \mathbf{v} + \mathbf{u}$ holds in W because it holds for all vectors in V including those in W. On the other hand, it *is* necessary to verify that W is closed under addition and scalar multiplication since it is possible that adding two vectors in W or multiplying a vector in W by a scalar produces a vector in V that is outside of W (Figure 4.2.1).

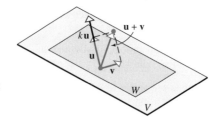

▶ Figure 4.2.1 The vectors **u** and **v** are in W, but the vectors **u** + **v** and k**u** are not.

Those axioms that are *not* inherited by W are

Axiom 1—Closure of W under addition

Axiom 4—Existence of a zero vector in W

Axiom 5—Existence of a negative in W for every vector in W

Axiom 6—Closure of W under scalar multiplication

so these must be verified to prove that it is a subspace of V. However, the following theorem shows that if Axiom 1 and Axiom 6 hold in W, then Axioms 4 and 5 hold in W as a consequence and hence need not be verified.

THEOREM 4.2.1 *If W is a set of one or more vectors in a vector space V, then W is a subspace of V if and only if the following conditions hold.*

*(a)   If **u** and **v** are vectors in W, then **u** + **v** is in W.*

*(b)   If k is any scalar and **u** is any vector in W, then k**u** is in W.*

Proof   If W is a subspace of V, then all the vector space axioms hold in W, including Axioms 1 and 6, which are precisely conditions (*a*) and (*b*).

Conversely, assume that conditions (*a*) and (*b*) hold. Since these are Axioms 1 and 6, and since Axioms 2, 3, 7, 8, 9, and 10 are inherited from V, we only need to show that Axioms 4 and 5 hold in W. For this purpose, let **u** be any vector in W. It follows from condition (*b*) that k**u** is a vector in W for every scalar k. In particular, 0**u** = **0** and (-1)**u** = −**u** are in W, which shows that Axioms 4 and 5 hold in W.   ◀

In words, Theorem 4.2.1 states that W is a subspace of V if and only if it is closed under addition and scalar multiplication.

▶ **EXAMPLE 1   The Zero Subspace**

If V is any vector space, and if $W = \{\mathbf{0}\}$ is the subset of V that consists of the zero vector only, then W is closed under addition and scalar multiplication since

$$\mathbf{0} + \mathbf{0} = \mathbf{0} \quad \text{and} \quad k\mathbf{0} = \mathbf{0}$$

for any scalar k. We call W the **zero subspace** of V.

Note that every vector space has at least two subspaces, itself and its zero subspace.

▶ **EXAMPLE 2   Lines Through the Origin Are Subspaces of R^2 and of R^3**

If W is a line through the origin of either R^2 or R^3, then adding two vectors on the line W or multiplying a vector on the line W by a scalar produces another vector on the line W, so W is closed under addition and scalar multiplication (see Figure 4.2.2 for an illustration in R^3).

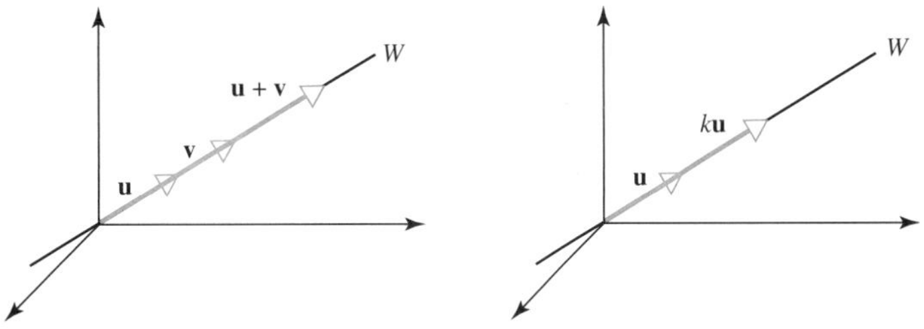

(a) W is closed under addition.

(b) W is closed under scalar multiplication.

Figure 4.2.2

▶ EXAMPLE 3 Planes Through the Origin Are Subspaces of R^3

If $\mathbf{u}$ and $\mathbf{v}$ are vectors in a plane W through the origin of R^3, then it is evident geometrically that $\mathbf{u} + \mathbf{v}$ and $k\mathbf{u}$ lie in the same plane W for any scalar k (Figure 4.2.3). Thus W is closed under addition and scalar multiplication. ◀

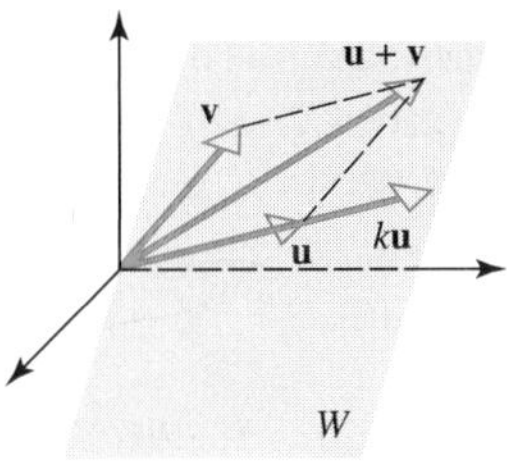

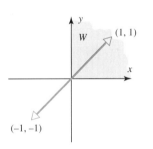

▲ Figure 4.2.3 The vectors $\mathbf{u} + \mathbf{v}$ and $k\mathbf{u}$ both lie in the same plane as $\mathbf{u}$ and $\mathbf{v}$.

Table 1 that follows gives a list of subspaces of R^2 and of R^3 that we have encountered thus far. We will see later that these are the only subspaces of R^2 and of R^3.

Table 1

Subspaces of R^2	Subspaces of R^3
• $\{0\}$	• $\{0\}$
• Lines through the origin	• Lines through the origin
• R^2	• Planes through the origin
	• R^3

▶ EXAMPLE 4 A Subset of R^2 That Is Not a Subspace

Let W be the set of all points (x, y) in R^2 for which $x \geq 0$ and $y \geq 0$ (the shaded region in Figure 4.2.4). This set is not a subspace of R^2 because it is not closed under scalar multiplication. For example, $\mathbf{v} = (1, 1)$ is a vector in W, but $(-1)\mathbf{v} = (-1, -1)$ is not.

▲ Figure 4.2.4 W is not closed under scalar multiplication.

▶ EXAMPLE 5 Subspaces of M_{nn}

We know from Theorem 1.7.2 that the sum of two symmetric $n \times n$ matrices is symmetric and that a scalar multiple of a symmetric $n \times n$ matrix is symmetric. Thus, the set of symmetric $n \times n$ matrices is closed under addition and scalar multiplication and hence is a subspace of M_{nn}. Similarly, the sets of upper triangular matrices, lower triangular matrices, and diagonal matrices are subspaces of M_{nn}.

▶ EXAMPLE 6 A Subset of M_{nn} That Is Not a Subspace

The set W of invertible $n \times n$ matrices is not a subspace of M_{nn}, failing on two counts—it is not closed under addition and not closed under scalar multiplication. We will illustrate this with an example in M_{22} that you can readily adapt to M_{nn}. Consider the matrices

$$ U = \begin{bmatrix} 1 & 2 \\ 2 & 5 \end{bmatrix} \quad \text{and} \quad V = \begin{bmatrix} -1 & 2 \\ -2 & 5 \end{bmatrix} $$

The matrix $0U$ is the 2×2 zero matrix and hence is not invertible, and the matrix $U + V$ has a column of zeros, so it also is not invertible.

▶ **EXAMPLE 7** The Subspace $C(-\infty, \infty)$

There is a theorem in calculus which states that a sum of continuous functions is continuous and that a constant times a continuous function is continuous. Rephrased in vector language, the set of continuous functions on $(-\infty, \infty)$ is a subspace of $F(-\infty, \infty)$. We will denote this subspace by $C(-\infty, \infty)$.

▶ **EXAMPLE 8** Functions with Continuous Derivatives

A function with a continuous derivative is said to be *continuously differentiable*. There is a theorem in calculus which states that the sum of two continuously differentiable functions is continuously differentiable and that a constant times a continuously differentiable function is continuously differentiable. Thus, the functions that are continuously differentiable on $(-\infty, \infty)$ form a subspace of $F(-\infty, \infty)$. We will denote this subspace by $C^1(-\infty, \infty)$, where the superscript emphasizes that the *first* derivative is continuous. To take this a step further, the set of functions with m continuous derivatives on $(-\infty, \infty)$ is a subspace of $F(-\infty, \infty)$ as is the set of functions with derivatives of all orders on $(-\infty, \infty)$. We will denote these subspaces by $C^m(-\infty, \infty)$ and $C^\infty(-\infty, \infty)$, respectively.

▶ **EXAMPLE 9** The Subspace of All Polynomials

Recall that a *polynomial* is a function that can be expressed in the form

$$p(x) = a_0 + a_1 x + \cdots + a_n x^n \tag{1}$$

where $a_0, a_1, \ldots, a_n$ are constants. It is evident that the sum of two polynomials is a polynomial and that a constant times a polynomial is a polynomial. Thus, the set W of all polynomials is closed under addition and scalar multiplication and hence is a subspace of $F(-\infty, \infty)$. We will denote this space by P_∞.

In this text we regard all constants to be polynomials of degree zero. Be aware, however, that some authors do not assign a degree to the constant 0.

▶ **EXAMPLE 10** The Subspace of Polynomials of Degree $\leq n$

Recall that the *degree* of a polynomial is the highest power of the variable that occurs with a nonzero coefficient. Thus, for example, if $a_n \neq 0$ in Formula (1), then that polynomial has degree n. It is *not* true that the set W of polynomials with positive degree n is a subspace of $F(-\infty, \infty)$ because that set is not closed under addition. For example, the polynomials

$$1 + 2x + 3x^2 \quad \text{and} \quad 5 + 7x - 3x^2$$

both have degree 2, but their sum has degree 1. What *is* true, however, is that for each nonnegative integer n the polynomials of degree n *or less* form a subspace of $F(-\infty, \infty)$. We will denote this space by P_n. ◀

The Hierarchy of Function Spaces

It is proved in calculus that polynomials are continuous functions and have continuous derivatives of all orders on $(-\infty, \infty)$. Thus, it follows that P_∞ is not only a subspace of $F(-\infty, \infty)$, as previously observed, but is also a subspace of $C^\infty(-\infty, \infty)$. We leave it for you to convince yourself that the vector spaces discussed in Examples 7 to 10 are "nested" one inside the other as illustrated in Figure 4.2.5.

Remark In our previous examples, and as illustrated in Figure 4.2.5, we have only considered functions that are defined at all points of the interval $(-\infty, \infty)$. Sometimes we will want to consider functions that are only defined on some subinterval of $(-\infty, \infty)$, say the closed interval $[a, b]$ or the open interval (a, b). In such cases we will make an appropriate notation change. For example, $C[a, b]$ is the space of continuous functions on $[a, b]$ and $C(a, b)$ is the space of continuous functions on (a, b).

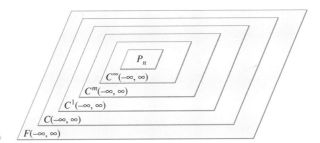

► Figure 4.2.5

Building Subspaces   The following theorem provides a useful way of creating a new subspace from known subspaces.

THEOREM 4.2.2 *If $W_1, W_2, \ldots, W_r$ are subspaces of a vector space V, then the intersection of these subspaces is also a subspace of V.*

Proof   Let W be the intersection of the subspaces $W_1, W_2, \ldots, W_r$. This set is not empty because each of these subspaces contains the zero vector of V, and hence so does their intersection. Thus, it remains to show that W is closed under addition and scalar multiplication.

Note that the first step in proving Theorem 4.2.2 was to establish that W contained at least one vector. This is important, for otherwise the subsequent argument might be logically correct but meaningless.

To prove closure under addition, let $\mathbf{u}$ and $\mathbf{v}$ be vectors in W. Since W is the intersection of $W_1, W_2, \ldots, W_r$, it follows that $\mathbf{u}$ and $\mathbf{v}$ also lie in each of these subspaces. Since these subspaces are all closed under addition, they all contain the vector $\mathbf{u} + \mathbf{v}$ and hence so does their intersection W. This proves that W is closed under addition. We leave the proof that W is closed under scalar multiplication to you.   ◄

Sometimes we will want to find the "smallest" subspace of a vector space V that contains all of the vectors in some set of interest. The following definition, which generalizes Definition 4 of Section 3.1, will help us to do that.

DEFINITION 2 If $\mathbf{w}$ is a vector in a vector space V, then $\mathbf{w}$ is said to be a **linear combination** of the vectors $\mathbf{v}_1, \mathbf{v}_2, \ldots, \mathbf{v}_r$ in V if $\mathbf{w}$ can be expressed in the form

$$\mathbf{w} = k_1\mathbf{v}_1 + k_2\mathbf{v}_2 + \cdots + k_r\mathbf{v}_r \tag{2}$$

where $k_1, k_2, \ldots, k_r$ are scalars. These scalars are called the **coefficients** of the linear combination.

If $k = 1$, then Equation (2) has the form $\mathbf{w} = k_1\mathbf{v}_1$, in which case the linear combination is just a scalar multiple of $\mathbf{v}_1$.

THEOREM 4.2.3 *If $S = \{\mathbf{w}_1, \mathbf{w}_2, \ldots, \mathbf{w}_r\}$ is a nonempty set of vectors in a vector space V, then*:

(a) *The set W of all possible linear combinations of the vectors in S is a subspace of V.*

(b) *The set W in part (a) is the "smallest" subspace of V that contains all of the vectors in S in the sense that any other subspace that contains those vectors contains W.*

Proof (a)   Let W be the set of all possible linear combinations of the vectors in S. We must show that S is closed under addition and scalar multiplication. To prove closure under addition, let

$$\mathbf{u} = c_1\mathbf{w}_1 + c_2\mathbf{w}_2 + \cdots + c_r\mathbf{w}_r \quad \text{and} \quad \mathbf{v} = k_1\mathbf{w}_1 + k_2\mathbf{w}_2 + \cdots + k_r\mathbf{w}_r$$

be two vectors in S. It follows that their sum can be written as

$$\mathbf{u} + \mathbf{v} = (c_1 + k_1)\mathbf{w}_1 + (c_2 + k_2)\mathbf{w}_2 + \cdots + (c_r + k_r)\mathbf{w}_r$$

which is a linear combination of the vectors in S. Thus, W is closed under addition. We leave it for you to prove that W is also closed under scalar multiplication and hence is a subspace of V.

Proof (b) Let W' be any subspace of V that contains all of the vectors in S. Since W' is closed under addition and scalar multiplication, it contains all linear combinations of the vectors in S and hence contains W. ◀

The following definition gives some important notation and terminology related to Theorem 4.2.3.

DEFINITION 3 The subspace of a vector space V that is formed from all possible linear combinations of the vectors in a nonempty set S is called the ***span of S***, and we say that the vectors in S ***span*** that subspace. If $S = \{\mathbf{w}_1, \mathbf{w}_2, \ldots, \mathbf{w}_r\}$, then we denote the span of S by

$$\text{span}\{\mathbf{w}_1, \mathbf{w}_2, \ldots, \mathbf{w}_r\} \quad \text{or} \quad \text{span}(S)$$

▶ **EXAMPLE 11 The Standard Unit Vectors Span R^n**

Recall that the standard unit vectors in R^n are

$$\mathbf{e}_1 = (1, 0, 0, \ldots, 0), \quad \mathbf{e}_2 = (0, 1, 0, \ldots, 0), \ldots, \quad \mathbf{e}_n = (0, 0, 0, \ldots, 1)$$

These vectors span R^n since every vector $\mathbf{v} = (v_1, v_2, \ldots, v_n)$ in R^n can be expressed as

$$\mathbf{v} = v_1\mathbf{e}_1 + v_2\mathbf{e}_2 + \cdots + v_n\mathbf{e}_n$$

which is a linear combination of $\mathbf{e}_1, \mathbf{e}_2, \ldots, \mathbf{e}_n$. Thus, for example, the vectors

$$\mathbf{i} = (1, 0, 0), \quad \mathbf{j} = (0, 1, 0), \quad \mathbf{k} = (0, 0, 1)$$

span R^3 since every vector $\mathbf{v} = (a, b, c)$ in this space can be expressed as

$$\mathbf{v} = (a, b, c) = a(1, 0, 0) + b(0, 1, 0) + c(0, 0, 1) = a\mathbf{i} + b\mathbf{j} + c\mathbf{k}.$$

▶ **EXAMPLE 12 A Geometric View of Spanning in R^2 and R^3**

(a)   If $\mathbf{v}$ is a nonzero vector in R^2 or R^3 that has its initial point at the origin, then span$\{\mathbf{v}\}$, which is the set of all scalar multiples of $\mathbf{v}$, is the line through the origin determined by $\mathbf{v}$. You should be able to visualize this from Figure 4.2.6a by observing that the

George William Hill
(1838–1914)

Historical Note   The terms *linearly independent* and *linearly dependent* were introduced by Maxime Bôcher (see p. 7) in his book *Introduction to Higher Algebra*, published in 1907. The term *linear combination* is due to the American mathematician G. W. Hill, who introduced it in a research paper on planetary motion published in 1900. Hill was a "loner" who preferred to work out of his home in West Nyack, New York, rather than in academia, though he did try lecturing at Columbia University for a few years. Interestingly, he apparently returned the teaching salary, indicating that he did not need the money and did not want to be bothered looking after it. Although technically a mathematician, Hill had little interest in modern developments of mathematics and worked almost entirely on the theory of planetary orbits.

[*Image: Courtesy of the American Mathematical Society*]

tip of the vector $k\mathbf{v}$ can be made to fall at any point on the line by choosing the value of k appropriately.

(b) If $\mathbf{v}_1$ and $\mathbf{v}_2$ are nonzero vectors in R^3 that have their initial points at the origin, then $\text{span}\{\mathbf{v}_1, \mathbf{v}_2\}$, which consists of all linear combinations of $\mathbf{v}_1$ and $\mathbf{v}_2$, is the plane through the origin determined by these two vectors. You should be able to visualize this from Figure 4.2.6b by observing that the tip of the vector $k_1\mathbf{v}_1 + k_2\mathbf{v}_2$ can be made to fall at any point in the plane by adjusting the scalars k_1 and k_2 to lengthen, shorten, or reverse the directions of the vectors $k_1\mathbf{v}_1$ and $k_2\mathbf{v}_2$ appropriately.

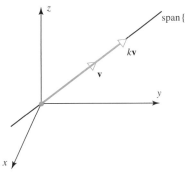

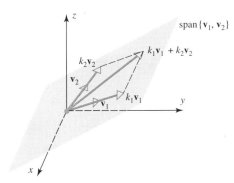

(a) Span $\{\mathbf{v}\}$ is the line through the origin determined by $\mathbf{v}$.

(b) Span $\{\mathbf{v}_1, \mathbf{v}_2\}$ is the plane through the origin determined by $\mathbf{v}_1$ and $\mathbf{v}_2$.

▶ Figure 4.2.6

▶ EXAMPLE 13 A Spanning Set for P_n

The polynomials $1, x, x^2, \ldots, x^n$ span the vector space P_n defined in Example 10 since each polynomial $\mathbf{p}$ in P_n can be written as

$$\mathbf{p} = a_0 + a_1x + \cdots + a_nx^n$$

which is a linear combination of $1, x, x^2, \ldots, x^n$. We can denote this by writing

$$P_n = \text{span}\{1, x, x^2, \ldots, x^n\} \quad ◀$$

The next two examples are concerned with two important types of problems:

- Given a set S of vectors in R^n and a vector $\mathbf{v}$ in R^n, determine whether $\mathbf{v}$ is a linear combination of the vectors in S.
- Given a set S of vectors in R^n, determine whether the vectors span R^n.

▶ EXAMPLE 14 Linear Combinations

Consider the vectors $\mathbf{u} = (1, 2, -1)$ and $\mathbf{v} = (6, 4, 2)$ in R^3. Show that $\mathbf{w} = (9, 2, 7)$ is a linear combination of $\mathbf{u}$ and $\mathbf{v}$ and that $\mathbf{w}' = (4, -1, 8)$ is *not* a linear combination of $\mathbf{u}$ and $\mathbf{v}$.

Solution In order for $\mathbf{w}$ to be a linear combination of $\mathbf{u}$ and $\mathbf{v}$, there must be scalars k_1 and k_2 such that $\mathbf{w} = k_1\mathbf{u} + k_2\mathbf{v}$; that is,

$$(9, 2, 7) = k_1(1, 2, -1) + k_2(6, 4, 2)$$

or

$$(9, 2, 7) = (k_1 + 6k_2, 2k_1 + 4k_2, -k_1 + 2k_2)$$

Equating corresponding components gives

$$k_1 + 6k_2 = 9$$
$$2k_1 + 4k_2 = 2$$
$$-k_1 + 2k_2 = 7$$

Solving this system using Gaussian elimination yields $k_1 = -3$, $k_2 = 2$, so

$$\mathbf{w} = -3\mathbf{u} + 2\mathbf{v}$$

Similarly, for $\mathbf{w}'$ to be a linear combination of $\mathbf{u}$ and $\mathbf{v}$, there must be scalars k_1 and k_2 such that $\mathbf{w}' = k_1\mathbf{u} + k_2\mathbf{v}$; that is,

$$(4, -1, 8) = k_1(1, 2, -1) + k_2(6, 4, 2)$$

or

$$(4, -1, 8) = (k_1 + 6k_2, 2k_1 + 4k_2, -k_1 + 2k_2)$$

Equating corresponding components gives

$$k_1 + 6k_2 = 4$$
$$2k_1 + 4k_2 = -1$$
$$-k_1 + 2k_2 = 8$$

This system of equations is inconsistent (verify), so no such scalars k_1 and k_2 exist. Consequently, $\mathbf{w}'$ is not a linear combination of $\mathbf{u}$ and $\mathbf{v}$.

▶ **EXAMPLE 15** Testing for Spanning

Determine whether $\mathbf{v}_1 = (1, 1, 2)$, $\mathbf{v}_2 = (1, 0, 1)$, and $\mathbf{v}_3 = (2, 1, 3)$ span the vector space R^3.

Solution We must determine whether an arbitrary vector $\mathbf{b} = (b_1, b_2, b_3)$ in R^3 can be expressed as a linear combination

$$\mathbf{b} = k_1\mathbf{v}_1 + k_2\mathbf{v}_2 + k_3\mathbf{v}_3$$

of the vectors $\mathbf{v}_1$, $\mathbf{v}_2$, and $\mathbf{v}_3$. Expressing this equation in terms of components gives

$$(b_1, b_2, b_3) = k_1(1, 1, 2) + k_2(1, 0, 1) + k_3(2, 1, 3)$$

or

$$(b_1, b_2, b_3) = (k_1 + k_2 + 2k_3, k_1 + k_3, 2k_1 + k_2 + 3k_3)$$

or

$$k_1 + k_2 + 2k_3 = b_1$$
$$k_1 + k_3 = b_2$$
$$2k_1 + k_2 + 3k_3 = b_3$$

Thus, our problem reduces to ascertaining whether this system is consistent for all values of b_1, b_2, and b_3. One way of doing this is to use parts (e) and (g) of Theorem 2.3.8, which state that the system is consistent if and only if its coefficient matrix

$$A = \begin{bmatrix} 1 & 1 & 2 \\ 1 & 0 & 1 \\ 2 & 1 & 3 \end{bmatrix}$$

has a nonzero determinant. But this is *not* the case here; we leave it for you to confirm that $\det(A) = 0$, so $\mathbf{v}_1$, $\mathbf{v}_2$, and $\mathbf{v}_3$ do not span R^3. ◀

Solution Spaces of Homogeneous Systems The solutions of a homogeneous linear system $A\mathbf{x} = \mathbf{0}$ of m equations in n unknowns can be viewed as vectors in R^n. The following theorem provides a useful insight into the geometric structure of the solution set.

THEOREM 4.2.4 *The solution set of a homogeneous linear system $A\mathbf{x} = \mathbf{0}$ in n unknowns is a subspace of R^n.*

Proof   Let W be the solution set for the system. The set W is not empty because it contains at least the trivial solution $\mathbf{x} = \mathbf{0}$.

To show that W is a subspace of R^n, we must show that it is closed under addition and scalar multiplication. To do this, let $\mathbf{x}_1$ and $\mathbf{x}_2$ be vectors in W. Since these vectors are solutions of $A\mathbf{x} = \mathbf{0}$, we have

$$A\mathbf{x}_1 = \mathbf{0} \quad \text{and} \quad A\mathbf{x}_2 = \mathbf{0}$$

It follows from these equations and the distributive property of matrix multiplication that

$$A(\mathbf{x}_1 + \mathbf{x}_2) = A\mathbf{x}_1 + A\mathbf{x}_2 = \mathbf{0} + \mathbf{0} = \mathbf{0}$$

so W is closed under addition. Similarly, if k is any scalar then

$$A(k\mathbf{x}_1) = kA\mathbf{x}_1 = k\mathbf{0} = \mathbf{0}$$

so W is also closed under scalar multiplication.   ◀

> Because the solution set of a homogeneous system in n unknowns is actually a subspace of R^n, we will generally refer to it as the **solution space** of the system.

▶ **EXAMPLE 16   Solution Spaces of Homogeneous Systems**

Consider the linear systems

(a) $\begin{bmatrix} 1 & -2 & 3 \\ 2 & -4 & 6 \\ 3 & -6 & 9 \end{bmatrix} \begin{bmatrix} x \\ y \\ z \end{bmatrix} = \begin{bmatrix} 0 \\ 0 \\ 0 \end{bmatrix}$ (b) $\begin{bmatrix} 1 & -2 & 3 \\ -3 & 7 & -8 \\ -2 & 4 & -6 \end{bmatrix} \begin{bmatrix} x \\ y \\ z \end{bmatrix} = \begin{bmatrix} 0 \\ 0 \\ 0 \end{bmatrix}$

(c) $\begin{bmatrix} 1 & -2 & 3 \\ -3 & 7 & -8 \\ 4 & 1 & 2 \end{bmatrix} \begin{bmatrix} x \\ y \\ z \end{bmatrix} = \begin{bmatrix} 0 \\ 0 \\ 0 \end{bmatrix}$ (d) $\begin{bmatrix} 0 & 0 & 0 \\ 0 & 0 & 0 \\ 0 & 0 & 0 \end{bmatrix} \begin{bmatrix} x \\ y \\ z \end{bmatrix} = \begin{bmatrix} 0 \\ 0 \\ 0 \end{bmatrix}$

Solution

(a)   We leave it for you to verify that the solutions are

$$x = 2s - 3t, \quad y = s, \quad z = t$$

from which it follows that

$$x = 2y - 3z \quad \text{or} \quad x - 2y + 3z = 0$$

This is the equation of a plane through the origin that has $\mathbf{n} = (1, -2, 3)$ as a normal.

(b)   We leave it for you to verify that the solutions are

$$x = -5t, \quad y = -t, \quad z = t$$

which are parametric equations for the line through the origin that is parallel to the vector $\mathbf{v} = (-5, -1, 1)$.

(c)   We leave it for you to verify that the only solution is $x = 0$, $y = 0$, $z = 0$, so the solution space is $\{\mathbf{0}\}$.

(d)   This linear system is satisfied by all real values of x, y, and z, so the solution space is all of R^3.   ◀

Remark Whereas the solution set of every *homogeneous* system of m equations in n unknowns is a subspace of R^n, it is *never* true that the solution set of a *nonhomogeneous* system of m equations in n unknowns is a subspace of R^n. There are two possible scenarios: first, the system may not have any solutions at all, and second, if there are solutions, then the solution set will not be closed under either addition or under scalar multiplication (Exercise 18).

A Concluding Observation   It is important to recognize that spanning sets are not unique. For example, any nonzero vector on the line in Figure 4.2.6a will span that line, and any two noncollinear vectors in the plane in Figure 4.2.6b will span that plane. The following theorem, whose proof we leave as an exercise, states conditions under which two sets of vectors will span the same space.

> **THEOREM 4.2.5** *If* $S = \{\mathbf{v}_1, \mathbf{v}_2, \ldots, \mathbf{v}_r\}$ *and* $S' = \{\mathbf{w}_1, \mathbf{w}_2, \ldots, \mathbf{w}_k\}$ *are nonempty sets of vectors in a vector space* V, *then*
>
> $$span\{\mathbf{v}_1, \mathbf{v}_2, \ldots, \mathbf{v}_r\} = span\{\mathbf{w}_1, \mathbf{w}_2, \ldots, \mathbf{w}_k\}$$
>
> *if and only if each vector in* S *is a linear combination of those in* S', *and each vector in* S' *is a linear combination of those in* S.

Concept Review

- Subspace
- Zero subspace
- Examples of subspaces
- Linear combination
- Span
- Solution space

Skills

- Determine whether a subset of a vector space is a subspace.
- Show that a subset of a vector space is a subspace.
- Show that a nonempty subset of a vector space is not a subspace by demonstrating that the set is either not closed under addition or not closed under scalar multiplication.

- Given a set S of vectors in R^n and a vector $\mathbf{v}$ in R^n, determine whether $\mathbf{v}$ is a linear combination of the vectors in S.
- Given a set S of vectors in R^n, determine whether the vectors in S span R^n.
- Determine whether two nonempty sets of vectors in a vector space V span the same subspace of V.

Exercise Set 4.2

1. Use Theorem 4.2.1 to determine which of the following are subspaces of R^3.

   (a) All vectors of the form $(a, 0, 0)$.

   (b) All vectors of the form $(a, 1, 1)$.

   (c) All vectors of the form (a, b, c), where $b = a + c$.

   (d) All vectors of the form (a, b, c), where $b = a + c + 1$.

   (e) All vectors of the form $(a, b, 0)$.

2. Use Theorem 4.2.1 to determine which of the following are subspaces of M_{nn}.

   (a) The set of all diagonal $n \times n$ matrices.

   (b) The set of all $n \times n$ matrices A such that $\det(A) = 0$.

   (c) The set of all $n \times n$ matrices A such that $\text{tr}(A) = 0$.

   (d) The set of all symmetric $n \times n$ matrices.

   (e) The set of all $n \times n$ matrices A such that $A^T = -A$.

   (f) The set of all $n \times n$ matrices A for which $A\mathbf{x} = \mathbf{0}$ has only the trivial solution.

   (g) The set of all $n \times n$ matrices A such that $AB = BA$ for some fixed $n \times n$ matrix B.

3. Use Theorem 4.2.1 to determine which of the following are subspaces of P_3.

   (a) All polynomials $a_0 + a_1x + a_2x^2 + a_3x^3$ for which $a_0 = 0$.

   (b) All polynomials $a_0 + a_1x + a_2x^2 + a_3x^3$ for which $a_0 + a_1 + a_2 + a_3 = 0$.

(c) All polynomials of the form $a_0 + a_1 x + a_2 x^2 + a_3 x^3$ in which a_0, a_1, a_2, and a_3 are integers.

(d) All polynomials of the form $a_0 + a_1 x$, where a_0 and a_1 are real numbers.

4. Which of the following are subspaces of $F(-\infty, \infty)$?

(a) All functions f in $F(-\infty, \infty)$ for which $f(0) = 0$.

(b) All functions f in $F(-\infty, \infty)$ for which $f(0) = 1$.

(c) All functions f in $F(-\infty, \infty)$ for which $f(-x) = f(x)$.

(d) All polynomials of degree 2.

5. Which of the following are subspaces of R^∞?

(a) All sequences $\mathbf{v}$ in R^∞ of the form
$\mathbf{v} = (v, 0, v, 0, v, 0, \ldots)$.

(b) All sequences $\mathbf{v}$ in R^∞ of the form
$\mathbf{v} = (v, 1, v, 1, v, 1, \ldots)$.

(c) All sequences $\mathbf{v}$ in R^∞ of the form
$\mathbf{v} = (v, 2v, 4v, 8v, 16v, \ldots)$.

(d) All sequences in R^∞ whose components are 0 from some point on.

6. A line L through the origin in R^3 can be represented by parametric equations of the form $x = at$, $y = bt$, and $z = ct$. Use these equations to show that L is a subspace of R^3 by showing that if $\mathbf{v}_1 = (x_1, y_1, z_1)$ and $\mathbf{v}_2 = (x_2, y_2, z_2)$ are points on L and k is any real number, then $k\mathbf{v}_1$ and $\mathbf{v}_1 + \mathbf{v}_2$ are also points on L.

7. Which of the following are linear combinations of $\mathbf{u} = (0, -2, 2)$ and $\mathbf{v} = (1, 3, -1)$?

(a) $(2, 2, 2)$ (b) $(3, 1, 5)$ (c) $(0, 4, 5)$ (d) $(0, 0, 0)$

8. Express the following as linear combinations of $\mathbf{u} = (2, 1, 4)$, $\mathbf{v} = (1, -1, 3)$, and $\mathbf{w} = (3, 2, 5)$.

(a) $(-9, -7, -15)$　　(b) $(6, 11, 6)$

(c) $(0, 0, 0)$　　(d) $(7, 8, 9)$

9. Which of the following are linear combinations of

$$A = \begin{bmatrix} 4 & 0 \\ -2 & -2 \end{bmatrix}, \quad B = \begin{bmatrix} 1 & -1 \\ 2 & 3 \end{bmatrix}, \quad C = \begin{bmatrix} 0 & 2 \\ 1 & 4 \end{bmatrix}?$$

(a) $\begin{bmatrix} 6 & -8 \\ -1 & -8 \end{bmatrix}$　　(b) $\begin{bmatrix} 0 & 0 \\ 0 & 0 \end{bmatrix}$

(c) $\begin{bmatrix} 6 & 0 \\ 3 & 8 \end{bmatrix}$　　(d) $\begin{bmatrix} -1 & 5 \\ 7 & 1 \end{bmatrix}$

10. In each part express the vector as a linear combination of $\mathbf{p}_1 = 2 + x + 4x^2$, $\mathbf{p}_2 = 1 - x + 3x^2$, and $\mathbf{p}_3 = 3 + 2x + 5x^2$.

(a) $-9 - 7x - 15x^2$　　(b) $6 + 11x + 6x^2$

(c) 0　　(d) $7 + 8x + 9x^2$

11. In each part, determine whether the given vectors span R^3.

(a) $\mathbf{v}_1 = (2, 2, 2)$, $\mathbf{v}_2 = (0, 0, 3)$, $\mathbf{v}_3 = (0, 1, 1)$

(b) $\mathbf{v}_1 = (2, -1, 3)$, $\mathbf{v}_2 = (4, 1, 2)$, $\mathbf{v}_3 = (8, -1, 8)$

(c) $\mathbf{v}_1 = (3, 1, 4)$, $\mathbf{v}_2 = (2, -3, 5)$, $\mathbf{v}_3 = (5, -2, 9)$, $\mathbf{v}_4 = (1, 4, -1)$

(d) $\mathbf{v}_1 = (1, 2, 6)$, $\mathbf{v}_2 = (3, 4, 1)$, $\mathbf{v}_3 = (4, 3, 1)$, $\mathbf{v}_4 = (3, 3, 1)$

12. Suppose that $\mathbf{v}_1 = (2, 1, 0, 3)$, $\mathbf{v}_2 = (3, -1, 5, 2)$, and $\mathbf{v}_3 = (-1, 0, 2, 1)$. Which of the following vectors are in span$\{\mathbf{v}_1, \mathbf{v}_2, \mathbf{v}_3\}$?

(a) $(2, 3, -7, 3)$　　　　(b) $(0, 0, 0, 0)$

(c) $(1, 1, 1, 1)$　　　　(d) $(-4, 6, -13, 4)$

13. Determine whether the following polynomials span P_2.

$$\mathbf{p}_1 = 1 - x + 2x^2, \quad \mathbf{p}_2 = 3 + x,$$
$$\mathbf{p}_3 = 5 - x + 4x^2, \quad \mathbf{p}_4 = -2 - 2x + 2x^2$$

14. Let $\mathbf{f} = \cos^2 x$ and $\mathbf{g} = \sin^2 x$. Which of the following lie in the space spanned by $\mathbf{f}$ and $\mathbf{g}$?

(a) $\cos 2x$ (b) $3 + x^2$ (c) 1 (d) $\sin x$ (e) 0

15. Determine whether the solution space of the system $A\mathbf{x} = \mathbf{0}$ is a line through the origin, a plane through the origin, or the origin only. If it is a plane, find an equation for it. If it is a line, find parametric equations for it.

(a) $A = \begin{bmatrix} -1 & 1 & 1 \\ 3 & -1 & 0 \\ 2 & -4 & -5 \end{bmatrix}$　(b) $A = \begin{bmatrix} 1 & -2 & 3 \\ -3 & 6 & 9 \\ -2 & 4 & -6 \end{bmatrix}$

(c) $A = \begin{bmatrix} 1 & 2 & 3 \\ 2 & 5 & 3 \\ 1 & 0 & 8 \end{bmatrix}$　(d) $A = \begin{bmatrix} 1 & 2 & -6 \\ 1 & 4 & 4 \\ 3 & 10 & 6 \end{bmatrix}$

(e) $A = \begin{bmatrix} 1 & -1 & 1 \\ 2 & -1 & 4 \\ 3 & 1 & 11 \end{bmatrix}$　(f) $A = \begin{bmatrix} 1 & -3 & 1 \\ 2 & -6 & 2 \\ 3 & -9 & 3 \end{bmatrix}$.

16. (**Calculus required**) Show that the following sets of functions are subspaces of $F(-\infty, \infty)$.

(a) All continuous functions on $(-\infty, \infty)$.

(b) All differentiable functions on $(-\infty, \infty)$.

(c) All differentiable functions on $(-\infty, \infty)$ that satisfy $\mathbf{f}' + 2\mathbf{f} = \mathbf{0}$.

17. (**Calculus required**) Show that the set of continuous functions $\mathbf{f} = f(x)$ on $[a, b]$ such that

$$\int_a^b f(x)\, dx = 0$$

is a subspace of $C[a, b]$.

18. Show that the solution vectors of a consistent nonhomogeneous system of m linear equations in n unknowns do not form a subspace of R^n.

19. Prove Theorem 4.2.5.

20. Use Theorem 4.2.5 to show that the vectors $\mathbf{v}_1 = (1, 6, 4)$, $\mathbf{v}_2 = (2, 4, -1)$, $\mathbf{v}_3 = (-1, 2, 5)$, and the vectors $\mathbf{w}_1 = (1, -2, -5)$, $\mathbf{w}_2 = (0, 8, 9)$ span the same subspace of R^3.

True-False Exercises

In parts (a)–(k) determine whether the statement is true or false, and justify your answer.

(a) Every subspace of a vector space is itself a vector space.

(b) Every vector space is a subspace of itself.

(c) Every subset of a vector space V that contains the zero vector in V is a subspace of V.

(d) The set R^2 is a subspace of R^3.

(e) The solution set of a consistent linear system $A\mathbf{x} = \mathbf{b}$ of m equations in n unknowns is a subspace of R^n.

(f) The span of any finite set of vectors in a vector space is closed under addition and scalar multiplication.

(g) The intersection of any two subspaces of a vector space V is a subspace of V.

(h) The union of any two subspaces of a vector space V is a subspace of V.

(i) Two subsets of a vector space V that span the same subspace of V must be equal.

(j) The set of upper triangular $n \times n$ matrices is a subspace of the vector space of all $n \times n$ matrices.

(k) The polynomials $x - 1$, $(x - 1)^2$, and $(x - 1)^3$ span P_3.

4.3 Linear Independence

In this section we will consider the question of whether the vectors in a given set are interrelated in the sense that one or more of them can be expressed as a linear combination of the others. This is important to know in applications because the existence of such relationships often signals that some kind of complication is likely to occur.

Extraneous Vectors

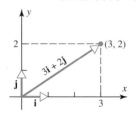

▲ Figure 4.3.1

In a rectangular xy-coordinate system every vector in the plane can be expressed in exactly one way as a linear combination of the standard unit vectors. For example, the only way to express the vector $(3, 2)$ as a linear combination of $\mathbf{i} = (1, 0)$ and $\mathbf{j} = (0, 1)$ is

$$(3, 2) = 3(1, 0) + 2(0, 1) = 3\mathbf{i} + 2\mathbf{j} \qquad (1)$$

(Figure 4.3.1). Suppose, however, that we were to introduce a third coordinate axis that makes an angle of $45°$ with the x-axis. Call it the w-axis. As illustrated in Figure 4.3.2, the unit vector along the w-axis is

$$\mathbf{w} = \left(\frac{1}{\sqrt{2}}, \frac{1}{\sqrt{2}} \right)$$

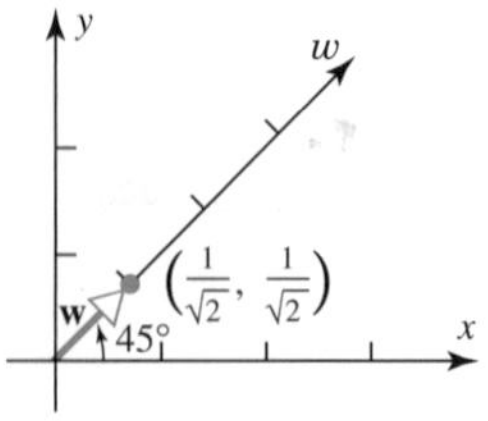

▲ Figure 4.3.2

Whereas Formula (1) shows the only way to express the vector $(3, 2)$ as a linear combination of $\mathbf{i}$ and $\mathbf{j}$, there are infinitely many ways to express this vector as a linear combination of $\mathbf{i}$, $\mathbf{j}$, and $\mathbf{w}$. Three possibilities are

$$(3, 2) = 3(1, 0) + 2(0, 1) + 0 \left(\frac{1}{\sqrt{2}}, \frac{1}{\sqrt{2}} \right) = 3\mathbf{i} + 2\mathbf{j} + 0\mathbf{w}$$

$$(3, 2) = 2(1, 0) + (0, 1) + \sqrt{2} \left(\frac{1}{\sqrt{2}}, \frac{1}{\sqrt{2}} \right) = 3\mathbf{i} + \mathbf{j} + \sqrt{2}\mathbf{w}$$

$$(3, 2) = 4(1, 0) + 3(0, 1) - \sqrt{2} \left(\frac{1}{\sqrt{2}}, \frac{1}{\sqrt{2}} \right) = 4\mathbf{i} + 3\mathbf{j} - \sqrt{2}\mathbf{w}$$

In short, by introducing a superfluous axis we created the complication of having multiple ways of assigning coordinates to points in the plane. What makes the vector $\mathbf{w}$ superfluous is the fact that it can be expressed as a linear combination of the vectors $\mathbf{i}$ and $\mathbf{j}$, namely,

$$\mathbf{w} = \left(\frac{1}{\sqrt{2}}, \frac{1}{\sqrt{2}} \right) = \frac{1}{\sqrt{2}}\mathbf{i} + \frac{1}{\sqrt{2}}\mathbf{j}$$

Thus, one of our main tasks in this section will be to develop ways of ascertaining whether one vector in a set S is a linear combination of other vectors in S.

Linear Independence and
Dependence

DEFINITION 1 If $S = \{\mathbf{v}_1, \mathbf{v}_2, \ldots, \mathbf{v}_r\}$ is a nonempty set of vectors in a vector space V, then the vector equation

$$k_1\mathbf{v}_1 + k_2\mathbf{v}_2 + \cdots + k_r\mathbf{v}_r = \mathbf{0}$$

has at least one solution, namely,

$$k_1 = 0, \quad k_2 = 0, \ldots, \quad k_r = 0$$

We call this the ***trivial solution***. If this is the only solution, then S is said to be a ***linearly independent set***. If there are solutions in addition to the trivial solution, then S is said to be a ***linearly dependent set***.

We will often apply the terms *linearly independent* and *linearly dependent* to the vectors themselves rather than to the set.

▶ **EXAMPLE 1 Linear Independence of the Standard Unit Vectors in R^n**
The most basic linearly independent set in R^n is the set of standard unit vectors

$$\mathbf{e}_1 = (1, 0, 0, \ldots, 0), \quad \mathbf{e}_2 = (0, 1, 0, \ldots, 0), \ldots, \quad \mathbf{e}_n = (0, 0, 0, \ldots, 1)$$

For notational simplicity, we will prove the linear independence in R^3 of

$$\mathbf{i} = (1, 0, 0), \quad \mathbf{j} = (0, 1, 0), \quad \mathbf{k} = (0, 0, 1)$$

The linear independence or linear dependence of these vectors is determined by whether there exist nontrivial solutions of the vector equation

$$k_1\mathbf{i} + k_2\mathbf{j} + k_3\mathbf{k} = \mathbf{0} \tag{2}$$

Since the component form of this equation is

$$(k_1, k_2, k_3) = (0, 0, 0)$$

it follows that $k_1 = k_2 = k_3 = 0$. This implies that (2) has only the trivial solution and hence that the vectors are linearly independent.

▶ **EXAMPLE 2 Linear Independence in R^3**
Determine whether the vectors

$$\mathbf{v}_1 = (1, -2, 3), \quad \mathbf{v}_2 = (5, 6, -1), \quad \mathbf{v}_3 = (3, 2, 1)$$

are linearly independent or linearly dependent in R^3.

Solution The linear independence or linear dependence of these vectors is determined by whether there exist nontrivial solutions of the vector equation

$$k_1\mathbf{v}_1 + k_2\mathbf{v}_2 + k_3\mathbf{v}_3 = \mathbf{0} \tag{3}$$

or, equivalently, of

$$k_1(1, -2, 3) + k_2(5, 6, -1) + k_3(3, 2, 1) = (0, 0, 0)$$

Equating corresponding components on the two sides yields the homogeneous linear system

$$\begin{aligned} k_1 + 5k_2 + 3k_3 &= 0 \\ -2k_1 + 6k_2 + 2k_3 &= 0 \\ 3k_1 - k_2 + k_3 &= 0 \end{aligned} \tag{4}$$

Thus, our problem reduces to determining whether this system has nontrivial solutions. There are various ways to do this; one possibility is to simply solve the system, which yields

$$k_1 = -\tfrac{1}{2}t, \quad k_2 = -\tfrac{1}{2}t, \quad k_3 = t$$

(we omit the details). This shows that the system has nontrivial solutions and hence that the vectors are linearly dependent. A second method for obtaining the same result is to compute the determinant of the coefficient matrix

$$A = \begin{bmatrix} 1 & 5 & 3 \\ -2 & 6 & 2 \\ 3 & -1 & 1 \end{bmatrix}.$$

In Example 2, what relationship do you see between the components of v_1, v_2, and v_3 and the columns of the coefficient matrix A?

and use parts (b) and (g) of Theorem 2.3.8. We leave it for you to verify that $\det(A) = 0$, from which it follows (3) has nontrivial solutions and the vectors are linearly dependent.

▶ **EXAMPLE 3 Linear Independence in R^4**

Determine whether the vectors

$$\mathbf{v}_1 = (1, 2, 2, -1), \quad \mathbf{v}_2 = (4, 9, 9, -4), \quad \mathbf{v}_3 = (5, 8, 9, -5)$$

in R^4 are linearly dependent or linearly independent.

Solution The linear independence or linear dependence of these vectors is determined by whether there exist nontrivial solutions of the vector equation

$$k_1 \mathbf{v}_1 + k_2 \mathbf{v}_2 + k_3 \mathbf{v}_3 = \mathbf{0}$$

or, equivalently, of

$$k_1(1, 2, 2, -1) + k_2(4, 9, 9, -4) + k_3(5, 8, 9, -5) = (0, 0, 0, 0)$$

Equating corresponding components on the two sides yields the homogeneous linear system

$$\begin{aligned} k_1 + 4k_2 + 5k_3 &= 0 \\ 2k_1 + 9k_2 + 8k_3 &= 0 \\ 2k_1 + 9k_2 + 9k_3 &= 0 \\ -k_1 - 4k_2 - 5k_3 &= 0 \end{aligned}$$

We leave it for you to show that this system has only the trivial solution

$$k_1 = 0, \quad k_2 = 0, \quad k_3 = 0$$

from which you can conclude that $\mathbf{v}_1$, $\mathbf{v}_2$, and $\mathbf{v}_3$ are linearly independent.

▶ **EXAMPLE 4 An Important Linearly Independent Set in P_n**

Show that the polynomials

$$1, \quad x, \quad x^2, \ldots, \quad x^n$$

form a linearly independent set in P_n.

Solution For convenience, let us denote the polynomials as

$$\mathbf{p}_0 = 1, \quad \mathbf{p}_1 = x, \quad \mathbf{p}_2 = x^2, \ldots, \quad \mathbf{p}_n = x^n$$

We must show that the vector equation

$$a_0 \mathbf{p}_0 + a_1 \mathbf{p}_1 + a_2 \mathbf{p}_2 + \cdots + a_n \mathbf{p}_n = \mathbf{0} \tag{5}$$

has only the trivial solution

$$a_0 = a_1 = a_2 = \cdots = a_n = 0$$

But (5) is equivalent to the statement that

$$a_0 + a_1 x + a_2 x^2 + \cdots + a_n x^n = 0 \tag{6}$$

for all x in $(-\infty, \infty)$, so we must show that this holds if and only if each coefficient in (6) is zero. To see that this is so, recall from algebra that a nonzero polynomial of degree n has at most n distinct roots. That being the case, each coefficient in (6) must be zero, for otherwise the left side of the equation would be a nonzero polynomial with infinitely many roots. Thus, (5) has only the trivial solution. ◀

The following example shows that the problem of determining whether a given set of vectors in P_n is linearly independent or linearly dependent can be reduced to determining whether a certain set of vectors in R^n is linearly dependent or independent.

▶ **EXAMPLE 5 Linear Independence of Polynomials**

Determine whether the polynomials

$$\mathbf{p}_1 = 1 - x, \quad \mathbf{p}_2 = 5 + 3x - 2x^2, \quad \mathbf{p}_3 = 1 + 3x - x^2$$

are linearly dependent or linearly independent in P_2.

Solution The linear independence or linear dependence of these vectors is determined by whether there exist nontrivial solutions of the vector equation

$$k_1 \mathbf{p}_1 + k_2 \mathbf{p}_2 + k_3 \mathbf{p}_3 = \mathbf{0} \tag{7}$$

This equation can be written as

$$k_1(1 - x) + k_2(5 + 3x - 2x^2) + k_3(1 + 3x - x^2) = 0 \tag{8}$$

or, equivalently, as

$$(k_1 + 5k_2 + k_3) + (-k_1 + 3k_2 + 3k_3)x + (-2k_2 - k_3)x^2 = 0$$

Since this equation must be satisfied by all x in $(-\infty, \infty)$, each coefficient must be zero (as explained in the previous example). Thus, the linear dependence or independence of the given polynomials hinges on whether the following linear system has a nontrivial solution:

$$\begin{array}{rcr} k_1 + 5k_2 + k_3 &=& 0 \\ -k_1 + 3k_2 + 3k_3 &=& 0 \\ -2k_2 - k_3 &=& 0 \end{array} \tag{9}$$

In Example 5, what relationship do you see between the coefficients of the given polynomials and the column vectors of the coefficient matrix of system (9)?

We leave it for you to show that this linear system has a nontrivial solutions either by solving it directly or by showing that the coefficient matrix has determinant zero. Thus, the set $\{\mathbf{p}_1, \mathbf{p}_2, \mathbf{p}_3\}$ is linearly dependent. ◀

An Alternative Interpretation of Linear Independence

The terms *linearly dependent* and *linearly independent* are intended to indicate whether the vectors in a given set are interrelated in some way. The following theorem, whose proof is deferred to the end of this section, makes this idea more precise.

THEOREM 4.3.1 *A set S with two or more vectors is*

(a) *Linearly dependent if and only if at least one of the vectors in S is expressible as a linear combination of the other vectors in S.*

(b) *Linearly independent if and only if no vector in S is expressible as a linear combination of the other vectors in S.*

▶ **EXAMPLE 6** **Example 1 Revisited**

In Example 1 we showed that the standard unit vectors in R^n are linearly independent. Thus, it follows from Theorem 4.3.1 that none of these vectors is expressible as a linear combination of the other two. To illustrate this in R^3, suppose, for example, that

$$\mathbf{k} = k_1 \mathbf{i} + k_2 \mathbf{j}$$

or in terms of components that

$$(0, 0, 1) = (k_1, k_2, 0)$$

Since this equation cannot be satisfied by any values of k_1 and k_2, there is no way to express $\mathbf{k}$ as a linear combination of $\mathbf{i}$ and $\mathbf{j}$. Similarly, $\mathbf{i}$ is not expressible as a linear combination of $\mathbf{j}$ and $\mathbf{k}$, and $\mathbf{j}$ is not expressible as a linear combination of $\mathbf{i}$ and $\mathbf{k}$.

▶ **EXAMPLE 7** **Example 2 Revisited**

In Example 2 we saw that the vectors

$$\mathbf{v}_1 = (1, -2, 3), \quad \mathbf{v}_2 = (5, 6, -1), \quad \mathbf{v}_3 = (3, 2, 1)$$

are linearly dependent. Thus, it follows from Theorem 4.3.1 that at least one of these vectors is expressible as a linear combination of the other two. We leave it for you to confirm that these vectors satisfy the equation

$$\tfrac{1}{2}\mathbf{v}_1 + \tfrac{1}{2}\mathbf{v}_2 - \mathbf{v}_3 = \mathbf{0}$$

from which it follows, for example, that

$$\mathbf{v}_3 = \tfrac{1}{2}\mathbf{v}_1 + \tfrac{1}{2}\mathbf{v}_2 \quad ◀$$

Sets with One or Two Vectors

The following basic theorem is concerned with the linear independence and linear dependence of sets with one or two vectors and sets that contain the zero vector.

THEOREM 4.3.2

(a) *A finite set that contains $\mathbf{0}$ is linearly dependent.*

(b) *A set with exactly one vector is linearly independent if and only if that vector is not $\mathbf{0}$.*

(c) *A set with exactly two vectors is linearly independent if and only if neither vector is a scalar multiple of the other.*

Historical Note The Polish-French mathematician Józef Hoëné de Wroński was born Józef Hoëné and adopted the name Wroński after he married. Wroński's life was fraught with controversy and conflict, which some say was due to his psychopathic tendencies and his exaggeration of the importance of his own work. Although Wroński's work was dismissed as rubbish for many years, and much of it was indeed erroneous, some of his ideas contained hidden brilliance and have survived. Among other things, Wroński designed a caterpillar vehicle to compete with trains (though it was never manufactured) and did research on the famous problem of determining the longitude of a ship at sea. His final years were spent in poverty.

[*Image: wikipedia*]

Józef Hoëné de Wroński
(1778–1853)

We will prove part (*a*) and leave the rest as exercises.

Proof (a) For any vectors $\mathbf{v}_1, \mathbf{v}_2, \ldots, \mathbf{v}_r$, the set $S = \{\mathbf{v}_1, \mathbf{v}_2, \ldots, \mathbf{v}_r, \mathbf{0}\}$ is linearly dependent since the equation

$$0\mathbf{v}_1 + 0\mathbf{v}_2 + \cdots + 0\mathbf{v}_r + 1(\mathbf{0}) = \mathbf{0}$$

expresses $\mathbf{0}$ as a linear combination of the vectors in S with coefficients that are not all zero. ◄

▶ **EXAMPLE 8 Linear Independence of Two Functions**

The functions $\mathbf{f}_1 = x$ and $\mathbf{f}_2 = \sin x$ are linearly independent vectors in $F(-\infty, \infty)$ since neither function is a scalar multiple of the other. On the other hand, the two functions $\mathbf{g}_1 = \sin 2x$ and $\mathbf{g}_2 = \sin x \cos x$ are linearly dependent because the trigonometric identity $\sin 2x = 2 \sin x \cos x$ reveals that $\mathbf{g}_1$ and $\mathbf{g}_2$ are scalar multiples of each other. ◄

A Geometric Interpretation of Linear Independence

Linear independence has the following useful geometric interpretations in R^2 and R^3:

- Two vectors in R^2 or R^3 are linearly independent if and only if they do not lie on the same line when they have their initial points at the origin. Otherwise one would be a scalar multiple of the other (Figure 4.3.3).

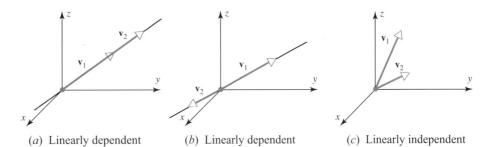

▶ Figure 4.3.3   (*a*) Linearly dependent   (*b*) Linearly dependent   (*c*) Linearly independent

- Three vectors in R^3 are linearly independent if and only if they do not lie in the same plane when they have their initial points at the origin. Otherwise at least one would be a linear combination of the other two (Figure 4.3.4).

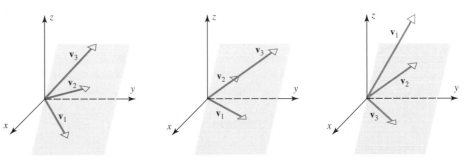

▶ Figure 4.3.4   (*a*) Linearly dependent   (*b*) Linearly dependent   (*c*) Linearly independent

At the beginning of this section we observed that a third coordinate axis in R^2 is superfluous by showing that a unit vector along such an axis would have to be expressible as a linear combination of unit vectors along the positive x- and y-axis. That result is a consequence of the next theorem, which shows that there can be at most n vectors in any linearly independent set R^n.

THEOREM 4.3.3 *Let $S = \{\mathbf{v}_1, \mathbf{v}_2, \ldots, \mathbf{v}_r\}$ be a set of vectors in R^n. If $r > n$, then S is linearly dependent.*

Proof Suppose that

$$\mathbf{v}_1 = (v_{11}, v_{12}, \ldots, v_{1n})$$
$$\mathbf{v}_2 = (v_{21}, v_{22}, \ldots, v_{2n})$$
$$\vdots \qquad\qquad \vdots$$
$$\mathbf{v}_r = (v_{r1}, v_{r2}, \ldots, v_{rn})$$

and consider the equation

$$k_1\mathbf{v}_1 + k_2\mathbf{v}_2 + \cdots + k_r\mathbf{v}_r = \mathbf{0}$$

It follows from Theorem 4.3.3, for example, that a set in R^2 with more than two vectors is linearly dependent and a set in R^3 with more than three vectors is linearly dependent.

If we express both sides of this equation in terms of components and then equate the corresponding components, we obtain the system

$$v_{11}k_1 + v_{21}k_2 + \cdots + v_{r1}k_r = 0$$
$$v_{12}k_1 + v_{22}k_2 + \cdots + v_{r2}k_r = 0$$
$$\vdots \qquad \vdots \qquad\qquad \vdots \qquad \vdots$$
$$v_{1n}k_1 + v_{2n}k_2 + \cdots + v_{rn}k_r = 0$$

This is a homogeneous system of n equations in the r unknowns $k_1, \ldots, k_r$. Since $r > n$, it follows from Theorem 1.2.2 that the system has nontrivial solutions. Therefore, $S = \{\mathbf{v}_1, \mathbf{v}_2, \ldots, \mathbf{v}_r\}$ is a linearly dependent set. ◀

CALCULUS REQUIRED
Linear Independence of Functions

Sometimes linear dependence of functions can be deduced from known identities. For example, the functions

$$\mathbf{f}_1 = \sin^2 x, \quad \mathbf{f}_2 = \cos^2 x, \quad \text{and} \quad \mathbf{f}_3 = 5$$

form a linearly dependent set in $F(-\infty, \infty)$, since the equation

$$5\mathbf{f}_1 + 5\mathbf{f}_2 - \mathbf{f}_3 = 5\sin^2 x + 5\cos^2 x - 5$$
$$= 5(\sin^2 x + \cos^2 x) - 5 = \mathbf{0}$$

expresses $\mathbf{0}$ as a linear combination of $\mathbf{f}_1$, $\mathbf{f}_2$, and $\mathbf{f}_3$ with coefficients that are not all zero.

Unfortunately, there is no *general* method that can be used to determine whether a set of functions is linearly independent or linearly dependent. However, there does exist a theorem that is useful for establishing linear independence in certain circumstances. The following definition will be useful for discussing that theorem.

DEFINITION 2 If $\mathbf{f}_1 = f_1(x), \mathbf{f}_2 = f_2(x), \ldots, \mathbf{f}_n = f_n(x)$ are functions that are $n - 1$ times differentiable on the interval $(-\infty, \infty)$, then the determinant

$$W(x) = \begin{vmatrix} f_1(x) & f_2(x) & \cdots & f_n(x) \\ f_1'(x) & f_2'(x) & \cdots & f_n'(x) \\ \vdots & \vdots & & \vdots \\ f_1^{(n-1)}(x) & f_2^{(n-1)}(x) & \cdots & f_n^{(n-1)}(x) \end{vmatrix}$$

is called the **Wronskian** of $f_1, f_2, \ldots, f_n$.

Suppose for the moment that $\mathbf{f}_1 = f_1(x), \mathbf{f}_2 = f_2(x), \ldots, \mathbf{f}_n = f_n(x)$ are *linearly dependent* vectors in $C^{(n-1)}(-\infty, \infty)$. This implies that for certain values of the coefficients the vector equation

$$k_1\mathbf{f}_1 + k_2\mathbf{f}_2 + \cdots + k_n\mathbf{f}_n = \mathbf{0}$$

has a nontrivial solution, or equivalently that the equation

$$k_1 f_1(x) + k_2 f_2(x) + \cdots + k_n f_n(x) = 0$$

is satisfied for all x in $(-\infty, \infty)$. Using this equation together with those that result by differentiating it $n - 1$ times yields the linear system

$$
\begin{array}{llll}
k_1 f_1(x) & + k_2 f_2(x) & + \cdots + k_n f_n(x) & = 0 \\
k_1 f_1'(x) & + k_2 f_2'(x) & + \cdots + k_n f_n'(x) & = 0 \\
\vdots & \vdots & \vdots & \vdots \\
k_1 f_1^{(n-1)}(x) & + k_2 f_2^{(n-1)}(x) + \cdots + k_n f_n^{(n-1)}(x) & = 0
\end{array}
$$

Thus, the linear dependence of $\mathbf{f}_1, \mathbf{f}_2, \ldots, \mathbf{f}_n$ implies that the linear system

$$
\begin{bmatrix}
f_1(x) & f_2(x) & \cdots & f_n(x) \\
f_1'(x) & f_2'(x) & \cdots & f_n'(x) \\
\vdots & \vdots & & \vdots \\
f_1^{(n-1)}(x) & f_2^{(n-1)}(x) & \cdots & f_n^{(n-1)}(x)
\end{bmatrix}
\begin{bmatrix}
k_1 \\
k_2 \\
\vdots \\
k_n
\end{bmatrix}
=
\begin{bmatrix}
0 \\
0 \\
\vdots \\
0
\end{bmatrix}
\tag{10}
$$

has a nontrivial solution. But this implies that the determinant of the coefficient matrix of (10) is zero for every such x. Since this determinant is the Wronskian of $f_1, f_2, \ldots, f_n$, we have established the following result.

THEOREM 4.3.4 *If the functions* $\mathbf{f}_1, \mathbf{f}_2, \ldots, \mathbf{f}_n$ *have* $n - 1$ *continuous derivatives on the interval* $(-\infty, \infty)$, *and if the Wronskian of these functions is not identically zero on* $(-\infty, \infty)$, *then these functions form a linearly independent set of vectors in* $C^{(n-1)}(-\infty, \infty)$.

In Example 8 we showed that x and $\sin x$ are linearly independent functions by observing that neither is a scalar multiple of the other. The following example shows how to obtain the same result using the Wronskian (though it is a more complicated procedure in this particular case).

▶ **EXAMPLE 9 Linear Independence Using the Wronskian**

Use the Wronskian to show that $\mathbf{f}_1 = x$ and $\mathbf{f}_2 = \sin x$ are linearly independent.

Solution The Wronskian is

$$W(x) = \begin{vmatrix} x & \sin x \\ 1 & \cos x \end{vmatrix} = x \cos x - \sin x$$

This function is not identically zero on the interval $(-\infty, \infty)$ since, for example,

$$W\left(\frac{\pi}{2}\right) = \frac{\pi}{2} \cos\left(\frac{\pi}{2}\right) - \sin\left(\frac{\pi}{2}\right) = \frac{\pi}{2}$$

Thus, the functions are linearly independent.

▶ **EXAMPLE 10 Linear Independence Using the Wronskian**

Use the Wronskian to show that $\mathbf{f}_1 = 1$, $\mathbf{f}_2 = e^x$, and $\mathbf{f}_3 = e^{2x}$ are linearly independent.

Solution The Wronskian is

$$W(x) = \begin{vmatrix} 1 & e^x & e^{2x} \\ 0 & e^x & 2e^{2x} \\ 0 & e^x & 4e^{2x} \end{vmatrix} = 2e^{3x}$$

This function is obviously not identically zero on $(-\infty, \infty)$, so $\mathbf{f}_1$, $\mathbf{f}_2$, and $\mathbf{f}_3$ form a linearly independent set. ◀

OPTIONAL

We will close this section by proving part (*a*) of Theorem 4.3.1. We will leave the proof of part (*b*) as an exercise.

Proof of Theorem 4.3.1(a) Let $S = \{\mathbf{v}_1, \mathbf{v}_2, \ldots, \mathbf{v}_r\}$ be a set with two or more vectors. If we assume that S is linearly dependent, then there are scalars $k_1, k_2, \ldots, k_r$, not all zero, such that

$$k_1\mathbf{v}_1 + k_2\mathbf{v}_2 + \cdots + k_r\mathbf{v}_r = \mathbf{0} \tag{11}$$

To be specific, suppose that $k_1 \neq 0$. Then (11) can be rewritten as

$$\mathbf{v}_1 = \left(-\frac{k_2}{k_1}\right)\mathbf{v}_2 + \cdots + \left(-\frac{k_r}{k_1}\right)\mathbf{v}_r$$

which expresses $\mathbf{v}_1$ as a linear combination of the other vectors in S. Similarly, if $k_j \neq 0$ in (11) for some $j = 2, 3, \ldots, r$, then $\mathbf{v}_j$ is expressible as a linear combination of the other vectors in S.

Conversely, let us assume that at least one of the vectors in S is expressible as a linear combination of the other vectors. To be specific, suppose that

$$\mathbf{v}_1 = c_2\mathbf{v}_2 + c_3\mathbf{v}_3 + \cdots + c_r\mathbf{v}_r$$

so

$$\mathbf{v}_1 - c_2\mathbf{v}_2 - c_3\mathbf{v}_3 - \cdots - c_r\mathbf{v}_r = \mathbf{0}$$

It follows that S is linearly dependent since the equation

$$k_1\mathbf{v}_1 + k_2\mathbf{v}_2 + \cdots + k_r\mathbf{v}_r = \mathbf{0}$$

is satisfied by

$$k_1 = 1, \quad k_2 = -c_2, \ldots, \quad k_r = -c_r$$

which are not all zero. The proof in the case where some vector other than $\mathbf{v}_1$ is expressible as a linear combination of the other vectors in S is similar. ◀

Concept Review

- Trivial solution
- Linearly independent set
- Linearly dependent set
- Wronskian

Skills

- Determine whether a set of vectors is linearly independent or linearly dependent.
- Express one vector in a linearly dependent set as a linear combination of the other vectors in the set.
- Use the Wronskian to show that a set of functions is linearly independent.

Exercise Set 4.3

1. Explain why the following are linearly dependent sets of vectors. (Solve this problem by inspection.)

   (a) $\mathbf{u}_1 = (-1, 2, 4)$ and $\mathbf{u}_2 = (5, -10, -20)$ in R^3

   (b) $\mathbf{u}_1 = (3, -1)$, $\mathbf{u}_2 = (4, 5)$, $\mathbf{u}_3 = (-4, 7)$ in R^2

   (c) $\mathbf{p}_1 = 3 - 2x + x^2$ and $\mathbf{p}_2 = 6 - 4x + 2x^2$ in P_2

   (d) $A = \begin{bmatrix} -3 & 4 \\ 2 & 0 \end{bmatrix}$ and $B = \begin{bmatrix} 3 & -4 \\ -2 & 0 \end{bmatrix}$ in M_{22}

2. Which of the following sets of vectors in R^3 are linearly dependent?

   (a) $(4, -1, 2)$, $(-4, 10, 2)$

   (b) $(-3, 0, 4)$, $(5, -1, 2)$, $(1, 1, 3)$

   (c) $(8, -1, 3)$, $(4, 0, 1)$

   (d) $(-2, 0, 1)$, $(3, 2, 5)$, $(6, -1, 1)$, $(7, 0, -2)$

3. Which of the following sets of vectors in R^4 are linearly dependent?

   (a) $(3, 8, 7, -3)$, $(1, 5, 3, -1)$, $(2, -1, 2, 6)$, $(1, 4, 0, 3)$

   (b) $(0, 0, 2, 2)$, $(3, 3, 0, 0)$, $(1, 1, 0, -1)$

   (c) $(0, 3, -3, -6)$, $(-2, 0, 0, -6)$, $(0, -4, -2, -2)$, $(0, -8, 4, -4)$

   (d) $(3, 0, -3, 6)$, $(0, 2, 3, 1)$, $(0, -2, -2, 0)$, $(-2, 1, 2, 1)$

4. Which of the following sets of vectors in P_2 are linearly dependent?

   (a) $2 - x + 4x^2$, $3 + 6x + 2x^2$, $2 + 10x - 4x^2$

   (b) $3 + x + x^2$, $2 - x + 5x^2$, $4 - 3x^2$

   (c) $6 - x^2$, $1 + x + 4x^2$

   (d) $1 + 3x + 3x^2$, $x + 4x^2$, $5 + 6x + 3x^2$, $7 + 2x - x^2$

5. Assume that $\mathbf{v}_1$, $\mathbf{v}_2$, and $\mathbf{v}_3$ are vectors in R^3 that have their initial points at the origin. In each part, determine whether the three vectors lie in a plane.

   (a) $\mathbf{v}_1 = (2, -2, 0)$, $\mathbf{v}_2 = (6, 1, 4)$, $\mathbf{v}_3 = (2, 0, -4)$

   (b) $\mathbf{v}_1 = (-6, 7, 2)$, $\mathbf{v}_2 = (3, 2, 4)$, $\mathbf{v}_3 = (4, -1, 2)$

6. Assume that $\mathbf{v}_1$, $\mathbf{v}_2$, and $\mathbf{v}_3$ are vectors in R^3 that have their initial points at the origin. In each part, determine whether the three vectors lie on the same line.

   (a) $\mathbf{v}_1 = (-1, 2, 3)$, $\mathbf{v}_2 = (2, -4, -6)$, $\mathbf{v}_3 = (-3, 6, 0)$

   (b) $\mathbf{v}_1 = (2, -1, 4)$, $\mathbf{v}_2 = (4, 2, 3)$, $\mathbf{v}_3 = (2, 7, -6)$

   (c) $\mathbf{v}_1 = (4, 6, 8)$, $\mathbf{v}_2 = (2, 3, 4)$, $\mathbf{v}_3 = (-2, -3, -4)$

7. (a) Show that the three vectors $\mathbf{v}_1 = (0, 3, 1, -1)$, $\mathbf{v}_2 = (6, 0, 5, 1)$, and $\mathbf{v}_3 = (4, -7, 1, 3)$ form a linearly dependent set in R^4.

   (b) Express each vector in part (a) as a linear combination of the other two.

8. (a) Show that the three vectors $\mathbf{v}_1 = (1, 2, 3, 4)$, $\mathbf{v}_2 = (0, 1, 0, -1)$, and $\mathbf{v}_3 = (1, 3, 3, 3)$ form a linearly dependent set in R^4.

   (b) Express each vector in part (a) as a linear combination of the other two.

9. For which real values of λ do the following vectors form a linearly dependent set in R^3?

   $$\mathbf{v}_1 = \left(\lambda, -\tfrac{1}{2}, -\tfrac{1}{2}\right), \quad \mathbf{v}_2 = \left(-\tfrac{1}{2}, \lambda, -\tfrac{1}{2}\right), \quad \mathbf{v}_3 = \left(-\tfrac{1}{2}, -\tfrac{1}{2}, \lambda\right)$$

10. Show that if $\{\mathbf{v}_1, \mathbf{v}_2, \mathbf{v}_3\}$ is a linearly independent set of vectors, then so are $\{\mathbf{v}_1, \mathbf{v}_2\}$, $\{\mathbf{v}_1, \mathbf{v}_3\}$, $\{\mathbf{v}_2, \mathbf{v}_3\}$, $\{\mathbf{v}_1\}$, $\{\mathbf{v}_2\}$, and $\{\mathbf{v}_3\}$.

11. Show that if $S = \{\mathbf{v}_1, \mathbf{v}_2, \dots, \mathbf{v}_r\}$ is a linearly independent set of vectors, then so is every nonempty subset of S.

12. Show that if $S = \{\mathbf{v}_1, \mathbf{v}_2, \mathbf{v}_3\}$ is a linearly dependent set of vectors in a vector space V, and $\mathbf{v}_4$ is any vector in V that is not in S, then $\{\mathbf{v}_1, \mathbf{v}_2, \mathbf{v}_3, \mathbf{v}_4\}$ is also linearly dependent.

13. Show that if $S = \{\mathbf{v}_1, \mathbf{v}_2, \dots, \mathbf{v}_r\}$ is a linearly dependent set of vectors in a vector space V, and if $\mathbf{v}_{r+1}, \dots, \mathbf{v}_n$ are any vectors in V that are not in S, then $\{\mathbf{v}_1, \mathbf{v}_2, \dots, \mathbf{v}_r, \mathbf{v}_{r+1}, \dots, \mathbf{v}_n\}$ is also linearly dependent.

14. Show that in P_2 every set with more than three vectors is linearly dependent.

15. Show that if $\{\mathbf{v}_1, \mathbf{v}_2\}$ is linearly independent and $\mathbf{v}_3$ does not lie in span$\{\mathbf{v}_1, \mathbf{v}_2\}$, then $\{\mathbf{v}_1, \mathbf{v}_2, \mathbf{v}_3\}$ is linearly independent.

16. Prove: For any vectors $\mathbf{u}$, $\mathbf{v}$, and $\mathbf{w}$ in a vector space V, the vectors $\mathbf{u} - \mathbf{v}$, $\mathbf{v} - \mathbf{w}$, and $\mathbf{w} - \mathbf{u}$ form a linearly dependent set.

17. Prove: The space spanned by two vectors in R^3 is a line through the origin, a plane through the origin, or the origin itself.

18. Under what conditions is a set with one vector linearly independent?

19. Are the vectors $\mathbf{v}_1$, $\mathbf{v}_2$, and $\mathbf{v}_3$ in part (a) of the accompanying figure linearly independent? What about those in part (b)? Explain.

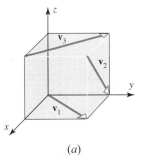

(a)

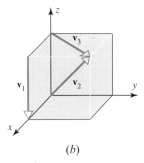

(b)

▲ Figure Ex-19

20. By using appropriate identities, where required, determine which of the following sets of vectors in $F(-\infty, \infty)$ are linearly dependent.

(a) $6,\ 3\sin^2 x,\ 2\cos^2 x$ (b) $x,\ \cos x$

(c) $1,\ \sin x,\ \sin 2x$ (d) $\cos 2x,\ \sin^2 x,\ \cos^2 x$

(e) $(3-x)^2,\ x^2 - 6x,\ 5$ (f) $0,\ \cos^3 \pi x,\ \sin^5 3\pi x$

21. The functions $f_1(x) = x$ and $f_2(x) = \cos x$ are linearly independent in $F(-\infty, \infty)$ because neither function is a scalar multiple of the other. Confirm the linear independence using Wroński's test.

22. The functions $f_1(x) = \sin x$ and $f_2(x) = \cos x$ are linearly independent in $F(-\infty, \infty)$ because neither function is a scalar multiple of the other. Confirm the linear independence using Wroński's test.

23. (*Calculus required*) Use the Wronskian to show that the following sets of vectors are linearly independent.

(a) $1,\ x,\ e^x$ (b) $1,\ x,\ x^2$

24. Use Wroński's test to show that the functions $f_1(x) = e^x$, $f_2(x) = xe^x$, and $f_3(x) = x^2 e^x$ are linearly independent vectors in $F(-\infty, \infty)$.

25. Use Wroński's test to show that the functions $f_1(x) = \sin x$, $f_2(x) = \cos x$, and $f_3(x) = x \cos x$ are linearly independent vectors in $F(-\infty, \infty)$.

26. Use part (*a*) of Theorem 4.3.1 to prove part (*b*).

27. Prove part (*b*) of Theorem 4.3.2.

28. (a) In Example 1 we showed that the mutually orthogonal vectors **i**, **j**, and **k** form a linearly independent set of vectors in R^3. Do you think that every set of three nonzero mutually orthogonal vectors in R^3 is linearly independent? Justify your conclusion with a geometric argument.

(b) Justify your conclusion with an algebraic argument. [*Hint:* Use dot products.]

True-False Exercises

In parts (a)–(h) determine whether the statement is true or false, and justify your answer.

(a) A set containing a single vector is linearly independent.

(b) The set of vectors $\{\mathbf{v}, k\mathbf{v}\}$ is linearly dependent for every scalar k.

(c) Every linearly dependent set contains the zero vector.

(d) If the set of vectors $\{\mathbf{v}_1, \mathbf{v}_2, \mathbf{v}_3\}$ is linearly independent, then $\{k\mathbf{v}_1, k\mathbf{v}_2, k\mathbf{v}_3\}$ is also linearly independent for every nonzero scalar k.

(e) If $\mathbf{v}_1, \ldots, \mathbf{v}_n$ are linearly dependent nonzero vectors, then at least one vector $\mathbf{v}_k$ is a unique linear combination of $\mathbf{v}_1, \ldots, \mathbf{v}_{k-1}$.

(f) The set of 2×2 matrices that contain exactly two 1's and two 0's is a linearly independent set in M_{22}.

(g) The three polynomials $(x-1)(x+2)$, $x(x+2)$, and $x(x-1)$ are linearly independent.

(h) The functions f_1 and f_2 are linearly dependent if there is a real number x so that $k_1 f_1(x) + k_2 f_2(x) = 0$ for some scalars k_1 and k_2.

4.4 Coordinates and Basis

We usually think of a line as being one-dimensional, a plane as two-dimensional, and the space around us as three-dimensional. It is the primary goal of this section and the next to make this intuitive notion of dimension precise. In this section we will discuss coordinate systems in general vector spaces and lay the groundwork for a precise definition of dimension in the next section.

Coordinate Systems in Linear Algebra

In analytic geometry we learned to use *rectangular* coordinate systems to create a one-to-one correspondence between points in 2-space and ordered pairs of real numbers and between points in 3-space and ordered triples of real numbers (Figure 4.4.1). Although rectangular coordinate systems are common, they are not essential. For example, Figure 4.4.2 shows coordinate systems in 2-space and 3-space in which the coordinate axes are not mutually perpendicular.

In linear algebra coordinate systems are commonly specified using vectors rather than coordinate axes. For example, in Figure 4.4.3 we have recreated the coordinate systems in Figure 4.4.2 by using unit vectors to identify the positive directions and then attaching coordinates to a point P using the scalar coefficients in the equations

$$\overrightarrow{OP} = a\mathbf{u}_1 + b\mathbf{u}_2 \quad \text{and} \quad \overrightarrow{OP} = a\mathbf{u}_1 + b\mathbf{u}_2 + c\mathbf{u}_3$$

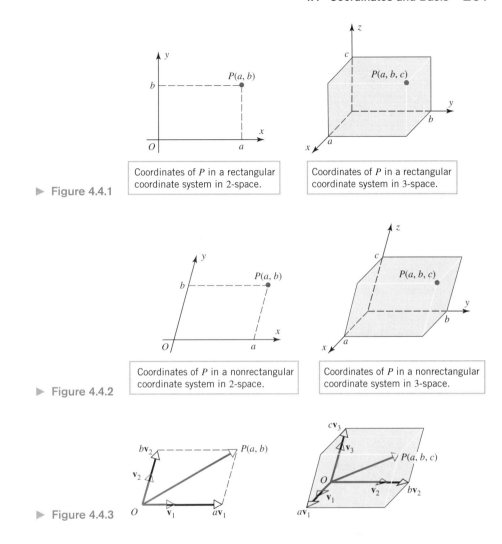

▶ Figure 4.4.1

Coordinates of P in a rectangular coordinate system in 2-space.

Coordinates of P in a rectangular coordinate system in 3-space.

▶ Figure 4.4.2

Coordinates of P in a nonrectangular coordinate system in 2-space.

Coordinates of P in a nonrectangular coordinate system in 3-space.

▶ Figure 4.4.3

Units of measurement are essential ingredients of any coordinate system. In geometry problems one tries to use the same unit of measurement on all axes to avoid distorting the shapes of figures. This is less important in applications where coordinates represent physical quantities with diverse units (for example, time in seconds on one axis and temperature in degrees Celsius on another axis). To allow for this level of generality, we will relax the requirement that *unit* vectors be used to identify the positive directions and require only that those vectors be linearly independent. We will refer to these as the "basis vectors" for the coordinate system. In summary, it is the directions of the basis vectors that establish the positive directions, and it is the lengths of the basis vectors that establish the spacing between the integer points on the axes (Figure 4.4.4).

Basis for a Vector Space

The following definition will make the preceding ideas more precise and will enable us to extend the concept of a coordinate system to general vector spaces.

Note that in Definition 1 we have required a basis to have finitely many vectors. Some authors call this a *finite basis*, but we will not use this terminology.

DEFINITION 1 If V is any vector space and $S = \{\mathbf{v}_1, \mathbf{v}_2, \dots, \mathbf{v}_n\}$ is a finite set of vectors in V, then S is called a **basis** for V if the following two conditions hold:

(a) S is linearly independent.

(b) S spans V.

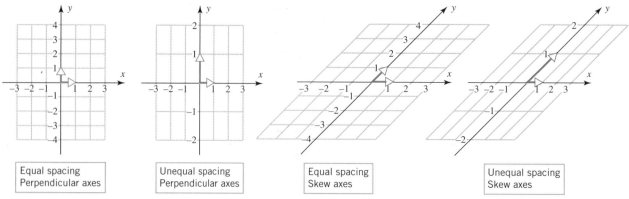

| Equal spacing
Perpendicular axes | Unequal spacing
Perpendicular axes | Equal spacing
Skew axes | Unequal spacing
Skew axes |

▲ Figure 4.4.4

If you think of a basis as describing a coordinate system for a vector space in V, then part (a) of this definition guarantees that there is no interrelationship between the basis vectors, and part (b) guarantees that there are enough basis vectors to provide coordinates for all vectors in V. Here are some examples.

▶ **EXAMPLE 1   The Standard Basis for R^n**

Recall from Example 11 of Section 4.2 that the standard unit vectors

$$\mathbf{e}_1 = (1, 0, 0, \ldots, 0), \quad \mathbf{e}_2 = (0, 1, 0, \ldots, 0), \ldots, \quad \mathbf{e}_n = (0, 0, 0, \ldots, 1)$$

span R^n and from Example 1 of Section 4.3 that they are linearly independent. Thus, they form a basis for R^n that we call the ***standard basis for R^n***. In particular,

$$\mathbf{i} = (1, 0, 0), \quad \mathbf{j} = (0, 1, 0), \quad \mathbf{k} = (0, 0, 1)$$

is the standard basis for R^3.

▶ **EXAMPLE 2   The Standard Basis for P_n**

Show that $S = \{1, x, x^2, \ldots, x^n\}$ is a basis for the vector space P_n of polynomials of degree n or less.

Solution   We must show that the polynomials in S are linearly independent and span P_n. Let us denote these polynomials by

$$\mathbf{p}_0 = 1, \quad \mathbf{p}_1 = x, \quad \mathbf{p}_2 = x^2, \ldots, \quad \mathbf{p}_n = x^n$$

We showed in Example 13 of Section 4.2 that these vectors span P_n and in Example 4 of Section 4.3 that they are linearly independent. Thus, they form a basis for P_n that we call the ***standard basis for P_n***.

▶ **EXAMPLE 3   Another Basis for R^3**

Show that the vectors $\mathbf{v}_1 = (1, 2, 1)$, $\mathbf{v}_2 = (2, 9, 0)$, and $\mathbf{v}_3 = (3, 3, 4)$ form a basis for R^3.

Solution   We must show that these vectors are linearly independent and span R^3. To prove linear independence we must show that the vector equation

$$c_1\mathbf{v}_1 + c_2\mathbf{v}_2 + c_3\mathbf{v}_3 = \mathbf{0} \tag{1}$$

has only the trivial solution; and to prove that the vectors span R^3 we must show that every vector $\mathbf{b} = (b_1, b_2, b_3)$ in R^3 can be expressed as

$$c_1\mathbf{v}_1 + c_2\mathbf{v}_2 + c_3\mathbf{v}_3 = \mathbf{b} \tag{2}$$

By equating corresponding components on the two sides, these two equations can be expressed as the linear systems

$$\begin{aligned} c_1 + 2c_2 + 3c_3 &= 0 \\ 2c_1 + 9c_2 + 3c_3 &= 0 \quad \text{and} \\ c_1 \qquad\quad + 4c_3 &= 0 \end{aligned} \qquad\qquad \begin{aligned} c_1 + 2c_2 + 3c_3 &= b_1 \\ 2c_1 + 9c_2 + 3c_3 &= b_2 \\ c_1 \qquad\quad + 4c_3 &= b_3 \end{aligned} \qquad (3)$$

(verify). Thus, we have reduced the problem to showing that in (3) the homogeneous system has only the trivial solution and that the nonhomogeneous system is consistent for all values of b_1, b_2, and b_3. But the two systems have the same coefficient matrix

$$A = \begin{bmatrix} 1 & 2 & 3 \\ 2 & 9 & 3 \\ 1 & 0 & 4 \end{bmatrix}$$

so it follows from parts (b), (e), and (g) of Theorem 2.3.8 that we can prove both results at the same time by showing that $\det(A) \neq 0$. We leave it for you to confirm that $\det(A) = -1$, which proves that the vectors $\mathbf{v}_1$, $\mathbf{v}_2$, and $\mathbf{v}_3$ form a basis for R^3.

▶ **EXAMPLE 4 The Standard Basis for M_{mn}**

Show that the matrices

$$M_1 = \begin{bmatrix} 1 & 0 \\ 0 & 0 \end{bmatrix}, \quad M_2 = \begin{bmatrix} 0 & 1 \\ 0 & 0 \end{bmatrix}, \quad M_3 = \begin{bmatrix} 0 & 0 \\ 1 & 0 \end{bmatrix}, \quad M_4 = \begin{bmatrix} 0 & 0 \\ 0 & 1 \end{bmatrix}$$

form a basis for the vector space M_{22} of 2×2 matrices.

Solution We must show that the matrices are linearly independent and span M_{22}. To prove linear independence we must show that the equation

$$c_1 M_1 + c_2 M_2 + c_3 M_3 + c_4 M_4 = 0 \qquad (4)$$

has only the trivial solution, where $\mathbf{0}$ is the 2×2 zero matrix; and to prove that the matrices span M_{22} we must show that every 2×2 matrix

$$B = \begin{bmatrix} a & b \\ c & d \end{bmatrix}$$

can be expressed as

$$c_1 M_1 + c_2 M_2 + c_3 M_3 + c_4 M_4 = B \qquad (5)$$

The matrix forms of Equations (4) and (5) are

$$c_1 \begin{bmatrix} 1 & 0 \\ 0 & 0 \end{bmatrix} + c_2 \begin{bmatrix} 0 & 1 \\ 0 & 0 \end{bmatrix} + c_3 \begin{bmatrix} 0 & 0 \\ 1 & 0 \end{bmatrix} + c_4 \begin{bmatrix} 0 & 0 \\ 0 & 1 \end{bmatrix} = \begin{bmatrix} 0 & 0 \\ 0 & 0 \end{bmatrix}$$

and

$$c_1 \begin{bmatrix} 1 & 0 \\ 0 & 0 \end{bmatrix} + c_2 \begin{bmatrix} 0 & 1 \\ 0 & 0 \end{bmatrix} + c_3 \begin{bmatrix} 0 & 0 \\ 1 & 0 \end{bmatrix} + c_4 \begin{bmatrix} 0 & 0 \\ 0 & 1 \end{bmatrix} = \begin{bmatrix} a & b \\ c & d \end{bmatrix}$$

which can be rewritten as

$$\begin{bmatrix} c_1 & c_2 \\ c_3 & c_4 \end{bmatrix} = \begin{bmatrix} 0 & 0 \\ 0 & 0 \end{bmatrix} \quad \text{and} \quad \begin{bmatrix} c_1 & c_2 \\ c_3 & c_4 \end{bmatrix} = \begin{bmatrix} a & b \\ c & d \end{bmatrix}$$

Since the first equation has only the trivial solution

$$c_1 = c_2 = c_3 = c_4 = 0$$

the matrices are linearly independent, and since the second equation has the solution

$$c_1 = a, \quad c_2 = b, \quad c_3 = c, \quad c_4 = d$$

the matrices span M_{22}. This proves that the matrices M_1, M_2, M_3, M_4 form a basis for M_{22}. More generally, the mn different matrices whose entries are zero except for a single entry of 1 form a basis for M_{mn} called the ***standard basis for*** M_{mn}. ◀

Some writers define the empty set to be a basis for the zero vector space, but we will not do so.

It is not true that every vector space has a basis in the sense of Definition 1. The simplest example is the zero vector space, which contains no linearly independent sets and hence no basis. The following is an example of a nonzero vector space that has no basis in the sense of Definition 1 because it cannot be spanned by finitely many vectors.

▶ **EXAMPLE 5** **A Vector Space That Has No Finite Spanning Set**

Show that the vector space of P_∞ of all polynomials with real coefficients has no finite spanning set.

Solution If there were a finite spanning set, say $S = \{\mathbf{p}_1, \mathbf{p}_2, \ldots, \mathbf{p}_r\}$, then the degrees of the polynomials in S would have a maximum value, say n; and this in turn would imply that any linear combination of the polynomials in S would have degree at most n. Thus, there would be no way to express the polynomial x^{n+1} as a linear combination of the polynomials in S, contradicting the fact that the vectors in S span P_∞. ◀

For reasons that will become clear shortly, a vector space that cannot be spanned by finitely many vectors is said to be ***infinite-dimensional***, whereas those that can are said to be ***finite-dimensional***.

▶ **EXAMPLE 6** **Some Finite- and Infinite-Dimensional Spaces**

In Examples 1, 2, and 4 we found bases for R^n, P_n, and M_{mn}, so these vector spaces are finite-dimensional. We showed in Example 5 that the vector space P_∞ is not spanned by finitely many vectors and hence is infinite-dimensional. In the exercises of this section and the next we will ask you to show that the vector spaces R^∞, $F(-\infty, \infty)$, $C(-\infty, \infty)$, $C^m(-\infty, \infty)$, and $C^\infty(-\infty, \infty)$ are infinite-dimensional. ◀

Coordinates Relative to a Basis

Earlier in this section we drew an informal analogy between basis vectors and coordinate systems. Our next goal is to make this informal idea precise by defining the notion of a coordinate system in a general vector space. The following theorem will be our first step in that direction.

THEOREM 4.4.1 **Uniqueness of Basis Representation**

If $S = \{\mathbf{v}_1, \mathbf{v}_2, \ldots, \mathbf{v}_n\}$ is a basis for a vector space V, then every vector $\mathbf{v}$ in V can be expressed in the form $\mathbf{v} = c_1\mathbf{v}_1 + c_2\mathbf{v}_2 + \cdots + c_n\mathbf{v}_n$ in exactly one way.

Proof Since S spans V, it follows from the definition of a spanning set that every vector in V is expressible as a linear combination of the vectors in S. To see that there is only *one* way to express a vector as a linear combination of the vectors in S, suppose that some vector $\mathbf{v}$ can be written as

$$\mathbf{v} = c_1\mathbf{v}_1 + c_2\mathbf{v}_2 + \cdots + c_n\mathbf{v}_n$$

and also as

$$\mathbf{v} = k_1\mathbf{v}_1 + k_2\mathbf{v}_2 + \cdots + k_n\mathbf{v}_n$$

Subtracting the second equation from the first gives

$$\mathbf{0} = (c_1 - k_1)\mathbf{v}_1 + (c_2 - k_2)\mathbf{v}_2 + \cdots + (c_n - k_n)\mathbf{v}_n$$

Since the right side of this equation is a linear combination of vectors in S, the linear independence of S implies that

$$c_1 - k_1 = 0, \quad c_2 - k_2 = 0, \ldots, \quad c_n - k_n = 0$$

that is,

$$c_1 = k_1, \quad c_2 = k_2, \ldots, \quad c_n = k_n$$

Thus, the two expressions for $\mathbf{v}$ are the same. ◀

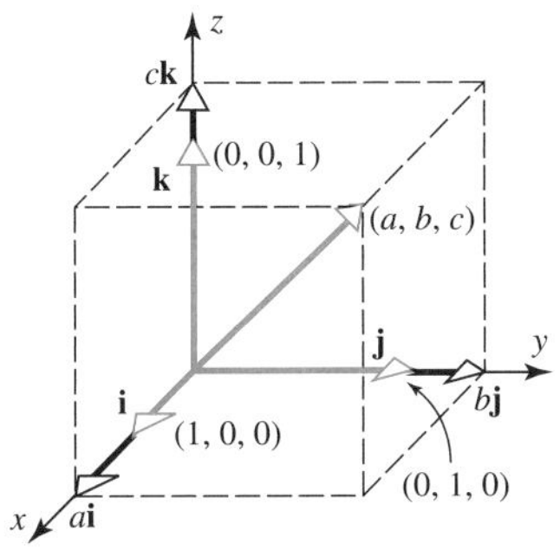

We now have all of the ingredients required to define the notion of "coordinates" in a general vector space V. For motivation, observe that in R^3, for example, the coordinates (a, b, c) of a vector $\mathbf{v}$ are precisely the coefficients in the formula

$$\mathbf{v} = a\mathbf{i} + b\mathbf{j} + c\mathbf{k}$$

that expresses $\mathbf{v}$ as a linear combination of the standard basis vectors for R^3 (see Figure 4.4.5). The following definition generalizes this idea.

▲ Figure 4.4.5

> Sometimes it will be desirable to write a coordinate vector as a column matrix, in which case we will denote it using square brackets as
>
> $$[\mathbf{v}]_S = \begin{bmatrix} c_1 \\ c_2 \\ \vdots \\ c_n \end{bmatrix}$$
>
> We will refer to $[\mathbf{v}]_S$ as a ***coordinate matrix*** and reserve the terminology ***coordinate vector*** for the comma delimited form $(\mathbf{v})_S$.

DEFINITION 2 If $S = \{\mathbf{v}_1, \mathbf{v}_2, \ldots, \mathbf{v}_n\}$ is a basis for a vector space V, and

$$\mathbf{v} = c_1\mathbf{v}_1 + c_2\mathbf{v}_2 + \cdots + c_n\mathbf{v}_n$$

is the expression for a vector $\mathbf{v}$ in terms of the basis S, then the scalars $c_1, c_2, \ldots, c_n$ are called the ***coordinates*** of $\mathbf{v}$ relative to the basis S. The vector $(c_1, c_2, \ldots, c_n)$ in R^n constructed from these coordinates is called the ***coordinate vector of $\mathbf{v}$ relative to S***; it is denoted by

$$(\mathbf{v})_S = (c_1, c_2, \ldots, c_n) \tag{6}$$

Remark Recall that two *sets* are considered to be the same if they have the same members, even if those members are written in a different order. However, if $S = \{\mathbf{v}_1, \mathbf{v}_2, \ldots, \mathbf{v}_n\}$ is a set of *basis vectors*, then changing the order in which the vectors are written would change the order of the entries in $(\mathbf{v})_S$, possibly producing a different coordinate vector. To avoid this complication, we will make the convention that in any discussion involving a basis S the order of the vectors in S remains fixed. Some authors call a set of basis vectors with this restriction an ***ordered basis***. However, we will use this terminology only when emphasis on the order is required for clarity.

Observe that $(\mathbf{v})_S$ is a vector in R^n, so that once basis S is given for a vector space V, Theorem 4.4.1 establishes a one-to-one correspondence between vectors in V and vectors in R^n (Figure 4.4.6).

A one-to-one correspondence

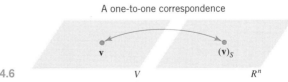

▶ Figure 4.4.6

▶ **EXAMPLE 7 Coordinates Relative to the Standard Basis for R^n**

In the special case where $V = R^n$ and S is the *standard basis*, the coordinate vector $(\mathbf{v})_S$ and the vector $\mathbf{v}$ are the same; that is,

$$\mathbf{v} = (\mathbf{v})_S$$

For example, in R^3 the representation of a vector $\mathbf{v} = (a, b, c)$ as a linear combination of the vectors in the standard basis $S = \{\mathbf{i}, \mathbf{j}, \mathbf{k}\}$ is

$$\mathbf{v} = a\mathbf{i} + b\mathbf{j} + c\mathbf{k}$$

so the coordinate vector relative to this basis is $(\mathbf{v})_S = (a, b, c)$, which is the same as the vector $\mathbf{v}$.

▶ **EXAMPLE 8 Coordinate Vectors Relative to Standard Bases**

(a) Find the coordinate vector for the polynomial

$$\mathbf{p}(x) = c_0 + c_1 x + c_2 x^2 + \cdots + c_n x^n$$

relative to the standard basis for the vector space P_n.

(b) Find the coordinate vector of

$$B = \begin{bmatrix} a & b \\ c & d \end{bmatrix}$$

relative to the standard basis for M_{22}.

Solution (a) The given formula for $\mathbf{p}(x)$ expresses this polynomial as a linear combination of the standard basis vectors $S = \{1, x, x^2, \ldots, x^n\}$. Thus, the coordinate vector for $\mathbf{p}$ relative to S is

$$(\mathbf{p})_S = (c_0, c_1, c_2, \ldots, c_n)$$

Solution (b) We showed in Example 4 that the representation of a vector

$$B = \begin{bmatrix} a & b \\ c & d \end{bmatrix}$$

as a linear combination of the standard basis vectors is

$$B = \begin{bmatrix} a & b \\ c & d \end{bmatrix} = a \begin{bmatrix} 1 & 0 \\ 0 & 0 \end{bmatrix} + b \begin{bmatrix} 0 & 1 \\ 0 & 0 \end{bmatrix} + c \begin{bmatrix} 0 & 0 \\ 1 & 0 \end{bmatrix} + d \begin{bmatrix} 0 & 0 \\ 0 & 1 \end{bmatrix}$$

so the coordinate vector of B relative to S is

$$(B)_S = (a, b, c, d)$$

▶ **EXAMPLE 9 Coordinates in R^3**

(a) We showed in Example 3 that the vectors

$$\mathbf{v}_1 = (1, 2, 1), \quad \mathbf{v}_2 = (2, 9, 0), \quad \mathbf{v}_3 = (3, 3, 4)$$

form a basis for R^3. Find the coordinate vector of $\mathbf{v} = (5, -1, 9)$ relative to the basis $S = \{\mathbf{v}_1, \mathbf{v}_2, \mathbf{v}_3\}$.

(b) Find the vector $\mathbf{v}$ in R^3 whose coordinate vector relative to S is $(\mathbf{v})_S = (-1, 3, 2)$.

Solution (a) To find $(\mathbf{v})_S$ we must first express $\mathbf{v}$ as a linear combination of the vectors in S; that is, we must find values of c_1, c_2, and c_3 such that

$$\mathbf{v} = c_1 \mathbf{v}_1 + c_2 \mathbf{v}_2 + c_3 \mathbf{v}_3$$

or, in terms of components,

$$(5, -1, 9) = c_1(1, 2, 1) + c_2(2, 9, 0) + c_3(3, 3, 4)$$

Equating corresponding components gives

$$\begin{aligned} c_1 + 2c_2 + 3c_3 &= 5 \\ 2c_1 + 9c_2 + 3c_3 &= -1 \\ c_1 + 4c_3 &= 9 \end{aligned}$$

Solving this system we obtain $c_1 = 1$, $c_2 = -1$, $c_3 = 2$ (verify). Therefore,

$$(\mathbf{v})_S = (1, -1, 2)$$

Solution (b) Using the definition of $(\mathbf{v})_S$, we obtain

$$\mathbf{v} = (-1)\mathbf{v}_1 + 3\mathbf{v}_2 + 2\mathbf{v}_3$$
$$= (-1)(1, 2, 1) + 3(2, 9, 0) + 2(3, 3, 4) = (11, 31, 7) \; \blacktriangleleft$$

Concept Review

- Basis
- Standard bases for R^n, P_n, M_{mn}
- Finite-dimensional
- Infinite-dimensional
- Coordinates
- Coordinate vector

Skills

- Show that a set of vectors is a basis for a vector space.
- Find the coordinates of a vector relative to a basis.
- Find the coordinate vector of a vector relative to a basis.

Exercise Set 4.4

1. In words, explain why the following sets of vectors are *not* bases for the indicated vector spaces.

(a) $\mathbf{u}_1 = (1, 2)$, $\mathbf{u}_2 = (0, 3)$, $\mathbf{u}_3 = (2, 7)$ for R^2

(b) $\mathbf{u}_1 = (-1, 3, 2)$, $\mathbf{u}_2 = (6, 1, 1)$ for R^3

(c) $\mathbf{p}_1 = 1 + x + x^2$, $\mathbf{p}_2 = x - 1$ for P_2

(d) $A = \begin{bmatrix} 1 & 1 \\ 2 & 3 \end{bmatrix}$, $B = \begin{bmatrix} 6 & 0 \\ -1 & 4 \end{bmatrix}$, $C = \begin{bmatrix} 3 & 0 \\ 1 & 7 \end{bmatrix}$,

$D = \begin{bmatrix} 5 & 1 \\ 4 & 2 \end{bmatrix}$, $E = \begin{bmatrix} 7 & 1 \\ 2 & 9 \end{bmatrix}$ for M_{22}

2. Which of the following sets of vectors are bases for R^2?

(a) $\{(2, 1), (3, 0)\}$

(b) $\{(4, 1), (-7, -8)\}$

(c) $\{(0, 0), (1, 3)\}$

(d) $\{(3, 9), (-4, -12)\}$

3. Which of the following sets of vectors are bases for R^3?

(a) $\{(1, 0, 0), (2, 2, 0), (3, 3, 3)\}$

(b) $\{(3, 1, -4), (2, 5, 6), (1, 4, 8)\}$

(c) $\{(2, -3, 1), (4, 1, 1), (0, -7, 1)\}$

(d) $\{(1, 6, 4), (2, 4, -1), (-1, 2, 5)\}$

4. Which of the following form bases for P_2?

(a) $1 - 3x + 2x^2$, $1 + x + 4x^2$, $1 - 7x$

(b) $4 + 6x + x^2$, $-1 + 4x + 2x^2$, $5 + 2x - x^2$

(c) $1 + x + x^2$, $x + x^2$, x^2

(d) $-4 + x + 3x^2$, $6 + 5x + 2x^2$, $8 + 4x + x^2$

5. Show that the following matrices form a basis for M_{22}.

$$\begin{bmatrix} 3 & 6 \\ 3 & -6 \end{bmatrix}, \quad \begin{bmatrix} 0 & -1 \\ -1 & 0 \end{bmatrix}, \quad \begin{bmatrix} 0 & -8 \\ -12 & -4 \end{bmatrix}, \quad \begin{bmatrix} 1 & 0 \\ -1 & 2 \end{bmatrix}$$

6. Let V be the space spanned by $\mathbf{v}_1 = \cos^2 x$, $\mathbf{v}_2 = \sin^2 x$, $\mathbf{v}_3 = \cos 2x$.

(a) Show that $S = \{\mathbf{v}_1, \mathbf{v}_2, \mathbf{v}_3\}$ is not a basis for V.

(b) Find a basis for V.

7. Find the coordinate vector of $\mathbf{w}$ relative to the basis $S = \{\mathbf{u}_1, \mathbf{u}_2\}$ for R^2.

(a) $\mathbf{u}_1 = (1, 0)$, $\mathbf{u}_2 = (0, 1)$; $\mathbf{w} = (3, -7)$

(b) $\mathbf{u}_1 = (2, -4)$, $\mathbf{u}_2 = (3, 8)$; $\mathbf{w} = (1, 1)$

(c) $\mathbf{u}_1 = (1, 1)$, $\mathbf{u}_2 = (0, 2)$; $\mathbf{w} = (a, b)$

8. Find the coordinate vector of $\mathbf{w}$ relative to the basis $S = \{\mathbf{u}_1, \mathbf{u}_2\}$ of R^2.

(a) $\mathbf{u}_1 = (1, -1)$, $\mathbf{u}_2 = (1, 1)$; $\mathbf{w} = (1, 0)$

(b) $\mathbf{u}_1 = (1, -1)$, $\mathbf{u}_2 = (1, 1)$; $\mathbf{w} = (0, 1)$

(c) $\mathbf{u}_1 = (1, -1)$, $\mathbf{u}_2 = (1, 1)$; $\mathbf{w} = (1, 1)$

9. Find the coordinate vector of $\mathbf{v}$ relative to the basis $S = \{\mathbf{v}_1, \mathbf{v}_2, \mathbf{v}_3\}$.

(a) $\mathbf{v} = (2, -1, 3)$; $\mathbf{v}_1 = (1, 0, 0)$, $\mathbf{v}_2 = (2, 2, 0)$, $\mathbf{v}_3 = (3, 3, 3)$

(b) $\mathbf{v} = (5, -12, 3)$; $\mathbf{v}_1 = (1, 2, 3)$, $\mathbf{v}_2 = (-4, 5, 6)$, $\mathbf{v}_3 = (7, -8, 9)$

10. Find the coordinate vector of $\mathbf{p}$ relative to the basis $S = \{\mathbf{p}_1, \mathbf{p}_2, \mathbf{p}_3\}$.

(a) $\mathbf{p} = 4 - 3x + x^2$; $\mathbf{p}_1 = 1$, $\mathbf{p}_2 = x$, $\mathbf{p}_3 = x^2$

(b) $\mathbf{p} = 2 - x + x^2$; $\mathbf{p}_1 = 1 + x$, $\mathbf{p}_2 = 1 + x^2$, $\mathbf{p}_3 = x + x^2$

11. Find the coordinate vector of A relative to the basis $S = \{A_1, A_2, A_3, A_4\}$.

$$A = \begin{bmatrix} 2 & 0 \\ -1 & 3 \end{bmatrix}; \quad A_1 = \begin{bmatrix} -1 & 1 \\ 0 & 0 \end{bmatrix}, \quad A_2 = \begin{bmatrix} 1 & 1 \\ 0 & 0 \end{bmatrix},$$

$$A_3 = \begin{bmatrix} 0 & 0 \\ 1 & 0 \end{bmatrix}, \quad A_4 = \begin{bmatrix} 0 & 0 \\ 0 & 1 \end{bmatrix}$$

In Exercises 12–13, show that $\{A_1, A_2, A_3, A_4\}$ is a basis for M_{22}, and express A as a linear combination of the basis vectors.

12. $A_1 = \begin{bmatrix} 1 & 0 \\ 1 & 0 \end{bmatrix}, \quad A_2 = \begin{bmatrix} 1 & 1 \\ 0 & 0 \end{bmatrix}, \quad A_3 = \begin{bmatrix} 1 & 0 \\ 0 & 1 \end{bmatrix},$

$A_4 = \begin{bmatrix} 0 & 0 \\ 1 & 0 \end{bmatrix}; \quad A = \begin{bmatrix} 6 & 2 \\ 5 & 3 \end{bmatrix}$

13. $A_1 = \begin{bmatrix} 1 & 1 \\ 1 & 1 \end{bmatrix}, \quad A_2 = \begin{bmatrix} 0 & 1 \\ 1 & 1 \end{bmatrix}, \quad A_3 = \begin{bmatrix} 0 & 0 \\ 1 & 1 \end{bmatrix},$

$A_4 = \begin{bmatrix} 0 & 0 \\ 0 & 1 \end{bmatrix}; \quad A = \begin{bmatrix} 1 & 0 \\ 1 & 0 \end{bmatrix}$

In Exercises 14–15, show that $\{\mathbf{p}_1, \mathbf{p}_2, \mathbf{p}_3\}$ is a basis for P_2, and express $\mathbf{p}$ as a linear combination of the basis vectors.

14. $\mathbf{p}_1 = 1 + 2x + x^2, \quad \mathbf{p}_2 = 2 + 9x, \quad \mathbf{p}_3 = 3 + 3x + 4x^2;$
$\mathbf{p} = 2 + 17x - 3x^2$

15. $\mathbf{p}_1 = 1 + x + x^2, \quad \mathbf{p}_2 = x + x^2, \quad \mathbf{p}_3 = x^2;$
$\mathbf{p} = 7 - x + 2x^2$

16. The accompanying figure shows a rectangular xy-coordinate system and an $x'y'$-coordinate system with skewed axes. Assuming that 1-unit scales are used on all the axes, find the $x'y'$-coordinates of the points whose xy-coordinates are given.

(a) (1, 1)   (b) (1, 0)   (c) (0, 1)   (d) (a, b)

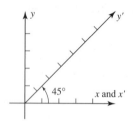

◀ Figure Ex-16

17. The accompanying figure shows a rectangular xy-coordinate system determined by the unit basis vectors $\mathbf{i}$ and $\mathbf{j}$ and an $x'y'$-coordinate system determined by unit basis vectors $\mathbf{u}_1$ and $\mathbf{u}_2$. Find the $x'y'$-coordinates of the points whose xy-coordinates are given.

(a) $(\sqrt{3}, 1)$ (b) (1, 0)   (c) (0, 1)   (d) (a, b)

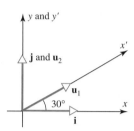

◀ Figure Ex-17

18. The basis that we gave for M_{22} in Example 4 consisted of noninvertible matrices. Do you think that there is a basis for M_{22} consisting of invertible matrices? Justify your answer.

19. Prove that R^∞ is infinite-dimensional.

True-False Exercises

In parts (a)–(e) determine whether the statement is true or false, and justify your answer.

(a) If $V = \text{span}\{\mathbf{v}_1, \ldots, \mathbf{v}_n\}$, then $\{\mathbf{v}_1, \ldots, \mathbf{v}_n\}$ is a basis for V.

(b) Every linearly independent subset of a vector space V is a basis for V.

(c) If $\{\mathbf{v}_1, \mathbf{v}_2, \ldots, \mathbf{v}_n\}$ is a basis for a vector space V, then every vector in V can be expressed as a linear combination of $\mathbf{v}_1, \mathbf{v}_2, \ldots, \mathbf{v}_n$.

(d) The coordinate vector of a vector $\mathbf{x}$ in R^n relative to the standard basis for R^n is $\mathbf{x}$.

(e) Every basis of P_4 contains at least one polynomial of degree 3 or less.

4.5 Dimension

We showed in the previous section that the standard basis R^n has n vectors and hence that the standard basis for R^3 has three vectors, the standard basis for R^2 has two vectors, and the standard basis for $R^1 (= R)$ has one vector. Since we think of space as three dimensional, a plane as two dimensional, and a line as one dimensional, there seems to be a link between the number of vectors in a basis and the dimension of a vector space. We will develop this idea in this section.

Number of Vectors in a Basis

Our first goal in this section is to establish the following fundamental theorem.

> **THEOREM 4.5.1** *All bases for a finite-dimensional vector space have the same number of vectors.*

To prove this theorem we will need the following preliminary result, whose proof is deferred to the end of the section.

> **THEOREM 4.5.2** *Let V be a finite-dimensional vector space, and let $\{\mathbf{v}_1, \mathbf{v}_2, \ldots, \mathbf{v}_n\}$ be any basis.*
> (a) *If a set has more than n vectors, then it is linearly dependent.*
> (b) *If a set has fewer than n vectors, then it does not span V.*

We can now see rather easily why Theorem 4.5.1 is true; for if

$$S = \{\mathbf{v}_1, \mathbf{v}_2, \ldots, \mathbf{v}_n\}$$

is an *arbitrary* basis for V, then the linear independence of S implies that any set in V with more than n vectors is linearly dependent and any set in V with fewer than n vectors does not span V. Thus, unless a set in V has exactly n vectors it cannot be a basis.

We noted in the introduction to this section that for certain familiar vector spaces the intuitive notion of dimension coincides with the number of vectors in a basis. The following definition makes this idea precise.

Some writers regard the empty set to be a basis for the zero vector space. This is consistent with our definition of dimension, since the empty set has no vectors and the zero vector space has dimension zero.

> **DEFINITION 1** The **dimension** of a finite-dimensional vector space V is denoted by $\dim(V)$ and is defined to be the number of vectors in a basis for V. In addition, the zero vector space is defined to have dimension zero.

Engineers often use the term **degrees of freedom** as a synonym for dimension.

▶ **EXAMPLE 1 Dimensions of Some Familiar Vector Spaces**

$$\dim(R^n) = n \qquad \text{The standard basis has } n \text{ vectors.}$$
$$\dim(P_n) = n + 1 \qquad \text{The standard basis has } n + 1 \text{ vectors.}$$
$$\dim(M_{mn}) = mn \qquad \text{The standard basis has } mn \text{ vectors.}$$

▶ **EXAMPLE 2 Dimension of Span(S)**

If $S = \{\mathbf{v}_1, \mathbf{v}_2, \ldots, \mathbf{v}_r\}$ is a *linearly independent* set in a vector space V, then S is automatically a basis for span(S) (why?), and this implies that

$$\dim[\text{span}(S)] = r$$

In words, the dimension of the space spanned by a linearly independent set of vectors is equal to the number of vectors in that set.

▶ **EXAMPLE 3 Dimension of a Solution Space**

Find a basis for and the dimension of the solution space of the homogeneous system

$$
\begin{aligned}
2x_1 + 2x_2 - \;\; x_3 \qquad\qquad + x_5 &= 0 \\
-x_1 - \;\; x_2 + 2x_3 - 3x_4 + x_5 &= 0 \\
x_1 + \;\; x_2 - 2x_3 \qquad\quad - x_5 &= 0 \\
x_3 + \;\; x_4 + x_5 &= 0
\end{aligned}
$$

Solution We leave it for you to solve this system by Gauss–Jordan elimination and show that its general solution is

$$
x_1 = -s - t, \quad x_2 = s, \quad x_3 = -t, \quad x_4 = 0, \quad x_5 = t
$$

which can be written in vector form as

$$
(x_1, x_2, x_3, x_4, x_5) = (-s - t, s, -t, 0, t)
$$

or, alternatively, as

$$
(x_1, x_2, x_3, x_4, x_5) = s(-1, 1, 0, 0, 0) + t(-1, 0, -1, 0, 1)
$$

This shows that the vectors $\mathbf{v}_1 = (-1, 1, 0, 0, 0)$ and $\mathbf{v}_2 = (-1, 0, -1, 0, 1)$ span the solution space. Since neither vector is a scalar multiple of the other, they are linearly independent and hence form a basis for the solution space. Thus, the solution space has dimension 2.

▶ **EXAMPLE 4 Dimension of a Solution Space**

Find a basis for and the dimension of the solution space of the homogeneous system

$$
\begin{aligned}
x_1 + 3x_2 - 2x_3 \qquad\qquad + 2x_5 \qquad\qquad &= 0 \\
2x_1 + 6x_2 - 5x_3 - \;\; 2x_4 + 4x_5 - \;\; 3x_6 &= 0 \\
5x_3 + 10x_4 \qquad\quad + 15x_6 &= 0 \\
2x_1 + 6x_2 \qquad\qquad + \;\; 8x_4 + 4x_5 + 18x_6 &= 0
\end{aligned}
$$

Solution In Example 6 of Section 1.2 we found the solution of this system to be

$$
x_1 = -3r - 4s - 2t, \quad x_2 = r, \quad x_3 = -2s, \quad x_4 = s, \quad x_5 = t, \quad x_6 = 0
$$

which can be written in vector form as

$$
(x_1, x_2, x_3, x_4, x_5, x_6) = (-3r - 4s - 2t, r, -2s, s, t, 0)
$$

or, alternatively, as

$$
(x_1, x_2, x_3, x_4, x_5) = r(-3, 1, 0, 0, 0, 0) + s(-4, 0, -2, 1, 0, 0) + t(-2, 0, 0, 0, 1, 0)
$$

This shows that the vectors

$$
\mathbf{v}_1 = (-3, 1, 0, 0, 0, 0), \quad \mathbf{v}_2 = (-4, 0, -2, 1, 0, 0), \quad \mathbf{v}_3 = (-2, 0, 0, 0, 1, 0)
$$

span the solution space. We leave it for you to check that these vectors are linearly independent by showing that none of them is a linear combination of the other two (but see the remark that follows). Thus, the solution space has dimension 3. ◀

Remark It can be shown that for a homogeneous linear system, the method of the last example *always* produces a basis for the solution space of the system. We omit the formal proof.

Some Fundamental
Theorems We will devote the remainder of this section to a series of theorems that reveal the subtle interrelationships among the concepts of linear independence, basis, and dimension. These theorems are not simply exercises in mathematical theory—they are essential to the understanding of vector spaces and the applications that build on them.

We will start with a theorem (proved at the end of this section) that is concerned with the effect on linear independence and spanning if a vector is added to or removed from a given nonempty set of vectors. Informally stated, if you start with a linearly independent set S and adjoin to it a vector that is not a linear combination of those in S, then the enlarged set will still be linearly independent. Also, if you start with a set S of two or more vectors in which one of the vectors is a linear combination of the others, then that vector can be removed from S without affecting span(S) (Figure 4.5.1).

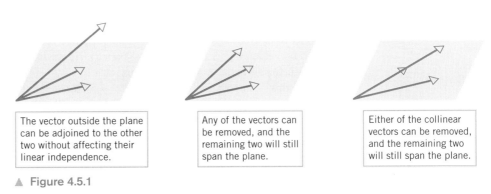

| The vector outside the plane can be adjoined to the other two without affecting their linear independence. | Any of the vectors can be removed, and the remaining two will still span the plane. | Either of the collinear vectors can be removed, and the remaining two will still span the plane. |

▲ Figure 4.5.1

THEOREM 4.5.3 **Plus/Minus Theorem**

Let S be a nonempty set of vectors in a vector space V.

(a) *If S is a linearly independent set, and if $\mathbf{v}$ is a vector in V that is outside of span(S), then the set $S \cup \{\mathbf{v}\}$ that results by inserting $\mathbf{v}$ into S is still linearly independent.*

(b) *If $\mathbf{v}$ is a vector in S that is expressible as a linear combination of other vectors in S, and if $S - \{\mathbf{v}\}$ denotes the set obtained by removing $\mathbf{v}$ from S, then S and $S - \{\mathbf{v}\}$ span the same space; that is,*

$$\text{span}(S) = \text{span}(S - \{\mathbf{v}\})$$

▶ **EXAMPLE 5** **Applying the Plus/Minus Theorem**

Show that $\mathbf{p}_1 = 1 - x^2$, $\mathbf{p}_2 = 2 - x^2$, and $\mathbf{p}_3 = x^3$ are linearly independent vectors.

Solution The set $S = \{\mathbf{p}_1, \mathbf{p}_2\}$ is linearly independent, since neither vector in S is a scalar multiple of the other. Since the vector $\mathbf{p}_3$ cannot be expressed as a linear combination of the vectors in S (why?), it can be adjoined to S to produce a linearly independent set $S' = \{\mathbf{p}_1, \mathbf{p}_2, \mathbf{p}_3\}$. ◀

In general, to show that a set of vectors $\{\mathbf{v}_1, \mathbf{v}_2, \ldots, \mathbf{v}_n\}$ is a basis for a vector space V, we must show that the vectors are linearly independent and span V. However, if we happen to know that V has dimension n (so that $\{\mathbf{v}_1, \mathbf{v}_2, \ldots, \mathbf{v}_n\}$ contains the right number of vectors for a basis), then it suffices to check *either* linear independence *or* spanning— the remaining condition will hold automatically. This is the content of the following theorem.

R^4

R^n n vectors indepe

THEOREM 4.5.4 *Let V be an n-dimensional vector space, and let S be a set in V with exactly n vectors. Then S is a basis for V if and only if S spans V or S is linearly independent.*

n

n − 1   span
        still be
        independent

However
impossible that
n+1 is independent
because theorem 4.5.2

Proof Assume that S has exactly n vectors and spans V. To prove that S is a basis, we must show that S is a linearly independent set. But if this is not so, then some vector $\mathbf{v}$ in S is a linear combination of the remaining vectors. If we remove this vector from S, then it follows from Theorem 4.5.3b that the remaining set of $n - 1$ vectors still spans V. But this is impossible, since it follows from Theorem 4.5.2b that no set with fewer than n vectors can span an n-dimensional vector space. Thus S is linearly independent.

Assume that S has exactly n vectors and is a linearly independent set. To prove that S is a basis, we must show that S spans V. But if this is not so, then there is some vector $\mathbf{v}$ in V that is not in span(S). If we insert this vector into S, then it follows from Theorem 4.5.3a that this set of $n + 1$ vectors is still linearly independent. But this is impossible, since Theorem 4.5.2a states that no set with more than n vectors in an n-dimensional vector space can be linearly independent. Thus S spans V. ◀

▶ **EXAMPLE 6 Bases by Inspection**

(a) By inspection, explain why $\mathbf{v}_1 = (-3, 7)$ and $\mathbf{v}_2 = (5, 5)$ form a basis for R^2.

(b) By inspection, explain why $\mathbf{v}_1 = (2, 0, -1)$, $\mathbf{v}_2 = (4, 0, 7)$, and $\mathbf{v}_3 = (-1, 1, 4)$ form a basis for R^3.

Solution (a) Since neither vector is a scalar multiple of the other, the two vectors form a linearly independent set in the two-dimensional space R^2, and hence they form a basis by Theorem 4.5.4.

Solution (b) The vectors $\mathbf{v}_1$ and $\mathbf{v}_2$ form a linearly independent set in the xz-plane (why?). The vector $\mathbf{v}_3$ is outside of the xz-plane, so the set $\{\mathbf{v}_1, \mathbf{v}_2, \mathbf{v}_3\}$ is also linearly independent. Since R^3 is three-dimensional, Theorem 4.5.4 implies that $\{\mathbf{v}_1, \mathbf{v}_2, \mathbf{v}_3\}$ is a basis for R^3. ◀

The next theorem (whose proof is deferred to the end of this section) reveals two important facts about the vectors in a finite-dimensional vector space V:

1. Every spanning set for a subspace is either a basis for that subspace or has a basis as a subset.

2. Every linearly independent set in a subspace is either a basis for that subspace or can be extended to a basis for it.

THEOREM 4.5.5 *Let S be a finite set of vectors in a finite-dimensional vector space V.*

(a) If S spans V but is not a basis for V, then S can be reduced to a basis for V by removing appropriate vectors from S.

(b) If S is a linearly independent set that is not already a basis for V, then S can be enlarged to a basis for V by inserting appropriate vectors into S.

We conclude this section with a theorem that relates the dimension of a vector space to the dimensions of its subspaces.

THEOREM 4.5.6 *If W is a subspace of a finite-dimensional vector space V, then*:

(a) *W is finite-dimensional.*

(b) *$\dim(W) \leq \dim(V)$.*

(c) *$W = V$ if and only if $\dim(W) = \dim(V)$.*

Proof (a)   We will leave the proof of this part for the exercises.

Proof (b)   Part (*a*) shows that W is finite-dimensional, so it has a basis

$$S = \{\mathbf{w}_1, \mathbf{w}_2, \ldots, \mathbf{w}_m\}$$

Either S is also a basis for V or it is not. If so, then $\dim(V) = m$, which means that $\dim(V) = \dim(W)$. If not, then because S is a linearly independent set it can be enlarged to a basis for V by part (*b*) of Theorem 4.5.5. But this implies that $\dim(W) < \dim(V)$, so we have shown that $\dim(W) \leq \dim(V)$ in all cases.

Proof (c)   Assume that $\dim(W) = \dim(V)$ and that

$$S = \{\mathbf{w}_1, \mathbf{w}_2, \ldots, \mathbf{w}_m\}$$

is a basis for W. If S is not also a basis for V, then being linearly independent S can be extended to a basis for V by part (*b*) of Theorem 4.5.5. But this would mean that $\dim(V) > \dim(W)$, which contradicts our hypothesis. Thus S must also be a basis for V, which means that $\dim(W) = \dim(V)$. ◀

Figure 4.5.2 illustrates the geometric relationship between the subspaces of R^3 in order of increasing dimension.

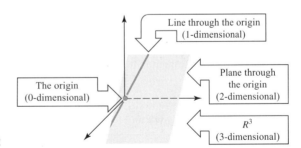

▶ Figure 4.5.2

OPTIONAL   We conclude this section with optional proofs of Theorems 4.5.2, 4.5.3, and 4.5.5.

Proof of Theorem 4.5.2(a)   Let $S' = \{\mathbf{w}_1, \mathbf{w}_2, \ldots, \mathbf{w}_m\}$ be any set of m vectors in V, where $m > n$. We want to show that S' is linearly dependent. Since $S = \{\mathbf{v}_1, \mathbf{v}_2, \ldots, \mathbf{v}_n\}$ is a basis, each $\mathbf{w}_i$ can be expressed as a linear combination of the vectors in S, say

$$
\begin{aligned}
\mathbf{w}_1 &= a_{11}\mathbf{v}_1 + a_{21}\mathbf{v}_2 + \cdots + a_{n1}\mathbf{v}_n \\
\mathbf{w}_2 &= a_{12}\mathbf{v}_1 + a_{22}\mathbf{v}_2 + \cdots + a_{n2}\mathbf{v}_n \\
&\;\;\vdots \qquad \vdots \qquad \vdots \qquad\qquad \vdots \\
\mathbf{w}_m &= a_{1m}\mathbf{v}_1 + a_{2m}\mathbf{v}_2 + \cdots + a_{nm}\mathbf{v}_n
\end{aligned}
\tag{1}
$$

To show that S' is linearly dependent, we must find scalars $k_1, k_2, \ldots, k_m$, not all zero, such that

$$k_1\mathbf{w}_1 + k_2\mathbf{w}_2 + \cdots + k_m\mathbf{w}_m = \mathbf{0} \tag{2}$$

Using the equations in (1), we can rewrite (2) as

$$
\begin{aligned}
(k_1 a_{11} + k_2 a_{12} + \cdots + k_m a_{1m})\mathbf{v}_1 & \\
+ (k_1 a_{21} + k_2 a_{22} + \cdots + k_m a_{2m})\mathbf{v}_2 & \\
\ddots \qquad\qquad & \\
+ (k_1 a_{n1} + k_2 a_{n2} + \cdots + k_m a_{nm})\mathbf{v}_n &= \mathbf{0}
\end{aligned}
$$

Thus, from the linear independence of S, the problem of proving that S' is a linearly dependent set reduces to showing there are scalars $k_1, k_2, \ldots, k_m$, not all zero, that satisfy

$$
\begin{aligned}
a_{11}k_1 + a_{12}k_2 + \cdots + a_{1m}k_m &= 0 \\
a_{21}k_1 + a_{22}k_2 + \cdots + a_{2m}k_m &= 0 \\
\vdots \qquad \vdots \qquad\qquad \vdots \qquad \vdots \ \\
a_{n1}k_1 + a_{n2}k_2 + \cdots + a_{nm}k_m &= 0
\end{aligned}
\tag{3}
$$

But (3) has more unknowns than equations, so the proof is complete since Theorem 1.2.2 guarantees the existence of nontrivial solutions.

Proof of Theorem 4.5.2(b)   Let $S' = \{\mathbf{w}_1, \mathbf{w}_2, \ldots, \mathbf{w}_m\}$ be any set of m vectors in V, where $m < n$. We want to show that S' does not span V. We will do this by showing that the assumption that S' spans V leads to a contradiction of the linear independence of $\{\mathbf{v}_1, \mathbf{v}_2, \ldots, \mathbf{v}_n\}$.

If S' spans V, then every vector in V is a linear combination of the vectors in S'. In particular, each basis vector $\mathbf{v}_i$ is a linear combination of the vectors in S', say

$$
\begin{aligned}
\mathbf{v}_1 &= a_{11}\mathbf{w}_1 + a_{21}\mathbf{w}_2 + \cdots + a_{m1}\mathbf{w}_m \\
\mathbf{v}_2 &= a_{12}\mathbf{w}_1 + a_{22}\mathbf{w}_2 + \cdots + a_{m2}\mathbf{w}_m \\
\vdots \qquad \vdots \qquad \vdots \qquad\qquad \vdots \ \\
\mathbf{v}_n &= a_{1n}\mathbf{w}_1 + a_{2n}\mathbf{w}_2 + \cdots + a_{mn}\mathbf{w}_m
\end{aligned}
\tag{4}
$$

To obtain our contradiction, we will show that there are scalars $k_1, k_2, \ldots, k_n$, not all zero, such that

$$k_1\mathbf{v}_1 + k_2\mathbf{v}_2 + \cdots + k_n\mathbf{v}_n = \mathbf{0} \tag{5}$$

But (4) and (5) have the same form as (1) and (2) except that m and n are interchanged and the $\mathbf{w}$'s and $\mathbf{v}$'s are interchanged. Thus, the computations that led to (3) now yield

$$
\begin{aligned}
a_{11}k_1 + a_{12}k_2 + \cdots + a_{1n}k_n &= 0 \\
a_{21}k_1 + a_{22}k_2 + \cdots + a_{2n}k_n &= 0 \\
\vdots \qquad \vdots \qquad\qquad \vdots \qquad \vdots \ \\
a_{m1}k_1 + a_{m2}k_2 + \cdots + a_{mn}k_n &= 0
\end{aligned}
$$

This linear system has more unknowns than equations and hence has nontrivial solutions by Theorem 1.2.2.

Proof of Theorem 4.5.3(a)   Assume that $S = \{\mathbf{v}_1, \mathbf{v}_2, \ldots, \mathbf{v}_r\}$ is a linearly independent set of vectors in V, and $\mathbf{v}$ is a vector in V outside of span(S). To show that $S' = \{\mathbf{v}_1, \mathbf{v}_2, \ldots, \mathbf{v}_r, \mathbf{v}\}$ is a linearly independent set, we must show that the only scalars that satisfy

$$k_1\mathbf{v}_1 + k_2\mathbf{v}_2 + \cdots + k_r\mathbf{v}_r + k_{r+1}\mathbf{v} = \mathbf{0} \tag{6}$$

are $k_1 = k_2 = \cdots = k_r = k_{r+1} = 0$. But it must be true that $k_{r+1} = 0$ for otherwise we could solve (6) for $\mathbf{v}$ as a linear combination of $\mathbf{v}_1, \mathbf{v}_2, \ldots, \mathbf{v}_r$, contradicting the assumption that $\mathbf{v}$ is outside of span(S). Thus, (6) simplifies to

$$k_1\mathbf{v}_1 + k_2\mathbf{v}_2 + \cdots + k_r\mathbf{v}_r = \mathbf{0} \tag{7}$$

which, by the linear independence of $\{\mathbf{v}_1, \mathbf{v}_2, \ldots, \mathbf{v}_r\}$, implies that

$$k_1 = k_2 = \cdots = k_r = 0$$

Proof Theorem 4.5.3(b) Assume that $S = \{\mathbf{v}_1, \mathbf{v}_2, \ldots, \mathbf{v}_r\}$ is a set of vectors in V, and (to be specific) suppose that $\mathbf{v}_r$ is a linear combination of $\mathbf{v}_1, \mathbf{v}_2, \ldots, \mathbf{v}_{r-1}$, say

$$\mathbf{v}_r = c_1\mathbf{v}_1 + c_2\mathbf{v}_2 + \cdots + c_{r-1}\mathbf{v}_{r-1} \tag{8}$$

We want to show that if $\mathbf{v}_r$ is removed from S, then the remaining set of vectors $\{\mathbf{v}_1, \mathbf{v}_2, \ldots, \mathbf{v}_{r-1}\}$ still spans S; that is, we must show that every vector $\mathbf{w}$ in span(S) is expressible as a linear combination of $\{\mathbf{v}_1, \mathbf{v}_2, \ldots, \mathbf{v}_{r-1}\}$. But if $\mathbf{w}$ is in span(S), then $\mathbf{w}$ is expressible in the form

$$\mathbf{w} = k_1\mathbf{v}_1 + k_2\mathbf{v}_2 + \cdots + k_{r-1}\mathbf{v}_{r-1} + k_r\mathbf{v}_r$$

or, on substituting (8),

$$\mathbf{w} = k_1\mathbf{v}_1 + k_2\mathbf{v}_2 + \cdots + k_{r-1}\mathbf{v}_{r-1} + k_r(c_1\mathbf{v}_1 + c_2\mathbf{v}_2 + \cdots + c_{r-1}\mathbf{v}_{r-1})$$

which expresses $\mathbf{w}$ as a linear combination of $\mathbf{v}_1, \mathbf{v}_2, \ldots, \mathbf{v}_{r-1}$.

Proof of Theorem 4.5.5(a) If S is a set of vectors that spans V but is not a basis for V, then S is a linearly dependent set. Thus some vector $\mathbf{v}$ in S is expressible as a linear combination of the other vectors in S. By the Plus/Minus Theorem (4.5.3b), we can remove $\mathbf{v}$ from S, and the resulting set S' will still span V. If S' is linearly independent, then S' is a basis for V, and we are done. If S' is linearly dependent, then we can remove some appropriate vector from S' to produce a set S'' that still spans V. We can continue removing vectors in this way until we finally arrive at a set of vectors in S that is linearly independent and spans V. This subset of S is a basis for V.

Proof of Theorem 4.5.5(b) Suppose that $\dim(V) = n$. If S is a linearly independent set that is not already a basis for V, then S fails to span V, so there is some vector $\mathbf{v}$ in V that is not in span(S). By the Plus/Minus Theorem (4.5.3a), we can insert $\mathbf{v}$ into S, and the resulting set S' will still be linearly independent. If S' spans V, then S' is a basis for V, and we are finished. If S' does not span V, then we can insert an appropriate vector into S' to produce a set S'' that is still linearly independent. We can continue inserting vectors in this way until we reach a set with n linearly independent vectors in V. This set will be a basis for V by Theorem 4.5.4. ◄

Concept Review

- Dimension

- Relationships among the concepts of linear independence, basis, and dimension

Skills

- Find a basis for and the dimension of the solution space of a homogeneous linear system.

- Use dimension to determine whether a set of vectors is a basis for a finite-dimensional vector space.

- Extend a linearly independent set to a basis.

Exercise Set 4.5

▶ In Exercises 1–6, find a basis for the solution space of the homogeneous linear system, and find the dimension of that space.

1. $\begin{aligned} x_1 + x_2 - x_3 &= 0 \\ -2x_1 - x_2 + 2x_3 &= 0 \\ -x_1 \quad\quad + x_3 &= 0 \end{aligned}$

2. $\begin{aligned} 3x_1 + x_2 + x_3 + x_4 &= 0 \\ 5x_1 - x_2 + x_3 - x_4 &= 0 \end{aligned}$

3. $\begin{aligned} x_1 - 4x_2 + 3x_3 - x_4 &= 0 \\ 2x_1 - 8x_2 + 6x_3 - 2x_4 &= 0 \end{aligned}$

4. $\begin{aligned} x_1 - 3x_2 + x_3 &= 0 \\ 2x_1 - 6x_2 + 2x_3 &= 0 \\ 3x_1 - 9x_2 + 3x_3 &= 0 \end{aligned}$

5. $\begin{aligned} 2x_1 + x_2 + 3x_3 &= 0 \\ x_1 \quad\quad + 5x_3 &= 0 \\ x_2 + x_3 &= 0 \end{aligned}$

6. $\begin{aligned} x + y + z &= 0 \\ 3x + 2y - 2z &= 0 \\ 4x + 3y - z &= 0 \\ 6x + 5y + z &= 0 \end{aligned}$

7. Find bases for the following subspaces of R^3.

   (a) The plane $3x - 2y + 5z = 0$.

   (b) The plane $x - y = 0$.

   (c) The line $x = 2t$, $y = -t$, $z = 4t$.

   (d) All vectors of the form (a, b, c), where $b = a + c$.

8. Find the dimensions of the following subspaces of R^4.

   (a) All vectors of the form $(a, b, c, 0)$.

   (b) All vectors of the form (a, b, c, d), where $d = a + b$ and $c = a - b$.

   (c) All vectors of the form (a, b, c, d), where $a = b = c = d$.

9. Find the dimension of each of the following vector spaces.

   (a) The vector space of all diagonal $n \times n$ matrices.

   (b) The vector space of all symmetric $n \times n$ matrices.

   (c) The vector space of all upper triangular $n \times n$ matrices.

10. Find the dimension of the subspace of P_3 consisting of all polynomials $a_0 + a_1x + a_2x^2 + a_3x^3$ for which $a_0 = 0$.

11. (a) Show that the set W of all polynomials in P_2 such that $p(1) = 0$ is a subspace of P_2.

   (b) Make a conjecture about the dimension of W.

   (c) Confirm your conjecture by finding a basis for W.

12. Find a standard basis vector for R^3 that can be added to the set $\{v_1, v_2\}$ to produce a basis for R^3.

   (a) $v_1 = (-1, 2, 3)$, $v_2 = (1, -2, -2)$

   (b) $v_1 = (1, -1, 0)$, $v_2 = (3, 1, -2)$

13. Find standard basis vectors for R^4 that can be added to the set $\{v_1, v_2\}$ to produce a basis for R^4.

$$v_1 = (1, -4, 2, -3), \quad v_2 = (-3, 8, -4, 6)$$

14. Let $\{v_1, v_2, v_3\}$ be a basis for a vector space V. Show that $\{u_1, u_2, u_3\}$ is also a basis, where $u_1 = v_1$, $u_2 = v_1 + v_2$, and $u_3 = v_1 + v_2 + v_3$.

15. The vectors $v_1 = (1, -2, 3)$ and $v_2 = (0, 5, -3)$ are linearly independent. Enlarge $\{v_1, v_2\}$ to a basis for R^3.

16. The vectors $v_1 = (1, -2, 3, -5)$ and $v_2 = (0, -1, 2, -3)$ are linearly independent. Enlarge $\{v_1, v_2\}$ to a basis for R^4.

17. (a) Show that for every positive integer n, one can find $n + 1$ linearly independent vectors in $F(-\infty, \infty)$. [*Hint:* Look for polynomials.]

   (b) Use the result in part (a) to prove that $F(-\infty, \infty)$ is infinite-dimensional.

   (c) Prove that $C(-\infty, \infty)$, $C^m(-\infty, \infty)$, and $C^\infty(-\infty, \infty)$ are infinite-dimensional vector spaces.

18. Let S be a basis for an n-dimensional vector space V. Show that if $v_1, v_2, \ldots, v_r$ form a linearly independent set of vectors in V, then the coordinate vectors $(v_1)_S$, $(v_2)_S$, $\ldots$, $(v_r)_S$ form a linearly independent set in R^n, and conversely.

19. Using the notation from Exercise 18, show that if the vectors $v_1, v_2, \ldots, v_r$ span V, then the coordinate vectors $(v_1)_S$, $(v_2)_S$, $\ldots$, $(v_r)_S$ span R^n, and conversely.

20. Find a basis for the subspace of P_2 spanned by the given vectors.

   (a) $-1 + x - 2x^2$, $3 + 3x + 6x^2$, 9

(b) $1 + x$, x^2, $-2 + 2x^2$, $-3x$

(c) $1 + x - 3x^2$, $2 + 2x - 6x^2$, $3 + 3x - 9x^2$

[*Hint:* Let S be the standard basis for P_2, and work with the coordinate vectors relative to S as in Exercises 18 and 19.]

21. Prove: A subspace of a finite-dimensional vector space is finite-dimensional.

22. State the two parts of Theorem 4.5.2 in contrapositive form.

True-False Exercises

In parts (a)–(j) determine whether the statement is true or false, and justify your answer.

(a) The zero vector space has dimension zero.

(b) There is a set of 17 linearly independent vectors in R^{17}.

(c) There is a set of 11 vectors that span R^{17}.

(d) Every linearly independent set of five vectors in R^5 is a basis for R^5.

(e) Every set of five vectors that spans R^5 is a basis for R^5.

(f) Every set of vectors that spans R^n contains a basis for R^n.

(g) Every linearly independent set of vectors in R^n is contained in some basis for R^n.

(h) There is a basis for M_{22} consisting of invertible matrices.

(i) If A has size $n \times n$ and I_n, A, A^2, ..., A^{n^2} are distinct matrices, then $\{I_n, A, A^2, \ldots, A^{n^2}\}$ is linearly dependent.

(j) There are at least two distinct three-dimensional subspaces of P_2.

4.6 Change of Basis

A basis that is suitable for one problem may not be suitable for another, so it is a common process in the study of vector spaces to change from one basis to another. Because a basis is the vector space generalization of a coordinate system, changing bases is akin to changing coordinate axes in R^2 and R^3. In this section we will study problems related to change of basis.

Coordinate Maps   If $S = \{v_1, v_2, \ldots, v_n\}$ is a basis for a finite-dimensional vector space V, and if

$$(\mathbf{v})_S = (c_1, c_2, \ldots, c_n)$$

is the coordinate vector of $\mathbf{v}$ relative to S, then, as observed in Section 4.4, the mapping

$$\mathbf{v} \rightarrow (\mathbf{v})_S \tag{1}$$

creates a connection (a one-to-one correspondence) between vectors in the *general* vector space V and vectors in the *familiar* vector space R^n. We call (1) the **coordinate map** from V to R^n. In this section we will find it convenient to express coordinate vectors in the matrix form

$$[\mathbf{v}]_S = \begin{bmatrix} c_1 \\ c_2 \\ \vdots \\ c_n \end{bmatrix} \tag{2}$$

Coordinate map

▲ Figure 4.6.1

where the square brackets emphasize the matrix notation (Figure 4.6.1).

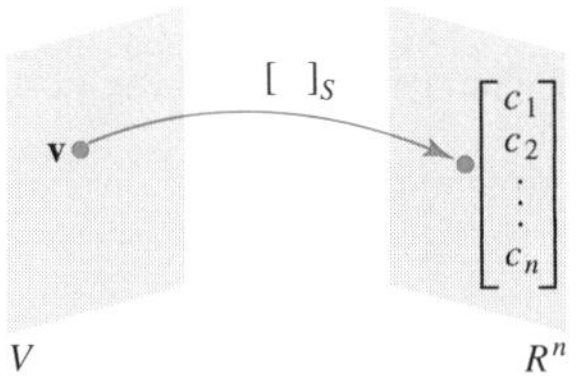

Change of Basis   There are many applications in which it is necessary to work with more than one coordinate system. In such cases it becomes important to know how the coordinates of a fixed vector relative to each coordinate system are related. This leads to the following problem.

> **The Change-of-Basis Problem**   If $\mathbf{v}$ is a vector in a finite-dimensional vector space V, and if we change the basis for V from a basis B to a basis B', how are the coordinate vectors $[\mathbf{v}]_B$ and $[\mathbf{v}]_{B'}$ related?

Remark To solve this problem, it will be convenient to refer to B as the "old basis" and B' as the "new basis." Thus, our objective is to find a relationship between the old and new coordinates of a fixed vector $\mathbf{v}$ in V.

For simplicity, we will solve this problem for two-dimensional spaces. The solution for n-dimensional spaces is similar. Let

$$B = \{\mathbf{u}_1, \mathbf{u}_2\} \quad \text{and} \quad B' = \{\mathbf{u}_1', \mathbf{u}_2'\}$$

be the old and new bases, respectively. We will need the coordinate vectors for the new basis vectors relative to the old basis. Suppose they are

$$[\mathbf{u}_1']_B = \begin{bmatrix} a \\ b \end{bmatrix} \quad \text{and} \quad [\mathbf{u}_2']_B = \begin{bmatrix} c \\ d \end{bmatrix} \tag{3}$$

That is,

$$\begin{aligned} \mathbf{u}_1' &= a\mathbf{u}_1 + b\mathbf{u}_2 \\ \mathbf{u}_2' &= c\mathbf{u}_1 + d\mathbf{u}_2 \end{aligned} \tag{4}$$

Now let $\mathbf{v}$ be any vector in V, and let

$$[\mathbf{v}]_{B'} = \begin{bmatrix} k_1 \\ k_2 \end{bmatrix} \tag{5}$$

be the new coordinate vector, so that

$$\mathbf{v} = k_1\mathbf{u}_1' + k_2\mathbf{u}_2' \tag{6}$$

In order to find the old coordinates of $\mathbf{v}$, we must express $\mathbf{v}$ in terms of the old basis B. To do this, we substitute (4) into (6). This yields

$$\mathbf{v} = k_1(a\mathbf{u}_1 + b\mathbf{u}_2) + k_2(c\mathbf{u}_1 + d\mathbf{u}_2)$$

or

$$\mathbf{v} = (k_1 a + k_2 c)\mathbf{u}_1 + (k_1 b + k_2 d)\mathbf{u}_2$$

Thus, the old coordinate vector for $\mathbf{v}$ is

$$[\mathbf{v}]_B = \begin{bmatrix} k_1 a + k_2 c \\ k_1 b + k_2 d \end{bmatrix}$$

which, by using (5), can be written as

$$[\mathbf{v}]_B = \begin{bmatrix} a & c \\ b & d \end{bmatrix} \begin{bmatrix} k_1 \\ k_2 \end{bmatrix} = \begin{bmatrix} a & c \\ b & d \end{bmatrix} [\mathbf{v}]_{B'}$$

This equation states that the old coordinate vector $[\mathbf{v}]_B$ results when we multiply the new coordinate vector $[\mathbf{v}]_{B'}$ on the left by the matrix

$$P = \begin{bmatrix} a & c \\ b & d \end{bmatrix}$$

Since the columns of this matrix are the coordinates of the new basis vectors relative to the old basis [see (3)] we have the following solution of the change-of-basis problem.

Solution of the Change-of-Basis Problem If we change the basis for a vector space V from an old basis $B = \{\mathbf{u}_1, \mathbf{u}_2, \ldots, \mathbf{u}_n\}$ to a new basis $B' = \{\mathbf{u}_1', \mathbf{u}_2', \ldots, \mathbf{u}_n'\}$, then for each vector $\mathbf{v}$ in V, the old coordinate vector $[\mathbf{v}]_B$ is related to the new coordinate vector $[\mathbf{v}]_{B'}$ by the equation

$$[\mathbf{v}]_B = P[\mathbf{v}]_{B'} \tag{7}$$

where the columns of P are the coordinate vectors of the new basis vectors relative to the old basis; that is, the column vectors of P are

$$[\mathbf{u}_1']_B, \quad [\mathbf{u}_2']_B, \ldots, \quad [\mathbf{u}_n']_B \tag{8}$$

Transition Matrices   The matrix P in Equation (7) is called the ***transition matrix*** from B' to B. For emphasis, we will often denote it by $P_{B' \to B}$. It follows from (8) that this matrix can be expressed in terms of its column vectors as

$$P_{B' \to B} = \left[[\mathbf{u}'_1]_B \mid [\mathbf{u}'_2]_B \mid \cdots \mid [\mathbf{u}'_n]_B \right] \qquad (9)$$

Similarly, the transition matrix from B to B' can be expressed in terms of its column vectors as

$$P_{B \to B'} = \left[[\mathbf{u}_1]_{B'} \mid [\mathbf{u}_2]_{B'} \mid \cdots \mid [\mathbf{u}_n]_{B'} \right] \qquad (10)$$

Remark   There is a simple way to remember both of these formulas using the terms "old basis" and "new basis" defined earlier in this section: In Formula (9) the old basis is B' and the new basis is B, whereas in Formula (10) the old basis is B and the new basis is B'. Thus, both formulas can be restated as follows:

The columns of the transition matrix from an old basis to a new basis are the coordinate vectors of the old basis relative to the new basis.

▶ **EXAMPLE 1   Finding Transition Matrices**

Consider the bases $B = \{\mathbf{u}_1, \mathbf{u}_2\}$ and $B' = \{\mathbf{u}'_1, \mathbf{u}'_2\}$ for R^2, where

$$\mathbf{u}_1 = (1, 0), \quad \mathbf{u}_2 = (0, 1), \quad \mathbf{u}'_1 = (1, 1), \quad \mathbf{u}'_2 = (2, 1)$$

(a)   Find the transition matrix $P_{B' \to B}$ from B' to B.

(b)   Find the transition matrix $P_{B \to B'}$ from B to B'.

Solution (a)   Here the old basis vectors are $\mathbf{u}'_1$ and $\mathbf{u}'_2$ and the new basis vectors are $\mathbf{u}_1$ and $\mathbf{u}_2$. We want to find the coordinate matrices of the old basis vectors $\mathbf{u}'_1$ and $\mathbf{u}'_2$ relative to the new basis vectors $\mathbf{u}_1$ and $\mathbf{u}_2$. To do this, first we observe that

$$\mathbf{u}'_1 = \mathbf{u}_1 + \mathbf{u}_2$$
$$\mathbf{u}'_2 = 2\mathbf{u}_1 + \mathbf{u}_2$$

from which it follows that

$$[\mathbf{u}'_1]_B = \begin{bmatrix} 1 \\ 1 \end{bmatrix} \quad \text{and} \quad [\mathbf{u}'_2]_B = \begin{bmatrix} 2 \\ 1 \end{bmatrix}$$

and hence that

$$P_{B' \to B} = \begin{bmatrix} 1 & 2 \\ 1 & 1 \end{bmatrix}$$

Solution (b)   Here the old basis vectors are $\mathbf{u}_1$ and $\mathbf{u}_2$ and the new basis vectors are $\mathbf{u}'_1$ and $\mathbf{u}'_2$. As in part (a), we want to find the coordinate matrices of the old basis vectors $\mathbf{u}'_1$ and $\mathbf{u}'_2$ relative to the new basis vectors $\mathbf{u}_1$ and $\mathbf{u}_2$. To do this, observe that

$$\mathbf{u}_1 = -\mathbf{u}'_1 + \mathbf{u}'_2$$
$$\mathbf{u}_2 = 2\mathbf{u}'_1 - \mathbf{u}'_2$$

from which it follows that

$$[\mathbf{u}_1]_{B'} = \begin{bmatrix} -1 \\ 1 \end{bmatrix} \quad \text{and} \quad [\mathbf{u}_2]_{B'} = \begin{bmatrix} 2 \\ -1 \end{bmatrix}$$

and hence that

$$P_{B \to B'} = \begin{bmatrix} -1 & 2 \\ 1 & -1 \end{bmatrix} \quad ◀$$

Suppose now that B and B' are bases for a finite-dimensional vector space V. Since multiplication by $P_{B'\to B}$ maps coordinate vectors relative to the basis B' into coordinate vectors relative to a basis B, and $P_{B\to B'}$ maps coordinate vectors relative to B into coordinate vectors relative to B', it follows that for every vector $\mathbf{v}$ in V we have

$$[\mathbf{v}]_B = P_{B'\to B}[\mathbf{v}]_{B'} \tag{11}$$

$$[\mathbf{v}]_{B'} = P_{B\to B'}[\mathbf{v}]_B \tag{12}$$

▶ **EXAMPLE 2 Computing Coordinate Vectors**

Let B and B' be the bases in Example 1. Use an appropriate formula to find $[\mathbf{v}]_B$ given that

$$[\mathbf{v}]_{B'} = \begin{bmatrix} -3 \\ 5 \end{bmatrix}$$

Solution To find $[\mathbf{v}]_B$ we need to make the transition from B' to B. It follows from Formula (11) and part (a) of Example 1 that

$$[\mathbf{v}]_B = P_{B'\to B}[\mathbf{v}]_{B'} = \begin{bmatrix} 1 & 2 \\ 1 & 1 \end{bmatrix}\begin{bmatrix} -3 \\ 5 \end{bmatrix} = \begin{bmatrix} 7 \\ 2 \end{bmatrix} \quad ◀$$

Invertibility of Transition Matrices

If B and B' are bases for a finite-dimensional vector space V, then

$$(P_{B'\to B})\,(P_{B\to B'}) = P_{B\to B}$$

because multiplication by $(P_{B'\to B})\,(P_{B\to B'})$ first maps B-coordinates of a vector into B'-coordinates, and then maps those B'-coordinates back into the original B-coordinates. Since the net effect of the two operations is to leave each coordinate vector unchanged, we are led to conclude that $P_{B\to B}$ must be the identity matrix, that is,

$$(P_{B'\to B})\,(P_{B\to B'}) = I \tag{13}$$

(we omit the formal proof). For example, for the transition matrices obtained in Example 1 we have

$$(P_{B'\to B})\,(P_{B\to B'}) = \begin{bmatrix} 1 & 2 \\ 1 & 1 \end{bmatrix}\begin{bmatrix} -1 & 2 \\ 1 & -1 \end{bmatrix} = \begin{bmatrix} 1 & 0 \\ 0 & 1 \end{bmatrix} = I$$

It follows from (13) that $P_{B'\to B}$ is invertible and that its inverse is $P_{B\to B'}$. Thus, we have the following theorem.

> **THEOREM 4.6.1** *If P is the transition matrix from a basis B' to a basis B for a finite-dimensional vector space V, then P is invertible and P^{-1} is the transition matrix from B to B'.*

An Efficient Method for Computing Transition Matrices for R^n

Our next objective is to develop an efficient procedure for computing transition matrices *between bases for R^n*. As illustrated in Example 1, the first step in computing a transition matrix is to express each new basis vector as a linear combination of the old basis vectors. For R^n this involves solving n linear systems of n equations in n unknowns, each of which has the same coefficient matrix (why?). An efficient way to do this is by the method illustrated in Example 2 of Section 1.6, which is as follows:

A Procedure for Computing $P_{B \to B'}$

Step 1. Form the matrix $[B' \mid B]$.

Step 2. Use elementary row operations to reduce the matrix in Step 1 to reduced row echelon form.

Step 3. The resulting matrix will be $[I \mid P_{B \to B'}]$.

Step 4. Extract the matrix $P_{B \to B'}$ from the right side of the matrix in Step 3.

This procedure is captured in the following diagram.

$$[\text{new basis} \mid \text{old basis}] \quad \xrightarrow{\text{row operations}} \quad [I \mid \text{transition from old to new}] \qquad (14)$$

▶ **EXAMPLE 3 Example 1 Revisited**

In Example 1 we considered the bases $B = \{\mathbf{u}_1, \mathbf{u}_2\}$ and $B' = \{\mathbf{u}'_1, \mathbf{u}'_2\}$ for R^2, where

$$\mathbf{u}_1 = (1, 0), \quad \mathbf{u}_2 = (0, 1), \quad \mathbf{u}'_1 = (1, 1), \quad \mathbf{u}'_2 = (2, 1)$$

(a) Use Formula (14) to find the transition matrix from B' to B.

(b) Use Formula (14) to find the transition matrix from B to B'.

Solution (a) Here B' is the old basis and B is the new basis, so

$$[\text{new basis} \mid \text{old basis}] = \begin{bmatrix} 1 & 0 & | & 1 & 2 \\ 0 & 1 & | & 1 & 1 \end{bmatrix}$$

Since the left side is already the identity matrix, no reduction is needed. We see by inspection that the transition matrix is

$$P_{B' \to B} = \begin{bmatrix} 1 & 2 \\ 1 & 1 \end{bmatrix}$$

which agrees with the result in Example 1.

Solution (b) Here B is the old basis and B' is the new basis, so

$$[\text{new basis} \mid \text{old basis}] = \begin{bmatrix} 1 & 2 & | & 1 & 0 \\ 1 & 1 & | & 0 & 1 \end{bmatrix}$$

By reducing this matrix, so the left side becomes the identity we obtain (verify)

$$[I \mid \text{transition from old to new}] = \begin{bmatrix} 1 & 0 & | & -1 & 2 \\ 0 & 1 & | & 1 & -1 \end{bmatrix}$$

so the transition matrix is

$$P_{B \to B'} = \begin{bmatrix} -1 & 2 \\ 1 & -1 \end{bmatrix}$$

which also agrees with the result in Example 1. ◀

Transition to the Standard Basis for R^n

Note that in part (a) of the last example the column vectors of the matrix that made the transition from the basis B' to the standard basis turned out to be the vectors in B' written in column form. This illustrates the following general result.

THEOREM 4.6.2 *Let* $B' = \{\mathbf{u}_1, \mathbf{u}_2, \ldots, \mathbf{u}_n\}$ *be any basis for the vector space* R^n *and let* $S = \{\mathbf{e}_1, \mathbf{e}_2, \ldots, \mathbf{e}_n\}$ *be the standard basis for* R^n. *If the vectors in these bases are written in column form, then*

$$P_{B' \to S} = [\mathbf{u}_1 \mid \mathbf{u}_2 \mid \cdots \mid \mathbf{u}_n] \tag{15}$$

It follows from this theorem that if

$$A = [\mathbf{u}_1 \mid \mathbf{u}_2 \mid \cdots \mid \mathbf{u}_n]$$

is *any* invertible $n \times n$ matrix, then A can be viewed as the transition matrix from the basis $\{\mathbf{u}_1, \mathbf{u}_2, \ldots, \mathbf{u}_n\}$ for R^n to the standard basis for R^n. Thus, for example, the matrix

$$A = \begin{bmatrix} 1 & 2 & 3 \\ 2 & 5 & 3 \\ 1 & 0 & 8 \end{bmatrix}$$

which was shown to be invertible in Example 4 of Section 1.5, is the transition matrix from the basis

$$\mathbf{u}_1 = (1, 2, 1), \quad \mathbf{u}_2 = (2, 5, 0), \quad \mathbf{u}_3 = (3, 3, 8)$$

to the basis

$$\mathbf{e}_1 = (1, 0, 0), \quad \mathbf{e}_2 = (0, 1, 0), \quad \mathbf{e}_3 = (0, 0, 1)$$

Concept Review

• Coordinate map
• Change-of-basis problem
• Transition matrix

Skills

• Find coordinate vectors relative to a given basis directly.
• Find the transition matrix from one basis to another.
• Use the transition matrix to compute coordinate vectors.

Exercise Set 4.6

1. Find the coordinate vector for $\mathbf{w}$ relative to the basis $S = \{\mathbf{u}_1, \mathbf{u}_2\}$ for R^2.

(a) $\mathbf{u}_1 = (1, 0)$, $\mathbf{u}_2 = (0, 1)$; $\mathbf{w} = (3, -7)$

(b) $\mathbf{u}_1 = (2, -4)$, $\mathbf{u}_2 = (3, 8)$; $\mathbf{w} = (1, 1)$

(c) $\mathbf{u}_1 = (1, 1)$, $\mathbf{u}_2 = (0, 2)$; $\mathbf{w} = (a, b)$

2. Find the coordinate vector for $\mathbf{v}$ relative to the basis $S = \{\mathbf{v}_1, \mathbf{v}_2, \mathbf{v}_3\}$ for R^3.

(a) $\mathbf{v} = (2, -1, 3)$; $\mathbf{v}_1 = (1, 0, 0)$, $\mathbf{v}_2 = (2, 2, 0)$, $\mathbf{v}_3 = (3, 3, 3)$

(b) $\mathbf{v} = (5, -12, 3)$; $\mathbf{v}_1 = (1, 2, 3)$, $\mathbf{v}_2 = (-4, 5, 6)$, $\mathbf{v}_3 = (7, -8, 9)$

3. Find the coordinate vector for $\mathbf{p}$ relative to the basis $S = \{\mathbf{p}_1, \mathbf{p}_2, \mathbf{p}_3\}$ for P_2.

(a) $\mathbf{p} = 4 - 3x + x^2$; $\mathbf{p}_1 = 1$, $\mathbf{p}_2 = x$, $\mathbf{p}_3 = x^2$

(b) $\mathbf{p} = 2 - x + x^2$; $\mathbf{p}_1 = 1 + x$, $\mathbf{p}_2 = 1 + x^2$, $\mathbf{p}_3 = x + x^2$

4. Find the coordinate vector for A relative to the basis $S = \{A_1, A_2, A_3, A_4\}$ for M_{22}.

$$A = \begin{bmatrix} 2 & 0 \\ -1 & 3 \end{bmatrix}, \quad A_1 = \begin{bmatrix} -1 & 1 \\ 0 & 0 \end{bmatrix}, \quad A_2 = \begin{bmatrix} 1 & 1 \\ 0 & 0 \end{bmatrix},$$

$$A_3 = \begin{bmatrix} 0 & 0 \\ 1 & 0 \end{bmatrix}, \quad A_4 = \begin{bmatrix} 0 & 0 \\ 0 & 1 \end{bmatrix}$$

5. Consider the coordinate vectors

$$[\mathbf{w}]_S = \begin{bmatrix} 6 \\ -1 \\ 4 \end{bmatrix}, \quad [\mathbf{q}]_S = \begin{bmatrix} 3 \\ 0 \\ 4 \end{bmatrix}, \quad [B]_S = \begin{bmatrix} -8 \\ 7 \\ 6 \\ 3 \end{bmatrix}$$

(a) Find $\mathbf{w}$ if S is the basis in Exercise 2(a).

(b) Find $\mathbf{q}$ if S is the basis in Exercise 3(a).

(c) Find B if S is the basis in Exercise 4.

6. Consider the bases $B = \{\mathbf{u}_1, \mathbf{u}_2\}$ and $B' = \{\mathbf{u}'_1, \mathbf{u}'_2\}$ for R^2, where

$$\mathbf{u}_1 = \begin{bmatrix} 1 \\ 0 \end{bmatrix}, \quad \mathbf{u}_2 = \begin{bmatrix} 0 \\ 1 \end{bmatrix}, \quad \mathbf{u}'_1 = \begin{bmatrix} 2 \\ 1 \end{bmatrix}, \quad \mathbf{u}'_2 = \begin{bmatrix} -3 \\ 4 \end{bmatrix}$$

(a) Find the transition matrix from B' to B.

(b) Find the transition matrix from B to B'.

(c) Compute the coordinate vector $[\mathbf{w}]_B$, where

$$\mathbf{w} = \begin{bmatrix} 3 \\ -5 \end{bmatrix}$$

and use (10) to compute $[\mathbf{w}]_{B'}$.

(d) Check your work by computing $[\mathbf{w}]_{B'}$ directly.

7. Repeat the directions of Exercise 6 with the same vector $\mathbf{w}$ but with

$$\mathbf{u}_1 = \begin{bmatrix} 2 \\ 2 \end{bmatrix}, \quad \mathbf{u}_2 = \begin{bmatrix} 4 \\ -1 \end{bmatrix}, \quad \mathbf{u}'_1 = \begin{bmatrix} 1 \\ 3 \end{bmatrix}, \quad \mathbf{u}'_2 = \begin{bmatrix} -1 \\ -1 \end{bmatrix}$$

8. Consider the bases $B = \{\mathbf{u}_1, \mathbf{u}_2, \mathbf{u}_3\}$ and $B' = \{\mathbf{u}'_1, \mathbf{u}'_2, \mathbf{u}'_3\}$ for R^3, where

$$\mathbf{u}_1 = \begin{bmatrix} -3 \\ 0 \\ -3 \end{bmatrix}, \quad \mathbf{u}_2 = \begin{bmatrix} -3 \\ 2 \\ -1 \end{bmatrix}, \quad \mathbf{u}_3 = \begin{bmatrix} 1 \\ 6 \\ -1 \end{bmatrix}$$

$$\mathbf{u}'_1 = \begin{bmatrix} -6 \\ -6 \\ 0 \end{bmatrix}, \quad \mathbf{u}'_2 = \begin{bmatrix} -2 \\ -6 \\ 4 \end{bmatrix}, \quad \mathbf{u}'_3 = \begin{bmatrix} -2 \\ -3 \\ 7 \end{bmatrix}$$

(a) Find the transition matrix from B to B'.

(b) Compute the coordinate vector $[\mathbf{w}]_B$, where

$$\mathbf{w} = \begin{bmatrix} -5 \\ 8 \\ -5 \end{bmatrix}$$

and use (12) to compute $[\mathbf{w}]_{B'}$.

(c) Check your work by computing $[\mathbf{w}]_{B'}$ directly.

9. Repeat the directions of Exercise 8 with the same vector $\mathbf{w}$, but with

$$\mathbf{u}_1 = \begin{bmatrix} 2 \\ 1 \\ 1 \end{bmatrix}, \quad \mathbf{u}_2 = \begin{bmatrix} 2 \\ -1 \\ 1 \end{bmatrix}, \quad \mathbf{u}_3 = \begin{bmatrix} 1 \\ 2 \\ 1 \end{bmatrix}$$

$$\mathbf{u}'_1 = \begin{bmatrix} 3 \\ 1 \\ -5 \end{bmatrix}, \quad \mathbf{u}'_2 = \begin{bmatrix} 1 \\ 1 \\ -3 \end{bmatrix}, \quad \mathbf{u}'_3 = \begin{bmatrix} -1 \\ 0 \\ 2 \end{bmatrix}$$

10. Consider the bases $B = \{\mathbf{p}_1, \mathbf{p}_2\}$ and $B' = \{\mathbf{q}_1, \mathbf{q}_2\}$ for P_1, where

$$\mathbf{p}_1 = 6 + 3x, \quad \mathbf{p}_2 = 10 + 2x, \quad \mathbf{q}_1 = 2, \quad \mathbf{q}_2 = 3 + 2x$$

(a) Find the transition matrix from B' to B.

(b) Find the transition matrix from B to B'.

(c) Compute the coordinate vector $[\mathbf{p}]_B$, where $\mathbf{p} = -4 + x$, and use (12) to compute $[\mathbf{p}]_{B'}$.

(d) Check your work by computing $[\mathbf{p}]_{B'}$ directly.

11. Let V be the space spanned by $\mathbf{f}_1 = \sin x$ and $\mathbf{f}_2 = \cos x$.

(a) Show that $\mathbf{g}_1 = 2\sin x + \cos x$ and $\mathbf{g}_2 = 3\cos x$ form a basis for V.

(b) Find the transition matrix from $B' = \{\mathbf{g}_1, \mathbf{g}_2\}$ to $B = \{\mathbf{f}_1, \mathbf{f}_2\}$.

(c) Find the transition matrix from B to B'.

(d) Compute the coordinate vector $[\mathbf{h}]_B$, where $\mathbf{h} = 2\sin x - 5\cos x$, and use (12) to obtain $[\mathbf{h}]_{B'}$.

(e) Check your work by computing $[\mathbf{h}]_{B'}$ directly.

12. Let S be the standard basis for R^2, and let $B = \{\mathbf{v}_1, \mathbf{v}_2\}$ be the basis in which $\mathbf{v}_1 = (2, 1)$ and $\mathbf{v}_2 = (-3, 4)$.

(a) Find the transition matrix $P_{B \to S}$ by inspection.

(b) Use Formula (14) to find the transition matrix $P_{S \to B}$.

(c) Confirm that $P_{B \to S}$ and $P_{S \to B}$ are inverses of one another.

(d) Let $\mathbf{w} = (5, -3)$. Find $[\mathbf{w}]_B$ and then use Formula (11) to compute $[\mathbf{w}]_S$.

(e) Let $\mathbf{w} = (3, -5)$. Find $[\mathbf{w}]_S$ and then use Formula (12) to compute $[\mathbf{w}]_B$.

13. Let S be the standard basis for R^3, and let $B = \{\mathbf{v}_1, \mathbf{v}_2, \mathbf{v}_3\}$ be the basis in which $\mathbf{v}_1 = (1, 2, 1)$, $\mathbf{v}_2 = (2, 5, 0)$, and $\mathbf{v}_3 = (3, 3, 8)$.

(a) Find the transition matrix $P_{B \to S}$ by inspection.

(b) Use Formula (14) to find the transition matrix $P_{S \to B}$.

(c) Confirm that $P_{B \to S}$ and $P_{S \to B}$ are inverses of one another.

(d) Let $\mathbf{w} = (5, -3, 1)$. Find $[\mathbf{w}]_B$ and then use Formula (11) to compute $[\mathbf{w}]_S$.

(e) Let $\mathbf{w} = (3, -5, 0)$. Find $[\mathbf{w}]_S$ and then use Formula (12) to compute $[\mathbf{w}]_B$.

14. Let $B_1 = \{\mathbf{u}_1, \mathbf{u}_2\}$ and $B_2 = \{\mathbf{v}_1, \mathbf{v}_2\}$ be the bases for R^2 in which $\mathbf{u}_1 = (2, 2)$, $\mathbf{u}_2 = (4, -1)$, $\mathbf{v}_1 = (1, 3)$, and $\mathbf{v}_2 = (-1, -1)$.

(a) Use Formula (14) to find the transition matrix $P_{B_2 \to B_1}$.

(b) Use Formula (14) to find the transition matrix $P_{B_1 \to B_2}$.

(c) Confirm that $P_{B_2 \to B_1}$ and $P_{B_1 \to B_2}$ are inverses of one another.

(d) Let $\mathbf{w} = (5, -3)$. Find $[\mathbf{w}]_{B_1}$ and then use the matrix $P_{B_1 \to B_2}$ to compute $[\mathbf{w}]_{B_2}$ from $[\mathbf{w}]_{B_1}$.

(e) Let $\mathbf{w} = (3, -5)$. Find $[\mathbf{w}]_{B_2}$ and then use the matrix $P_{B_2 \to B_1}$ to compute $[\mathbf{w}]_{B_1}$ from $[\mathbf{w}]_{B_2}$.

15. Let $B_1 = \{\mathbf{u}_1, \mathbf{u}_2\}$ and $B_2 = \{\mathbf{v}_1, \mathbf{v}_2\}$ be the bases for R^2 in which $\mathbf{u}_1 = (1, 2)$, $\mathbf{u}_2 = (2, 3)$, $\mathbf{v}_1 = (1, 3)$, and $\mathbf{v}_2 = (1, 4)$.

(a) Use Formula (14) to find the transition matrix $P_{B_2 \to B_1}$.

(b) Use Formula (14) to find the transition matrix $P_{B_1 \to B_2}$.

(c) Confirm that $P_{B_2 \to B_1}$ and $P_{B_1 \to B_2}$ are inverses of one another.

(d) Let $\mathbf{w} = (0, 1)$. Find $[\mathbf{w}]_{B_1}$ and then use the matrix $P_{B_1 \to B_2}$ to compute $[\mathbf{w}]_{B_2}$ from $[\mathbf{w}]_{B_1}$.

(e) Let $\mathbf{w} = (2, 5)$. Find $[\mathbf{w}]_{B_2}$ and then use the matrix $P_{B_2 \to B_1}$ to compute $[\mathbf{w}]_{B_1}$ from $[\mathbf{w}]_{B_2}$.

16. Let $B_1 = \{\mathbf{u}_1, \mathbf{u}_2, \mathbf{u}_3\}$ and $B_2 = \{\mathbf{v}_1, \mathbf{v}_2, \mathbf{v}_3\}$ be the bases for R^3 in which $\mathbf{u}_1 = (-3, 0, -3)$, $\mathbf{u}_2 = (-3, 2, -1)$, $\mathbf{u}_3 = (1, 6, -1)$, $\mathbf{v}_1 = (-6, -6, 0)$, $\mathbf{v}_2 = (-2, -6, 4)$, and $\mathbf{v}_3 = (-2, -3, 7)$.

(a) Find the transition matrix $P_{B_1 \to B_2}$.

(b) Let $\mathbf{w} = (-5, 8, -5)$. Find $[\mathbf{w}]_{B_1}$ and then use the transition matrix obtained in part (a) to compute $[\mathbf{w}]_{B_2}$ by matrix multiplication.

(c) Check the result in part (b) by computing $[\mathbf{w}]_{B_2}$ directly.

17. Follow the directions of Exercise 16 with the same vector $\mathbf{w}$ but with $\mathbf{u}_1 = (2, 1, 1)$, $\mathbf{u}_2 = (2, -1, 1)$, $\mathbf{u}_3 = (1, 2, 1)$, $\mathbf{v}_1 = (3, 1, -5)$, $\mathbf{v}_2 = (1, 1, -3)$, and $\mathbf{v}_3 = (-1, 0, 2)$.

18. Let $S = \{\mathbf{e}_1, \mathbf{e}_2\}$ be the standard basis for R^2, and let $B = \{\mathbf{v}_1, \mathbf{v}_2\}$ be the basis that results when the vectors in S are reflected about the line $y = x$.

(a) Find the transition matrix $P_{B \to S}$.

(b) Let $P = P_{B \to S}$ and show that $P^T = P_{S \to B}$.

19. Let $S = \{\mathbf{e}_1, \mathbf{e}_2\}$ be the standard basis for R^2, and let $B = \{\mathbf{v}_1, \mathbf{v}_2\}$ be the basis that results when the vectors in S are reflected about the line that makes an angle θ with the positive x-axis.

(a) Find the transition matrix $P_{B \to S}$.

(b) Let $P = P_{B \to S}$ and show that $P^T = P_{S \to B}$.

20. If B_1, B_2, and B_3 are bases for R^2, and if

$$P_{B_1 \to B_2} = \begin{bmatrix} 3 & 1 \\ 5 & 2 \end{bmatrix} \quad \text{and} \quad P_{B_2 \to B_3} = \begin{bmatrix} 7 & 2 \\ 4 & -1 \end{bmatrix}$$

then $P_{B_3 \to B_1} = $ _____.

21. If P is the transition matrix from a basis B' to a basis B, and Q is the transition matrix from B to a basis C, what is the transition matrix from B' to C? What is the transition matrix from C to B'?

22. To write the coordinate vector for a vector, it is necessary to specify an order for the vectors in the basis. If P is the transition matrix from a basis B' to a basis B, what is the effect

on P if we reverse the order of vectors in B from $\mathbf{v}_1, \ldots, \mathbf{v}_n$ to $\mathbf{v}_n, \ldots, \mathbf{v}_1$? What is the effect on P if we reverse the order of vectors in both B' and B?

23. Consider the matrix

$$P = \begin{bmatrix} 1 & 1 & 0 \\ 1 & 0 & 2 \\ 0 & 2 & 1 \end{bmatrix}$$

(a) P is the transition matrix from what basis B to the standard basis $S = \{\mathbf{e}_1, \mathbf{e}_2, \mathbf{e}_3\}$ for R^3?

(b) P is the transition matrix from the standard basis $S = \{\mathbf{e}_1, \mathbf{e}_2, \mathbf{e}_3\}$ to what basis B for R^3?

24. The matrix

$$P = \begin{bmatrix} 1 & 0 & 0 \\ 0 & 3 & 2 \\ 0 & 1 & 1 \end{bmatrix}$$

is the transition matrix from what basis B to the basis $\{(1, 1, 1), (1, 1, 0), (1, 0, 0)\}$ for R^3?

25. Let B be a basis for R^n. Prove that the vectors $\mathbf{v}_1, \mathbf{v}_2, \ldots, \mathbf{v}_k$ form a linearly independent set in R^n if and only if the vectors $[\mathbf{v}_1]_B, [\mathbf{v}_2]_B, \ldots, [\mathbf{v}_k]_B$ form a linearly independent set in R^n.

26. Let B be a basis for R^n. Prove that the vectors $\mathbf{v}_1, \mathbf{v}_2, \ldots, \mathbf{v}_k$ span R^n if and only if the vectors $[\mathbf{v}_1]_B, [\mathbf{v}_2]_B, \ldots, [\mathbf{v}_k]_B$ span R^n.

27. If $[\mathbf{w}]_B = \mathbf{w}$ holds for all vectors $\mathbf{w}$ in R^n, what can you say about the basis B?

True-False Exercises

In parts (a)–(f) determine whether the statement is true or false, and justify your answer.

(a) If B_1 and B_2 are bases for a vector space V, then there exists a transition matrix from B_1 to B_2.

(b) Transition matrices are invertible.

(c) If B is a basis for a vector space R^n, then $P_{B \to B}$ is the identity matrix.

(d) If $P_{B_1 \to B_2}$ is a diagonal matrix, then each vector in B_2 is a scalar multiple of some vector in B_1.

(e) If each vector in B_2 is a scalar multiple of some vector in B_1, then $P_{B_1 \to B_2}$ is a diagonal matrix.

(f) If A is a square matrix, then $A = P_{B_1 \to B_2}$ for some bases B_1 and B_2 for R^n.

4.7 Row Space, Column Space, and Null Space

In this section we will study some important vector spaces that are associated with matrices. Our work here will provide us with a deeper understanding of the relationships between the solutions of a linear system and properties of its coefficient matrix.

Row Space, Column Space, and Null Space

Recall that vectors can be written in comma-delimited form or in matrix form as either row vectors or column vectors. In this section we will use the latter two.

> **DEFINITION 1** For an $m \times n$ matrix
>
> $$A = \begin{bmatrix} a_{11} & a_{12} & \cdots & a_{1n} \\ a_{21} & a_{22} & \cdots & a_{2n} \\ \vdots & \vdots & & \vdots \\ a_{m1} & a_{m2} & \cdots & a_{mn} \end{bmatrix}$$
>
> the vectors
>
> $$\mathbf{r}_1 = [a_{11} \quad a_{12} \quad \cdots \quad a_{1n}]$$
> $$\mathbf{r}_2 = [a_{21} \quad a_{22} \quad \cdots \quad a_{2n}]$$
> $$\vdots \qquad\qquad \vdots$$
> $$\mathbf{r}_m = [a_{m1} \quad a_{m2} \quad \cdots \quad a_{mn}]$$
>
> in R^n that are formed from the rows of A are called the ***row vectors*** of A, and the vectors
>
> $$\mathbf{c}_1 = \begin{bmatrix} a_{11} \\ a_{21} \\ \vdots \\ a_{m1} \end{bmatrix}, \quad \mathbf{c}_2 = \begin{bmatrix} a_{12} \\ a_{22} \\ \vdots \\ a_{m2} \end{bmatrix}, \ldots, \quad \mathbf{c}_n = \begin{bmatrix} a_{1n} \\ a_{2n} \\ \vdots \\ a_{mn} \end{bmatrix}$$
>
> in R^m formed from the columns of A are called the ***column vectors*** of A.

▶ **EXAMPLE 1 Row and Column Vectors of a 2 × 3 Matrix**

Let

$$A = \begin{bmatrix} 2 & 1 & 0 \\ 3 & -1 & 4 \end{bmatrix}$$

The row vectors of A are

$$\mathbf{r}_1 = [2 \quad 1 \quad 0] \quad \text{and} \quad \mathbf{r}_2 = [3 \quad -1 \quad 4]$$

and the column vectors of A are

$$\mathbf{c}_1 = \begin{bmatrix} 2 \\ 3 \end{bmatrix}, \quad \mathbf{c}_2 = \begin{bmatrix} 1 \\ -1 \end{bmatrix}, \quad \text{and} \quad \mathbf{c}_3 = \begin{bmatrix} 0 \\ 4 \end{bmatrix} \quad ◀$$

The following definition defines three important vector spaces associated with a matrix.

> **DEFINITION 2** If A is an $m \times n$ matrix, then the subspace of R^n spanned by the row vectors of A is called the ***row space*** of A, and the subspace of R^m spanned by the column vectors of A is called the ***column space*** of A. The solution space of the homogeneous system of equations $A\mathbf{x} = \mathbf{0}$, which is a subspace of R^n, is called the ***null space*** of A.

In this section and the next we will be concerned with two general questions:

Question 1. What relationships exist among the solutions of a linear system $A\mathbf{x} = \mathbf{b}$ and the row space, column space, and null space of the coefficient matrix A?

Question 2. What relationships exist among the row space, column space, and null space of a matrix?

Starting with the first question, suppose that

$$A = \begin{bmatrix} a_{11} & a_{12} & \cdots & a_{1n} \\ a_{21} & a_{22} & \cdots & a_{2n} \\ \vdots & \vdots & & \vdots \\ a_{m1} & a_{m2} & \cdots & a_{mn} \end{bmatrix} \quad \text{and} \quad \mathbf{x} = \begin{bmatrix} x_1 \\ x_2 \\ \vdots \\ x_n \end{bmatrix}$$

It follows from Formula (10) of Section 1.3 that if $\mathbf{c}_1, \mathbf{c}_2, \ldots, \mathbf{c}_n$ denote the column vectors of A, then the product $A\mathbf{x}$ can be expressed as a linear combination of these vectors with coefficients from $\mathbf{x}$; that is,

$$A\mathbf{x} = x_1\mathbf{c}_1 + x_2\mathbf{c}_2 + \cdots + x_n\mathbf{c}_n \tag{1}$$

Thus, a linear system, $A\mathbf{x} = \mathbf{b}$, of m equations in n unknowns can be written as

$$x_1\mathbf{c}_1 + x_2\mathbf{c}_2 + \cdots + x_n\mathbf{c}_n = \mathbf{b} \tag{2}$$

from which we conclude that $A\mathbf{x} = \mathbf{b}$ is consistent if and only if $\mathbf{b}$ is expressible as a linear combination of the column vectors of A. This yields the following theorem.

THEOREM 4.7.1 *A system of linear equations $A\mathbf{x} = \mathbf{b}$ is consistent if and only if $\mathbf{b}$ is in the column space of A.*

▶ **EXAMPLE 2** A Vector b in the Column Space of A

Let $A\mathbf{x} = \mathbf{b}$ be the linear system

$$\begin{bmatrix} -1 & 3 & 2 \\ 1 & 2 & -3 \\ 2 & 1 & -2 \end{bmatrix} \begin{bmatrix} x_1 \\ x_2 \\ x_3 \end{bmatrix} = \begin{bmatrix} 1 \\ -9 \\ -3 \end{bmatrix}$$

Show that $\mathbf{b}$ is in the column space of A by expressing it as a linear combination of the column vectors of A.

Solution Solving the system by Gaussian elimination yields (verify)

$$x_1 = 2, \quad x_2 = -1, \quad x_3 = 3$$

It follows from this and Formula (2) that

$$2 \begin{bmatrix} -1 \\ 1 \\ 2 \end{bmatrix} - \begin{bmatrix} 3 \\ 2 \\ 1 \end{bmatrix} + 3 \begin{bmatrix} 2 \\ -3 \\ -2 \end{bmatrix} = \begin{bmatrix} 1 \\ -9 \\ -3 \end{bmatrix} \quad ◀$$

Recall from Theorem 3.4.4 that the general solution of a consistent linear system $A\mathbf{x} = \mathbf{b}$ can be obtained by adding any specific solution of this system to the general solution of the corresponding homogeneous system $A\mathbf{x} = \mathbf{0}$. Keeping in mind that the null space of A is the same as the solution space of $A\mathbf{x} = \mathbf{0}$, we can rephrase that theorem in the following vector form.

THEOREM 4.7.2 *If $\mathbf{x}_0$ is any solution of a consistent linear system $A\mathbf{x} = \mathbf{b}$, and if $S = \{\mathbf{v}_1, \mathbf{v}_2, \ldots, \mathbf{v}_k\}$ is a basis for the null space of A, then every solution of $A\mathbf{x} = \mathbf{b}$ can be expressed in the form*

$$\mathbf{x} = \mathbf{x}_0 + c_1\mathbf{v}_1 + c_2\mathbf{v}_2 + \cdots + c_k\mathbf{v}_k \tag{3}$$

Conversely, for all choices of scalars $c_1, c_2, \ldots, c_k$, the vector $\mathbf{x}$ in this formula is a solution of $A\mathbf{x} = \mathbf{b}$.

Equation (3) gives a formula for the *general solution of* $A\mathbf{x} = \mathbf{b}$. The vector $\mathbf{x}_0$ in that formula is called a *particular solution of* $A\mathbf{x} = \mathbf{b}$, and the remaining part of the formula is called the *general solution of* $A\mathbf{x} = \mathbf{0}$. In words, this formula tells us that:

The general solution of a consistent linear system can be expressed as the sum of a particular solution of that system and the general solution of the corresponding homogeneous system.

Geometrically, the solution set of $A\mathbf{x} = \mathbf{b}$ can be viewed as the translation by $\mathbf{x}_0$ of the solution space of $A\mathbf{x} = \mathbf{0}$ (Figure 4.7.1).

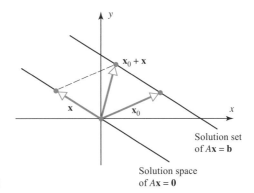

▶ Figure 4.7.1

▶ **EXAMPLE 3 General Solution of a Linear System $A\mathbf{x} = \mathbf{b}$**

In the concluding subsection of Section 3.4 we compared solutions of the linear systems

$$\begin{bmatrix} 1 & 3 & -2 & 0 & 2 & 0 \\ 2 & 6 & -5 & -2 & 4 & -3 \\ 0 & 0 & 5 & 10 & 0 & 15 \\ 2 & 6 & 0 & 8 & 4 & 18 \end{bmatrix} \begin{bmatrix} x_1 \\ x_2 \\ x_3 \\ x_4 \\ x_5 \\ x_6 \end{bmatrix} = \begin{bmatrix} 0 \\ 0 \\ 0 \\ 0 \end{bmatrix} \quad \text{and} \quad \begin{bmatrix} 1 & 3 & -2 & 0 & 2 & 0 \\ 2 & 6 & -5 & -2 & 4 & -3 \\ 0 & 0 & 5 & 10 & 0 & 15 \\ 2 & 6 & 0 & 8 & 4 & 18 \end{bmatrix} \begin{bmatrix} x_1 \\ x_2 \\ x_3 \\ x_4 \\ x_5 \\ x_6 \end{bmatrix} = \begin{bmatrix} 0 \\ -1 \\ 5 \\ 6 \end{bmatrix}$$

and deduced that the general solution $\mathbf{x}$ of the nonhomogeneous system and the general solution $\mathbf{x}_h$ of the corresponding homogeneous system (when written in column-vector form) are related by

$$
\underbrace{\begin{bmatrix} x_1 \\ x_2 \\ x_3 \\ x_4 \\ x_5 \\ x_6 \end{bmatrix}}_{\mathbf{x}} = \begin{bmatrix} -3r - 4s - 2t \\ r \\ -2s \\ s \\ t \\ \frac{1}{3} \end{bmatrix} = \underbrace{\begin{bmatrix} 0 \\ 0 \\ 0 \\ 0 \\ 0 \\ \frac{1}{3} \end{bmatrix}}_{\mathbf{x}_0} + \underbrace{r\begin{bmatrix} -3 \\ 1 \\ 0 \\ 0 \\ 0 \\ 0 \end{bmatrix} + s\begin{bmatrix} -4 \\ 0 \\ -2 \\ 1 \\ 0 \\ 0 \end{bmatrix} + t\begin{bmatrix} -2 \\ 0 \\ 0 \\ 0 \\ 1 \\ 0 \end{bmatrix}}_{\mathbf{x}_h}
$$

◀

Recall from the Remark following Example 4 of Section 4.5 that the vectors in $\mathbf{x}_h$ form a basis for the solution space of $A\mathbf{x} = \mathbf{0}$.

Bases for Row Spaces, Column Spaces, and Null Spaces   We first developed elementary row operations for the purpose of solving linear systems, and we know from that work that performing an elementary row operation on an augmented matrix does not change the solution set of the corresponding linear system. It follows that applying an elementary row operation to a matrix A does not change the solution set of the corresponding linear system $A\mathbf{x} = \mathbf{0}$, or, stated another way, it does not change the null space of A. Thus we have the following theorem.

THEOREM 4.7.3 *Elementary row operations do not change the null space of a matrix.*

The following theorem, whose proof is left as an exercise, is a companion to Theorem 4.7.3.

THEOREM 4.7.4 *Elementary row operations do not change the row space of a matrix.*

Theorems 4.7.3 and 4.7.4 might tempt you into *incorrectly* believing that elementary row operations do not change the column space of a matrix. To see why this is *not* true, compare the matrices

$$
A = \begin{bmatrix} 1 & 3 \\ 2 & 6 \end{bmatrix} \quad \text{and} \quad B = \begin{bmatrix} 1 & 3 \\ 0 & 0 \end{bmatrix}
$$

The matrix B can be obtained from A by adding -2 times the first row to the second. However, this operation has changed the column space of A, since that column space consists of all scalar multiples of

$$
\begin{bmatrix} 1 \\ 2 \end{bmatrix}
$$

whereas the column space of B consists of all scalar multiples of

$$
\begin{bmatrix} 1 \\ 0 \end{bmatrix}
$$

and the two are different spaces.

▶ EXAMPLE 4   **Finding a Basis for the Null Space of a Matrix**

Find a basis for the null space of the matrix

$$
A = \begin{bmatrix} 1 & 3 & -2 & 0 & 2 & 0 \\ 2 & 6 & -5 & -2 & 4 & -3 \\ 0 & 0 & 5 & 10 & 0 & 15 \\ 2 & 6 & 0 & 8 & 4 & 18 \end{bmatrix}
$$

Solution The null space of A is the solution space of the homogeneous linear system $A\mathbf{x} = \mathbf{0}$, which, as shown in Example 3, has the basis

$$
\mathbf{v}_1 = \begin{bmatrix} -3 \\ 1 \\ 0 \\ 0 \\ 0 \\ 0 \end{bmatrix}, \quad \mathbf{v}_2 = \begin{bmatrix} -4 \\ 0 \\ -2 \\ 1 \\ 0 \\ 0 \end{bmatrix}, \quad \mathbf{v}_3 = \begin{bmatrix} -2 \\ 0 \\ 0 \\ 0 \\ 1 \\ 0 \end{bmatrix} \quad ◀
$$

Remark Observe that the basis vectors $\mathbf{v}_1$, $\mathbf{v}_2$, and $\mathbf{v}_3$ in the last example are the vectors that result by successively setting one of the parameters in the general solution equal to 1 and the others equal to 0.

The following theorem makes it possible to find bases for the row and column spaces of a matrix in row echelon form by inspection.

THEOREM 4.7.5 *If a matrix R is in row echelon form, then the row vectors with the leading 1's (the nonzero row vectors) form a basis for the row space of R, and the column vectors with the leading 1's of the row vectors form a basis for the column space of R.*

The proof involves little more than an analysis of the positions of the 0's and 1's of R. We omit the details.

▶ EXAMPLE 5   **Bases for Row and Column Spaces**

The matrix

$$
R = \begin{bmatrix} 1 & -2 & 5 & 0 & 3 \\ 0 & 1 & 3 & 0 & 0 \\ 0 & 0 & 0 & 1 & 0 \\ 0 & 0 & 0 & 0 & 0 \end{bmatrix}
$$

is in row echelon form. From Theorem 4.7.5, the vectors

$$
\begin{aligned}
\mathbf{r}_1 &= \begin{bmatrix} 1 & -2 & 5 & 0 & 3 \end{bmatrix} \\
\mathbf{r}_2 &= \begin{bmatrix} 0 & 1 & 3 & 0 & 0 \end{bmatrix} \\
\mathbf{r}_3 &= \begin{bmatrix} 0 & 0 & 0 & 1 & 0 \end{bmatrix}
\end{aligned}
$$

form a basis for the row space of R, and the vectors

$$\mathbf{c}_1 = \begin{bmatrix} 1 \\ 0 \\ 0 \\ 0 \end{bmatrix}, \quad \mathbf{c}_2 = \begin{bmatrix} -2 \\ 1 \\ 0 \\ 0 \end{bmatrix}, \quad \mathbf{c}_4 = \begin{bmatrix} 0 \\ 0 \\ 1 \\ 0 \end{bmatrix}$$

form a basis for the column space of R.

▶ **EXAMPLE 6 Basis for a Row Space by Row Reduction**

Find a basis for the row space of the matrix

$$A = \begin{bmatrix} 1 & -3 & 4 & -2 & 5 & 4 \\ 2 & -6 & 9 & -1 & 8 & 2 \\ 2 & -6 & 9 & -1 & 9 & 7 \\ -1 & 3 & -4 & 2 & -5 & -4 \end{bmatrix}$$

Solution Since elementary row operations do not change the row space of a matrix, we can find a basis for the row space of A by finding a basis for the row space of any row echelon form of A. Reducing A to row echelon form, we obtain (verify)

$$R = \begin{bmatrix} 1 & -3 & 4 & -2 & 5 & 4 \\ 0 & 0 & 1 & 3 & -2 & -6 \\ 0 & 0 & 0 & 0 & 1 & 5 \\ 0 & 0 & 0 & 0 & 0 & 0 \end{bmatrix}$$

By Theorem 4.7.5, the nonzero row vectors of R form a basis for the row space of R and hence form a basis for the row space of A. These basis vectors are

$$\mathbf{r}_1 = \begin{bmatrix} 1 & -3 & 4 & -2 & 5 & 4 \end{bmatrix}$$
$$\mathbf{r}_2 = \begin{bmatrix} 0 & 0 & 1 & 3 & -2 & -6 \end{bmatrix}$$
$$\mathbf{r}_3 = \begin{bmatrix} 0 & 0 & 0 & 0 & 1 & 5 \end{bmatrix} \quad ◀$$

The problem of finding a basis for the column space of a matrix A in Example 6 is complicated by the fact that an elementary row operation can alter its column space. However, the good news is that *elementary row operations do not alter dependence relationships among the column vectors*. To make this more precise, suppose that $\mathbf{w}_1, \mathbf{w}_2, \ldots, \mathbf{w}_k$ are linearly dependent column vectors of A, so there are scalars $c_1, c_2, \ldots, c_k$ that are not all zero and such that

$$c_1\mathbf{w}_1 + c_2\mathbf{w}_2 + \cdots + c_k\mathbf{w}_k = \mathbf{0} \tag{4}$$

If we perform an elementary row operation on A, then these vectors will be changed into new column vectors $\mathbf{w}_1', \mathbf{w}_2', \ldots, \mathbf{w}_k'$. At first glance it would seem possible that the transformed vectors might be linearly independent. However, this is not so, since it can be proved that these new column vectors will be linear dependent and, in fact, related by an equation

$$c_1\mathbf{w}_1' + c_2\mathbf{w}_2' + \cdots + c_k\mathbf{w}_k' = \mathbf{0}$$

that has exactly the same coefficients as (4). It follows from the fact that elementary row operations are reversible that they also preserve linear independence among column vectors (why?). The following theorem summarizes all of these results.

> **THEOREM 4.7.6** *If A and B are row equivalent matrices, then:*
>
> (a) *A given set of column vectors of A is linearly independent if and only if the corresponding column vectors of B are linearly independent.*
>
> (b) *A given set of column vectors of A forms a basis for the column space of A if and only if the corresponding column vectors of B form a basis for the column space of B.*

▶ **EXAMPLE 7 Basis for a Column Space by Row Reduction**

Find a basis for the column space of the matrix

$$A = \begin{bmatrix} 1 & -3 & 4 & -2 & 5 & 4 \\ 2 & -6 & 9 & -1 & 8 & 2 \\ 2 & -6 & 9 & -1 & 9 & 7 \\ -1 & 3 & -4 & 2 & -5 & -4 \end{bmatrix}$$

Solution We observed in Example 6 that the matrix

$$R = \begin{bmatrix} 1 & -3 & 4 & -2 & 5 & 4 \\ 0 & 0 & 1 & 3 & -2 & -6 \\ 0 & 0 & 0 & 0 & 1 & 5 \\ 0 & 0 & 0 & 0 & 0 & 0 \end{bmatrix}$$

is a row echelon form of A. Keeping in mind that A and R can have different column spaces, we cannot find a basis for the column space of A directly from the column vectors of R. However, it follows from Theorem 4.7.6b that if we can find a set of column vectors of R that forms a basis for the column space of R, then the *corresponding* column vectors of A will form a basis for the column space of A.

Since the first, third, and fifth columns of R contain the leading 1's of the row vectors, the vectors

$$\mathbf{c}_1' = \begin{bmatrix} 1 \\ 0 \\ 0 \\ 0 \end{bmatrix}, \quad \mathbf{c}_3' = \begin{bmatrix} 4 \\ 1 \\ 0 \\ 0 \end{bmatrix}, \quad \mathbf{c}_5' = \begin{bmatrix} 5 \\ -2 \\ 1 \\ 0 \end{bmatrix}$$

form a basis for the column space of R. Thus, the corresponding column vectors of A, which are

$$\mathbf{c}_1 = \begin{bmatrix} 1 \\ 2 \\ 2 \\ -1 \end{bmatrix}, \quad \mathbf{c}_3 = \begin{bmatrix} 4 \\ 9 \\ 9 \\ -4 \end{bmatrix}, \quad \mathbf{c}_5 = \begin{bmatrix} 5 \\ 8 \\ 9 \\ -5 \end{bmatrix}$$

form a basis for the column space of A. ◀

Up to now we have focused on methods for finding bases associated with matrices. Those methods can readily be adapted to the more general problem of finding a basis for the space spanned by a set of vectors in R^n.

▶ **EXAMPLE 8** **Basis for a Vector Space Using Row Operations**

Find a basis for the subspace of R^5 spanned by the vectors

$$\mathbf{v}_1 = (1, -2, 0, 0, 3), \quad \mathbf{v}_2 = (2, -5, -3, -2, 6),$$
$$\mathbf{v}_3 = (0, 5, 15, 10, 0), \quad \mathbf{v}_4 = (2, 6, 18, 8, 6)$$

Solution The space spanned by these vectors is the row space of the matrix

$$\begin{bmatrix} 1 & -2 & 0 & 0 & 3 \\ 2 & -5 & -3 & -2 & 6 \\ 0 & 5 & 15 & 10 & 0 \\ 2 & 6 & 18 & 8 & 6 \end{bmatrix}$$

Reducing this matrix to row echelon form, we obtain

$$\begin{bmatrix} 1 & -2 & 0 & 0 & 3 \\ 0 & 1 & 3 & 2 & 0 \\ 0 & 0 & 1 & 1 & 0 \\ 0 & 0 & 0 & 0 & 0 \end{bmatrix}$$

The nonzero row vectors in this matrix are

$$\mathbf{w}_1 = (1, -2, 0, 0, 3), \quad \mathbf{w}_2 = (0, 1, 3, 2, 0), \quad \mathbf{w}_3 = (0, 0, 1, 1, 0)$$

These vectors form a basis for the row space and consequently form a basis for the subspace of R^5 spanned by $\mathbf{v}_1$, $\mathbf{v}_2$, $\mathbf{v}_3$, and $\mathbf{v}_4$. ◀

Bases Formed from Row and Column Vectors of a Matrix

In all of the examples we have considered thus far we have looked for bases in which no restrictions were imposed on the individual vectors in the basis. We now want to focus on the problem of finding a basis for the row space of a matrix A consisting entirely of row vectors from A and a basis for the column space of A consisting entirely of column vectors of A.

Looking back on our earlier work, we see that the procedure followed in Example 7 did, in fact, produce a basis for the column space of A consisting of column vectors of A, whereas the procedure used in Example 6 produced a basis for the row space of A, but that basis did not consist of row vectors of A. The following example shows how to adapt the procedure from Example 7 to find a basis for the row space of a matrix that is formed from its row vectors.

▶ **EXAMPLE 9** **Basis for the Row Space of a Matrix**

Find a basis for the row space of

$$A = \begin{bmatrix} 1 & -2 & 0 & 0 & 3 \\ 2 & -5 & -3 & -2 & 6 \\ 0 & 5 & 15 & 10 & 0 \\ 2 & 6 & 18 & 8 & 6 \end{bmatrix}$$

consisting entirely of row vectors from A.

Solution We will transpose A, thereby converting the row space of A into the column space of A^T; then we will use the method of Example 7 to find a basis for the column space of A^T; and then we will transpose again to convert column vectors back to row vectors. Transposing A yields

$$A^T = \begin{bmatrix} 1 & 2 & 0 & 2 \\ -2 & -5 & 5 & 6 \\ 0 & -3 & 15 & 18 \\ 0 & -2 & 10 & 8 \\ 3 & 6 & 0 & 6 \end{bmatrix}$$

Reducing this matrix to row echelon form yields

$$\begin{bmatrix} 1 & 2 & 0 & 2 \\ 0 & 1 & -5 & -10 \\ 0 & 0 & 0 & 1 \\ 0 & 0 & 0 & 0 \\ 0 & 0 & 0 & 0 \end{bmatrix}$$

The first, second, and fourth columns contain the leading 1's, so the corresponding column vectors in A^T form a basis for the column space of A^T; these are

$$\mathbf{c}_1 = \begin{bmatrix} 1 \\ -2 \\ 0 \\ 0 \\ 3 \end{bmatrix}, \quad \mathbf{c}_2 = \begin{bmatrix} 2 \\ -5 \\ -3 \\ -2 \\ 6 \end{bmatrix}, \quad \text{and} \quad \mathbf{c}_4 = \begin{bmatrix} 2 \\ 6 \\ 18 \\ 8 \\ 6 \end{bmatrix}$$

Transposing again and adjusting the notation appropriately yields the basis vectors

$$\mathbf{r}_1 = \begin{bmatrix} 1 & -2 & 0 & 0 & 3 \end{bmatrix}, \qquad \mathbf{r}_2 = \begin{bmatrix} 2 & -5 & -3 & -2 & 6 \end{bmatrix},$$

and

$$\mathbf{r}_4 = \begin{bmatrix} 2 & 6 & 18 & 8 & 6 \end{bmatrix}$$

for the row space of A. ◀

Next, we will give an example that adapts the methods we have developed above to solve the following general problem in R^n:

Problem   Given a set of vectors $S = \{\mathbf{v}_1, \mathbf{v}_2, \ldots, \mathbf{v}_k\}$ in R^n, find a subset of these vectors that forms a basis for span(S), and express those vectors that are not in that basis as a linear combination of the basis vectors.

▶ **EXAMPLE 10 Basis and Linear Combinations**

(a)   Find a subset of the vectors

$$\mathbf{v}_1 = (1, -2, 0, 3), \quad \mathbf{v}_2 = (2, -5, -3, 6),$$
$$\mathbf{v}_3 = (0, 1, 3, 0), \quad \mathbf{v}_4 = (2, -1, 4, -7), \quad \mathbf{v}_5 = (5, -8, 1, 2)$$

that forms a basis for the space spanned by these vectors.

(b)   Express each vector not in the basis as a linear combination of the basis vectors.

Solution (a) We begin by constructing a matrix that has $\mathbf{v}_1, \mathbf{v}_2, \ldots, \mathbf{v}_5$ as its column vectors:

$$\begin{bmatrix} 1 & 2 & 0 & 2 & 5 \\ -2 & -5 & 1 & -1 & -8 \\ 0 & -3 & 3 & 4 & 1 \\ 3 & 6 & 0 & -7 & 2 \end{bmatrix} \tag{5}$$
$$\begin{array}{ccccc} \uparrow & \uparrow & \uparrow & \uparrow & \uparrow \\ \mathbf{v}_1 & \mathbf{v}_2 & \mathbf{v}_3 & \mathbf{v}_4 & \mathbf{v}_5 \end{array}$$

The first part of our problem can be solved by finding a basis for the column space of this matrix. Reducing the matrix to *reduced* row echelon form and denoting the column vectors of the resulting matrix by $\mathbf{w}_1, \mathbf{w}_2, \mathbf{w}_3, \mathbf{w}_4$, and $\mathbf{w}_5$ yields

$$\begin{bmatrix} 1 & 0 & 2 & 0 & 1 \\ 0 & 1 & -1 & 0 & 1 \\ 0 & 0 & 0 & 1 & 1 \\ 0 & 0 & 0 & 0 & 0 \end{bmatrix} \tag{6}$$
$$\begin{array}{ccccc} \uparrow & \uparrow & \uparrow & \uparrow & \uparrow \\ \mathbf{w}_1 & \mathbf{w}_2 & \mathbf{w}_3 & \mathbf{w}_4 & \mathbf{w}_5 \end{array}$$

The leading 1's occur in columns 1, 2, and 4, so by Theorem 4.7.5,

$$\{\mathbf{w}_1, \mathbf{w}_2, \mathbf{w}_4\}$$

is a basis for the column space of (6), and consequently,

$$\{\mathbf{v}_1, \mathbf{v}_2, \mathbf{v}_4\}$$

is a basis for the column space of (5).

Solution (b) We will start by expressing $\mathbf{w}_3$ and $\mathbf{w}_5$ as linear combinations of the basis vectors $\mathbf{w}_1, \mathbf{w}_2, \mathbf{w}_4$. The simplest way of doing this is to express $\mathbf{w}_3$ and $\mathbf{w}_5$ in terms of basis vectors with smaller subscripts. Accordingly, we will express $\mathbf{w}_3$ as a linear combination of $\mathbf{w}_1$ and $\mathbf{w}_2$, and we will express $\mathbf{w}_5$ as a linear combination of $\mathbf{w}_1, \mathbf{w}_2$, and $\mathbf{w}_4$. By inspection of (6), these linear combinations are

$$\mathbf{w}_3 = 2\mathbf{w}_1 - \mathbf{w}_2$$
$$\mathbf{w}_5 = \mathbf{w}_1 + \mathbf{w}_2 + \mathbf{w}_4$$

We call these the ***dependency equations***. The corresponding relationships in (5) are

$$\mathbf{v}_3 = 2\mathbf{v}_1 - \mathbf{v}_2$$
$$\mathbf{v}_5 = \mathbf{v}_1 + \mathbf{v}_2 + \mathbf{v}_4 \quad \blacktriangleleft$$

The following is a summary of the steps that we followed in our last example to solve the problem posed above.

Basis for Span(S)

Step 1. Form the matrix A having vectors in $S = \{\mathbf{v}_1, \mathbf{v}_2, \ldots, \mathbf{v}_k\}$ as column vectors.

Step 2. Reduce the matrix A to reduced row echelon form R.

Step 3. Denote the column vectors of R by $\mathbf{w}_1, \mathbf{w}_2, \ldots, \mathbf{w}_k$.

Step 4. Identify the columns of R that contain the leading 1's. The corresponding column vectors of A form a basis for span(S).

This completes the first part of the problem.

Step 5. Obtain a set of dependency equations by expressing each column vector of *R* that does not contain a leading 1 as a linear combination of preceding column vectors that do contain leading 1's.

Step 6. Replace the column vectors of *R* that appear in the dependency equations by the corresponding column vectors of *A*.

This completes the second part of the problem.

Concept Review

- Row vectors
- Column vectors
- Row space
- Column space
- Null space

- General solution
- Particular solution
- Relationships among linear systems and row spaces, column spaces, and null spaces

- Relationships among the row space, column space, and null space of a matrix
- Dependency equations

Skills

- Determine whether a given vector is in the column space of a matrix; if it is, express it as a linear combination of the column vectors of the matrix.
- Find a basis for the null space of a matrix.

- Find a basis for the row space of a matrix.
- Find a basis for the column space of a matrix.
- Find a basis for the span of a set of vectors in R^n.

Exercise Set 4.7

1. List the row vectors and column vectors of the matrix

$$\begin{bmatrix} 2 & -1 & 0 & 1 \\ 3 & 5 & 7 & -1 \\ 1 & 4 & 2 & 7 \end{bmatrix}$$

2. Express the product $A\mathbf{x}$ as a linear combination of the column vectors of A.

(a) $\begin{bmatrix} 2 & 3 \\ -1 & 4 \end{bmatrix}\begin{bmatrix} 1 \\ 2 \end{bmatrix}$

(b) $\begin{bmatrix} 4 & 0 & -1 \\ 3 & 6 & 2 \\ 0 & -1 & 4 \end{bmatrix}\begin{bmatrix} -2 \\ 3 \\ 5 \end{bmatrix}$

(c) $\begin{bmatrix} -3 & 6 & 2 \\ 5 & -4 & 0 \\ 2 & 3 & -1 \\ 1 & 8 & 3 \end{bmatrix}\begin{bmatrix} -1 \\ 2 \\ 5 \end{bmatrix}$

(d) $\begin{bmatrix} 2 & 1 & 5 \\ 6 & 3 & -8 \end{bmatrix}\begin{bmatrix} 3 \\ 0 \\ -5 \end{bmatrix}$

3. Determine whether **b** is in the column space of *A*, and if so, express **b** as a linear combination of the column vectors of *A*.

(a) $A = \begin{bmatrix} 1 & 3 \\ 4 & -6 \end{bmatrix}$; $\mathbf{b} = \begin{bmatrix} -2 \\ 10 \end{bmatrix}$

(b) $A = \begin{bmatrix} 1 & 1 & 2 \\ 1 & 0 & 1 \\ 2 & 1 & 3 \end{bmatrix}$; $\mathbf{b} = \begin{bmatrix} -1 \\ 0 \\ 2 \end{bmatrix}$

(c) $A = \begin{bmatrix} 1 & -1 & 1 \\ 9 & 3 & 1 \\ 1 & 1 & 1 \end{bmatrix}$; $\mathbf{b} = \begin{bmatrix} 5 \\ 1 \\ -1 \end{bmatrix}$

(d) $A = \begin{bmatrix} 1 & -1 & 1 \\ 1 & 1 & -1 \\ -1 & -1 & 1 \end{bmatrix}$; $\mathbf{b} = \begin{bmatrix} 2 \\ 0 \\ 0 \end{bmatrix}$

(e) $A = \begin{bmatrix} 1 & 2 & 0 & 1 \\ 0 & 1 & 2 & 1 \\ 1 & 2 & 1 & 3 \\ 0 & 1 & 2 & 2 \end{bmatrix}$; $\mathbf{b} = \begin{bmatrix} 4 \\ 3 \\ 5 \\ 7 \end{bmatrix}$

4. Suppose that $x_1 = -1$, $x_2 = 2$, $x_3 = 4$, $x_4 = -3$ is a solution of a nonhomogeneous linear system $A\mathbf{x} = \mathbf{b}$ and that the solution set of the homogeneous system $A\mathbf{x} = \mathbf{0}$ is given by the formulas

$$x_1 = -3r + 4s, \quad x_2 = r - s, \quad x_3 = r, \quad x_4 = s$$

(a) Find a vector form of the general solution of $A\mathbf{x} = \mathbf{0}$.

(b) Find a vector form of the general solution of $A\mathbf{x} = \mathbf{b}$.

5. In parts (a)–(d), find the vector form of the general solution of the given linear system $A\mathbf{x} = \mathbf{b}$; then use that result to find the vector form of the general solution of $A\mathbf{x} = \mathbf{0}$.

(a) $\begin{aligned} x_1 - 3x_2 &= 1 \\ 2x_1 - 6x_2 &= 2 \end{aligned}$

(b) $\begin{aligned} x_1 + x_2 + 2x_3 &= 5 \\ x_1 \qquad + x_3 &= -2 \\ 2x_1 + x_2 + 3x_3 &= 3 \end{aligned}$

(c) $\begin{aligned} x_1 - 2x_2 + x_3 + 2x_4 &= -1 \\ 2x_1 - 4x_2 + 2x_3 + 4x_4 &= -2 \\ -x_1 + 2x_2 - x_3 - 2x_4 &= 1 \\ 3x_1 - 6x_2 + 3x_3 + 6x_4 &= -3 \end{aligned}$

(d)
$$
\begin{aligned}
x_1 + 2x_2 - 3x_3 + x_4 &= 4 \\
-2x_1 + x_2 + 2x_3 + x_4 &= -1 \\
-x_1 + 3x_2 - x_3 + 2x_4 &= 3 \\
4x_1 - 7x_2 \qquad\quad - 5x_4 &= -5
\end{aligned}
$$

6. Find a basis for the null space of A.

(a) $A = \begin{bmatrix} 1 & -1 & 3 \\ 5 & -4 & -4 \\ 7 & -6 & 2 \end{bmatrix}$ (b) $A = \begin{bmatrix} 2 & 0 & -1 \\ 4 & 0 & -2 \\ 0 & 0 & 0 \end{bmatrix}$

(c) $A = \begin{bmatrix} 1 & 4 & 5 & 2 \\ 2 & 1 & 3 & 0 \\ -1 & 3 & 2 & 2 \end{bmatrix}$

(d) $A = \begin{bmatrix} 1 & 4 & 5 & 6 & 9 \\ 3 & -2 & 1 & 4 & -1 \\ -1 & 0 & -1 & -2 & -1 \\ 2 & 3 & 5 & 7 & 8 \end{bmatrix}$

(e) $A = \begin{bmatrix} 1 & -3 & 2 & 2 & 1 \\ 0 & 3 & 6 & 0 & -3 \\ 2 & -3 & -2 & 4 & 4 \\ 3 & -6 & 0 & 6 & 5 \\ -2 & 9 & 2 & -4 & -5 \end{bmatrix}$

7. In each part, a matrix in row echelon form is given. By inspection, find bases for the row and column spaces of A.

(a) $\begin{bmatrix} 1 & 0 & 2 \\ 0 & 0 & 1 \\ 0 & 0 & 0 \end{bmatrix}$ (b) $\begin{bmatrix} 1 & -3 & 0 & 0 \\ 0 & 1 & 0 & 0 \\ 0 & 0 & 0 & 0 \\ 0 & 0 & 0 & 0 \end{bmatrix}$

(c) $\begin{bmatrix} 1 & 2 & 4 & 5 \\ 0 & 1 & -3 & 0 \\ 0 & 0 & 1 & -3 \\ 0 & 0 & 0 & 1 \\ 0 & 0 & 0 & 0 \end{bmatrix}$ (d) $\begin{bmatrix} 1 & 2 & -1 & 5 \\ 0 & 1 & 4 & 3 \\ 0 & 0 & 1 & -7 \\ 0 & 0 & 0 & 1 \end{bmatrix}$

8. For the matrices in Exercise 6, find a basis for the row space of A by reducing the matrix to row echelon form.

9. By inspection, find a basis for the row space and a basis for the column space of each matrix.

(a) $\begin{bmatrix} 1 & 0 & 2 \\ 0 & 0 & 1 \\ 0 & 0 & 0 \end{bmatrix}$ (b) $\begin{bmatrix} 1 & -3 & 0 & 0 \\ 0 & 1 & 0 & 0 \\ 0 & 0 & 0 & 0 \\ 0 & 0 & 0 & 0 \end{bmatrix}$

(c) $\begin{bmatrix} 1 & 2 & 4 & 5 \\ 0 & 1 & -3 & 0 \\ 0 & 0 & 1 & -3 \\ 0 & 0 & 0 & 1 \\ 0 & 0 & 0 & 0 \end{bmatrix}$ (d) $\begin{bmatrix} 1 & 2 & -1 & 5 \\ 0 & 1 & 4 & 3 \\ 0 & 0 & 1 & -7 \\ 0 & 0 & 0 & 1 \end{bmatrix}$

10. For the matrices in Exercise 6, find a basis for the row space of A consisting entirely of row vectors of A.

11. Find a basis for the subspace of R^4 spanned by the given vectors.

(a) $(1, 1, -4, -3),\ (2, 0, 2, -2),\ (2, -1, 3, 2)$

(b) $(-1, 1, -2, 0),\ (3, 3, 6, 0),\ (9, 0, 0, 3)$

(c) $(1, 1, 0, 0),\ (0, 0, 1, 1),\ (-2, 0, 2, 2),\ (0, -3, 0, 3)$

12. Find a subset of the vectors that forms a basis for the space spanned by the vectors; then express each vector that is not in the basis as a linear combination of the basis vectors.

(a) $\mathbf{v}_1 = (1, 0, 1, 1),\ \mathbf{v}_2 = (-3, 3, 7, 1),$
$\mathbf{v}_3 = (-1, 3, 9, 3),\ \mathbf{v}_4 = (-5, 3, 5, -1)$

(b) $\mathbf{v}_1 = (1, -2, 0, 3),\ \mathbf{v}_2 = (2, -4, 0, 6),$
$\mathbf{v}_3 = (-1, 1, 2, 0),\ \mathbf{v}_4 = (0, -1, 2, 3)$

(c) $\mathbf{v}_1 = (1, -1, 5, 2),\ \mathbf{v}_2 = (-2, 3, 1, 0),$
$\mathbf{v}_3 = (4, -5, 9, 4),\ \mathbf{v}_4 = (0, 4, 2, -3),$
$\mathbf{v}_5 = (-7, 18, 2, -8)$

13. Prove that the row vectors of an $n \times n$ invertible matrix A form a basis for R^n.

14. Construct a matrix whose null space consists of all linear combinations of the vectors

$$
\mathbf{v}_1 = \begin{bmatrix} 1 \\ -1 \\ 3 \\ 2 \end{bmatrix} \quad \text{and} \quad \mathbf{v}_2 = \begin{bmatrix} 2 \\ 0 \\ -2 \\ 4 \end{bmatrix}
$$

15. (a) Let

$$
A = \begin{bmatrix} 0 & 1 & 0 \\ 1 & 0 & 0 \\ 0 & 0 & 0 \end{bmatrix}
$$

Show that relative to an xyz-coordinate system in 3-space the null space of A consists of all points on the z-axis and that the column space consists of all points in the xy-plane (see the accompanying figure).

(b) Find a 3×3 matrix whose null space is the x-axis and whose column space is the yz-plane.

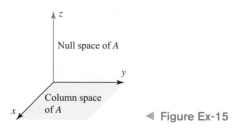

Figure Ex-15

16. Find a 3×3 matrix whose null space is

(a) a point.   (b) a line.   (c) a plane.

17. (a) Find all 2×2 matrices whose null space is the line $3x - 5y = 0$.

(b) Sketch the null spaces of the following matrices:

$$A = \begin{bmatrix} 1 & 4 \\ 0 & 5 \end{bmatrix}, \quad B = \begin{bmatrix} 1 & 0 \\ 0 & 5 \end{bmatrix},$$

$$C = \begin{bmatrix} 6 & 2 \\ 3 & 1 \end{bmatrix}, \quad D = \begin{bmatrix} 0 & 0 \\ 0 & 0 \end{bmatrix}$$

18. The equation $x_1 + x_2 + x_3 = 1$ can be viewed as a linear system of one equation in three unknowns. Express its general solution as a particular solution plus the general solution of the corresponding homogeneous system. [*Suggestion:* Write the vectors in column form.]

19. Suppose that A and B are $n \times n$ matrices and A is invertible. Invent and prove a theorem that describes how the row spaces of AB and B are related.

True-False Exercises

In parts (a)–(j) determine whether the statement is true or false, and justify your answer.

(a) The span of $v_1, \ldots, v_n$ is the column space of the matrix whose column vectors are $v_1, \ldots, v_n$.

(b) The column space of a matrix A is the set of solutions of $Ax = b$.

(c) If R is the reduced row echelon form of A, then those column vectors of R that contain the leading 1's form a basis for the column space of A.

(d) The set of nonzero row vectors of a matrix A is a basis for the row space of A.

(e) If A and B are $n \times n$ matrices that have the same row space, then A and B have the same column space.

(f) If E is an $m \times m$ elementary matrix and A is an $m \times n$ matrix, then the null space of EA is the same as the null space of A.

(g) If E is an $m \times m$ elementary matrix and A is an $m \times n$ matrix, then the row space of EA is the same as the row space of A.

(h) If E is an $m \times m$ elementary matrix and A is an $m \times n$ matrix, then the column space of EA is the same as the column space of A.

(i) The system $Ax = b$ is inconsistent if and only if b is not in the column space of A.

(j) There is an invertible matrix A and a singular matrix B such that the row spaces of A and B are the same.

4.8 Rank, Nullity, and the Fundamental Matrix Spaces

In the last section we investigated relationships between a system of linear equations and the row space, column space, and null space of its coefficient matrix. In this section we will be concerned with the dimensions of those spaces. The results we obtain will provide a deeper insight into the relationship between a linear system and its coefficient matrix.

Row and Column Spaces Have Equal Dimensions

In Examples 6 and 7 of Section 4.7 we found that the row and column spaces of the matrix

$$A = \begin{bmatrix} 1 & -3 & 4 & -2 & 5 & 4 \\ 2 & -6 & 9 & -1 & 8 & 2 \\ 2 & -6 & 9 & -1 & 9 & 7 \\ -1 & 3 & -4 & 2 & -5 & -4 \end{bmatrix}$$

both have three basis vectors and hence are both three-dimensional. The fact that these spaces have the same dimension is not accidental, but rather a consequence of the following theorem.

THEOREM 4.8.1 *The row space and column space of a matrix A have the same dimension.*

Proof Let R be any row echelon form of A. It follows from Theorems 4.7.4 and 4.7.6*b* that

$$\dim(\text{row space of } A) = \dim(\text{row space of } R)$$
$$\dim(\text{column space of } A) = \dim(\text{column space of } R)$$

so it suffices to show that the row and column spaces of R have the same dimension. But the dimension of the row space of R is the number of nonzero rows, and by Theorem 4.7.5 the dimension of the column space of R is the number of leading 1's. Since these two numbers are the same, the row and column space have the same dimension. ◀

Rank and Nullity The dimensions of the row space, column space, and null space of a matrix are such important numbers that there is some notation and terminology associated with them.

The proof of Theorem 4.8.1 shows that the rank of A can be interpreted as the number of leading 1's in any row echelon form of A.

> **DEFINITION 1** The common dimension of the row space and column space of a matrix A is called the **rank** of A and is denoted by $\text{rank}(A)$; the dimension of the null space of A is called the **nullity** of A and is denoted by $\text{nullity}(A)$.

▶ **EXAMPLE 1 Rank and Nullity of a 4 × 6 Matrix**

Find the rank and nullity of the matrix

$$A = \begin{bmatrix} -1 & 2 & 0 & 4 & 5 & -3 \\ 3 & -7 & 2 & 0 & 1 & 4 \\ 2 & -5 & 2 & 4 & 6 & 1 \\ 4 & -9 & 2 & -4 & -4 & 7 \end{bmatrix}$$

Solution The reduced row echelon form of A is

$$\begin{bmatrix} 1 & 0 & -4 & -28 & -37 & 13 \\ 0 & 1 & -2 & -12 & -16 & 5 \\ 0 & 0 & 0 & 0 & 0 & 0 \\ 0 & 0 & 0 & 0 & 0 & 0 \end{bmatrix} \tag{1}$$

(verify). Since this matrix has two leading 1's, its row and column spaces are two-dimensional and $\text{rank}(A) = 2$. To find the nullity of A, we must find the dimension of the solution space of the linear system $A\mathbf{x} = \mathbf{0}$. This system can be solved by reducing its augmented matrix to reduced row echelon form. The resulting matrix will be identical to (1), except that it will have an additional last column of zeros, and hence the corresponding system of equations will be

$$x_1 - 4x_3 - 28x_4 - 37x_5 + 13x_6 = 0$$
$$x_2 - 2x_3 - 12x_4 - 16x_5 + 5x_6 = 0$$

Solving these equations for the leading variables yields

$$x_1 = 4x_3 + 28x_4 + 37x_5 - 13x_6$$
$$x_2 = 2x_3 + 12x_4 + 16x_5 - 5x_6 \tag{2}$$

from which we obtain the general solution

$$x_1 = 4r + 28s + 37t - 13u$$
$$x_2 = 2r + 12s + 16t - 5u$$
$$x_3 = r$$
$$x_4 = s$$
$$x_5 = t$$
$$x_6 = u$$

or in column vector form

$$
\begin{bmatrix} x_1 \\ x_2 \\ x_3 \\ x_4 \\ x_5 \\ x_6 \end{bmatrix} = r \begin{bmatrix} 4 \\ 2 \\ 1 \\ 0 \\ 0 \\ 0 \end{bmatrix} + s \begin{bmatrix} 28 \\ 12 \\ 0 \\ 1 \\ 0 \\ 0 \end{bmatrix} + t \begin{bmatrix} 37 \\ 16 \\ 0 \\ 0 \\ 1 \\ 0 \end{bmatrix} + u \begin{bmatrix} -13 \\ -5 \\ 0 \\ 0 \\ 0 \\ 1 \end{bmatrix} \tag{3}
$$

Because the four vectors on the right side of (3) form a basis for the solution space, $\text{nullity}(A) = 4$.

▶ **EXAMPLE 2 Maximum Value for Rank**

What is the maximum possible rank of an $m \times n$ matrix A that is not square?

Solution Since the row vectors of A lie in R^n and the column vectors in R^m, the row space of A is at most n-dimensional and the column space is at most m-dimensional. Since the rank of A is the common dimension of its row and column space, it follows that the rank is at most the smaller of m and n. We denote this by writing

$$
\text{rank}(A) \le \min(m, n)
$$

in which $\min(m, n)$ is the minimum of m and n. ◀

The following theorem establishes an important relationship between the rank and nullity of a matrix.

THEOREM 4.8.2 Dimension Theorem for Matrices

If A is a matrix with n columns, then

$$
rank(A) + nullity(A) = n \tag{4}
$$

Proof Since A has n columns, the homogeneous linear system $A\mathbf{x} = \mathbf{0}$ has n unknowns (variables). These fall into two distinct categories: the leading variables and the free variables. Thus,

$$
\begin{bmatrix} \text{number of leading} \\ \text{variables} \end{bmatrix} + \begin{bmatrix} \text{number of free} \\ \text{variables} \end{bmatrix} = n
$$

But the number of leading variables is the same as the number of leading 1's in the reduced row echelon form of A, which is the rank of A; and the number of free variables is the same as the number of parameters in the general solution of $A\mathbf{x} = \mathbf{0}$, which is the nullity of A. This yields Formula (4). ◀

▶ **EXAMPLE 3 The Sum of Rank and Nullity**

The matrix

$$
A = \begin{bmatrix} -1 & 2 & 0 & 4 & 5 & -3 \\ 3 & -7 & 2 & 0 & 1 & 4 \\ 2 & -5 & 2 & 4 & 6 & 1 \\ 4 & -9 & 2 & -4 & -4 & 7 \end{bmatrix}
$$

has 6 columns, so

$$\text{rank}(A) + \text{nullity}(A) = 6$$

This is consistent with Example 1, where we showed that

$$\text{rank}(A) = 2 \quad \text{and} \quad \text{nullity}(A) = 4 \quad \blacktriangleleft$$

The following theorem, which summarizes results already obtained, interprets rank and nullity in the context of a homogeneous linear system.

> **THEOREM 4.8.3** *If A is an $m \times n$ matrix, then*
>
> *(a)* *rank(A) = the number of leading variables in the general solution of $A\mathbf{x} = \mathbf{0}$.*
> *(b)* *nullity(A) = the number of parameters in the general solution of $A\mathbf{x} = \mathbf{0}$.*

▶ **EXAMPLE 4 Number of Parameters in a General Solution**

Find the number of parameters in the general solution of $A\mathbf{x} = \mathbf{0}$ if A is a 5×7 matrix of rank 3.

Solution From (4),

$$\text{nullity}(A) = n - \text{rank}(A) = 7 - 3 = 4$$

Thus there are four parameters. ◀

Equivalence Theorem   In Theorem 2.3.8 we listed seven results that are equivalent to the invertibility of a square matrix A. We are now in a position to add eight more results to that list to produce a single theorem that summarizes most of the topics we have covered thus far.

> **THEOREM 4.8.4 Equivalent Statements**
>
> *If A is an $n \times n$ matrix, then the following statements are equivalent.*
>
> *(a)* *A is invertible.*
> *(b)* *$A\mathbf{x} = \mathbf{0}$ has only the trivial solution.*
> *(c)* *The reduced row echelon form of A is I_n.*
> *(d)* *A is expressible as a product of elementary matrices.*
> *(e)* *$A\mathbf{x} = \mathbf{b}$ is consistent for every $n \times 1$ matrix $\mathbf{b}$.*
> *(f)* *$A\mathbf{x} = \mathbf{b}$ has exactly one solution for every $n \times 1$ matrix $\mathbf{b}$.*
> *(g)* *$\det(A) \neq 0$.*
> *(h)* *The column vectors of A are linearly independent.*
> *(i)* *The row vectors of A are linearly independent.*
> *(j)* *The column vectors of A span R^n.*
> *(k)* *The row vectors of A span R^n.*
> *(l)* *The column vectors of A form a basis for R^n.*
> *(m)* *The row vectors of A form a basis for R^n.*
> *(n)* *A has rank n.*
> *(o)* *A has nullity 0.*

Proof The equivalence of (h) through (m) follows from Theorem 4.5.4 (we omit the details). To complete the proof we will show that (b), (n), and (o) are equivalent by proving the chain of implications (b) $\Rightarrow$ (o) $\Rightarrow$ (n) $\Rightarrow$ (b).

(b) $\Rightarrow$ (o) If $A\mathbf{x} = \mathbf{0}$ has only the trivial solution, then there are no parameters in that solution, so nullity(A) = 0 by Theorem 4.8.3b.

(o) $\Rightarrow$ (n) Theorem 4.8.2.

(n) $\Rightarrow$ (b) If A has rank n, then Theorem 4.8.3a implies that there are n leading variables (hence no free variables) in the general solution of $A\mathbf{x} = \mathbf{0}$. This leaves the trivial solution as the only possibility. ◄

Overdetermined and Underdetermined Systems

In engineering and other applications, the occurrence of an overdetermined or underdetermined linear system often signals that one or more variables were omitted in formulating the problem or that extraneous variables were included. This often leads to some kind of undesirable physical result.

In many applications the equations in a linear system correspond to physical constraints or conditions that must be satisfied. In general, the most desirable systems are those that have the same number of constraints as unknowns, since such systems often have a unique solution. Unfortunately, it is not always possible to match the number of constraints and unknowns, so researchers are often faced with linear systems that have more constraints than unknowns, called ***overdetermined systems***, or with fewer constraints than unknowns, called ***underdetermined systems***. The following two theorems will help us to analyze both overdetermined and underdetermined systems.

THEOREM 4.8.5 *If $A\mathbf{x} = \mathbf{b}$ is a consistent linear system of m equations in n unknowns, and if A has rank r, then the general solution of the system contains $n - r$ parameters.*

Proof It follows from Theorem 4.7.2 that the number of parameters is equal to the nullity of A, which, by Theorem 4.8.2, is $n - r$. ◄

THEOREM 4.8.6 *Let A be an $m \times n$ matrix.*

(a) (***Overdetermined Case***). *If $m > n$, then the linear system $A\mathbf{x} = \mathbf{b}$ is inconsistent for at least one vector $\mathbf{b}$ in R^n.*

(b) (***Underdetermined Case***). *If $m < n$, then for each vector $\mathbf{b}$ in R^m the linear system $A\mathbf{x} = \mathbf{b}$ is either inconsistent or has infinitely many solutions.*

Proof (a) Assume that $m > n$, in which case the column vectors of A cannot span R^m (fewer vectors than the dimension of R^m). Thus, there is at least one vector $\mathbf{b}$ in R^m that is not in the column space of A, and for that $\mathbf{b}$ the system $A\mathbf{x} = \mathbf{b}$ is inconsistent by Theorem 4.7.1.

Proof (b) Assume that $m < n$. For each vector $\mathbf{b}$ in R^n there are two possibilities: either the system $A\mathbf{x} = \mathbf{b}$ is consistent or it is inconsistent. If it is inconsistent, then the proof is complete. If it is consistent, then Theorem 4.8.5 implies that the general solution has $n - r$ parameters, where $r = \text{rank}(A)$. But rank(A) is the smaller of m and n, so

$$n - r = n - m > 0$$

This means that the general solution has at least one parameter and hence there are infinitely many solutions. ◄

▶ **EXAMPLE 5** **Overdetermined and Underdetermined Systems**

(a)   What can you say about the solutions of an overdetermined system $A\mathbf{x} = \mathbf{b}$ of 7 equations in 5 unknowns in which A has rank $r = 4$?

(b)   What can you say about the solutions of an underdetermined system $A\mathbf{x} = \mathbf{b}$ of 5 equations in 7 unknowns in which A has rank $r = 4$?

Solution (a)   The system is consistent for some vector $\mathbf{b}$ in R^7, and for any such $\mathbf{b}$ the number of parameters in the general solution is $n - r = 5 - 4 = 1$.

Solution (b)   The system may be consistent or inconsistent, but if it is consistent for the vector $\mathbf{b}$ in R^5, then the general solution has $n - r = 7 - 4 = 3$ parameters.

▶ **EXAMPLE 6** **An Overdetermined System**

The linear system

$$\begin{aligned}
x_1 - 2x_2 &= b_1 \\
x_1 - x_2 &= b_2 \\
x_1 + x_2 &= b_3 \\
x_1 + 2x_2 &= b_4 \\
x_1 + 3x_2 &= b_5
\end{aligned}$$

is overdetermined, so it cannot be consistent for all possible values of b_1, b_2, b_3, b_4, and b_5. Exact conditions under which the system is consistent can be obtained by solving the linear system by Gauss–Jordan elimination. We leave it for you to show that the augmented matrix is row equivalent to

$$\begin{bmatrix}
1 & 0 & 2b_2 - b_1 \\
0 & 1 & b_2 - b_1 \\
0 & 0 & b_3 - 3b_2 + 2b_1 \\
0 & 0 & b_4 - 4b_2 + 3b_1 \\
0 & 0 & b_5 - 5b_2 + 4b_1
\end{bmatrix} \tag{5}$$

Thus, the system is consistent if and only if b_1, b_2, b_3, b_4, and b_5 satisfy the conditions

$$\begin{aligned}
2b_1 - 3b_2 + b_3 &= 0 \\
3b_1 - 4b_2 + b_4 &= 0 \\
4b_1 - 5b_2 + b_5 &= 0
\end{aligned}$$

Solving this homogeneous linear system yields

$$b_1 = 5r - 4s, \quad b_2 = 4r - 3s, \quad b_3 = 2r - s, \quad b_4 = r, \quad b_5 = s$$

where r and s are arbitrary. ◀

Remark   The coefficient matrix for the linear system in the last example has $n = 2$ columns, and it has rank $r = 2$ because there are two nonzero rows in its reduced row echelon form. This implies that when the system is consistent its general solution will contain $n - r = 0$ parameters; that is, the solution will be unique. With a moment's thought, you should be able to see that this is so from (5).

The Fundamental Spaces of a Matrix

There are six important vector spaces associated with a matrix A and its transpose A^T:

row space of A	row space of A^T
column space of A	column space of A^T
null space of A	null space of A^T

However, transposing a matrix converts row vectors into column vectors and conversely, so except for a difference in notation, the row space of A^T is the same as the column space of A, and the column space of A^T is the same as the row space of A. Thus, of the six spaces listed above, only the following four are distinct:

If A is an $m \times n$ matrix, then the row space and null space of A are subspaces of R^n, and the column space of A and the null space of A^T are subspaces of R^m.

row space of A column space of A

null space of A null space of A^T

These are called the ***fundamental spaces*** of a matrix A. We will conclude this section by discussing how these four subspaces are related.

Let us focus for a moment on the matrix A^T. Since the row space and column space of a matrix have the same dimension, and since transposing a matrix converts its columns to rows and its rows to columns, the following result should not be surprising.

THEOREM 4.8.7 *If A is any matrix, then $rank(A) = rank(A^T)$.*

Proof

$$rank(A) = \dim(\text{row space of } A) = \dim(\text{column space of } A^T) = rank(A^T). \ \blacktriangleleft$$

This result has some important implications. For example, if A is an $m \times n$ matrix, then applying Formula (4) to the matrix A^T and using the fact that this matrix has m columns yields

$$rank(A^T) + \text{nullity}(A^T) = m$$

which, by virtue of Theorem 4.8.7, can be rewritten as

$$rank(A) + \text{nullity}(A^T) = m \tag{6}$$

This alternative form of Formula (4) in Theorem 4.8.2 makes it possible to express the dimensions of all four fundamental spaces in terms of the size and rank of A. Specifically, if $rank(A) = r$, then

$$\dim[\text{row}(A)] = r \qquad \dim[\text{col}(A)] = r$$
$$\dim[\text{null}(A)] = n - r \quad \dim[\text{null}(A^T)] = m - r \tag{7}$$

The four formulas in (7) provide an *algebraic* relationship between the size of a matrix and the dimensions of its fundamental spaces. Our next objective is to find a *geometric* relationship between the fundamental spaces themselves. For this purpose recall from Theorem 3.4.3 that if A is an $m \times n$ matrix, then the null space of A consists of those vectors that are orthogonal to each of the row vectors of A. To develop that idea in more detail, we make the following definition.

DEFINITION 2 If W is a subspace of R^n, then the set of all vectors in R^n that are orthogonal to every vector in W is called the ***orthogonal complement*** of W and is denoted by the symbol $W^{\perp}$.

The following theorem lists three basic properties of orthogonal complements. We will omit the formal proof because a more general version of this theorem will be given later in the text.

THEOREM 4.8.8 *If W is a subspace of R^n, then:*

(a) $W^\perp$ *is a subspace of R^n.*

(b) *The only vector common to W and $W^\perp$ is $\mathbf{0}$.*

(c) *The orthogonal complement of $W^\perp$ is W.*

▶ **EXAMPLE 7 Orthogonal Complements**

Explain why $\{\mathbf{0}\}$ and R^n are orthogonal complements.

In R^2 the orthogonal complement of a line W through the origin is the line through the origin that is perpendicular to W (Figure 4.8.1a); and in R^3 the orthogonal complement of a plane W through the origin is the line through the origin that is perpendicular to that plane (Figure 4.8.1b). ◀

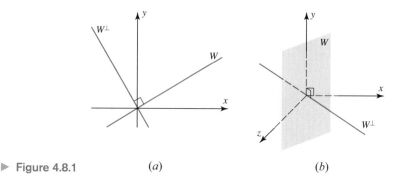

▶ **Figure 4.8.1**   (a)   (b)

A Geometric Link Between the Fundamental Spaces

The following theorem provides a geometric link between the fundamental spaces of a matrix. Part (a) is essentially a restatement of Theorem 3.4.3 in the language of orthogonal complements, and part (b), whose proof is left as an exercise, follows from part (a). The essential idea of the theorem is illustrated in Figure 4.8.2.

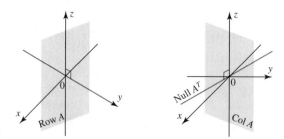

▶ **Figure 4.8.2**

THEOREM 4.8.9 *If A is an $m \times n$ matrix, then:*

(a) *The null space of A and the row space of A are orthogonal complements in R^n.*

(b) *The null space of A^T and the column space of A are orthogonal complements in R^m.*

More on the Equivalence Theorem

As our final result in this section, we will add two more statements to Theorem 4.8.4. We leave the proof that those statements are equivalent to the rest as an exercise.

THEOREM 4.8.10 Equivalent Statements

If A is an $n \times n$ matrix, then the following statements are equivalent.

(a) *A is invertible.*

(b) $A\mathbf{x} = \mathbf{0}$ *has only the trivial solution.*

(c) *The reduced row echelon form of A is I_n.*

(d) *A is expressible as a product of elementary matrices.*

(e) $A\mathbf{x} = \mathbf{b}$ *is consistent for every $n \times 1$ matrix $\mathbf{b}$.*

(f) $A\mathbf{x} = \mathbf{b}$ *has exactly one solution for every $n \times 1$ matrix $\mathbf{b}$.*

(g) $\det(A) \neq 0$.

(h) *The column vectors of A are linearly independent.*

(i) *The row vectors of A are linearly independent.*

(j) *The column vectors of A span R^n.*

(k) *The row vectors of A span R^n.*

(l) *The column vectors of A form a basis for R^n.*

(m) *The row vectors of A form a basis for R^n.*

(n) *A has rank n.*

(o) *A has nullity 0.*

(p) *The orthogonal complement of the null space of A is R^n.*

(q) *The orthogonal complement of the row space of A is $\{\mathbf{0}\}$.*

Applications of Rank

The advent of the Internet has stimulated research on finding efficient methods for transmitting large amounts of digital data over communications lines with limited bandwidths. Digital data are commonly stored in matrix form, and many techniques for improving transmission speed use the rank of a matrix in some way. Rank plays a role because it measures the "redundancy" in a matrix in the sense that if A is an $m \times n$ matrix of rank k, then $n - k$ of the column vectors and $m - k$ of the row vectors can be expressed in terms of k linearly independent column or row vectors. The essential idea in many data compression schemes is to approximate the original data set by a data set with smaller rank that conveys nearly the same information, then eliminate redundant vectors in the approximating set to speed up the transmission time.

Concept Review

- Rank
- Nullity
- Dimension Theorem
- Overdetermined system
- Underdetermined system
- Fundamental spaces of a matrix
- Relationships among the fundamental spaces
- Orthogonal complement
- Equivalent characterizations of invertible matrices

Skills

- Find the rank and nullity of a matrix.
- Find the dimension of the row space of a matrix.

Exercise Set 4.8

1. Verify that $\text{rank}(A) = \text{rank}(A^T)$.

$$A = \begin{bmatrix} 1 & 2 & 4 & 0 \\ -3 & 1 & 5 & 2 \\ -2 & 3 & 9 & 2 \end{bmatrix}$$

2. Find the rank and nullity of the matrix; then verify that the values obtained satisfy Formula (4) in the Dimension Theorem.

(a) $A = \begin{bmatrix} 1 & -1 & 3 \\ 5 & -4 & -4 \\ 7 & -6 & 2 \end{bmatrix}$ (b) $A = \begin{bmatrix} 2 & 0 & -1 \\ 4 & 0 & -2 \\ 0 & 0 & 0 \end{bmatrix}$

(c) $A = \begin{bmatrix} 1 & 4 & 5 & 2 \\ 2 & 1 & 3 & 0 \\ -1 & 3 & 2 & 2 \end{bmatrix}$

(d) $A = \begin{bmatrix} 1 & 4 & 5 & 6 & 9 \\ 3 & -2 & 1 & 4 & -1 \\ -1 & 0 & -1 & -2 & -1 \\ 2 & 3 & 5 & 7 & 8 \end{bmatrix}$

(e) $A = \begin{bmatrix} 1 & -3 & 2 & 2 & 1 \\ 0 & 3 & 6 & 0 & -3 \\ 2 & -3 & -2 & 4 & 4 \\ 3 & -6 & 0 & 6 & 5 \\ -2 & 9 & 2 & -4 & -5 \end{bmatrix}$

3. In each part of Exercise 2, use the results obtained to find the number of leading variables and the number of parameters in the solution of $A\mathbf{x} = \mathbf{0}$ without solving the system.

4. In each part, use the information in the table to find the dimension of the row space of A, column space of A, null space of A, and null space of A^T.

	(a)	(b)	(c)	(d)	(e)	(f)	(g)
Size of A	3×3	3×3	3×3	5×9	9×5	4×4	6×2
Rank(A)	3	2	1	2	2	0	2

5. In each part, find the largest possible value for the rank of A and the smallest possible value for the nullity of A.

(a) A is 4×4 (b) A is 3×5 (c) A is 5×3

6. If A is an $m \times n$ matrix, what is the largest possible value for its rank and the smallest possible value for its nullity?

7. In each part, use the information in the table to determine whether the linear system $A\mathbf{x} = \mathbf{b}$ is consistent. If so, state the number of parameters in its general solution.

	(a)	(b)	(c)	(d)	(e)	(f)	(g)
Size of A	3×3	3×3	3×3	5×9	5×9	4×4	6×2
Rank(A)	3	2	1	2	2	0	2
Rank[$A \mid \mathbf{b}$]	3	3	1	2	3	0	2

8. For each of the matrices in Exercise 7, find the nullity of A, and determine the number of parameters in the general solution of the homogeneous linear system $A\mathbf{x} = \mathbf{0}$.

9. What conditions must be satisfied by b_1, b_2, b_3, b_4, and b_5 for the overdetermined linear system

$$x_1 - 3x_2 = b_1$$
$$x_1 - 2x_2 = b_2$$
$$x_1 + x_2 = b_3$$
$$x_1 - 4x_2 = b_4$$
$$x_1 + 5x_2 = b_5$$

to be consistent?

10. Let

$$A = \begin{bmatrix} a_{11} & a_{12} & a_{13} \\ a_{21} & a_{22} & a_{23} \end{bmatrix}$$

Show that A has rank 2 if and only if one or more of the determinants

$$\begin{vmatrix} a_{11} & a_{12} \\ a_{21} & a_{22} \end{vmatrix}, \quad \begin{vmatrix} a_{11} & a_{13} \\ a_{21} & a_{23} \end{vmatrix}, \quad \begin{vmatrix} a_{12} & a_{13} \\ a_{22} & a_{23} \end{vmatrix}$$

is nonzero.

11. Suppose that A is a 3×3 matrix whose null space is a line through the origin in 3-space. Can the row or column space of A also be a line through the origin? Explain.

12. Discuss how the rank of A varies with t.

(a) $A = \begin{bmatrix} 1 & 1 & t \\ 1 & t & 1 \\ t & 1 & 1 \end{bmatrix}$ (b) $A = \begin{bmatrix} t & 3 & -1 \\ 3 & 6 & -2 \\ -1 & -3 & t \end{bmatrix}$

13. Are there values of r and s for which

$$\begin{bmatrix} 1 & 0 & 0 \\ 0 & r-2 & 2 \\ 0 & s-1 & r+2 \\ 0 & 0 & 3 \end{bmatrix}$$

has rank 1? Has rank 2? If so, find those values.

14. Use the result in Exercise 10 to show that the set of points (x, y, z) in R^3 for which the matrix

$$\begin{bmatrix} x & y & z \\ 1 & x & y \end{bmatrix}$$

has rank 1 is the curve with parametric equations $x = t$, $y = t^2, z = t^3$.

15. Prove: If $k \neq 0$, then A and kA have the same rank.

16. (a) Give an example of a 3×3 matrix whose column space is a plane through the origin in 3-space.

(b) What kind of geometric object is the null space of your matrix?

(c) What kind of geometric object is the row space of your matrix?

17. (a) If A is a 3×5 matrix, then the number of leading 1's in the reduced row echelon form of A is at most _____. Why?

   (b) If A is a 3×5 matrix, then the number of parameters in the general solution of $A\mathbf{x} = \mathbf{0}$ is at most _____. Why?

   (c) If A is a 5×3 matrix, then the number of leading 1's in the reduced row echelon form of A is at most _____. Why?

   (d) If A is a 5×3 matrix, then the number of parameters in the general solution of $A\mathbf{x} = \mathbf{0}$ is at most _____. Why?

18. (a) If A is a 3×5 matrix, then the rank of A is at most _____. Why?

   (b) If A is a 3×5 matrix, then the nullity of A is at most _____. Why?

   (c) If A is a 3×5 matrix, then the rank of A^T is at most _____. Why?

   (d) If A is a 3×5 matrix, then the nullity of A^T is at most _____. Why?

19. Find matrices A and B for which $\text{rank}(A) = \text{rank}(B)$, but $\text{rank}(A^2) \neq \text{rank}(B^2)$.

20. Prove: If a matrix A is not square, then either the row vectors or the column vectors of A are linearly dependent.

True-False Exercises

In parts (a)–(j) determine whether the statement is true or false, and justify your answer.

(a) Either the row vectors or the column vectors of a square matrix are linearly independent.

(b) A matrix with linearly independent row vectors and linearly independent column vectors is square.

(c) The nullity of a nonzero $m \times n$ matrix is at most m.

(d) Adding one additional column to a matrix increases its rank by one.

(e) The nullity of a square matrix with linearly dependent rows is at least one.

(f) If A is square and $A\mathbf{x} = \mathbf{b}$ is inconsistent for some vector $\mathbf{b}$, then the nullity of A is zero.

(g) If a matrix A has more rows than columns, then the dimension of the row space is greater than the dimension of the column space.

(h) If $\text{rank}(A^T) = \text{rank}(A)$, then A is square.

(i) There is no 3×3 matrix whose row space and null space are both lines in 3-space.

(j) If V is a subspace of R^n and W is a subspace of V, then $W^\perp$ is a subspace of $V^\perp$.

4.9 Matrix Transformations from R^n to R^m

In this section we will study functions of the form $\mathbf{w} = F(\mathbf{x})$, where the independent variable $\mathbf{x}$ is a vector in R^n and the dependent variable $\mathbf{w}$ is a vector in R^m. We will concentrate on a special class of such functions called "matrix transformations." Such transformations are fundamental in the study of linear algebra and have important applications in physics, engineering, social sciences, and various branches of mathematics.

Functions and Transformations

Recall that a **function** is a rule that associates with each element of a set A one and only one element in a set B. If f associates the element b with the element a, then we write

$$b = f(a)$$

and we say that b is the **image** of a under f or that $f(a)$ is the **value** of f at a. The set A is called the **domain** of f and the set B the **codomain** of f (Figure 4.9.1). The subset of the codomain that consists of all images of points in the domain is called the **range** of f.

For many common functions the domain and codomain are sets of real numbers, but in this text we will be concerned with functions for which the domain and codomain are vector spaces.

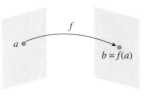

Domain
A

Codomain
B

▲ Figure 4.9.1

DEFINITION 1 If V and W are vector spaces, and if f is a function with domain V and codomain W, then we say that f is a ***transformation*** from V to W or that f ***maps*** V to W, which we denote by writing

$$f: V \rightarrow W$$

In the special case where $V = W$, the transformation is also called an ***operator*** on V.

In this section we will be concerned exclusively with transformations from R^n to R^m; transformations of general vector spaces will be considered in a later section. To illustrate one way in which such transformations can arise, suppose that $f_1, f_2, \ldots, f_m$ are real-valued functions of n variables, say

$$
\begin{aligned}
w_1 &= f_1(x_1, x_2, \ldots, x_n) \\
w_2 &= f_2(x_1, x_2, \ldots, x_n) \\
&\vdots \qquad\qquad \vdots \\
w_m &= f_m(x_1, x_2, \ldots, x_n)
\end{aligned}
\tag{1}
$$

These m equations assign a unique point $(w_1, w_2, \ldots, w_m)$ in R^m to each point $(x_1, x_2, \ldots, x_n)$ in R^n and thus define a transformation from R^n to R^m. If we denote this transformation by T, then $T: R^n \rightarrow R^m$ and

$$T(x_1, x_2, \ldots, x_n) = (w_1, w_2, \ldots, w_m)$$

Matrix Transformations   In the special case where the equations in (1) are linear, they can be expressed in the form

$$
\begin{aligned}
w_1 &= a_{11}x_1 + a_{12}x_2 + \cdots + a_{1n}x_n \\
w_2 &= a_{21}x_1 + a_{22}x_2 + \cdots + a_{2n}x_n \\
&\vdots \qquad \vdots \qquad \vdots \qquad\qquad \vdots \\
w_m &= a_{m1}x_1 + a_{m2}x_2 + \cdots + a_{mn}x_n
\end{aligned}
\tag{2}
$$

which we can write in matrix notation as

$$
\begin{bmatrix} w_1 \\ w_2 \\ \vdots \\ w_m \end{bmatrix} =
\begin{bmatrix}
a_{11} & a_{12} & \cdots & a_{1n} \\
a_{21} & a_{22} & \cdots & a_{2n} \\
\vdots & \vdots & & \vdots \\
a_{m1} & a_{m2} & \cdots & a_{mn}
\end{bmatrix}
\begin{bmatrix} x_1 \\ x_2 \\ \vdots \\ x_n \end{bmatrix}
\tag{3}
$$

or more briefly as

$$\mathbf{w} = A\mathbf{x} \tag{4}$$

Although we could view this as a linear system, we will view it instead as a transformation that maps the column vector $\mathbf{x}$ in R^n into the column vector $\mathbf{w}$ in R^m by multiplying $\mathbf{x}$ on the left by A. We call this a ***matrix transformation*** (or ***matrix operator*** if $m = n$), and we denote it by $T_A: R^n \rightarrow R^m$. With this notation, Equation (4) can be expressed as

$$\mathbf{w} = T_A(\mathbf{x}) \tag{5}$$

The matrix transformation T_A is called ***multiplication by A***, and the matrix A is called the ***standard matrix*** for the transformation.

We will also find it convenient, on occasion, to express (5) in the schematic form

$$\mathbf{x} \xrightarrow{\;T_A\;} \mathbf{w} \tag{6}$$

which is read "T_A maps $\mathbf{x}$ into $\mathbf{w}$."

▶ **EXAMPLE 1** **A Matrix Transformation from R^4 to R^3**

The matrix transformation $T: R^4 \rightarrow R^3$ defined by the equations

$$
\begin{aligned}
w_1 &= 2x_1 - 3x_2 + x_3 - 5x_4 \\
w_2 &= 4x_1 + x_2 - 2x_3 + x_4 \\
w_3 &= 5x_1 - x_2 + 4x_3
\end{aligned}
\tag{7}
$$

can be expressed in matrix form as

$$
\begin{bmatrix} w_1 \\ w_2 \\ w_3 \end{bmatrix} =
\begin{bmatrix} 2 & -3 & 1 & -5 \\ 4 & 1 & -2 & 1 \\ 5 & -1 & 4 & 0 \end{bmatrix}
\begin{bmatrix} x_1 \\ x_2 \\ x_3 \\ x_4 \end{bmatrix}
\tag{8}
$$

so the standard matrix for T is

$$
A = \begin{bmatrix} 2 & -3 & 1 & -5 \\ 4 & 1 & -2 & 1 \\ 5 & -1 & 4 & 0 \end{bmatrix}
$$

The image of a point (x_1, x_2, x_3, x_4) can be computed directly from the defining equations (7) or from (8) by matrix multiplication. For example, if

$$(x_1, x_2, x_3, x_4) = (1, -3, 0, 2)$$

then substituting in (7) yields $w_1 = 1$, $w_2 = 3$, $w_3 = 8$ (verify), or alternatively from (8),

$$
\begin{bmatrix} w_1 \\ w_2 \\ w_3 \end{bmatrix} =
\begin{bmatrix} 2 & -3 & 1 & -5 \\ 4 & 1 & -2 & 1 \\ 5 & -1 & 4 & 0 \end{bmatrix}
\begin{bmatrix} 1 \\ -3 \\ 0 \\ 2 \end{bmatrix} =
\begin{bmatrix} 1 \\ 3 \\ 8 \end{bmatrix} ◀
$$

Some Notational Matters

Sometimes we will want to denote a matrix transformation without giving a name to the matrix itself. In such cases we will denote the standard matrix for $T: R^n \rightarrow R^m$ by the symbol $[T]$. Thus, the equation

$$T(\mathbf{x}) = [T]\mathbf{x} \tag{9}$$

is simply the statement that T is a matrix transformation with standard matrix $[T]$, and the image of $\mathbf{x}$ under this transformation is the product of the matrix $[T]$ and the column vector $\mathbf{x}$.

Properties of Matrix Transformations

The following theorem lists four basic properties of matrix transformations that follow from properties of matrix multiplication.

THEOREM 4.9.1 *For every matrix A the matrix transformation $T_A: R^n \rightarrow R^m$ has the following properties for all vectors $\mathbf{u}$ and $\mathbf{v}$ in R^n and for every scalar k:*

(a) $T_A(\mathbf{0}) = \mathbf{0}$

(b) $T_A(k\mathbf{u}) = kT_A(\mathbf{u})$ **[Homogeneity property]**

(c) $T_A(\mathbf{u} + \mathbf{v}) = T_A(\mathbf{u}) + T_A(\mathbf{v})$ **[Additivity property]**

(d) $T_A(\mathbf{u} - \mathbf{v}) = T_A(\mathbf{u}) - T_A(\mathbf{v})$

Proof All four parts are restatements of familiar properties of matrix multiplication:

$$A\mathbf{0} = \mathbf{0}, \quad A(k\mathbf{u}) = k(A\mathbf{u}), \quad A(\mathbf{u} + \mathbf{v}) = A\mathbf{u} + A\mathbf{v}, \quad A(\mathbf{u} - \mathbf{v}) = A\mathbf{u} - A\mathbf{v} \quad \blacktriangleleft$$

It follows from Theorem 4.9.1 that a matrix transformation maps linear combinations of vectors in R^n into the corresponding linear combinations in R^m in the sense that

$$T_A(k_1\mathbf{u}_1 + k_2\mathbf{u}_2 + \cdots + k_r\mathbf{u}_r) = k_1 T_A(\mathbf{u}_1) + k_2 T_A(\mathbf{u}_2) + \cdots + k_r T_A(\mathbf{u}_r) \tag{10}$$

Depending on whether n-tuples and m-tuples are regarded as vectors or points, the geometric effect of a matrix transformation $T_A: R^n \to R^m$ is to map each vector (point) in R^n into a vector (point) in R^m (Figure 4.9.2).

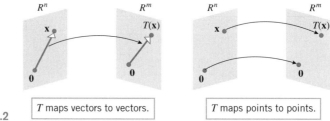

| T maps vectors to vectors. | T maps points to points. |

▶ Figure 4.9.2

The following theorem states that if two matrix transformations from R^n to R^m have the same image at each point of R^n, then the matrices themselves must be the same.

THEOREM 4.9.2 *If $T_A: R^n \to R^m$ and $T_B: R^n \to R^m$ are matrix transformations, and if $T_A(\mathbf{x}) = T_B(\mathbf{x})$ for every vector $\mathbf{x}$ in R^n, then $A = B$.*

Proof To say that $T_A(\mathbf{x}) = T_B(\mathbf{x})$ for every vector in R^n is the same as saying that

$$A\mathbf{x} = B\mathbf{x}$$

for every vector $\mathbf{x}$ in R^n. This is true, in particular, if $\mathbf{x}$ is any of the standard basis vectors $\mathbf{e}_1, \mathbf{e}_2, \ldots, \mathbf{e}_n$ for R^n; that is,

$$A\mathbf{e}_j = B\mathbf{e}_j \quad (j = 1, 2, \ldots, n) \tag{11}$$

Since every entry of $\mathbf{e}_j$ is 0 except for the jth, which is 1, it follows from Theorem 1.3.1 that $A\mathbf{e}_j$ is the jth column of A and $B\mathbf{e}_j$ is the jth column of B. Thus, it follows from (11) that corresponding columns of A and B are the same, and hence that $A = B$. $\blacktriangleleft$

▶ **EXAMPLE 2 Zero Transformations**

If 0 is the $m \times n$ zero matrix, then

$$T_0(\mathbf{x}) = 0\mathbf{x} = \mathbf{0}$$

so multiplication by zero maps every vector in R^n into the zero vector in R^m. We call T_0 the *zero transformation* from R^n to R^m.

▶ **EXAMPLE 3 Identity Operators**

If I is the $n \times n$ identity matrix, then

$$T_I(\mathbf{x}) = I\mathbf{x} = \mathbf{x}$$

so multiplication by I maps every vector in R^n into itself. We call T_I the *identity operator* on R^n. ◀

A Procedure for Finding Standard Matrices

There is a way of finding the standard matrix for a matrix transformation from R^n to R^m by considering the effect of that transformation on the standard basis vectors for R^n. To explain the idea, suppose that A is unknown and that

$$\mathbf{e}_1, \quad \mathbf{e}_2, \ldots, \quad \mathbf{e}_n$$

are the standard basis vectors for R^n. Suppose also that the images of these vectors under the transformation T_A are

$$T_A(\mathbf{e}_1) = A\mathbf{e}_1, \quad T_A(\mathbf{e}_2) = A\mathbf{e}_2, \ldots, \quad T_A(\mathbf{e}_n) = A\mathbf{e}_n$$

It follows from Theorem 1.3.1 that $A\mathbf{e}_j$ is a linear combination of the columns of A in which the successive coefficients are the entries of $\mathbf{e}_j$. But all entries of $\mathbf{e}_j$ are zero except the jth, so the product $A\mathbf{e}_j$ is just the jth column of the matrix A. Thus,

$$A = [T_A(\mathbf{e}_1) \mid T_A(\mathbf{e}_2) \mid \cdots \mid T_A(\mathbf{e}_n)] \tag{12}$$

In summary, we have the following procedure for finding the standard matrix for a matrix transformation:

Finding the Standard Matrix for a Matrix Transformation

Step 1.   Find the images of the standard basis vectors $\mathbf{e}_1, \mathbf{e}_2, \ldots, \mathbf{e}_n$ for R^n in column form.

Step 2.   Construct the matrix that has the images obtained in Step 1 as its successive columns. This matrix is the standard matrix for the transformation.

Reflection Operators

Some of the most basic matrix operators on R^2 and R^3 are those that map each point into its symmetric image about a fixed line or a fixed plane; these are called *reflection operators*. Table 1 shows the standard matrices for the reflections about the coordinate axes in R^2, and Table 2 shows the standard matrices for the reflections about the coordinate planes in R^3. In each case the standard matrix was obtained by finding the images of the standard basis vectors, converting those images to column vectors, and then using those column vectors as successive columns of the standard matrix.

Projection Operators

Matrix operators on R^2 and R^3 that map each point into its orthogonal projection on a fixed line or plane are called *projection operators* (or more precisely, *orthogonal projection* operators). Table 3 shows the standard matrices for the orthogonal projections on the coordinate axes in R^2, and Table 4 shows the standard matrices for the orthogonal projections on the coordinate planes in R^3.

Table 1

Operator	Illustration	Images of e_1 and e_2	Standard Matrix
Reflection about the y-axis $T(x, y) = (-x, y)$		$T(e_1) = T(1, 0) = (-1, 0)$ $T(e_2) = T(0, 1) = (0, 1)$	$\begin{bmatrix} -1 & 0 \\ 0 & 1 \end{bmatrix}$
Reflection about the x-axis $T(x, y) = (x, -y)$		$T(e_1) = T(1, 0) = (1, 0)$ $T(e_2) = T(0, 1) = (0, -1)$	$\begin{bmatrix} 1 & 0 \\ 0 & -1 \end{bmatrix}$
Reflection about the line $y = x$ $T(x, y) = (y, x)$		$T(e_1) = T(1, 0) = (0, 1)$ $T(e_2) = T(0, 1) = (1, 0)$	$\begin{bmatrix} 0 & 1 \\ 1 & 0 \end{bmatrix}$

Table 2

Operator	Illustration	e_1, e_2, e_3	Standard Matrix
Reflection about the xy-plane $T(x, y, z) = (x, y, -z)$		$T(e_1) = T(1, 0, 0) = (1, 0, 0)$ $T(e_2) = T(0, 1, 0) = (0, 1, 0)$ $T(e_3) = T(0, 0, 1) = (0, 0, -1)$	$\begin{bmatrix} 1 & 0 & 0 \\ 0 & 1 & 0 \\ 0 & 0 & -1 \end{bmatrix}$
Reflection about the xz-plane $T(x, y, z) = (x, -y, z)$		$T(e_1) = T(1, 0, 0) = (1, 0, 0)$ $T(e_2) = T(0, 1, 0) = (0, -1, 0)$ $T(e_3) = T(0, 0, 1) = (0, 0, 1)$	$\begin{bmatrix} 1 & 0 & 0 \\ 0 & -1 & 0 \\ 0 & 0 & 1 \end{bmatrix}$
Reflection about the yz-plane $T(x, y, z) = (-x, y, z)$		$T(e_1) = T(1, 0, 0) = (-1, 0, 0)$ $T(e_2) = T(0, 1, 0) = (0, 1, 0)$ $T(e_3) = T(0, 0, 1) = (0, 0, 1)$	$\begin{bmatrix} -1 & 0 & 0 \\ 0 & 1 & 0 \\ 0 & 0 & 1 \end{bmatrix}$

Table 3

Operator	Illustration	Images of e_1 and e_2	Standard Matrix
Orthogonal projection on the x-axis $T(x, y) = (x, 0)$		$T(\mathbf{e}_1) = T(1, 0) = (1, 0)$ $T(\mathbf{e}_2) = T(0, 1) = (0, 0)$	$\begin{bmatrix} 1 & 0 \\ 0 & 0 \end{bmatrix}$
Orthogonal projection on the y-axis $T(x, y) = (0, y)$		$T(\mathbf{e}_1) = T(1, 0) = (0, 0)$ $T(\mathbf{e}_2) = T(0, 1) = (0, 1)$	$\begin{bmatrix} 0 & 0 \\ 0 & 1 \end{bmatrix}$

Table 4

Operator	Illustration	Images of e_1, e_2, e_3	Standard Matrix
Orthogonal projection on the xy-plane $T(x, y, z) = (x, y, 0)$		$T(\mathbf{e}_1) = T(1, 0, 0) = (1, 0, 0)$ $T(\mathbf{e}_2) = T(0, 1, 0) = (0, 1, 0)$ $T(\mathbf{e}_3) = T(0, 0, 1) = (0, 0, 0)$	$\begin{bmatrix} 1 & 0 & 0 \\ 0 & 1 & 0 \\ 0 & 0 & 0 \end{bmatrix}$
Orthogonal projection on the xz-plane $T(x, y, z) = (x, 0, z)$		$T(\mathbf{e}_1) = T(1, 0, 0) = (1, 0, 0)$ $T(\mathbf{e}_2) = T(0, 1, 0) = (0, 0, 0)$ $T(\mathbf{e}_3) = T(0, 0, 1) = (0, 0, 1)$	$\begin{bmatrix} 1 & 0 & 0 \\ 0 & 0 & 0 \\ 0 & 0 & 1 \end{bmatrix}$
Orthogonal projection on the yz-plane $T(x, y, z) = (0, y, z)$		$T(\mathbf{e}_1) = T(1, 0, 0) = (0, 0, 0)$ $T(\mathbf{e}_2) = T(0, 1, 0) = (0, 1, 0)$ $T(\mathbf{e}_3) = T(0, 0, 1) = (0, 0, 1)$	$\begin{bmatrix} 0 & 0 & 0 \\ 0 & 1 & 0 \\ 0 & 0 & 1 \end{bmatrix}$

Rotation Operators　　Matrix operators on R^2 and R^3 that move points along circular arcs are called ***rotation operators***. Let us consider how to find the standard matrix for the rotation operator $T: R^2 \to R^2$ that moves points *counterclockwise* about the origin through an angle θ (Figure 4.9.3). As illustrated in Figure 4.9.3, the images of the standard basis vectors are

$$T(\mathbf{e}_1) = T(1, 0) = (\cos\theta, \sin\theta) \quad \text{and} \quad T(\mathbf{e}_2) = T(0, 1) = (-\sin\theta, \cos\theta)$$

so the standard matrix for T is

$$[T(\mathbf{e}_1) \mid T(\mathbf{e}_2)] = \begin{bmatrix} \cos\theta & -\sin\theta \\ \sin\theta & \cos\theta \end{bmatrix}$$

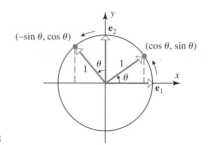

► Figure 4.9.3

In keeping with common usage we will denote this operator by R_θ and call

$$R_\theta = \begin{bmatrix} \cos\theta & -\sin\theta \\ \sin\theta & \cos\theta \end{bmatrix} \tag{13}$$

the *rotation matrix* for R^2. If $\mathbf{x} = (x, y)$ is a vector in R^2, and if $\mathbf{w} = (w_1, w_2)$ is its image under the rotation, then the relationship $\mathbf{w} = R_\theta \mathbf{x}$ can be written in component form as

$$\begin{aligned} w_1 &= x\cos\theta - y\sin\theta \\ w_2 &= x\sin\theta + y\cos\theta \end{aligned} \tag{14}$$

These are called the *rotation equations* for R^2. These ideas are summarized in Table 5.

Table 5

Operator	Illustration	Rotation Equations	Standard Matrix
Rotation through an angle θ		$\begin{aligned} w_1 &= x\cos\theta - y\sin\theta \\ w_2 &= x\sin\theta + y\cos\theta \end{aligned}$	$\begin{bmatrix} \cos\theta & -\sin\theta \\ \sin\theta & \cos\theta \end{bmatrix}$

In the plane, counterclockwise angles are positive and clockwise angles are negative. The rotation matrix for a *clockwise* rotation of $-\theta$ radians can be obtained by replacing θ by $-\theta$ in (12). After simplification this yields

$$R_{-\theta} = \begin{bmatrix} \cos\theta & \sin\theta \\ -\sin\theta & \cos\theta \end{bmatrix}$$

► **EXAMPLE 4 A Rotation Operator**

Find the image of $\mathbf{x} = (1, 1)$ under a rotation of $\pi/6$ radians ($= 30°$) about the origin.

Solution It follows from (13) with $\theta = \pi/6$ that

$$R_{\pi/6}\mathbf{x} = \begin{bmatrix} \frac{\sqrt{3}}{2} & -\frac{1}{2} \\ \frac{1}{2} & \frac{\sqrt{3}}{2} \end{bmatrix} \begin{bmatrix} 1 \\ 1 \end{bmatrix} = \begin{bmatrix} \frac{\sqrt{3}-1}{2} \\ \frac{1+\sqrt{3}}{2} \end{bmatrix} \approx \begin{bmatrix} 0.37 \\ 1.37 \end{bmatrix}$$

or in comma-delimited notation, $R_{\pi/6}(1, 1) = (0.37, 1.37)$. ◄

Rotations in R^3 A rotation of vectors in R^3 is usually described in relation to a ray emanating from the origin, called the *axis of rotation*. As a vector revolves around the axis of rotation, it sweeps out some portion of a cone (Figure 4.9.4a). The *angle of rotation*, which is measured in the base of the cone, is described as "clockwise" or "counterclockwise" in relation to a viewpoint that is along the axis of rotation *looking toward the origin*. For example, in Figure 4.9.4a the vector $\mathbf{w}$ results from rotating the vector $\mathbf{x}$ counterclockwise around the axis l through an angle θ. As in R^2, angles are *positive* if they are generated by counterclockwise rotations and *negative* if they are generated by clockwise rotations.

The most common way of describing a general axis of rotation is to specify a nonzero vector $\mathbf{u}$ that runs along the axis of rotation and has its initial point at the origin. The counterclockwise direction for a rotation about the axis can then be determined by a "right-hand rule" (Figure 4.9.4b): If the thumb of the right hand points in the direction of $\mathbf{u}$, then the cupped fingers point in a counterclockwise direction.

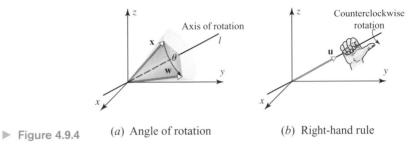

▶ Figure 4.9.4   (a) Angle of rotation   (b) Right-hand rule

A ***rotation operator*** on R^3 is a matrix operator that rotates each vector in R^3 about some rotation axis through a fixed angle θ. In Table 6 we have described the rotation operators on R^3 whose axes of rotation are the positive coordinate axes. For each of these rotations one of the components is unchanged, and the relationships between the other components can be derived by the same procedure used to derive (14). For example, in the rotation about the z-axis, the z-components of $\mathbf{x}$ and $\mathbf{w} = T(\mathbf{x})$ are the same, and the x- and y-components are related as in (14). This yields the rotation equation shown in the last row of Table 6.

Table 6

Operator	Illustration	Rotation Equations	Standard Matrix
Counterclockwise rotation about the positive x-axis through an angle θ		$w_1 = x$ $w_2 = y \cos\theta - z \sin\theta$ $w_3 = y \sin\theta + z \cos\theta$	$\begin{bmatrix} 1 & 0 & 0 \\ 0 & \cos\theta & -\sin\theta \\ 0 & \sin\theta & \cos\theta \end{bmatrix}$
Counterclockwise rotation about the positive y-axis through an angle θ		$w_1 = x \cos\theta + z \sin\theta$ $w_2 = y$ $w_3 = -x \sin\theta + z \cos\theta$	$\begin{bmatrix} \cos\theta & 0 & \sin\theta \\ 0 & 1 & 0 \\ -\sin\theta & 0 & \cos\theta \end{bmatrix}$
Counterclockwise rotation about the positive z-axis through an angle θ		$w_1 = x \cos\theta - y \sin\theta$ $w_2 = x \sin\theta + y \cos\theta$ $w_3 = z$	$\begin{bmatrix} \cos\theta & -\sin\theta & 0 \\ \sin\theta & \cos\theta & 0 \\ 0 & 0 & 1 \end{bmatrix}$

For completeness, we note that the standard matrix for a counterclockwise rotation through an angle θ about an axis in R^3, which is determined by an arbitrary *unit vector* $\mathbf{u} = (a, b, c)$ that has its initial point at the origin, is

$$\begin{bmatrix} a^2(1 - \cos\theta) + \cos\theta & ab(1 - \cos\theta) - c\sin\theta & ac(1 - \cos\theta) + b\sin\theta \\ ab(1 - \cos\theta) + c\sin\theta & b^2(1 - \cos\theta) + \cos\theta & bc(1 - \cos\theta) - a\sin\theta \\ ac(1 - \cos\theta) - b\sin\theta & bc(1 - \cos\theta) + a\sin\theta & c^2(1 - \cos\theta) + \cos\theta \end{bmatrix} \quad (15)$$

The derivation can be found in the book *Principles of Interactive Computer Graphics*, by W. M. Newman and R. F. Sproull (New York: McGraw-Hill, 1979). You may find it instructive to derive the results in Table 6 as special cases of this more general result.

Dilations and Contractions If k is a nonnegative scalar, then the operator $T(\mathbf{x}) = k\mathbf{x}$ on R^2 or R^3 has the effect of increasing or decreasing the length of each vector by a factor of k. If $0 \le k < 1$ the operator is called a ***contraction*** with factor k, and if $k > 1$ it is called a ***dilation*** with factor k (Figure 4.9.5). If $k = 1$, then T is the identity operator and can be regarded either as a contraction or a dilation. Tables 7 and 8 illustrate these operators.

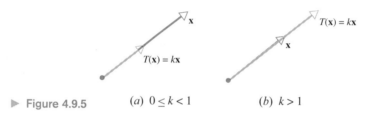

▶ **Figure 4.9.5**   (*a*) $0 \le k < 1$ (*b*) $k > 1$

Table 7

Operator	Illustration $T(x, y) = (kx, ky)$	Effect on the Standard Basis	Standard Matrix
Contraction with factor k on R^2 $(0 \le k < 1)$			$\begin{bmatrix} k & 0 \\ 0 & k \end{bmatrix}$
Dilation with factor k on R^2 $(k > 1)$			

Yaw, Pitch, and Roll

In aeronautics and astronautics, the orientation of an aircraft or space shuttle relative to an xyz-coordinate system is often described in terms of angles called ***yaw***, ***pitch***, and ***roll***. If, for example, an aircraft is flying along the y-axis and the xy-plane defines the horizontal, then the aircraft's angle of rotation about the z-axis is called the ***yaw***, its angle of rotation about the x-axis is called the ***pitch***, and its angle of rotation about the y-axis is called the ***roll***. A combination of yaw, pitch, and roll can be achieved by a single rotation about some axis through the origin. This is, in fact, how a space shuttle makes attitude adjustments—it doesn't perform each rotation separately; it calculates one axis, and rotates about that axis to get the correct orientation. Such rotation maneuvers are used to align an an-

tenna, point the nose toward a celestial object, or position a payload bay for docking.

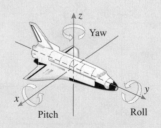

Table 8

Operator	Illustration $T(x, y, z) = (kx, ky, kz)$	Standard Matrix
Contraction with factor k on R^3 $(0 \leq k \leq 1)$		$\begin{bmatrix} k & 0 & 0 \\ 0 & k & 0 \\ 0 & 0 & k \end{bmatrix}$
Dilation with factor k on R^3 $(k \geq 1)$		

Expansion and Compressions

In a dilation or contraction of R^2 or R^3, all coordinates are multiplied by a factor k. If only one of the coordinates is multiplied by k, then the resulting operator is called an *expansion* or *compression* with factor k. This is illustrated in Table 9 for R^2. You should have no trouble extending these results to R^3.

Table 9

Operator	Illustration $T(x, y) = (kx, y)$	Effect on the Standard Basis	Standard Matrix
Compression of R^2 in the x-direction with factor k $(0 \leq k < 1)$			$\begin{bmatrix} k & 0 \\ 0 & 1 \end{bmatrix}$
Expansion of R^2 in the x-direction with factor k $(k > 1)$			

Operator	Illustration $T(x, y) = (x, ky)$	Effect on the Standard Basis	Standard Matrix
Compression of R^2 in the y-direction with factor k $(0 \leq k < 1)$			$\begin{bmatrix} 1 & 0 \\ 0 & k \end{bmatrix}$
Expansion of R^2 in the y-direction with factor k $(k > 1)$			

Shears A matrix operator of the form $T(x, y) = (x + ky, y)$ translates a point (x, y) in the xy-plane parallel to the x-axis by an amount ky that is proportional to the y-coordinate of the point. This operator leaves the points on the x-axis fixed (since $y = 0$), but as we progress away from the x-axis, the translation distance increases. We call this operator the **shear in the x-direction with factor k.** Similarly, a matrix operator of the form $T(x, y) = (x, y + kx)$ is called the **shear in the y-direction with factor k.** Table 10 illustrates the basic information about shears in R^2.

Table 10

Operator	Effect on the Standard Basis			Standard Matrix
Shear of R^2 in the x-direction with factor k $T(x, y) = (x + ky, y)$	$(0, 1)$ $(1, 0)$	$(k, 1)$ $(1, 0)$ $(k > 0)$	$(k, 1)$ $(1, 0)$ $(k < 0)$	$\begin{bmatrix} 1 & k \\ 0 & 1 \end{bmatrix}$
Shear of R^2 in the y-direction with factor k $T(x, y) = (x, y + kx)$	$(0, 1)$ $(1, 0)$	$(0, 1)$ $(1, k)$ $(k > 0)$	$(0, 1)$ $(1, k)$ $(k < 0)$	$\begin{bmatrix} 1 & 0 \\ k & 1 \end{bmatrix}$

▶ **EXAMPLE 5 Some Basic Matrix Operators on R^2**

In each part describe the matrix operator corresponding to A, and show its effect on the unit square.

(a) $A_1 = \begin{bmatrix} 1 & 2 \\ 0 & 1 \end{bmatrix}$ (b) $A_2 = \begin{bmatrix} 2 & 0 \\ 0 & 2 \end{bmatrix}$ (c) $A_3 = \begin{bmatrix} 2 & 0 \\ 0 & 1 \end{bmatrix}$

Solution By comparing the forms of these matrices to those in Tables 7, 9, and 10, we see that the matrix A_1 corresponds to a shear in the x-direction with factor 2, the matrix A_2 corresponds to a dilation with factor 2, and A_3 corresponds to an expansion in the x-direction with factor 2. The effects of these operators on the unit square are shown in Figure 4.9.6. ◀

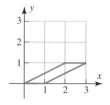 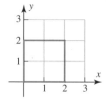 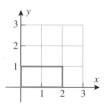

▶ Figure 4.9.6

In Table 3 we listed the standard matrices for the orthogonal projections on the coordinate axes in R^2. These are special cases of the more general operator $T: R^2 \rightarrow R^2$ that maps each point into its orthogonal projection on a line L through the origin that makes an

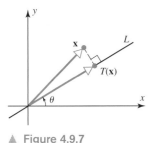

▲ Figure 4.9.7

angle θ with the positive x-axis (Figure 4.9.7). In Example 4 of Section 3.3 we used Formula (10) of that section to find the orthogonal projections of the standard basis vectors for R^2 on that line. Expressed in matrix form, we found those projections to be

$$T(\mathbf{e}_1) = \begin{bmatrix} \cos^2 \theta \\ \sin \theta \cos \theta \end{bmatrix} \quad \text{and} \quad T(\mathbf{e}_2) = \begin{bmatrix} \sin \theta \cos \theta \\ \sin^2 \theta \end{bmatrix}$$

Thus, the standard matrix for T is

$$[T] = [T(\mathbf{e}_1) \mid T(\mathbf{e}_2)] = \begin{bmatrix} \cos^2 \theta & \sin \theta \cos \theta \\ \sin \theta \cos \theta & \sin^2 \theta \end{bmatrix} = \begin{bmatrix} \cos^2 \theta & \frac{1}{2} \sin 2\theta \\ \frac{1}{2} \sin 2\theta & \sin^2 \theta \end{bmatrix}$$

In keeping with common usage, we will denote this operator by

We have included two versions of Formula (16) because both are commonly used. Whereas the first version involves only the angle θ, the second involves both θ and 2θ.

$$P_\theta = \begin{bmatrix} \cos^2 \theta & \sin \theta \cos \theta \\ \sin \theta \cos \theta & \sin^2 \theta \end{bmatrix} = \begin{bmatrix} \cos^2 \theta & \frac{1}{2} \sin 2\theta \\ \frac{1}{2} \sin 2\theta & \sin^2 \theta \end{bmatrix} \tag{16}$$

▶ **EXAMPLE 6 Orthogonal Projection on a Line Through the Origin**

Use Formula (16) to find the orthogonal projection of the vector $\mathbf{x} = (1, 5)$ on the line through the origin that makes an angle of $\pi/6 \,(= 30°)$ with the x-axis.

Solution Since $\sin(\pi/6) = 1/2$ and $\cos(\pi/6) = \sqrt{3}/2$, it follows from (16) that the standard matrix for this projection is

$$P_{\pi/6} = \begin{bmatrix} \cos^2(\pi/6) & \sin(\pi/6)\cos(\pi/6) \\ \sin(\pi/6)\cos(\pi/6) & \sin^2(\pi/6) \end{bmatrix} = \begin{bmatrix} \frac{3}{4} & \frac{\sqrt{3}}{4} \\ \frac{\sqrt{3}}{4} & \frac{1}{4} \end{bmatrix}$$

Thus,

$$P_{\pi/6}\mathbf{x} = \begin{bmatrix} \frac{3}{4} & \frac{\sqrt{3}}{4} \\ \frac{\sqrt{3}}{4} & \frac{1}{4} \end{bmatrix} \begin{bmatrix} 1 \\ 5 \end{bmatrix} = \begin{bmatrix} \frac{3+5\sqrt{3}}{4} \\ \frac{\sqrt{3}+5}{4} \end{bmatrix} \approx \begin{bmatrix} 2.91 \\ 1.68 \end{bmatrix}$$

or in comma-delimited notation, $P_{\pi/6}(1, 5) \approx (2.91, 1.68)$ ◀

Reflections About Lines Through the Origin

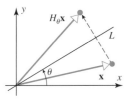

▲ Figure 4.9.8

In Table 1 we listed the reflections about the coordinate axes in R^2. These are special cases of the more general operator $H_\theta: R^2 \to R^2$ that maps each point into its reflection about a line L through the origin that makes an angle θ with the positive x-axis (Figure 4.9.8). We could find the standard matrix for H_θ by finding the images of the standard basis vectors, but instead we will take advantage of our work on orthogonal projections by using the Formula (16) for P_θ to find a formula for H_θ.

You should be able to see from Figure 4.9.9 that for every vector $\mathbf{x}$ in R^n

$$P_\theta \mathbf{x} - \mathbf{x} = \tfrac{1}{2}(H_\theta \mathbf{x} - \mathbf{x}) \quad \text{or equivalently} \quad H_\theta \mathbf{x} = (2P_\theta - I)\mathbf{x}$$

Thus, it follows from Theorem 4.9.2 that

$$H_\theta = 2P_\theta - I \tag{17}$$

and hence from (16) that

$$H_\theta = \begin{bmatrix} \cos 2\theta & \sin 2\theta \\ \sin 2\theta & -\cos 2\theta \end{bmatrix} \tag{18}$$

▲ Figure 4.9.9

▶ **EXAMPLE 7 Reflection About a Line Through the Origin**

Find the reflection of the vector $\mathbf{x} = (1, 5)$ on the line through the origin that makes an angle of $\pi/6 (= 30°)$ with the x-axis.

Solution Since $\sin(\pi/3) = \sqrt{3}/2$ and $\cos(\pi/3) = 1/2$, it follows from (18) that the standard matrix for this reflection is

$$H_{\pi/6} = \begin{bmatrix} \cos(\pi/3) & \sin(\pi/3) \\ \sin(\pi/3) & -\cos(\pi/3) \end{bmatrix} = \begin{bmatrix} \frac{1}{2} & \frac{\sqrt{3}}{2} \\ \frac{\sqrt{3}}{2} & -\frac{1}{2} \end{bmatrix}$$

Thus,

$$H_{\pi/6}\mathbf{x} = \begin{bmatrix} \frac{1}{2} & \frac{\sqrt{3}}{2} \\ \frac{\sqrt{3}}{2} & -\frac{1}{2} \end{bmatrix}\begin{bmatrix} 1 \\ 5 \end{bmatrix} = \begin{bmatrix} \frac{1+5\sqrt{3}}{2} \\ \frac{\sqrt{3}-5}{2} \end{bmatrix} \approx \begin{bmatrix} 4.83 \\ -1.63 \end{bmatrix}$$

or in comma-delimited notation, $H_{\pi/6}(1, 5) \approx (4.83, -1.63)$ ◀

Show that the standard matrices in Tables 1 and 3 are special cases of (18) and (16).

Concept Review

- Function
- Image
- Value
- Domain
- Codomain
- Transformation
- Operator
- Matrix transformation
- Matrix operator

- Standard matrix
- Properties of matrix transformations
- Zero transformation
- Identity operator
- Reflection operator
- Projection operator
- Rotation operator
- Rotation matrix

- Rotation equations
- Axis of rotation in 3-space
- Angle of rotation in 3-space
- Expansion operator
- Compression operator
- Shear
- Dilation
- Contraction

Skills

- Find the domain and codomain of a transformation, and determine whether the transformation is linear.
- Find the standard matrix for a matrix transformation.

- Describe the effect of a matrix operator on the standard basis in R^n.

Exercise Set 4.9

▷ In Exercises 1–2, find the domain and codomain of the transformation $T_A(\mathbf{x}) = A\mathbf{x}$. ◀

1. (a) A has size 3×2.   (b) A has size 2×3.
    (c) A has size 3×3.   (d) A has size 1×6.

2. (a) A has size 4×5.   (b) A has size 5×4.
    (c) A has size 4×4.   (d) A has size 3×1.

3. If $T(x_1, x_2) = (x_1 + x_2, -x_2, 3x_1)$, then the domain of T is _____, the codomain of T is _____, and the image of $\mathbf{x} = (1, -2)$ under T is _____.

4. If $T(x_1, x_2, x_3) = (x_1 + 2x_2, x_1 - 2x_2)$, then the domain of T is _____, the codomain of T is _____, and the image of $\mathbf{x} = (0, -1, 4)$ under T is _____.

5. In each part, find the domain and codomain of the transformation defined by the equations, and determine whether the transformation is linear.

(a) $w_1 = 3x_1 - 2x_2 + 4x_3$
    $w_2 = 5x_1 - 8x_2 + x_3$

(b) $w_1 = 2x_1x_2 - x_2$
    $w_2 = x_1 + 3x_1x_2$
    $w_3 = x_1 + x_2$

(c) $w_1 = 5x_1 - x_2 + x_3$
    $w_2 = -x_1 + x_2 + 7x_3$
    $w_3 = 2x_1 - 4x_2 - x_3$

(d) $w_1 = x_1^2 - 3x_2 + x_3 - 2x_4$
    $w_2 = 3x_1 - 4x_2 - x_3^2 + x_4$

6. In each part, determine whether T is a matrix transformation.

(a) $T(x, y) = (2x, y)$ (b) $T(x, y) = (-y, x)$

(c) $T(x, y) = (2x + y, x - y)$

(d) $T(x, y) = (x^2, y)$ (e) $T(x, y) = (x, y + 1)$

7. In each part, determine whether T is a matrix transformation.

(a) $T(x, y, z) = (0, 0)$ (b) $T(x, y, z) = (1, 1)$

(c) $T(x, y, z) = (3x - 4y, 2x - 5z)$

(d) $T(x, y, z) = (y^2, z)$ (e) $T(x, y, z) = (y - 1, x)$

8. Find the standard matrix for the transformation defined by the equations.

(a) $w_1 = 2x_1 - 3x_2 + x_4$ (b) $w_1 = 7x_1 + 2x_2 - 8x_3$
    $w_2 = 3x_1 + 5x_2 - x_4$ $w_2 = - x_2 + 5x_3$
                                           $w_3 = 4x_1 + 7x_2 - x_3$

(c) $w_1 = -x_1 + x_2$ (d) $w_1 = x_1$
    $w_2 = 3x_1 - 2x_2$ $w_2 = x_1 + x_2$
    $w_3 = 5x_1 - 7x_2$ $w_3 = x_1 + x_2 + x_3$
                                              $w_4 = x_1 + x_2 + x_3 + x_4$

9. Find the standard matrix for the operator $T: R^3 \to R^3$ defined by

$$w_1 = 3x_1 + 5x_2 - x_3$$
$$w_2 = 4x_1 - x_2 + x_3$$
$$w_3 = 3x_1 + 2x_2 - x_3$$

and then calculate $T(-1, 2, 4)$ by directly substituting in the equations and also by matrix multiplication.

10. Find the standard matrix for the operator T defined by the formula.

(a) $T(x_1, x_2) = (2x_1 - x_2, x_1 + x_2)$

(b) $T(x_1, x_2) = (x_1, x_2)$

(c) $T(x_1, x_2, x_3) = (x_1 + 2x_2 + x_3, x_1 + 5x_2, x_3)$

(d) $T(x_1, x_2, x_3) = (4x_1, 7x_2, -8x_3)$

11. Find the standard matrix for the transformation T defined by the formula.

(a) $T(x_1, x_2) = (x_2, -x_1, x_1 + 3x_2, x_1 - x_2)$

(b) $T(x_1, x_2, x_3, x_4) = (7x_1 + 2x_2 - x_3 + x_4, x_2 + x_3, -x_1)$

(c) $T(x_1, x_2, x_3) = (0, 0, 0, 0, 0)$

(d) $T(x_1, x_2, x_3, x_4) = (x_4, x_1, x_3, x_2, x_1 - x_3)$

12. In each part, find $T(\mathbf{x})$, and express the answer in matrix form.

(a) $[T] = \begin{bmatrix} 1 & 2 \\ 3 & 4 \end{bmatrix}; \ \mathbf{x} = \begin{bmatrix} 3 \\ -2 \end{bmatrix}$

(b) $[T] = \begin{bmatrix} -1 & 2 & 0 \\ 3 & 1 & 5 \end{bmatrix}; \ \mathbf{x} = \begin{bmatrix} -1 \\ 1 \\ 3 \end{bmatrix}$

(c) $[T] = \begin{bmatrix} -2 & 1 & 4 \\ 3 & 5 & 7 \\ 6 & 0 & -1 \end{bmatrix}; \ \mathbf{x} = \begin{bmatrix} x_1 \\ x_2 \\ x_3 \end{bmatrix}$

(d) $[T] = \begin{bmatrix} -1 & 1 \\ 2 & 4 \\ 7 & 8 \end{bmatrix}; \ \mathbf{x} = \begin{bmatrix} x_1 \\ x_2 \end{bmatrix}$

13. In each part, use the standard matrix for T to find $T(\mathbf{x})$; then check the result by calculating $T(\mathbf{x})$ directly.

(a) $T(x_1, x_2) = (-x_1 + x_2, x_2); \ \mathbf{x} = (-1, 4)$

(b) $T(x_1, x_2, x_3) = (2x_1 - x_2 + x_3, x_2 + x_3, 0);$
    $\mathbf{x} = (2, 1, -3)$

14. Use matrix multiplication to find the reflection of $(-1, 2)$ about

(a) the x-axis.   (b) the y-axis.   (c) the line $y = x$.

15. Use matrix multiplication to find the reflection of $(2, -5, 3)$ about

(a) the xy-plane.   (b) the xz-plane.   (c) the yz-plane.

16. Use matrix multiplication to find the orthogonal projection of $(2, -5)$ on

(a) the x-axis.   (b) the y-axis.

17. Use matrix multiplication to find the orthogonal projection of $(-2, 1, 3)$ on

(a) the xy-plane.   (b) the xz-plane.   (c) the yz-plane.

18. Use matrix multiplication to find the image of the vector $(3, -4)$ when it is rotated through an angle of

(a) $\theta = 30°$.   (b) $\theta = -60°$.

(c) $\theta = 45°$.   (d) $\theta = 90°$.

19. Use matrix multiplication to find the image of the vector $(-2, 1, 2)$ if it is rotated

(a) $30°$ about the x-axis.   (b) $45°$ about the y-axis.

(c) $90°$ about the z-axis.

20. Find the standard matrix for the operator that rotates a vector in R^3 through an angle of $-60°$ about

(a) the x-axis.   (b) the y-axis.   (c) the z-axis.

21. Use matrix multiplication to find the image of the vector $(-2, 1, 2)$ if it is rotated

(a) $-30°$ about the x-axis.   (b) $-45°$ about the y-axis.

(c) $-90°$ about the z-axis.

22. In R^3 the *orthogonal projections* on the x-axis, y-axis, and z-axis are defined by

$$T_1(x, y, z) = (x, 0, 0), \quad T_2(x, y, z) = (0, y, 0),$$
$$T_3(x, y, z) = (0, 0, z)$$

respectively.

(a) Show that the orthogonal projections on the coordinate axes are matrix operators, and find their standard matrices.

(b) Show that if $T: R^3 \to R^3$ is an orthogonal projection on one of the coordinate axes, then for every vector $\mathbf{x}$ in R^3, the vectors $T(\mathbf{x})$ and $\mathbf{x} - T(\mathbf{x})$ are orthogonal.

(c) Make a sketch showing $\mathbf{x}$ and $\mathbf{x} - T(\mathbf{x})$ in the case where T is the orthogonal projection on the x-axis.

23. Use Formula (15) to derive the standard matrices for the rotations about the x-axis, y-axis, and z-axis in R^3.

24. Use Formula (15) to find the standard matrix for a rotation of $\pi/2$ radians about the axis determined by the vector $\mathbf{v} = (1, 1, 1)$. [*Note:* Formula (15) requires that the vector defining the axis of rotation have length 1.]

25. Use Formula (15) to find the standard matrix for a rotation of $180°$ about the axis determined by the vector $\mathbf{v} = (2, 2, 1)$. [*Note:* Formula (15) requires that the vector defining the axis of rotation have length 1.]

26. It can be proved that if A is a 2×2 matrix with orthonormal column vectors and for which $\det(A) = 1$, then multiplication by A is a rotation through some angle θ. Verify that

$$A = \begin{bmatrix} -\frac{1}{\sqrt{2}} & -\frac{1}{\sqrt{2}} \\ \frac{1}{\sqrt{2}} & -\frac{1}{\sqrt{2}} \end{bmatrix}$$

satisfies the stated conditions and find the angle of rotation.

27. The result stated in Exercise 26 can be extended to R^3; that is, it can be proved that if A is a 3×3 matrix with orthonormal column vectors and for which $\det(A) = 1$, then multiplication by A is a rotation about some axis through some angle θ. Use Formula (15) to show that the angle of rotation satisfies the equation

$$\cos \theta = \frac{\text{tr}(A) - 1}{2}$$

28. Let A be a 3×3 matrix (other than the identity matrix) satisfying the conditions stated in Exercise 27. It can be shown that if $\mathbf{x}$ is any nonzero vector in R^3, then the vector $\mathbf{u} = A\mathbf{x} + A^T\mathbf{x} + [1 - \text{tr}(A)]\mathbf{x}$ determines an axis of rotation when $\mathbf{u}$ is positioned with its initial point at the origin. [See "The Axis of Rotation: Analysis, Algebra, Geometry," by Dan Kalman, *Mathematics Magazine*, Vol. 62, No. 4, October 1989.]

(a) Show that multiplication by

$$A = \begin{bmatrix} \frac{1}{9} & -\frac{4}{9} & \frac{8}{9} \\ \frac{8}{9} & \frac{4}{9} & \frac{1}{9} \\ -\frac{4}{9} & \frac{7}{9} & \frac{4}{9} \end{bmatrix}$$

is a rotation.

(b) Find a vector of length 1 that defines an axis for the rotation.

(c) Use the result in Exercise 27 to find the angle of rotation about the axis obtained in part (b).

29. In words, describe the geometric effect of multiplying a vector $\mathbf{x}$ by the matrix A.

(a) $A = \begin{bmatrix} 2 & 0 \\ 0 & 0 \end{bmatrix}$ (b) $A = \begin{bmatrix} 2 & 0 \\ 0 & -2 \end{bmatrix}$

30. In words, describe the geometric effect of multiplying a vector $\mathbf{x}$ by the matrix A.

(a) $A = \begin{bmatrix} 2 & 0 \\ 0 & 3 \end{bmatrix}$ (b) $A = \begin{bmatrix} \frac{\sqrt{3}}{2} & -\frac{1}{2} \\ \frac{1}{2} & \frac{\sqrt{3}}{2} \end{bmatrix}$

31. In words, describe the geometric effect of multiplying a vector $\mathbf{x}$ by the matrix

$$A = \begin{bmatrix} \cos^2 \theta - \sin^2 \theta & -2\sin\theta\cos\theta \\ 2\sin\theta\cos\theta & \cos^2\theta - \sin^2\theta \end{bmatrix}$$

32. If multiplication by A rotates a vector $\mathbf{x}$ in the xy-plane through an angle θ, what is the effect of multiplying $\mathbf{x}$ by A^T? Explain your reasoning.

33. Let $\mathbf{x}_0$ be a nonzero column vector in R^2, and suppose that $T: R^2 \to R^2$ is the transformation defined by the formula $T(\mathbf{x}) = \mathbf{x}_0 + R_\theta \mathbf{x}$, where R_θ is the standard matrix of the rotation of R^2 about the origin through the angle θ. Give a geometric description of this transformation. Is it a matrix transformation? Explain.

34. A function of the form $f(x) = mx + b$ is commonly called a "linear function" because the graph of $y = mx + b$ is a line. Is f a matrix transformation on R?

35. Let $\mathbf{x} = \mathbf{x}_0 + t\mathbf{v}$ be a line in R^n, and let $T: R^n \to R^n$ be a matrix operator on R^n. What kind of geometric object is the image of this line under the operator T? Explain your reasoning.

True-False Exercises

In parts (a)–(i) determine whether the statement is true or false, and justify your answer.

(a) If A is a 2×3 matrix, then the domain of the transformation T_A is R^2.

(b) If A is an $m \times n$ matrix, then the codomain of the transformation T_A is R^n.

(c) If $T: R^n \to R^m$ and $T(\mathbf{0}) = \mathbf{0}$, then T is a matrix transformation.

(d) If $T: R^n \to R^m$ and $T(c_1\mathbf{x} + c_2\mathbf{y}) = c_1 T(\mathbf{x}) + c_2 T(\mathbf{y})$ for all scalars c_1 and c_2 and all vectors $\mathbf{x}$ and $\mathbf{y}$ in R^n, then T is a matrix transformation.

(e) There is only one matrix transformation $T: R^n \to R^m$ such that $T(-\mathbf{x}) = -T(\mathbf{x})$ for every vector $\mathbf{x}$ in R^n.

(f) There is only one matrix transformation $T: R^n \to R^m$ such that $T(\mathbf{x} + \mathbf{y}) = T(\mathbf{x} - \mathbf{y})$ for all vectors $\mathbf{x}$ and $\mathbf{y}$ in R^n.

(g) If $\mathbf{b}$ is a nonzero vector in R^n, then $T(\mathbf{x}) = \mathbf{x} + \mathbf{b}$ is a matrix operator on R^n.

(h) The matrix $\begin{bmatrix} \frac{1}{2} & -\frac{1}{2} \\ \frac{1}{2} & \frac{1}{2} \end{bmatrix}$ is the standard matrix for a rotation.

(i) The standard matrices of the reflections about the coordinate axes in 2-space have the form $\begin{bmatrix} a & 0 \\ 0 & -a \end{bmatrix}$, where $a = \pm 1$.

4.10 Properties of Matrix Transformations

In this section we will discuss properties of matrix transformations. We will show, for example, that if several matrix transformations are performed in succession, then the same result can be obtained by a single matrix transformation that is chosen appropriately. We will also explore the relationship between the invertibility of a matrix and properties of the corresponding transformation.

Compositions of Matrix Transformations

Suppose that T_A is a matrix transformation from R^n to R^k and T_B is a matrix transformation from R^k to R^m. If $\mathbf{x}$ is a vector in R^n, then T_A maps this vector into a vector $T_A(\mathbf{x})$ in R^k, and T_B, in turn, maps that vector into the vector $T_B(T_A(\mathbf{x}))$ in R^m. This process creates a transformation from R^n to R^m that we call the **composition of T_B with T_A** and denote by the symbol

$$T_B \circ T_A$$

which is read "T_B circle T_A". As illustrated in Figure 4.10.1, the transformation T_A in the formula is performed first; that is,

$$(T_B \circ T_A)(\mathbf{x}) = T_B(T_A(\mathbf{x})) \tag{1}$$

This composition is itself a matrix transformation since

$$(T_B \circ T_A)(\mathbf{x}) = T_B(T_A(\mathbf{x})) = B(T_A(\mathbf{x})) = B(A\mathbf{x}) = (BA)\mathbf{x}$$

which shows that it is multiplication by BA. This is expressed by the formula

$$T_B \circ T_A = T_{BA} \tag{2}$$

> **WARNING** Just as it is *not* true, in general, that
> $$AB = BA$$
> so it is *not* true, in general, that
> $$T_B \circ T_A = T_A \circ T_B$$
> That is, *order matters when matrix transformations are composed.*

Compositions can be defined for any finite succession of matrix transformations whose domains and ranges have the appropriate dimensions. For example, to extend Formula (2) to three factors, consider the matrix transformations

$$T_A: R^n \rightarrow R^k, \quad T_B: R^k \rightarrow R^l, \quad T_C: R^l \rightarrow R^m$$

We define the composition $(T_C \circ T_B \circ T_A): R^n \rightarrow R^m$ by

$$(T_C \circ T_B \circ T_A)(\mathbf{x}) = T_C(T_B(T_A(\mathbf{x})))$$

As above, it can be shown that this is a matrix transformation whose standard matrix is CBA and that

$$T_C \circ T_B \circ T_A = T_{CBA} \tag{3}$$

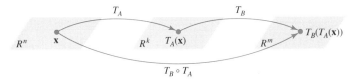

▷ **Figure 4.10.1**

As in Formula (9) of Section 4.9, we can use square brackets to denote a matrix transformation without referencing a specific matrix. Thus, for example, the formula

$$[T_2 \circ T_1] = [T_2][T_1] \tag{4}$$

is a restatement of Formula (2) which states that the standard matrix for a composition is the product of the standard matrices in the appropriate order. Similarly,

$$[T_3 \circ T_2 \circ T_1] = [T_3][T_2][T_1] \tag{5}$$

is a restatement of Formula (3).

▶ **EXAMPLE 1 Composition of Two Rotations**

Let $T_1: R^2 \to R^2$ and $T_2: R^2 \to R^2$ be the matrix operators that rotate vectors through the angles θ_1 and θ_2, respectively. Thus the operation

$$(T_2 \circ T_1)(\mathbf{x}) = T_2(T_1(\mathbf{x}))$$

first rotates $\mathbf{x}$ through the angle θ_1, then rotates $T_1(\mathbf{x})$ through the angle θ_2. It follows that the net effect of $T_2 \circ T_1$ is to rotate each vector in R^2 through the angle $\theta_1 + \theta_2$ (Figure 4.10.2). Thus, the standard matrices for these matrix operators are

$$[T_1] = \begin{bmatrix} \cos\theta_1 & -\sin\theta_1 \\ \sin\theta_1 & \cos\theta_1 \end{bmatrix}, \quad [T_2] = \begin{bmatrix} \cos\theta_2 & -\sin\theta_2 \\ \sin\theta_2 & \cos\theta_2 \end{bmatrix},$$

$$[T_2 \circ T_1] = \begin{bmatrix} \cos(\theta_1 + \theta_2) & -\sin(\theta_1 + \theta_2) \\ \sin(\theta_1 + \theta_2) & \cos(\theta_1 + \theta_2) \end{bmatrix}$$

These matrices should satisfy (4). With the help of some basic trigonometric identities, we can confirm that this is so as follows:

$$[T_2][T_1] = \begin{bmatrix} \cos\theta_2 & -\sin\theta_2 \\ \sin\theta_2 & \cos\theta_2 \end{bmatrix} \begin{bmatrix} \cos\theta_1 & -\sin\theta_1 \\ \sin\theta_1 & \cos\theta_1 \end{bmatrix}$$

$$= \begin{bmatrix} \cos\theta_2 \cos\theta_1 - \sin\theta_2 \sin\theta_1 & -(\cos\theta_2 \sin\theta_1 + \sin\theta_2 \cos\theta_1) \\ \sin\theta_2 \cos\theta_1 + \cos\theta_2 \sin\theta_1 & -\sin\theta_2 \sin\theta_1 + \cos\theta_2 \cos\theta_1 \end{bmatrix}$$

$$= \begin{bmatrix} \cos(\theta_1 + \theta_2) & -\sin(\theta_1 + \theta_2) \\ \sin(\theta_1 + \theta_2) & \cos(\theta_1 + \theta_2) \end{bmatrix}$$

$$= [T_2 \circ T_1]$$

▶ **EXAMPLE 2 Composition Is Not Commutative**

Let $T_1: R^2 \to R^2$ be the reflection about the line $y = x$, and let $T_2: R^2 \to R^2$ be the orthogonal projection on the y-axis. Figure 4.10.3 illustrates graphically that $T_1 \circ T_2$ and $T_2 \circ T_1$ have different effects on a vector $\mathbf{x}$. This same conclusion can be reached by showing that the standard matrices for T_1 and T_2 do not commute:

$$[T_1 \circ T_2] = [T_1][T_2] = \begin{bmatrix} 0 & 1 \\ 1 & 0 \end{bmatrix} \begin{bmatrix} 0 & 0 \\ 0 & 1 \end{bmatrix} = \begin{bmatrix} 0 & 1 \\ 0 & 0 \end{bmatrix}$$

$$[T_2 \circ T_1] = [T_2][T_1] = \begin{bmatrix} 0 & 0 \\ 0 & 1 \end{bmatrix} \begin{bmatrix} 0 & 1 \\ 1 & 0 \end{bmatrix} = \begin{bmatrix} 0 & 0 \\ 1 & 0 \end{bmatrix}$$

so $[T_2 \circ T_1] \neq [T_1 \circ T_2]$.

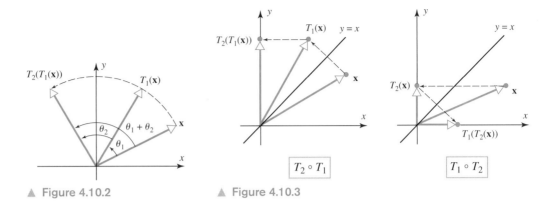

▲ Figure 4.10.2   ▲ Figure 4.10.3

▶ **EXAMPLE 3 Composition of Two Reflections**

Let $T_1: R^2 \to R^2$ be the reflection about the y-axis, and let $T_2: R^2 \to R^2$ be the reflection about the x-axis. In this case $T_1 \circ T_2$ and $T_2 \circ T_1$ are the same; both map every vector $\mathbf{x} = (x, y)$ into its negative $-\mathbf{x} = (-x, -y)$ (Figure 4.10.4):

$$(T_1 \circ T_2)(x, y) = T_1(x, -y) = (-x, -y)$$
$$(T_2 \circ T_1)(x, y) = T_2(-x, y) = (-x, -y)$$

The equality of $T_1 \circ T_2$ and $T_2 \circ T_1$ can also be deduced by showing that the standard matrices for T_1 and T_2 commute:

$$[T_1 \circ T_2] = [T_1][T_2] = \begin{bmatrix} -1 & 0 \\ 0 & 1 \end{bmatrix} \begin{bmatrix} 1 & 0 \\ 0 & -1 \end{bmatrix} = \begin{bmatrix} -1 & 0 \\ 0 & -1 \end{bmatrix}$$

$$[T_2 \circ T_1] = [T_2][T_1] = \begin{bmatrix} 1 & 0 \\ 0 & -1 \end{bmatrix} \begin{bmatrix} -1 & 0 \\ 0 & 1 \end{bmatrix} = \begin{bmatrix} -1 & 0 \\ 0 & -1 \end{bmatrix}$$

The operator $T(\mathbf{x}) = -\mathbf{x}$ on R^2 or R^3 is called the ***reflection about the origin***. As the foregoing computations show, the standard matrix for this operator on R^2 is

$$[T] = \begin{bmatrix} -1 & 0 \\ 0 & -1 \end{bmatrix}$$

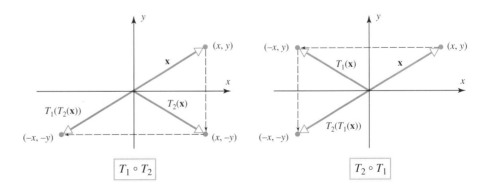

▶ Figure 4.10.4

▶ **EXAMPLE 4 Composition of Three Transformations**

Find the standard matrix for the operator $T: R^3 \to R^3$ that first rotates a vector counterclockwise about the z-axis through an angle θ, then reflects the resulting vector about the yz-plane, and then projects that vector orthogonally onto the xy-plane.

Solution The operator T can be expressed as the composition

$$T = T_3 \circ T_2 \circ T_1$$

where T_1 is the rotation about the z-axis, T_2 is the reflection about the yz-plane, and T_3 is the orthogonal projection on the xy-plane. From Tables 6, 2, and 4 of Section 4.9, the standard matrices for these operators are

$$[T_1] = \begin{bmatrix} \cos\theta & -\sin\theta & 0 \\ \sin\theta & \cos\theta & 0 \\ 0 & 0 & 1 \end{bmatrix}, \quad [T_2] = \begin{bmatrix} -1 & 0 & 0 \\ 0 & 1 & 0 \\ 0 & 0 & 1 \end{bmatrix}, \quad [T_3] = \begin{bmatrix} 1 & 0 & 0 \\ 0 & 1 & 0 \\ 0 & 0 & 0 \end{bmatrix}$$

Thus, it follows from (5) that the standard matrix for T is

$$[T] = \begin{bmatrix} 1 & 0 & 0 \\ 0 & 1 & 0 \\ 0 & 0 & 0 \end{bmatrix} \begin{bmatrix} -1 & 0 & 0 \\ 0 & 1 & 0 \\ 0 & 0 & 1 \end{bmatrix} \begin{bmatrix} \cos\theta & -\sin\theta & 0 \\ \sin\theta & \cos\theta & 0 \\ 0 & 0 & 1 \end{bmatrix}$$

$$= \begin{bmatrix} -\cos\theta & \sin\theta & 0 \\ \sin\theta & \cos\theta & 0 \\ 0 & 0 & 0 \end{bmatrix} \blacktriangleleft$$

One-to-One Matrix Transformations
Our next objective is to establish a link between the invertibility of a matrix A and properties of the corresponding matrix transformation T_A.

DEFINITION 1 A matrix transformation $T_A : R^n \to R^m$ is said to be ***one-to-one*** if T_A maps distinct vectors (points) in R^n into distinct vectors (points) in R^m.

(See Figure 4.10.5). This idea can be expressed in various ways. For example, you should be able to see that the following are just restatements of Definition 1:

1. T_A is one-to-one if for each vector **b** in the range of A there is exactly one vector **x** in R^n such that $T_A \mathbf{x} = \mathbf{b}$.

2. T_A is one-to-one if the equality $T_A(\mathbf{u}) = T_A(\mathbf{v})$ implies that $\mathbf{u} = \mathbf{v}$.

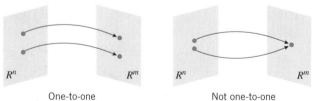

R^n R^m R^n R^m

▶ **Figure 4.10.5**   One-to-one   Not one-to-one

Rotation operators on R^2 are one-to-one since distinct vectors that are rotated through the same angle have distinct images (Figure 4.10.6). In contrast, the orthogonal projection of R^3 on the xy-plane is not one-to-one because it maps distinct points on the same vertical line into the same point (Figure 4.10.7).

The following theorem establishes a fundamental relationship between the invertibility of a matrix and properties of the corresponding matrix transformation.

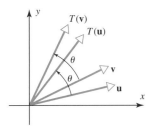

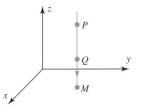

▲ **Figure 4.10.6** Distinct vectors **u** and **v** are rotated into distinct vectors $T(\mathbf{u})$ and $T(\mathbf{v})$.

▲ **Figure 4.10.7** The distinct points P and Q are mapped into the same point M.

THEOREM 4.10.1 *If A is an $n \times n$ matrix and $T_A \colon R^n \to R^n$ is the corresponding matrix operator, then the following statements are equivalent.*

(a) *A is invertible.*

(b) *The range of T_A is R^n.*

(c) *T_A is one-to-one.*

Proof We will establish the chain of implications $(a) \Rightarrow (b) \Rightarrow (c) \Rightarrow (a)$.

(a) ⇒ (b) Assume that A is invertible. By parts (a) and (e) of Theorem 4.8.10, the system $A\mathbf{x} = \mathbf{b}$ is consistent for every $n \times 1$ matrix $\mathbf{b}$ in R^n. This implies that T_A maps $\mathbf{x}$ into the arbitrary vector $\mathbf{b}$ in R^n, which in turn implies that the range of T_A is all of R^n.

(b) ⇒ (c) Assume that the range of T_A is R^n. This implies that for every vector $\mathbf{b}$ in R^n there is some vector $\mathbf{x}$ in R^n for which $T_A(\mathbf{x}) = \mathbf{b}$ and hence that the linear system $A\mathbf{x} = \mathbf{b}$ is consistent for every vector $\mathbf{b}$ in R^n. But the equivalence of parts (e) and (f) of Theorem 4.8.10 implies that $A\mathbf{x} = \mathbf{b}$ has a unique solution for every vector $\mathbf{b}$ in R^n and hence that for every vector $\mathbf{b}$ in the range of T_A there is exactly one vector $\mathbf{x}$ in R^n such that $T_A\mathbf{x} = \mathbf{b}$.

(c) ⇒ (a) Assume that T_A is one-to-one. Thus, if $\mathbf{b}$ is a vector in the range of T_A, there is a unique vector $\mathbf{x}$ in R^n for which $T_A(\mathbf{x}) = \mathbf{b}$. We leave it for you to complete the proof using Exercise 30. ◀

▶ **EXAMPLE 5 Properties of a Rotation Operator**

As indicated in Figure 4.10.6, the operator $T \colon R^n \to R^n$ that rotates vectors in R^2 through an angle θ is one-to-one. Confirm that $[T]$ is invertible in accordance with Theorem 4.10.1.

Solution From Table 5 of Section 4.9 the standard matrix for T is

$$[T] = \begin{bmatrix} \cos\theta & -\sin\theta \\ \sin\theta & \cos\theta \end{bmatrix}$$

This matrix is invertible because

$$\det[T] = \begin{vmatrix} \cos\theta & -\sin\theta \\ \sin\theta & \cos\theta \end{vmatrix} = \cos^2\theta + \sin^2\theta = 1 \neq 0$$

▶ **EXAMPLE 6 Properties of a Projection Operator**

As indicated in Figure 4.10.7, the operator $T \colon R^n \to R^n$ that projects each vector in R^3 orthogonally on the xy-plane is not one-to-one. Confirm that $[T]$ is not invertible in accordance with Theorem 4.10.1.

Solution From Table 4 of Section 4.9 the standard matrix for T is

$$[T] = \begin{bmatrix} 1 & 0 & 0 \\ 0 & 1 & 0 \\ 0 & 0 & 0 \end{bmatrix}$$

This matrix is not invertible since $\det[T] = 0$. ◀

Inverse of a One-to-One Matrix Operator

If $T_A: R^n \to R^n$ is a one-to-one matrix operator, then it follows from Theorem 4.10.1 that A is invertible. The matrix operator

$$T_{A^{-1}}: R^n \to R^n$$

that corresponds to A^{-1} is called the **inverse operator** or (more simply) the **inverse** of T_A. This terminology is appropriate because T_A and $T_{A^{-1}}$ cancel the effect of each other in the sense that if $\mathbf{x}$ is any vector in R^n, then

$$T_A(T_{A^{-1}}(\mathbf{x})) = AA^{-1}\mathbf{x} = I\mathbf{x} = \mathbf{x}$$
$$T_{A^{-1}}(T_A(\mathbf{x})) = A^{-1}A\mathbf{x} = I\mathbf{x} = \mathbf{x}$$

or, equivalently,

$$T_A \circ T_{A^{-1}} = T_{AA^{-1}} = T_I$$
$$T_{A^{-1}} \circ T_A = T_{A^{-1}A} = T_I$$

From a more geometric viewpoint, if $\mathbf{w}$ is the image of $\mathbf{x}$ under T_A, then $T_{A^{-1}}$ maps $\mathbf{w}$ back into $\mathbf{x}$, since

$$T_{A^{-1}}(\mathbf{w}) = T_{A^{-1}}(T_A(\mathbf{x})) = \mathbf{x}$$

(Figure 4.10.8).

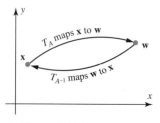

▲ Figure 4.10.8

Before considering examples, it will be helpful to touch on some notational matters. If $T_A: R^n \to R^n$ is a one-to-one matrix operator, and if $T_{A^{-1}}: R^n \to R^n$ is its inverse, then the standard matrices for these operators are related by the equation

$$T_{A^{-1}} = T_A^{-1} \tag{6}$$

In cases where it is preferable not to assign a name to the matrix, we will write this equation as

$$[T^{-1}] = [T]^{-1} \tag{7}$$

▶ **EXAMPLE 7 Standard Matrix for T^{-1}**

Let $T: R^2 \to R^2$ be the operator that rotates each vector in R^2 through the angle θ, so from Table 5 of Section 4.9,

$$[T] = \begin{bmatrix} \cos\theta & -\sin\theta \\ \sin\theta & \cos\theta \end{bmatrix} \tag{8}$$

It is evident geometrically that to undo the effect of T, one must rotate each vector in R^2 through the angle $-\theta$. But this is exactly what the operator T^{-1} does, since the standard matrix for T^{-1} is

$$[T^{-1}] = [T]^{-1} = \begin{bmatrix} \cos\theta & \sin\theta \\ -\sin\theta & \cos\theta \end{bmatrix} = \begin{bmatrix} \cos(-\theta) & -\sin(-\theta) \\ \sin(-\theta) & \cos(-\theta) \end{bmatrix}$$

(verify), which is the standard matrix for a rotation through the angle $-\theta$.

▶ **EXAMPLE 8 Finding T^{-1}**

Show that the operator $T: R^2 \to R^2$ defined by the equations

$$w_1 = 2x_1 + \ x_2$$
$$w_2 = 3x_1 + 4x_2$$

is one-to-one, and find $T^{-1}(w_1, w_2)$.

Solution The matrix form of these equations is

$$\begin{bmatrix} w_1 \\ w_2 \end{bmatrix} = \begin{bmatrix} 2 & 1 \\ 3 & 4 \end{bmatrix} \begin{bmatrix} x_1 \\ x_2 \end{bmatrix}$$

so the standard matrix for T is

$$[T] = \begin{bmatrix} 2 & 1 \\ 3 & 4 \end{bmatrix}$$

This matrix is invertible (so T is one-to-one) and the standard matrix for T^{-1} is

$$[T^{-1}] = [T]^{-1} = \begin{bmatrix} \frac{4}{5} & -\frac{1}{5} \\ -\frac{3}{5} & \frac{2}{5} \end{bmatrix}$$

Thus

$$[T^{-1}] \begin{bmatrix} w_1 \\ w_2 \end{bmatrix} = \begin{bmatrix} \frac{4}{5} & -\frac{1}{5} \\ -\frac{3}{5} & \frac{2}{5} \end{bmatrix} \begin{bmatrix} w_1 \\ w_2 \end{bmatrix} = \begin{bmatrix} \frac{4}{5}w_1 - \frac{1}{5}w_2 \\ -\frac{3}{5}w_1 + \frac{2}{5}w_2 \end{bmatrix}$$

from which we conclude that

$$T^{-1}(w_1, w_2) = \left(\tfrac{4}{5}w_1 - \tfrac{1}{5}w_2, -\tfrac{3}{5}w_1 + \tfrac{2}{5}w_2 \right) \blacktriangleleft$$

Linearity Properties Up to now we have focused exclusively on matrix transformations from R^n to R^m. However, these are not the only kinds of transformations from R^n to R^m. For example, if $f_1, f_2, \ldots, f_m$ are any functions of the n variables $x_1, x_2, \ldots, x_n$, then the equations

$$w_1 = f_1(x_1, x_2, \ldots, x_n)$$
$$w_2 = f_2(x_1, x_2, \ldots, x_n)$$
$$\vdots$$
$$w_m = f_m(x_1, x_2, \ldots, x_n)$$

define a transformation $T: R^n \to R^m$ that maps the vector $\mathbf{x} = (x_1, x_2, \ldots, x_n)$ into the vector $(w_1, w_2, \ldots, w_m)$. But it is only in the case where these equations are *linear* that T is a matrix transformation. The question that we will now consider is this:

> **Question** Are there algebraic properties of a transformation $T: R^n \to R^m$ that can be used to determine whether T is a matrix transformation?

The answer is provided by the following theorem.

THEOREM 4.10.2 *$T: R^n \to R^m$ is a matrix transformation if and only if the following relationships hold for all vectors $\mathbf{u}$ and $\mathbf{v}$ in R^n and for every scalar k:*

(i) $T(\mathbf{u} + \mathbf{v}) = T(\mathbf{u}) + T(\mathbf{v})$ [Additivity property]

(ii) $T(k\mathbf{u}) = kT(\mathbf{u})$ [Homogeneity property]

Proof If T is a matrix transformation, then properties (i) and (ii) follow respectively from parts (*c*) and (*b*) of Theorem 4.9.1.

Conversely, assume that properties (i) and (ii) hold. We must show that there exists an $m \times n$ matrix A such that

$$T(\mathbf{x}) = A\mathbf{x}$$

for every vector $\mathbf{x}$ in R^n. As a first step, recall from Formula (10) of Section 4.9 that the additivity and homogeneity properties imply that

$$T(k_1\mathbf{u}_1 + k_2\mathbf{u}_2 + \cdots + k_r\mathbf{u}_r) = k_1T(\mathbf{u}_1) + k_2T(\mathbf{u}_2) + \cdots + k_rT(\mathbf{u}_r) \qquad (9)$$

for all scalars $k_1, k_2, \ldots, k_r$ and all vectors $\mathbf{u}_1, \mathbf{u}_2, \ldots, \mathbf{u}_r$ in R^n. Let A be the matrix

$$A = [T(\mathbf{e}_1) \mid T(\mathbf{e}_2) \mid \cdots \mid T(\mathbf{e}_n)]$$

in which $\mathbf{e}_1, \mathbf{e}_2, \ldots, \mathbf{e}_n$ are the standard basis vectors for R^n. It follows from Theorem 1.3.1 that $A\mathbf{x}$ is a linear combination of the columns of A in which the successive coefficients are the entries $x_1, x_2, \ldots, x_n$ of $\mathbf{x}$. That is,

$$A\mathbf{x} = x_1 T(\mathbf{e}_1) + x_2 T(\mathbf{e}_2) + \cdots + x_n T(\mathbf{e}_n)$$

Using (9) we can rewrite this as

$$A\mathbf{x} = T(x_1\mathbf{e}_1 + x_2\mathbf{e}_2 + \cdots + x_n\mathbf{e}_n) = T(\mathbf{x})$$

which completes the proof. ◀

The additivity and homogeneity properties in Theorem 4.10.2 are called *linearity conditions*, and a transformation that satisfies these conditions is called a *linear transformation*. Using this terminology Theorem 4.10.2 can be restated as follows.

> **THEOREM 4.10.3** *Every linear transformation from R^n to R^m is a matrix transformation, and conversely, every matrix transformation from R^n to R^m is a linear transformation.*

More on the Equivalence Theorem

As our final result in this section, we will add parts (b) and (c) of Theorem 4.10.1 to Theorem 4.8.10.

> **THEOREM 4.10.4 Equivalent Statements**
>
> *If A is an $n \times n$ matrix, then the following statements are equivalent.*
>
> (a) *A is invertible.*
> (b) *$A\mathbf{x} = \mathbf{0}$ has only the trivial solution.*
> (c) *The reduced row echelon form of A is I_n.*
> (d) *A is expressible as a product of elementary matrices.*
> (e) *$A\mathbf{x} = \mathbf{b}$ is consistent for every $n \times 1$ matrix $\mathbf{b}$.*
> (f) *$A\mathbf{x} = \mathbf{b}$ has exactly one solution for every $n \times 1$ matrix $\mathbf{b}$.*
> (g) *$\det(A) \neq 0$.*
> (h) *The column vectors of A are linearly independent.*
> (i) *The row vectors of A are linearly independent.*
> (j) *The column vectors of A span R^n.*
> (k) *The row vectors of A span R^n.*
> (l) *The column vectors of A form a basis for R^n.*
> (m) *The row vectors of A form a basis for R^n.*
> (n) *A has rank n.*
> (o) *A has nullity 0.*
> (p) *The orthogonal complement of the null space of A is R^n.*
> (q) *The orthogonal complement of the row space of A is $\{\mathbf{0}\}$.*
> (r) *The range of T_A is R^n.*
> (s) *T_A is one-to-one.*

Concept Review

- Composition of matrix transformations
- Reflection about the origin
- One-to-one transformation
- Inverse of a matrix operator
- Linearity conditions
- Linear transformation
- Equivalent characterizations of invertible matrices

Skills

- Find the standard matrix for a composition of matrix transformations.
- Determine whether a matrix operator is one-to-one; if it is, then find the inverse operator.
- Determine whether a transformation is a linear transformation.

Exercise Set 4.10

In Exercises 1–2, let T_A and T_B be the operators whose standard matrices are given. Find the standard matrices for $T_B \circ T_A$ and $T_A \circ T_B$.

1. $A = \begin{bmatrix} 1 & -2 & 0 \\ 4 & 1 & -3 \\ 5 & 2 & 4 \end{bmatrix}$, $B = \begin{bmatrix} 2 & -3 & 3 \\ 5 & 0 & 1 \\ 6 & 1 & 7 \end{bmatrix}$

2. $A = \begin{bmatrix} 6 & 3 & -1 \\ 2 & 0 & 1 \\ 4 & -3 & 6 \end{bmatrix}$, $B = \begin{bmatrix} 4 & 0 & 4 \\ -1 & 5 & 2 \\ 2 & -3 & 8 \end{bmatrix}$

3. Let $T_1(x_1, x_2) = (x_1 + x_2, x_1 - x_2)$ and $T_2(x_1, x_2) = (3x_1, 2x_1 + 4x_2)$.
   (a) Find the standard matrices for T_1 and T_2.
   (b) Find the standard matrices for $T_2 \circ T_1$ and $T_1 \circ T_2$.
   (c) Use the matrices obtained in part (b) to find formulas for $T_1(T_2(x_1, x_2))$ and $T_2(T_1(x_1, x_2))$.

4. Let $T_1(x_1, x_2, x_3) = (4x_1, -2x_1 + x_2, -x_1 - 3x_2)$ and $T_2(x_1, x_2, x_3) = (x_1 + 2x_2, -x_3, 4x_1 - x_3)$.
   (a) Find the standard matrices for T_1 and T_2.
   (b) Find the standard matrices for $T_2 \circ T_1$ and $T_1 \circ T_2$.
   (c) Use the matrices obtained in part (b) to find formulas for $T_1(T_2(x_1, x_2, x_3))$ and $T_2(T_1(x_1, x_2, x_3))$.

5. Find the standard matrix for the stated composition in R^2.
   (a) A rotation of $90°$, followed by a reflection about the line $y = x$.
   (b) An orthogonal projection on the y-axis, followed by a contraction with factor $k = \frac{1}{2}$.
   (c) A reflection about the x-axis, followed by a dilation with factor $k = 3$.

6. Find the standard matrix for the stated composition in R^2.
   (a) A rotation of $60°$, followed by an orthogonal projection on the x-axis, followed by a reflection about the line $y = x$.
   (b) A dilation with factor $k = 2$, followed by a rotation of $45°$, followed by a reflection about the y-axis.

   (c) A rotation of $15°$, followed by a rotation of $105°$, followed by a rotation of $60°$.

7. Find the standard matrix for the stated composition in R^3.
   (a) A reflection about the yz-plane, followed by an orthogonal projection on the xz-plane.
   (b) A rotation of $45°$ about the y-axis, followed by a dilation with factor $k = \sqrt{2}$.
   (c) An orthogonal projection on the xy-plane, followed by a reflection about the yz-plane.

8. Find the standard matrix for the stated composition in R^3.
   (a) A rotation of $30°$ about the x-axis, followed by a rotation of $30°$ about the z-axis, followed by a contraction with factor $k = \frac{1}{4}$.
   (b) A reflection about the xy-plane, followed by a reflection about the xz-plane, followed by an orthogonal projection on the yz-plane.
   (c) A rotation of $270°$ about the x-axis, followed by a rotation of $90°$ about the y-axis, followed by a rotation of $180°$ about the z-axis.

9. Determine whether $T_1 \circ T_2 = T_2 \circ T_1$.
   (a) $T_1: R^2 \to R^2$ is the orthogonal projection on the x-axis, and $T_2: R^2 \to R^2$ is the orthogonal projection on the y-axis.
   (b) $T_1: R^2 \to R^2$ is the rotation through an angle θ_1, and $T_2: R^2 \to R^2$ is the rotation through an angle θ_2.
   (c) $T_1: R^2 \to R^2$ is the orthogonal projection on the x-axis, and $T_2: R^2 \to R^2$ is the rotation through an angle θ.

10. Determine whether $T_1 \circ T_2 = T_2 \circ T_1$.
    (a) $T_1: R^3 \to R^3$ is a dilation by a factor k, and $T_2: R^3 \to R^3$ is the rotation about the z-axis through an angle θ.
    (b) $T_1: R^3 \to R^3$ is the rotation about the x-axis t' angle θ_1, and $T_2: R^3 \to R^3$ is the rotation about through an angle θ_2.

11. By inspection, determine whether the matrix operator is one-to-one.

(a) the orthogonal projection on the x-axis in R^2

(b) the reflection about the y-axis in R^2

(c) the reflection about the line $y = x$ in R^2

(d) a contraction with factor $k > 0$ in R^2

(e) a rotation about the z-axis in R^3

(f) a reflection about the xy-plane in R^3

(g) a dilation with factor $k > 0$ in R^3

12. Find the standard matrix for the matrix operator defined by the equations, and use Theorem 4.10.4 to determine whether the operator is one-to-one.

(a) $\begin{aligned} w_1 &= 8x_1 + 4x_2 \\ w_2 &= 2x_1 + x_2 \end{aligned}$

(b) $\begin{aligned} w_1 &= 2x_1 - 3x_2 \\ w_2 &= 5x_1 + x_2 \end{aligned}$

(c) $\begin{aligned} w_1 &= -x_1 + 3x_2 + 2x_3 \\ w_2 &= 2x_1 \quad\quad + 4x_3 \\ w_3 &= x_1 + 3x_2 + 6x_3 \end{aligned}$

(d) $\begin{aligned} w_1 &= x_1 + 2x_2 + 3x_3 \\ w_2 &= 2x_1 + 5x_2 + 3x_3 \\ w_3 &= x_1 \quad\quad + 8x_3 \end{aligned}$

13. Determine whether the matrix operator $T: R^2 \to R^2$ defined by the equations is one-to-one; if so, find the standard matrix for the inverse operator, and find $T^{-1}(w_1, w_2)$.

(a) $\begin{aligned} w_1 &= x_1 + 2x_2 \\ w_2 &= -x_1 + x_2 \end{aligned}$

(b) $\begin{aligned} w_1 &= 4x_1 - 6x_2 \\ w_2 &= -2x_1 + 3x_2 \end{aligned}$

(c) $\begin{aligned} w_1 &= -x_2 \\ w_2 &= -x_1 \end{aligned}$

(d) $\begin{aligned} w_1 &= 3x_1 \\ w_2 &= -5x_1 \end{aligned}$

14. Determine whether the matrix operator $T: R^3 \to R^3$ defined by the equations is one-to-one; if so, find the standard matrix for the inverse operator, and find $T^{-1}(w_1, w_2, w_3)$.

(a) $\begin{aligned} w_1 &= x_1 - 2x_2 + 2x_3 \\ w_2 &= 2x_1 + x_2 + x_3 \\ w_3 &= x_1 + x_2 \end{aligned}$

(b) $\begin{aligned} w_1 &= x_1 - 3x_2 + 4x_3 \\ w_2 &= -x_1 + x_2 + x_3 \\ w_3 &= \quad\quad - 2x_2 + 5x_3 \end{aligned}$

(c) $\begin{aligned} w_1 &= x_1 + 4x_2 - x_3 \\ w_2 &= 2x_1 + 7x_2 + x_3 \\ w_3 &= x_1 + 3x_2 \end{aligned}$

(d) $\begin{aligned} w_1 &= x_1 + 2x_2 + x_3 \\ w_2 &= -2x_1 + x_2 + 4x_3 \\ w_3 &= 7x_1 + 4x_2 - 5x_3 \end{aligned}$

15. By inspection, find the inverse of the given one-to-one matrix operator.

(a) The reflection about the x-axis in R^2.

(b) The rotation through an angle of $\pi/4$ in R^2.

(c) The dilation by a factor of 3 in R^2.

(d) The reflection about the yz-plane in R^3.

(e) The contraction by a factor of $\frac{1}{5}$ in R^3.

▷ In Exercises 16–17, use Theorem 4.10.2 to determine whether $T: R^2 \to R^2$ is a matrix operator. ◁

16. (a) $T(x, y) = (2x, y)$

(b) $T(x, y) = (x^2, y)$

(c) $T(x, y) = (-y, x)$

(d) $T(x, y) = (x, 0)$

17. (a) $T(x, y) = (2x + y, x - y)$

(b) $T(x, y) = (x + 1, y)$

(c) $T(x, y) = (y, y)$

(d) $T(x, y) = (\sqrt[3]{x}, \sqrt[3]{y})$

▷ In Exercises 18–19, use Theorem 4.10.2 to determine whether $T: R^3 \to R^2$ is a matrix transformation. ◁

18. (a) $T(x, y, z) = (x, x + y + z)$

(b) $T(x, y, z) = (1, 1)$

19. (a) $T(x, y, z) = (0, 0)$

(b) $T(x, y, z) = (3x - 4y, 2x - 5z)$

20. In each part, use Theorem 4.10.3 to find the standard matrix for the matrix operator from the images of the standard basis vectors.

(a) The reflection operators on R^2 in Table 1 of Section 4.9.

(b) The reflection operators on R^3 in Table 2 of Section 4.9.

(c) The projection operators on R^2 in Table 3 of Section 4.9.

(d) The projection operators on R^3 in Table 4 of Section 4.9.

(e) The rotation operators on R^2 in Table 5 of Section 4.9.

(f) The dilation and contraction operators on R^3 in Table 8 of Section 4.9.

21. Find the standard matrix for the given matrix operator.

(a) $T: R^2 \to R^2$ projects a vector orthogonally onto the x-axis and then reflects that vector about the y-axis.

(b) $T: R^2 \to R^2$ reflects a vector about the line $y = x$ and then reflects that vector about the x-axis.

(c) $T: R^2 \to R^2$ dilates a vector by a factor of 3, then reflects that vector about the line $y = x$, and then projects that vector orthogonally onto the y-axis.

22. Find the standard matrix for the given matrix operator.

(a) $T: R^3 \to R^3$ reflects a vector about the xz-plane and then contracts that vector by a factor of $\frac{1}{5}$.

(b) $T: R^3 \to R^3$ projects a vector orthogonally onto the xz-plane and then projects that vector orthogonally onto the xy-plane.

(c) $T: R^3 \to R^3$ reflects a vector about the xy-plane, then reflects that vector about the xz-plane, and then reflects that vector about the yz-plane.

23. Let $T_A: R^3 \to R^3$ be multiplication by

$$A = \begin{bmatrix} -1 & 3 & 0 \\ 2 & 1 & 2 \\ 4 & 5 & -3 \end{bmatrix}$$

and let $e_1, e_2,$ and e_3 be the standard basis vectors for R^3. Find the following vectors by inspection.

(a) $T_A(e_1), T_A(e_2),$ and $T_A(e_3)$

(b) $T_A(e_1 + e_2 + e_3)$

(c) $T_A(7e_3)$

24. Determine whether multiplication by A is a one-to-one matrix transformation.

(a) $A = \begin{bmatrix} 1 & -1 \\ 2 & 0 \\ 3 & -4 \end{bmatrix}$ (b) $A = \begin{bmatrix} 1 & 2 & 3 \\ -1 & 0 & -4 \end{bmatrix}$

(c) $A = \begin{bmatrix} 1 & 2 & 1 \\ 0 & 1 & 1 \\ 1 & 1 & 0 \\ 1 & 0 & -1 \end{bmatrix}$

25. (a) Is a composition of one-to-one matrix transformations one-to-one? Justify your conclusion.

(b) Can the composition of a one-to-one matrix transformation and a matrix transformation that is not one-to-one be one-to-one? Account for both possible orders of composition and justify your conclusion.

26. Show that $T(x, y) = (0, 0)$ defines a matrix operator on R^2 but $T(x, y) = (1, 1)$ does not.

27. (a) Prove: If $T: R^n \rightarrow R^m$ is a matrix transformation, then $T(\mathbf{0}) = \mathbf{0}$; that is, T maps the zero vector in R^n into the zero vector in R^m.

(b) The converse of this is not true. Find an example of a function that satisfies $T(0) = 0$ but is not a matrix transformation.

28. Prove: An $n \times n$ matrix A is invertible if and only if the linear system $A\mathbf{x} = \mathbf{w}$ has exactly one solution for every vector $\mathbf{w}$ in R^n for which the system is consistent.

29. Let A be an $n \times n$ matrix such that $\det(A) = 0$, and let $T: R^n \rightarrow R^n$ be multiplication by A.

(a) What can you say about the range of the matrix operator T? Give an example that illustrates your conclusion.

(b) What can you say about the number of vectors that T maps into $\mathbf{0}$?

30. Prove: If the matrix transformation $T_A: R^n \rightarrow R^n$ is one-to-one, then A is invertible.

True-False Exercises

In parts (a)–(f) determine whether the statement is true or false, and justify your answer.

(a) If $T: R^n \rightarrow R^m$ and $T(\mathbf{0}) = \mathbf{0}$, then T is a matrix transformation.

(b) If $T: R^n \rightarrow R^m$ and $T(c_1\mathbf{x} + c_2\mathbf{y}) = c_1 T(\mathbf{x}) + c_2 T(\mathbf{y})$ for all scalars c_1 and c_2 and all vectors $\mathbf{x}$ and $\mathbf{y}$ in R^n, then T is a matrix transformation.

(c) If $T: R^n \rightarrow R^m$ is a one-to-one matrix transformation, then there are no distinct vectors $\mathbf{x}$ and $\mathbf{y}$ for which $T(\mathbf{x} - \mathbf{y}) = \mathbf{0}$.

(d) If $T: R^n \rightarrow R^m$ is a matrix transformation and $m > n$, then T is one-to-one.

(e) If $T: R^n \rightarrow R^m$ is a matrix transformation and $m = n$, then T is one-to-one.

(f) If $T: R^n \rightarrow R^m$ is a matrix transformation and $m < n$, then T is one-to-one.

4.11 Geometry of Matrix Operators on R^2

In this optional section we will discuss matrix operators on R^2 in a little more depth. The ideas that we will develop here have important applications to computer graphics.

Transformations of Regions

In Section 4.9 we focused on the effect that a matrix operator has on individual vectors in R^2 and R^3. However, it is also important to understand how such operators affect the shapes of regions. For example, Figure 4.11.1 shows a famous picture of Albert Einstein and three computer-generated modifications of that image that result from matrix operators on R^2. The original picture was scanned and then digitized to decompose it into a rectangular array of pixels. The pixels were then transformed as follows:

- The program MATLAB was used to assign coordinates and a gray level to each pixel.
- The coordinates of the pixels were transformed by matrix multiplication.
- The pixels were then assigned their original gray levels to produce the transformed picture.

The overall effect of a matrix operator on R^2 can often be ascertained by graphing the images of the vertices $(0, 0)$, $(1, 0)$, $(0, 1)$, and $(1, 1)$ of the unit square (Figure 4.11.2).

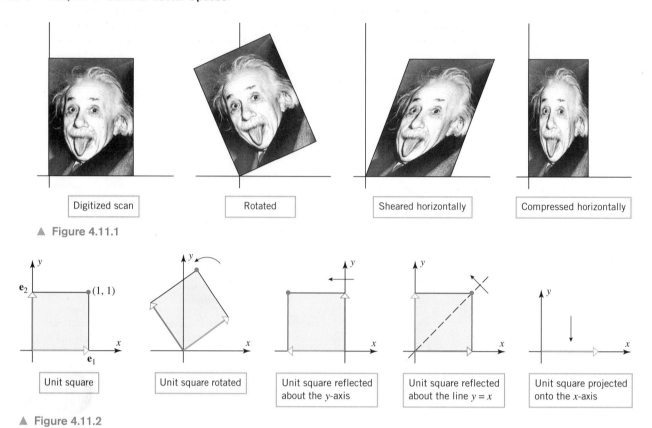

▲ Figure 4.11.1

▲ Figure 4.11.2

Table 1 shows the effect that some of the matrix operators studied in Section 4.9 have on the unit square. For clarity, we have shaded a portion of the original square and its corresponding image.

▶ **EXAMPLE 1 Transforming with Diagonal Matrices**

Suppose that the xy-plane first is compressed or expanded by a factor of k_1 in the x-direction and then is compressed or expanded by a factor of k_2 in the y-direction. Find a single matrix operator that performs both operations.

Solution The standard matrices for the two operations are

$$\begin{bmatrix} k_1 & 0 \\ 0 & 1 \end{bmatrix} \qquad \begin{bmatrix} 1 & 0 \\ 0 & k_2 \end{bmatrix}$$

x-**compression (expansion)** y-**compression (expansion)**

Thus, the standard matrix for the composition of the x-operation followed by the y-operation is

$$A = \begin{bmatrix} 1 & 0 \\ 0 & k_2 \end{bmatrix} \begin{bmatrix} k_1 & 0 \\ 0 & 1 \end{bmatrix} = \begin{bmatrix} k_1 & 0 \\ 0 & k_2 \end{bmatrix} \tag{1}$$

This shows that multiplication by a diagonal 2×2 matrix compresses or expands the plane in the x-direction and also in the y-direction. In the special case where k_1 and k_2 are the same, say $k_1 = k_2 = k$, Formula (1) simplifies to

$$A = \begin{bmatrix} k & 0 \\ 0 & k \end{bmatrix}$$

which is a contraction or a dilation (Table 7 of Section 4.9). ◀

Table 1

Operator	Standard Matrix	Effect on the Unit Square
Reflection about the y-axis	$\begin{bmatrix} -1 & 0 \\ 0 & 1 \end{bmatrix}$	
Reflection about the x-axis	$\begin{bmatrix} 1 & 0 \\ 0 & -1 \end{bmatrix}$	
Reflection about the line $y = x$	$\begin{bmatrix} 0 & 1 \\ 1 & 0 \end{bmatrix}$	
Counterclockwise rotation through an angle θ	$\begin{bmatrix} \cos\theta & -\sin\theta \\ \sin\theta & \cos\theta \end{bmatrix}$	
Compression in the x-direction by a factor of k $(0 < k < 1)$	$\begin{bmatrix} k & 0 \\ 0 & 1 \end{bmatrix}$	
Expansion in the x-direction by a factor of k $(k > 1)$	$\begin{bmatrix} k & 0 \\ 0 & 1 \end{bmatrix}$	
Shear in the x-direction with factor $k > 0$	$\begin{bmatrix} 1 & k \\ 0 & 1 \end{bmatrix}$	
Shear in the x-direction with factor $k < 0$	$\begin{bmatrix} 1 & k \\ 0 & 1 \end{bmatrix}$	

▶ **EXAMPLE 2 Finding Matrix Operators**

(a)   Find the standard matrix for the operator on R^2 that first shears by a factor of 2 in the x-direction and then reflects the result about the line $y = x$. Sketch the image of the unit square under this operator.

(b)   Find the standard matrix for the operator on R^2 that first reflects about $y = x$ and then shears by a factor of 2 in the x-direction. Sketch the image of the unit square under this operator.

(c)   Confirm that the shear and the reflection in parts (a) and (b) do not commute.

Solution (a) The standard matrix for the shear is

$$A_1 = \begin{bmatrix} 1 & 2 \\ 0 & 1 \end{bmatrix}$$

and for the reflection is

$$A_2 = \begin{bmatrix} 0 & 1 \\ 1 & 0 \end{bmatrix}$$

Thus, the standard matrix for the shear followed by the reflection is

$$A_2 A_1 = \begin{bmatrix} 0 & 1 \\ 1 & 0 \end{bmatrix} \begin{bmatrix} 1 & 2 \\ 0 & 1 \end{bmatrix} = \begin{bmatrix} 0 & 1 \\ 1 & 2 \end{bmatrix}$$

Solution (b) The standard matrix for the reflection followed by the shear is

$$A_1 A_2 = \begin{bmatrix} 1 & 2 \\ 0 & 1 \end{bmatrix} \begin{bmatrix} 0 & 1 \\ 1 & 0 \end{bmatrix} = \begin{bmatrix} 2 & 1 \\ 1 & 0 \end{bmatrix}$$

Solution (c) The computations in Solutions (a) and (b) show that $A_1 A_2 \neq A_2 A_1$, so the standard matrices, and hence the operators, do not commute. The same conclusion follows from Figures 4.11.3 and 4.11.4, since the two operators produce different images of the unit square. ◀

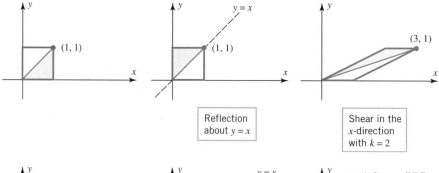

▶ Figure 4.11.3

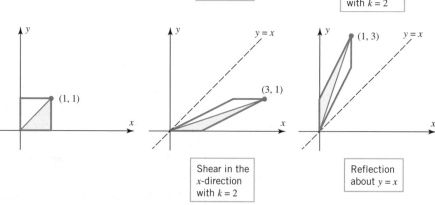

▶ Figure 4.11.4

Geometry of One-to-One Matrix Operators

We will now turn our attention to one-to-one matrix operators on R^2, which are important because they map distinct points into distinct points. Recall from Theorem 4.10.4 (the Equivalence Theorem) that a matrix transformation T_A is one-to-one if and only if A can be expressed as a product of elementary matrices. Thus, we can analyze the effect of any one-to-one transformation T_A by first factoring the matrix A into a product of elementary matrices, say

$$A = E_1 E_2 \cdots E_r$$

and then expressing T_A as the composition

$$T_A = T_{E_1 E_2 \cdots E_r} = T_{E_1} \circ T_{E_2} \circ \cdots \circ T_{E_r} \tag{2}$$

The following theorem explains the geometric effect of matrix operators corresponding to elementary matrices.

THEOREM 4.11.1 *If E is an elementary matrtix, then $T_E \colon R^2 \to R^2$ is one of the following*:

(a) *A shear along a coordinate axis.*

(b) *A reflection about $y = x$.*

(c) *A compression along a coordinate axis.*

(d) *An expansion along a coordinate axis.*

(e) *A reflection about a coordinate axis.*

(f) *A compression or expansion along a coordinate axis followed by a reflection about a coordinate axis.*

Proof   Because a 2×2 elementary matrix results from performing a single elementary row operation on the 2×2 identity matrix, such a matrix must have one of the following forms (verify):

$$\begin{bmatrix} 1 & 0 \\ k & 1 \end{bmatrix}, \quad \begin{bmatrix} 1 & k \\ 0 & 1 \end{bmatrix}, \quad \begin{bmatrix} 0 & 1 \\ 1 & 0 \end{bmatrix}, \quad \begin{bmatrix} k & 0 \\ 0 & 1 \end{bmatrix}, \quad \begin{bmatrix} 1 & 0 \\ 0 & k \end{bmatrix}$$

The first two matrices represent shears along coordinate axes, and the third represents a reflection about $y = x$. If $k > 0$, the last two matrices represent compressions or expansions along coordinate axes, depending on whether $0 \leq k < 1$ or $k > 1$. If $k < 0$, and if we express k in the form $k = -k_1$, where $k_1 > 0$, then the last two matrices can be written as

$$\begin{bmatrix} k & 0 \\ 0 & 1 \end{bmatrix} = \begin{bmatrix} -k_1 & 0 \\ 0 & 1 \end{bmatrix} = \begin{bmatrix} -1 & 0 \\ 0 & 1 \end{bmatrix} \begin{bmatrix} k_1 & 0 \\ 0 & 1 \end{bmatrix} \tag{3}$$

$$\begin{bmatrix} 1 & 0 \\ 0 & k \end{bmatrix} = \begin{bmatrix} 1 & 0 \\ 0 & -k_1 \end{bmatrix} = \begin{bmatrix} 1 & 0 \\ 0 & -1 \end{bmatrix} \begin{bmatrix} 1 & 0 \\ 0 & k_1 \end{bmatrix} \tag{4}$$

Since $k_1 > 0$, the product in (3) represents a compression or expansion along the x-axis followed by a reflection about the y-axis, and (4) represents a compression or expansion along the y-axis followed by a reflection about the x-axis. In the case where $k = -1$, transformations (3) and (4) are simply reflections about the y-axis and x-axis, respectively. ◄

Since every invertible matrix is a product of elementary matrices, the following result follows from Theorem 4.11.1 and Formula (2).

> **THEOREM 4.11.2** *If $T_A: R^2 \to R^2$ is multiplication by an invertible matrix A, then the geometric effect of T_A is the same as an appropriate succession of shears, compressions, expansions, and reflections.*

▶ **EXAMPLE 3 Analyzing the Geometric Effect of a Matrix Operator**

Assuming that k_1 and k_2 are positive, express the diagonal matrix

$$A = \begin{bmatrix} k_1 & 0 \\ 0 & k_2 \end{bmatrix}$$

as a product of elementary matrices, and describe the geometric effect of multiplication by A in terms of compressions and expansions.

Solution From Example 1 we have

$$A = \begin{bmatrix} k_1 & 0 \\ 0 & k_2 \end{bmatrix} = \begin{bmatrix} 1 & 0 \\ 0 & k_2 \end{bmatrix} \begin{bmatrix} k_1 & 0 \\ 0 & 1 \end{bmatrix}$$

which shows that multiplication by A has the geometric effect of compressing or expanding by a factor of k_1 in the x-direction and then compressing or expanding by a factor of k_2 in the y-direction. ◀

▶ **EXAMPLE 4 Analyzing the Geometric Effect of a Matrix Operator**

Express

$$A = \begin{bmatrix} 1 & 2 \\ 3 & 4 \end{bmatrix}$$

as a product of elementary matrices, and then describe the geometric effect of multiplication by A in terms of shears, compressions, expansions, and reflections.

Solution A can be reduced to I as follows:

$$\begin{bmatrix} 1 & 2 \\ 3 & 4 \end{bmatrix} \longrightarrow \begin{bmatrix} 1 & 2 \\ 0 & -2 \end{bmatrix} \longrightarrow \begin{bmatrix} 1 & 2 \\ 0 & 1 \end{bmatrix} \longrightarrow \begin{bmatrix} 1 & 0 \\ 0 & 1 \end{bmatrix}$$

| Add -3 times the first row to the second. | Multiply the second row by $-\frac{1}{2}$. | Add -2 times the second row to the first. |

The three successive row operations can be performed by multiplying A on the left successively by

$$E_1 = \begin{bmatrix} 1 & 0 \\ -3 & 1 \end{bmatrix}, \quad E_2 = \begin{bmatrix} 1 & 0 \\ 0 & -\frac{1}{2} \end{bmatrix}, \quad E_3 = \begin{bmatrix} 1 & -2 \\ 0 & 1 \end{bmatrix}$$

Inverting these matrices and using Formula (4) of Section 1.5 yields

$$A = E_1^{-1} E_2^{-1} E_3^{-1} = \begin{bmatrix} 1 & 0 \\ 3 & 1 \end{bmatrix} \begin{bmatrix} 1 & 0 \\ 0 & -2 \end{bmatrix} \begin{bmatrix} 1 & 2 \\ 0 & 1 \end{bmatrix}$$

Reading from right to left and noting that

$$\begin{bmatrix} 1 & 0 \\ 0 & -2 \end{bmatrix} = \begin{bmatrix} 1 & 0 \\ 0 & -1 \end{bmatrix} \begin{bmatrix} 1 & 0 \\ 0 & 2 \end{bmatrix}$$

it follows that the effect of multiplying by A is equivalent to

1. shearing by a factor of 2 in the x-direction,
2. then expanding by a factor of 2 in the y-direction,
3. then reflecting about the x-axis,
4. then shearing by a factor of 3 in the y-direction. ◄

Images of Lines Under Matrix Operators   Many images in computer graphics are constructed by connecting points with line segments. The following theorem, some of whose parts are proved in the exercises, is helpful for understanding how matrix operators transform such figures.

> **THEOREM 4.11.3** *If $T: R^2 \rightarrow R^2$ is multiplication by an invertible matrix, then:*
>
> (a) *The image of a straight line is a straight line.*
>
> (b) *The image of a straight line through the origin is a straight line through the origin.*
>
> (c) *The images of parallel straight lines are parallel straight lines.*
>
> (d) *The image of the line segment joining points P and Q is the line segment joining the images of P and Q.*
>
> (e) *The images of three points lie on a line if and only if the points themselves lie on a line.*

Note that it follows from Theorem 4.11.3 that if A is an invertible 2×2 matrix, then multiplication by A maps triangles into triangles and parallelograms into parallelograms.

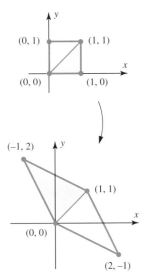

▲ Figure 4.11.5

► **EXAMPLE 5 Image of a Square**

Sketch the image of the square with vertices $(0, 0)$, $(1, 0)$, $(1, 1)$, and $(0, 1)$ under multiplication by

$$A = \begin{bmatrix} -1 & 2 \\ 2 & -1 \end{bmatrix}$$

Solution Since

$$\begin{bmatrix} -1 & 2 \\ 2 & -1 \end{bmatrix}\begin{bmatrix} 0 \\ 0 \end{bmatrix} = \begin{bmatrix} 0 \\ 0 \end{bmatrix}, \quad \begin{bmatrix} -1 & 2 \\ 2 & -1 \end{bmatrix}\begin{bmatrix} 1 \\ 0 \end{bmatrix} = \begin{bmatrix} -1 \\ 2 \end{bmatrix}$$

$$\begin{bmatrix} -1 & 2 \\ 2 & -1 \end{bmatrix}\begin{bmatrix} 0 \\ 1 \end{bmatrix} = \begin{bmatrix} 2 \\ -1 \end{bmatrix}, \quad \begin{bmatrix} -1 & 2 \\ 2 & -1 \end{bmatrix}\begin{bmatrix} 1 \\ 1 \end{bmatrix} = \begin{bmatrix} 1 \\ 1 \end{bmatrix}$$

the image of the square is a parallelogram with vertices $(0, 0)$, $(-1, 2)$, $(2, -1)$, and $(1, 1)$ (Figure 4.11.5).

► **EXAMPLE 6 Image of a Line**

According to Theorem 4.11.3, the invertible matrix

$$A = \begin{bmatrix} 3 & 1 \\ 2 & 1 \end{bmatrix}$$

maps the line $y = 2x + 1$ into another line. Find its equation.

Solution Let (x, y) be a point on the line $y = 2x + 1$, and let (x', y') be its image under multiplication by A. Then

$$\begin{bmatrix} x' \\ y' \end{bmatrix} = \begin{bmatrix} 3 & 1 \\ 2 & 1 \end{bmatrix}\begin{bmatrix} x \\ y \end{bmatrix} \quad \text{and} \quad \begin{bmatrix} x \\ y \end{bmatrix} = \begin{bmatrix} 3 & 1 \\ 2 & 1 \end{bmatrix}^{-1}\begin{bmatrix} x' \\ y' \end{bmatrix} = \begin{bmatrix} 1 & -1 \\ -2 & 3 \end{bmatrix}\begin{bmatrix} x' \\ y' \end{bmatrix}$$

so

$$x = x' - y'$$
$$y = -2x' + 3y'$$

Substituting in $y = 2x + 1$ yields

$$-2x' + 3y' = 2(x' - y') + 1 \quad \text{or equivalently} \quad y' = \tfrac{4}{5}x' + \tfrac{1}{5}$$

Thus (x', y') satisfies

$$y = \tfrac{4}{5}x + \tfrac{1}{5}$$

which is the equation we want. ◀

Concept Review

- Effect of a matrix operator on the unit square
- Geometry of one-to-one matrix operators
- Images of lines under matrix operators

Skills

- Find standard matrices for geometric transformations of R^2.
- Describe the geometric effect of an invertible matrix operator.
- Find the image of the unit square under a matrix operator.
- Find the image of a line under a matrix operator.

Exercise Set 4.11

1. Find the standard matrix for the operator $T: R^2 \to R^2$ that maps a point (x, y) into

  (a) its reflection about the line $y = -x$.

  (b) its reflection through the origin.

  (c) its orthogonal projection on the x-axis.

  (d) its orthogonal projection on the y-axis.

2. For each part of Exercise 1, use the matrix you have obtained to compute $T(2, 1)$. Check your answers geometrically by plotting the points $(2, 1)$ and $T(2, 1)$.

3. Find the standard matrix for the operator $T: R^3 \to R^3$ that maps a point (x, y, z) into

  (a) its reflection through the xy-plane.

  (b) its reflection through the xz-plane.

  (c) its reflection through the yz-plane.

4. For each part of Exercise 3, use the matrix you have obtained to compute $T(1, 1, 1)$. Check your answers geometrically by plotting the points $(1, 1, 1)$ and $T(1, 1, 1)$.

5. Find the standard matrix for the operator $T: R^3 \to R^3$ that

  (a) rotates each vector $90°$ counterclockwise about the z-axis (looking along the positive z-axis toward the origin).

  (b) rotates each vector $90°$ counterclockwise about the x-axis (looking along the positive x-axis toward the origin).

  (c) rotates each vector $90°$ counterclockwise about the y-axis (looking along the positive y-axis toward the origin).

6. Sketch the image of the rectangle with vertices $(0, 0)$, $(1, 0)$, $(1, 2)$, and $(0, 2)$ under

  (a) a reflection about the x-axis.

  (b) a reflection about the y-axis.

  (c) a compression of factor $k = \tfrac{1}{4}$ in the y-direction.

  (d) an expansion of factor $k = 2$ in the x-direction.

  (e) a shear of factor $k = 3$ in the x-direction.

  (f) a shear of factor $k = 2$ in the y-direction.

7. Sketch the image of the square with vertices $(0, 0)$, $(1, 0)$, $(0, 1)$, and $(1, 1)$ under multiplication by

$$A = \begin{bmatrix} -3 & 0 \\ 0 & 1 \end{bmatrix}$$

8. Find the matrix that rotates a point (x, y) about the origin through

  (a) $45°$ (b) $90°$ (c) $180°$ (d) $270°$ (e) $-30°$

9. Find the matrix that shears by

  (a) a factor of $k = 4$ in the y-direction.

  (b) a factor of $k = -2$ in the x-direction.

10. Find the matrix that compresses or expands by

  (a) a factor of $\tfrac{1}{3}$ in the y-direction.

  (b) a factor of 6 in the x-direction.

11. In each part, describe the geometric effect on R^2 of multiplication by A.

(a) $A = \begin{bmatrix} 3 & 0 \\ 0 & 1 \end{bmatrix}$ (b) $A = \begin{bmatrix} 1 & 0 \\ 0 & -5 \end{bmatrix}$ (c) $A = \begin{bmatrix} 1 & 4 \\ 0 & 1 \end{bmatrix}$

12. In each part, express the matrix as a product of elementary matrices, and then describe the effect on R^2 of multiplication by A in terms of compressions, expansions, reflections, and shears.

(a) $A = \begin{bmatrix} 2 & 0 \\ 0 & 3 \end{bmatrix}$ (b) $A = \begin{bmatrix} 1 & 4 \\ 2 & 9 \end{bmatrix}$

(c) $A = \begin{bmatrix} 0 & -2 \\ 4 & 0 \end{bmatrix}$ (d) $A = \begin{bmatrix} 1 & -3 \\ 4 & 6 \end{bmatrix}$

13. In each part, find a single matrix that performs the indicated succession of operations on R^2.

(a) Compresses by a factor of $\frac{1}{2}$ in the x-direction, then expands by a factor of 5 in the y-direction.

(b) Expands by a factor of 5 in the y-direction, then shears by a factor of 2 in the y-direction.

(c) Reflects about $y = x$, then rotates through an angle of $180°$ about the origin.

14. In each part, find a single matrix that performs the indicated succession of operations on R^2.

(a) Reflects about the y-axis, then expands by a factor of 5 in the x-direction, and then reflects about $y = x$.

(b) Rotates through $30°$ about the origin, then shears by a factor of -2 in the y-direction, and then expands by a factor of 3 in the y-direction.

15. Use matrix inversion to show the following.

(a) The inverse transformation for a reflection about $y = x$ is a reflection about $y = x$.

(b) The inverse transformation for a compression along an axis is an expansion along that axis.

(c) The inverse transformation for a reflection about a coordinate axis is a reflection about that axis.

(d) The inverse transformation for a shear along a coordinate axis is a shear along that axis.

16. Find an equation of the image of the line $y = -4x + 3$ under multiplication by

$$A = \begin{bmatrix} 4 & -3 \\ 3 & -2 \end{bmatrix}$$

17. In parts (a) through (e), find an equation of the image of the line $y = 2x$ in R^2 under

(a) a shear of factor 3 in the x-direction.

(b) a compression of factor $\frac{1}{2}$ in the y-direction.

(c) a reflection about $y = x$.

(d) a reflection about the y-axis.

(e) a rotation of $60°$ about the origin.

18. Find the matrix for a shear in the x-direction that transforms the triangle with vertices $(0, 0)$, $(2, 1)$, and $(3, 0)$ into a right triangle with the right angle at the origin.

19. (a) Show that multiplication by

$$A = \begin{bmatrix} 3 & 1 \\ 6 & 2 \end{bmatrix}$$

maps each point in the plane onto the line $y = 2x$.

(b) It follows from part (a) that the noncollinear points $(1, 0)$, $(0, 1)$, $(-1, 0)$ are mapped onto a line. Does this violate part (e) of Theorem 4.11.3?

20. Prove part (a) of Theorem 4.11.3. [*Hint:* A line in the plane has an equation of the form $Ax + By + C = 0$, where A and B are not both zero. Use the method of Example 6 to show that the image of this line under multiplication by the invertible matrix

$$\begin{bmatrix} a & b \\ c & d \end{bmatrix}$$

has the equation $A'x + B'y + C = 0$, where

$$A' = (dA - cB)/(ad - bc)$$

and

$$B' = (-bA + aB)/(ad - bc)$$

Then show that A' and B' are not both zero to conclude that the image is a line.]

21. Use the hint in Exercise 20 to prove parts (b) and (c) of Theorem 4.11.3.

22. In each part of the accompanying figure, find the standard matrix for the operator described.

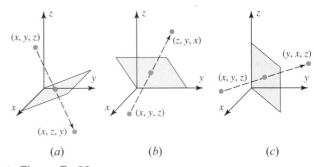

(a)   (b)   (c)

▲ **Figure Ex-22**

23. In R^3 the **shear in the xy-direction with factor k** is the matrix transformation that moves each point (x, y, z) parallel to the xy-plane to the new position $(x + kz, y + kz, z)$. (See the accompanying figure.)

(a) Find the standard matrix for the shear in the xy-direction with factor k.

(b) How would you define the shear in the xz-direction with factor k and the shear in the yz-direction with factor k?

Find the standard matrices for these matrix transformations.

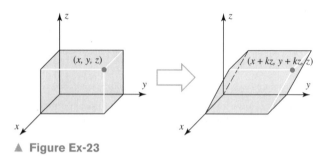

▲ **Figure Ex-23**

True-False Exercises

In parts (a)–(g) determine whether the statement is true or false, and justify your answer.

(a) The image of the unit square under a one-to-one matrix operator is a square.

(b) A 2×2 invertible matrix operator has the geometric effect of a succession of shears, compressions, expansions, and reflections.

(c) The image of a line under a one-to-one matrix operator is a line.

(d) Every reflection operator on R^2 is its own inverse.

(e) The matrix $\begin{bmatrix} 1 & 1 \\ 1 & -1 \end{bmatrix}$ represents reflection about a line.

(f) The matrix $\begin{bmatrix} 1 & -2 \\ 2 & 1 \end{bmatrix}$ represents a shear.

(g) The matrix $\begin{bmatrix} 1 & 0 \\ 0 & 3 \end{bmatrix}$ represents an expansion.

4.12 Dynamical Systems and Markov Chains

In this optional section we will show how matrix methods can be used to analyze the behavior of physical systems that evolve over time. The methods that we will study here have been applied to problems in business, ecology, demographics, sociology, and most of the physical sciences.

Dynamical Systems   A ***dynamical system*** is a finite set of variables whose values change with time. The value of a variable at a point in time is called the ***state of the variable*** at that time, and the vector formed from these states is called the ***state of the dynamical system*** at that time. Our primary objective in this section is to analyze how the state of a dynamical system changes with time. Let us begin with an example.

▶ **EXAMPLE 1 Market Share as a Dynamical System**

Suppose that two competing television channels, channel 1 and channel 2, each have 50% of the viewer market at some initial point in time. Assume that over each one-year period channel 1 captures 10% of channel 2's share, and channel 2 captures 20% of channel 1's share (see Figure 4.12.1). What is each channel's market share after one year?

Solution   Let us begin by introducing the time-dependent variables

$$x_1(t) = \text{fraction of the market held by channel 1 at time } t$$
$$x_2(t) = \text{fraction of the market held by channel 2 at time } t$$

and the column vector

$$\mathbf{x}(t) = \begin{bmatrix} x_1(t) \\ x_2(t) \end{bmatrix} \quad \begin{matrix} \leftarrow \textbf{Channel 1's fraction of the market at time } t \textbf{ in years} \\ \leftarrow \textbf{Channel 2's fraction of the market at time } t \textbf{ in years} \end{matrix}$$

The variables $x_1(t)$ and $x_2(t)$ form a dynamical system whose state at time t is the vector $\mathbf{x}(t)$. If we take $t = 0$ to be the starting point at which the two channels had 50% of the

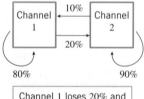

Channel 1 loses 20% and holds 80%.
Channel 2 loses 10% and holds 90%.

▲ **Figure 4.12.1**

market, then the state of the system at that time is

$$\mathbf{x}(0) = \begin{bmatrix} x_1(0) \\ x_2(0) \end{bmatrix} = \begin{bmatrix} 0.5 \\ 0.5 \end{bmatrix} \quad \begin{matrix} \leftarrow \text{ Channel 1's fraction of the market at time } t = 0 \\ \leftarrow \text{ Channel 2's fraction of the market at time } t = 0 \end{matrix} \tag{1}$$

Now let us try to find the state of the system at time $t = 1$ (one year later). Over the one-year period, channel 1 retains 80% of its initial 50%, and it gains 10% of channel 2's initial 50%. Thus,

$$x_1(1) = 0.8(0.5) + 0.1(0.5) = 0.45 \tag{2}$$

Similarly, channel 2 gains 20% of channel 1's initial 50%, and retains 90% of its initial 50%. Thus,

$$x_2(1) = 0.2(0.5) + 0.9(0.5) = 0.55 \tag{3}$$

Therefore, the state of the system at time $t = 1$ is

$$\mathbf{x}(1) = \begin{bmatrix} x_1(1) \\ x_2(1) \end{bmatrix} = \begin{bmatrix} 0.45 \\ 0.55 \end{bmatrix} \quad \begin{matrix} \leftarrow \text{ Channel 1's fraction of the market at time } t = 1 \\ \leftarrow \text{ Channel 2's fraction of the market at time } t = 1 \end{matrix} \tag{4}$$

▶ **EXAMPLE 2 Evolution of Market Share over Five Years**

Track the market shares of channels 1 and 2 in Example 1 over a five-year period.

Solution To solve this problem suppose that we have already computed the market share of each channel at time $t = k$ and we are interested in using the known values of $x_1(k)$ and $x_2(k)$ to compute the market shares $x_1(k + 1)$ and $x_2(k + 1)$ one year later. The analysis is exactly the same as that used to obtain Equations (2) and (3). Over the one-year period, channel 1 retains 80% of its starting fraction $x_1(k)$ and gains 10% of channel 2's starting fraction $x_2(k)$. Thus,

$$x_1(k + 1) = (0.8)x_1(k) + (0.1)x_2(k) \tag{5}$$

Similarly, channel 2 gains 20% of channel 1's starting fraction $x_1(k)$ and retains 90% of its own starting fraction $x_2(k)$. Thus,

$$x_2(k + 1) = (0.2)x_1(k) + (0.9)x_2(k) \tag{6}$$

Equations (5) and (6) can be expressed in matrix form as

$$\begin{bmatrix} x_1(k + 1) \\ x_2(k + 1) \end{bmatrix} = \begin{bmatrix} 0.8 & 0.1 \\ 0.2 & 0.9 \end{bmatrix} \begin{bmatrix} x_1(k) \\ x_2(k) \end{bmatrix} \tag{7}$$

which provides a way of using matrix multiplication to compute the state of the system at time $t = k + 1$ from the state at time $t = k$. For example, using (1) and (7) we obtain

$$\mathbf{x}(1) = \begin{bmatrix} 0.8 & 0.1 \\ 0.2 & 0.9 \end{bmatrix} \mathbf{x}(0) = \begin{bmatrix} 0.8 & 0.1 \\ 0.2 & 0.9 \end{bmatrix} \begin{bmatrix} 0.5 \\ 0.5 \end{bmatrix} = \begin{bmatrix} 0.45 \\ 0.55 \end{bmatrix}$$

which agrees with (4). Similarly,

$$\mathbf{x}(2) = \begin{bmatrix} 0.8 & 0.1 \\ 0.2 & 0.9 \end{bmatrix} \mathbf{x}(1) = \begin{bmatrix} 0.8 & 0.1 \\ 0.2 & 0.9 \end{bmatrix} \begin{bmatrix} 0.45 \\ 0.55 \end{bmatrix} = \begin{bmatrix} 0.415 \\ 0.585 \end{bmatrix}$$

We can now continue this process, using Formula (7) to compute $\mathbf{x}(3)$ from $\mathbf{x}(2)$, then $\mathbf{x}(4)$ from $\mathbf{x}(3)$, and so on. This yields (verify)

$$\mathbf{x}(3) = \begin{bmatrix} 0.3905 \\ 0.6095 \end{bmatrix}, \quad \mathbf{x}(4) = \begin{bmatrix} 0.37335 \\ 0.62665 \end{bmatrix}, \quad \mathbf{x}(5) = \begin{bmatrix} 0.361345 \\ 0.638655 \end{bmatrix} \tag{8}$$

Thus, after five years, channel 1 will hold about 36% of the market and channel 2 will hold about 64% of the market. ◀

If desired, we can continue the market analysis in the last example beyond the five-year period and explore what happens to the market share over the long term. We did so, using a computer, and obtained the following state vectors (rounded to six decimal places):

$$\mathbf{x}(10) \approx \begin{bmatrix} 0.338041 \\ 0.661959 \end{bmatrix}, \quad \mathbf{x}(20) \approx \begin{bmatrix} 0.333466 \\ 0.666534 \end{bmatrix}, \quad \mathbf{x}(40) \approx \begin{bmatrix} 0.333333 \\ 0.666667 \end{bmatrix} \quad (9)$$

All subsequent state vectors, when rounded to six decimal places, are the same as $\mathbf{x}(40)$, so we see that the market shares eventually stabilize with channel 1 holding about one-third of the market and channel 2 holding about two-thirds. Later in this section, we will explain why this stabilization occurs.

Markov Chains   In many dynamical systems the states of the variables are not known with certainty but can be expressed as probabilities; such dynamical systems are called ***stochastic processes*** (from the Greek word *stokastikos*, meaning "proceeding by guesswork"). A detailed study of stochastic processes requires a precise definition of the term *probability*, which is outside the scope of this course. However, the following interpretation will suffice for our present purposes:

> Stated informally, the ***probability*** that an experiment or observation will have a certain outcome is approximately the fraction of the time that the outcome would occur if the experiment were to be repeated many times under constant conditions—the greater the number of repetitions, the more accurately the probability describes the fraction of occurrences.

For example, when we say that the probability of tossing heads with a fair coin is $\frac{1}{2}$, we mean that if the coin were tossed many times under constant conditions, then we would expect about half of the outcomes to be heads. Probabilities are often expressed as decimals or percentages. Thus, the probability of tossing heads with a fair coin can also be expressed as 0.5 or 50%.

If an experiment or observation has n possible outcomes, then the probabilities of those outcomes must be nonnegative fractions whose sum is 1. The probabilities are nonnegative because each describes the fraction of occurrences of an outcome over the long term, and the sum is 1 because they account for all possible outcomes. For example, if a box containing 10 balls has one red ball, three green balls, and six yellow balls, and if a ball is drawn at random from the box, then the probabilities of the various outcomes are

$$p_1 = \text{prob(red)} = 1/10 = 0.1$$
$$p_2 = \text{prob(green)} = 3/10 = 0.3$$
$$p_3 = \text{prob(yellow)} = 6/10 = 0.6$$

Each probability is a nonnegative fraction and

$$p_1 + p_2 + p_3 = 0.1 + 0.3 + 0.6 = 1$$

In a stochastic process with n possible states, the state vector at each time t has the form

$$\mathbf{x}(t) = \begin{bmatrix} x_1(t) \\ x_2(t) \\ \vdots \\ x_n(t) \end{bmatrix} \quad \begin{matrix} \textbf{Probability that the system is in state 1} \\ \textbf{Probability that the system is in state 2} \\ \vdots \\ \textbf{Probability that the system is in state } n \end{matrix}$$

The entries in this vector must add up to 1 since they account for all n possibilities. In general, a vector with nonnegative entries that add up to 1 is called a ***probability vector***.

▶ EXAMPLE 3 **Example 1 Revisited from the Probability Viewpoint**

Observe that the state vectors in Examples 1 and 2 are all probability vectors. This is to be expected since the entries in each state vector are the fractional market shares of the channels, and together they account for the entire market. In practice, it is preferable to interpret the entries in the state vectors as probabilities rather than exact market fractions, since market information is usually obtained by statistical sampling procedures with intrinsic uncertainties. Thus, for example, the state vector

$$\mathbf{x}(1) = \begin{bmatrix} x_1(1) \\ x_2(1) \end{bmatrix} = \begin{bmatrix} 0.45 \\ 0.55 \end{bmatrix}$$

which we interpreted in Example 1 to mean that channel 1 has 45% of the market and channel 2 has 55%, can also be interpreted to mean that an individual picked at random from the market will be a channel 1 viewer with probability 0.45 and a channel 2 viewer with probability 0.55. ◀

A square matrix, each of whose columns is a probability vector, is called a ***stochastic matrix***. Such matrices commonly occur in formulas that relate successive states of a stochastic process. For example, the state vectors $\mathbf{x}(k + 1)$ and $\mathbf{x}(k)$ in (7) are related by an equation of the form $\mathbf{x}(k + 1) = P\mathbf{x}(k)$ in which

$$P = \begin{bmatrix} 0.8 & 0.1 \\ 0.2 & 0.9 \end{bmatrix} \tag{10}$$

is a stochastic matrix. It should not be surprising that the column vectors of P are probability vectors, since the entries in each column provide a breakdown of what happens to each channel's market share over the year—the entries in column 1 convey that each year channel 1 retains 80% of its market share and loses 20%; and the entries in column 2 convey that each year channel 2 retains 90% of its market share and loses 10%. The entries in (10) can also be viewed as probabilities:

$p_{11} = 0.8 =$ probability that a channel 1 viewer remains a channel 1 viewer

$p_{21} = 0.2 =$ probability that a channel 1 viewer becomes a channel 2 viewer

$p_{12} = 0.1 =$ probability that a channel 2 viewer becomes a channel 1 viewer

$p_{22} = 0.9 =$ probability that a channel 2 viewer remains a channel 2 viewer

Example 1 is a special case of a large class of stochastic processes, called *Markov chains*.

Andrei Andreyevich Markov (1856–1922)

Historical Note Markov chains are named in honor of the Russian mathematician A. A. Markov, a lover of poetry, who used them to analyze the alternation of vowels and consonants in the poem *Eugene Onegin* by Pushkin. Markov believed that the only applications of his chains were to the analysis of literary works, so he would be astonished to learn that his discovery is used today in the social sciences, quantum theory, and genetics!

[*Image: wikipedia*]

State at time $t = k$

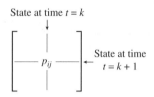

State at time $t = k + 1$

The entry p_{ij} is the probability that the system is in state i at time $t = k + 1$ if it is in state j at time $t = k$.

▲ **Figure 4.12.2**

DEFINITION 1 A ***Markov chain*** is a dynamical system whose state vectors at a succession of time intervals are probability vectors and for which the state vectors at successive time intervals are related by an equation of the form

$$\mathbf{x}(k + 1) = P\mathbf{x}(k)$$

in which $P = [p_{ij}]$ is a stochastic matrix and p_{ij} is the probability that the system will be in state i at time $t = k + 1$ if it is in state j at time $t = k$. The matrix P is called the ***transition matrix*** for the system.

Remark Note that in this definition the row index i corresponds to the later state and the column index j to the earlier state (Figure 4.12.2).

▶ **EXAMPLE 4 Wildlife Migration as a Markov Chain**

Suppose that a tagged lion can migrate over three adjacent game reserves in search of food, reserve 1, reserve 2, and reserve 3. Based on data about the food resources, researchers conclude that the monthly migration pattern of the lion can be modeled by a Markov chain with transition matrix

Reserve at time $t = k$

$$P = \begin{bmatrix} 0.5 & 0.4 & 0.6 \\ 0.2 & 0.2 & 0.3 \\ 0.3 & 0.4 & 0.1 \end{bmatrix} \begin{matrix} 1 \\ 2 \\ 3 \end{matrix} \quad \text{Reserve at time } t = k + 1$$

(see Figure 4.12.3). That is,

$p_{11} = 0.5 =$ probability that the lion will stay in reserve 1 when it is in reserve 1
$p_{12} = 0.4 =$ probability that the lion will move from reserve 2 to reserve 1
$p_{13} = 0.6 =$ probability that the lion will move from reserve 3 to reserve 1
$p_{21} = 0.2 =$ probability that the lion will move from reserve 1 to reserve 2
$p_{22} = 0.2 =$ probability that the lion will stay in reserve 2 when it is in reserve 2
$p_{23} = 0.3 =$ probability that the lion will move from reserve 3 to reserve 2
$p_{31} = 0.3 =$ probability that the lion will move from reserve 1 to reserve 3
$p_{32} = 0.4 =$ probability that the lion will move from reserve 2 to reserve 3
$p_{33} = 0.1 =$ probability that the lion will stay in reserve 3 when it is in reserve 3

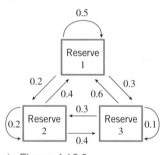

▲ **Figure 4.12.3**

Assuming that t is in months and the lion is released in reserve 2 at time $t = 0$, track its probable locations over a six-month period.

Solution Let $x_1(k)$, $x_2(k)$, and $x_3(k)$ be the probabilities that the lion is in reserve 1, 2, or 3, respectively, at time $t = k$, and let

$$\mathbf{x}(k) = \begin{bmatrix} x_1(k) \\ x_2(k) \\ x_3(k) \end{bmatrix}$$

be the state vector at that time. Since we know with certainty that the lion is in reserve 2 at time $t = 0$, the initial state vector is

$$\mathbf{x}(0) = \begin{bmatrix} 0 \\ 1 \\ 0 \end{bmatrix}$$

We leave it for you to show that the state vectors over a six-month period are

$$\mathbf{x}(1) = P\mathbf{x}(0) = \begin{bmatrix} 0.400 \\ 0.200 \\ 0.400 \end{bmatrix}, \quad \mathbf{x}(2) = P\mathbf{x}(1) = \begin{bmatrix} 0.520 \\ 0.240 \\ 0.240 \end{bmatrix}, \quad \mathbf{x}(3) = P\mathbf{x}(2) = \begin{bmatrix} 0.500 \\ 0.224 \\ 0.276 \end{bmatrix}$$

$$\mathbf{x}(4) = P\mathbf{x}(3) \approx \begin{bmatrix} 0.505 \\ 0.228 \\ 0.267 \end{bmatrix}, \quad \mathbf{x}(5) = P\mathbf{x}(4) \approx \begin{bmatrix} 0.504 \\ 0.227 \\ 0.269 \end{bmatrix}, \quad \mathbf{x}(6) = P\mathbf{x}(5) \approx \begin{bmatrix} 0.504 \\ 0.227 \\ 0.269 \end{bmatrix}$$

As in Example 2, the state vectors here seem to stabilize over time with a probability of approximately 0.504 that the lion is in reserve 1, a probability of approximately 0.227 that it is in reserve 2, and a probability of approximately 0.269 that it is in reserve 3. ◄

Markov Chains in Terms of Powers of the Transition Matrix

In a Markov chain with an initial state of $\mathbf{x}(0)$, the successive state vectors are

$$\mathbf{x}(1) = P\mathbf{x}(0), \quad \mathbf{x}(2) = P\mathbf{x}(1), \quad \mathbf{x}(3) = P\mathbf{x}(2), \quad \mathbf{x}(4) = P\mathbf{x}(3), \dots$$

For brevity, it is common to denote $\mathbf{x}(k)$ by $\mathbf{x}_k$, which allows us to write the successive state vectors more briefly as

$$\mathbf{x}_1 = P\mathbf{x}_0, \quad \mathbf{x}_2 = P\mathbf{x}_1, \quad \mathbf{x}_3 = P\mathbf{x}_2, \quad \mathbf{x}_4 = P\mathbf{x}_3, \dots \tag{11}$$

Alternatively, these state vectors can be expressed in terms of the initial state vector $\mathbf{x}_0$ as

$$\mathbf{x}_1 = P\mathbf{x}_0, \quad \mathbf{x}_2 = P(P\mathbf{x}_0) = P^2\mathbf{x}_0, \quad \mathbf{x}_3 = P(P^2\mathbf{x}_0) = P^3\mathbf{x}_0, \quad \mathbf{x}_4 = P(P^3\mathbf{x}_0) = P^4\mathbf{x}_0, \dots$$

from which it follows that

$$\mathbf{x}_k = P^k\mathbf{x}_0 \tag{12}$$

> Note that Formula (12) makes it possible to compute the state vector $\mathbf{x}_k$ without first computing the earlier state vectors as required in Formula (11).

▶ **EXAMPLE 5 Finding a State Vector Directly from $\mathbf{x}_0$**

Use Formula (12) to find the state vector $\mathbf{x}(3)$ in Example 2.

Solution From (1) and (7), the initial state vector and transition matrix are

$$\mathbf{x}_0 = \mathbf{x}(0) = \begin{bmatrix} 0.5 \\ 0.5 \end{bmatrix} \quad \text{and} \quad P = \begin{bmatrix} 0.8 & 0.1 \\ 0.2 & 0.9 \end{bmatrix}$$

We leave it for you to calculate P^3 and show that

$$\mathbf{x}(3) = \mathbf{x}_3 = P^3\mathbf{x}_0 = \begin{bmatrix} 0.562 & 0.219 \\ 0.438 & 0.781 \end{bmatrix} \begin{bmatrix} 0.5 \\ 0.5 \end{bmatrix} = \begin{bmatrix} 0.3905 \\ 0.6095 \end{bmatrix}$$

which agrees with the result in (8). ◄

Long-Term Behavior of a Markov Chain

We have seen two examples of Markov chains in which the state vectors seem to stabilize after a period of time. Thus, it is reasonable to ask whether all Markov chains have this property. The following example shows that this is not the case.

▶ **EXAMPLE 6 A Markov Chain That Does Not Stabilize**

The matrix

$$P = \begin{bmatrix} 0 & 1 \\ 1 & 0 \end{bmatrix}$$

is stochastic and hence can be regarded as the transition matrix for a Markov chain. A simple calculation shows that $P^2 = I$, from which it follows that

$$I = P^2 = P^4 = P^6 = \cdots \quad \text{and} \quad P = P^3 = P^5 = P^7 = \cdots$$

Thus, the successive states in the Markov chain with initial vector $\mathbf{x}_0$ are

$$\mathbf{x}_0, \quad P\mathbf{x}_0, \quad \mathbf{x}_0, \quad P\mathbf{x}_0, \quad \mathbf{x}_0, \ldots$$

which oscillate between $\mathbf{x}_0$ and $P\mathbf{x}_0$. Thus, the Markov chain does not stabilize unless both components of $\mathbf{x}_0$ are $\frac{1}{2}$ (verify). ◀

A precise definition of what it means for a sequence of numbers or vectors to stabilize is given in calculus; however, that level of precision will not be needed here. Stated informally, we will say that a sequence of vectors

$$\mathbf{x}_1, \quad \mathbf{x}_2, \ldots, \quad \mathbf{x}_k, \ldots$$

approaches a *limit* $\mathbf{q}$ or that it *converges* to $\mathbf{q}$ if all entries in $\mathbf{x}_k$ can be made as close as we like to the corresponding entries in the vector $\mathbf{q}$ by taking k sufficiently large. We denote this by writing $\mathbf{x}_k \to \mathbf{q}$ as $k \to \infty$.

We saw in Example 6 that the state vectors of a Markov chain need not approach a limit in all cases. However, by imposing a mild condition on the transition matrix of a Markov chain, we can guarantee that the state vectors will approach a limit.

DEFINITION 2 A stochastic matrix P is said to be *regular* if P or some positive power of P has all positive entries, and a Markov chain whose transition matrix is regular is said to be a *regular Markov chain*.

▶ **EXAMPLE 7** Regular Stochastic Matrices

The transition matrices in Examples 2 and 4 are regular because their entries are positive. The matrix

$$P = \begin{bmatrix} 0.5 & 1 \\ 0.5 & 0 \end{bmatrix}$$

is regular because

$$P^2 = \begin{bmatrix} 0.75 & 0.5 \\ 0.25 & 0.5 \end{bmatrix}$$

has positive entries. The matrix P in Example 6 is not regular because P and every positive power of P have some zero entries (verify). ◀

The following theorem, which we state without proof, is the fundamental result about the long-term behavior of Markov chains.

THEOREM 4.12.1 *If P is the transition matrix for a regular Markov chain, then:*

(a) *There is a unique probability vector $\mathbf{q}$ such that $P\mathbf{q} = \mathbf{q}$.*

(b) *For any initial probability vector $\mathbf{x}_0$, the sequence of state vectors*

$$\mathbf{x}_0, \quad P\mathbf{x}_0, \ldots, \quad P^k\mathbf{x}_0, \ldots$$

converges to $\mathbf{q}$.

The vector $\mathbf{q}$ in this theorem is called the *steady-state* vector of the Markov chain. It can be found by rewriting the equation in part (a) as

$$(I - P)\mathbf{q} = \mathbf{0}$$

and then solving this equation for $\mathbf{q}$ subject to the requirement that $\mathbf{q}$ be a probability vector. Here are some examples.

▶ EXAMPLE 8 **Examples 1 and 2 Revisited**

The transition matrix for the Markov chain in Example 2 is

$$P = \begin{bmatrix} 0.8 & 0.1 \\ 0.2 & 0.9 \end{bmatrix}$$

Since the entries of P are positive, the Markov chain is regular and hence has a unique steady-state vector $\mathbf{q}$. To find $\mathbf{q}$ we will solve the system $(I - P)\mathbf{q} = \mathbf{0}$, which we can write as

$$\begin{bmatrix} 0.2 & -0.1 \\ -0.2 & 0.1 \end{bmatrix} \begin{bmatrix} q_1 \\ q_2 \end{bmatrix} = \begin{bmatrix} 0 \\ 0 \end{bmatrix}$$

The general solution of this system is

$$q_1 = 0.5s, \quad q_2 = s$$

(verify), which we can write in vector form as

$$\mathbf{q} = \begin{bmatrix} q_1 \\ q_2 \end{bmatrix} = \begin{bmatrix} 0.5s \\ s \end{bmatrix} = \begin{bmatrix} \frac{1}{2}s \\ s \end{bmatrix} \tag{13}$$

For $\mathbf{q}$ to be a probability vector, we must have

$$1 = q_1 + q_2 = \tfrac{3}{2}s$$

which implies that $s = \tfrac{2}{3}$. Substituting this value in (13) yields the steady-state vector

$$\mathbf{q} = \begin{bmatrix} \frac{1}{3} \\ \frac{2}{3} \end{bmatrix}$$

which is consistent with the numerical results obtained in (9).

▶ EXAMPLE 9 **Example 4 Revisited**

The transition matrix for the Markov chain in Example 4 is

$$P = \begin{bmatrix} 0.5 & 0.4 & 0.6 \\ 0.2 & 0.2 & 0.3 \\ 0.3 & 0.4 & 0.1 \end{bmatrix}$$

Since the entries of P are positive, the Markov chain is regular and hence has a unique steady-state vector $\mathbf{q}$. To find $\mathbf{q}$ we will solve the system $(I - P)\mathbf{q} = \mathbf{0}$, which we can write (using fractions) as

$$\begin{bmatrix} \frac{1}{2} & -\frac{2}{5} & -\frac{3}{5} \\ -\frac{1}{5} & \frac{4}{5} & -\frac{3}{10} \\ -\frac{3}{10} & -\frac{2}{5} & \frac{9}{10} \end{bmatrix} \begin{bmatrix} q_1 \\ q_2 \\ q_3 \end{bmatrix} = \begin{bmatrix} 0 \\ 0 \\ 0 \end{bmatrix} \tag{14}$$

(We have converted to fractions to avoid roundoff error in this illustrative example.) We leave it for you to confirm that the reduced row echelon form of the coefficient matrix is

$$\begin{bmatrix} 1 & 0 & -\frac{15}{8} \\ 0 & 1 & -\frac{27}{32} \\ 0 & 0 & 0 \end{bmatrix}$$

and that the general solution of (14) is

$$q_1 = \tfrac{15}{8}s, \quad q_2 = \tfrac{27}{32}s, \quad q_3 = s \tag{15}$$

For $\mathbf{q}$ to be a probability vector we must have $q_1 + q_2 + q_3 = 1$, from which it follows that $s = \frac{32}{119}$ (verify). Substituting this value in (15) yields the steady-state vector

$$\mathbf{q} = \begin{bmatrix} \frac{60}{119} \\ \frac{27}{119} \\ \frac{32}{119} \end{bmatrix} \approx \begin{bmatrix} 0.5042 \\ 0.2269 \\ 0.2689 \end{bmatrix}$$

(verify), which is consistent with the results obtained in Example 4. ◀

Concept Review

- Dynamical system
- State of a variable
- State of a dynamical system
- Stochastic process

- Probability
- Probability vector
- Stochastic matrix
- Markov chain

- Transition matrix
- Regular stochastic matrix
- Regular Markov chain
- Steady-state vector

Skills

- Determine whether a matrix is stochastic.
- Compute the state vectors from a transition matrix and an initial state.
- Determine whether a stochastic matrix is regular.

- Determine whether a Markov chain is regular.
- Find the steady-state vector for a regular transition matrix.

Exercise Set 4.12

▶ In Exercises 1–2, determine whether A is a stochastic matrix. If A is not stochastic, then explain why not. ◀

1. (a) $A = \begin{bmatrix} 0.4 & 0.3 \\ 0.6 & 0.7 \end{bmatrix}$ (b) $A = \begin{bmatrix} 0.4 & 0.6 \\ 0.3 & 0.7 \end{bmatrix}$

(c) $A = \begin{bmatrix} 1 & \frac{1}{2} & \frac{1}{3} \\ 0 & 0 & \frac{1}{3} \\ 0 & \frac{1}{2} & \frac{1}{3} \end{bmatrix}$ (d) $A = \begin{bmatrix} \frac{1}{3} & \frac{1}{3} & \frac{1}{2} \\ \frac{1}{6} & \frac{1}{3} & -\frac{1}{2} \\ \frac{1}{2} & \frac{1}{3} & 1 \end{bmatrix}$

2. (a) $A = \begin{bmatrix} 0.2 & 0.9 \\ 0.8 & 0.1 \end{bmatrix}$ (b) $A = \begin{bmatrix} 0.2 & 0.8 \\ 0.9 & 0.1 \end{bmatrix}$

(c) $A = \begin{bmatrix} \frac{1}{12} & \frac{1}{9} & \frac{1}{6} \\ \frac{1}{2} & 0 & \frac{5}{6} \\ \frac{5}{12} & \frac{8}{9} & 0 \end{bmatrix}$ (d) $A = \begin{bmatrix} -1 & \frac{1}{3} & \frac{1}{2} \\ 0 & \frac{1}{3} & \frac{1}{2} \\ 2 & \frac{1}{3} & 0 \end{bmatrix}$

▶ In Exercises 3–4, use Formulas (11) and (12) to compute the state vector $\mathbf{x}_4$ in two different ways. ◀

3. $P = \begin{bmatrix} 0.5 & 0.6 \\ 0.5 & 0.4 \end{bmatrix}$; $\mathbf{x}_0 = \begin{bmatrix} 0.5 \\ 0.5 \end{bmatrix}$

4. $P = \begin{bmatrix} 0.8 & 0.5 \\ 0.2 & 0.5 \end{bmatrix}$; $\mathbf{x}_0 = \begin{bmatrix} 1 \\ 0 \end{bmatrix}$

▶ In Exercises 5–6, determine whether P is a regular stochastic matrix. ◀

5. (a) $P = \begin{bmatrix} \frac{1}{5} & \frac{1}{7} \\ \frac{4}{5} & \frac{6}{7} \end{bmatrix}$ (b) $P = \begin{bmatrix} \frac{1}{5} & 0 \\ \frac{4}{5} & 1 \end{bmatrix}$ (c) $P = \begin{bmatrix} \frac{1}{5} & 1 \\ \frac{4}{5} & 0 \end{bmatrix}$

6. (a) $P = \begin{bmatrix} \frac{1}{2} & 1 \\ \frac{1}{2} & 0 \end{bmatrix}$ (b) $P = \begin{bmatrix} 1 & \frac{2}{3} \\ 0 & \frac{1}{3} \end{bmatrix}$ (c) $P = \begin{bmatrix} \frac{3}{4} & \frac{1}{3} \\ \frac{1}{4} & \frac{2}{3} \end{bmatrix}$

▷ In Exercises 7–10, verify that P is a regular stochastic matrix, and find the steady-state vector for the associated Markov chain.

7. $P = \begin{bmatrix} \frac{1}{4} & \frac{2}{3} \\ \frac{3}{4} & \frac{1}{3} \end{bmatrix}$

8. $P = \begin{bmatrix} 0.2 & 0.6 \\ 0.8 & 0.4 \end{bmatrix}$

9. $P = \begin{bmatrix} \frac{1}{2} & \frac{1}{2} & 0 \\ \frac{1}{4} & \frac{1}{2} & \frac{1}{3} \\ \frac{1}{4} & 0 & \frac{2}{3} \end{bmatrix}$

10. $P = \begin{bmatrix} \frac{1}{3} & \frac{1}{4} & \frac{2}{5} \\ 0 & \frac{3}{4} & \frac{2}{5} \\ \frac{2}{3} & 0 & \frac{1}{5} \end{bmatrix}$

11. Consider a Markov process with transition matrix

$$\begin{array}{cc} & \begin{array}{cc} \text{State 1} & \text{State 2} \end{array} \\ \begin{array}{c} \text{State 1} \\ \text{State 2} \end{array} & \begin{bmatrix} 0.2 & 0.1 \\ 0.8 & 0.9 \end{bmatrix} \end{array}$$

(a) What does the entry 0.2 represent?

(b) What does the entry 0.1 represent?

(c) If the system is in state 1 initially, what is the probability that it will be in state 2 at the next observation?

(d) If the system has a 50% chance of being in state 1 initially, what is the probability that it will be in state 2 at the next observation?

12. Consider a Markov process with transition matrix

$$\begin{array}{cc} & \begin{array}{cc} \text{State 1} & \text{State 2} \end{array} \\ \begin{array}{c} \text{State 1} \\ \text{State 2} \end{array} & \begin{bmatrix} 0 & \frac{1}{7} \\ 1 & \frac{6}{7} \end{bmatrix} \end{array}$$

(a) What does the entry $\frac{6}{7}$ represent?

(b) What does the entry 0 represent?

(c) If the system is in state 1 initially, what is the probability that it will be in state 1 at the next observation?

(d) If the system has a 50% chance of being in state 1 initially, what is the probability that it will be in state 2 at the next observation?

13. On a given day the air quality in a certain city is either good or bad. Records show that when the air quality is good on one day, then there is a 95% chance that it will be good the next day, and when the air quality is bad on one day, then there is a 45% chance that it will be bad the next day.

(a) Find a transition matrix for this phenomenon.

(b) If the air quality is good today, what is the probability that it will be good two days from now?

(c) If the air quality is bad today, what is the probability that it will be bad three days from now?

(d) If there is a 20% chance that the air quality will be good today, what is the probability that it will be good tomorrow?

14. In a laboratory experiment, a mouse can choose one of two food types each day, type I or type II. Records show that if the mouse chooses type I on a given day, then there is a 75%

chance that it will choose type I the next day, and if it chooses type II on one day, then there is a 50% chance that it will choose type II the next day.

(a) Find a transition matrix for this phenomenon.

(b) If the mouse chooses type I today, what is the probability that it will choose type I two days from now?

(c) If the mouse chooses type II today, what is the probability that it will choose type II three days from now?

(d) If there is a 10% chance that the mouse will choose type I today, what is the probability that it will choose type I tomorrow?

15. Suppose that at some initial point in time 100,000 people live in a certain city and 25,000 people live in its suburbs. The Regional Planning Commission determines that each year 5% of the city population moves to the suburbs and 3% of the suburban population moves to the city.

(a) Assuming that the total population remains constant, make a table that shows the populations of the city and its suburbs over a five-year period (round to the nearest integer).

(b) Over the long term, how will the population be distributed between the city and its suburbs?

16. Suppose that two competing television stations, station 1 and station 2, each have 50% of the viewer market at some initial point in time. Assume that over each one-year period station 1 captures 5% of station 2's market share and station 2 captures 10% of station 1's market share.

(a) Make a table that shows the market share of each station over a five-year period.

(b) Over the long term, how will the market share be distributed between the two stations?

17. Suppose that a car rental agency has three locations, numbered 1, 2, and 3. A customer may rent a car from any of the three locations and return it to any of the three locations. Records show that cars are rented and returned in accordance with the following probabilities:

| | | Rented from Location | |
		1	2	3
Returned to Location	1	$\frac{1}{10}$	$\frac{1}{5}$	$\frac{3}{5}$
	2	$\frac{4}{5}$	$\frac{3}{10}$	$\frac{1}{5}$
	3	$\frac{1}{10}$	$\frac{1}{2}$	$\frac{1}{5}$

(a) Assuming that a car is rented from location 1, what is the probability that it will be at location 1 after two rentals?

(b) Assuming that this dynamical system can be modeled as a Markov chain, find the steady-state vector.

(c) If the rental agency owns 120 cars, how many parking spaces should it allocate at each location to be reasonably certain that it will have enough spaces for the cars over the long term? Explain your reasoning.

18. Physical traits are determined by the genes that an offspring receives from its parents. In the simplest case a trait in the offspring is determined by one pair of genes, one member of the pair inherited from the male parent and the other from the female parent. Typically, each gene in a pair can assume one of two forms, called *alleles*, denoted by A and a. This leads to three possible pairings:

$$AA, \quad Aa, \quad aa$$

called *genotypes* (the pairs Aa and aA determine the same trait and hence are not distinguished from one another). It is shown in the study of heredity that if a parent of known genotype is crossed with a random parent of unknown genotype, then the offspring will have the genotype probabilities given in the following table, which can be viewed as a transition matrix for a Markov process:

		Genotype of Parent		
		AA	Aa	aa
Genotype of Offspring	AA	$\frac{1}{2}$	$\frac{1}{4}$	0
	Aa	$\frac{1}{2}$	$\frac{1}{2}$	$\frac{1}{2}$
	aa	0	$\frac{1}{4}$	$\frac{1}{2}$

Thus, for example, the offspring of a parent of genotype AA that is crossed at random with a parent of unknown genotype will have a 50% chance of being AA, a 50% chance of being Aa, and no chance of being aa.

(a) Show that the transition matrix is regular.

(b) Find the steady-state vector, and discuss its physical interpretation.

19. Fill in the missing entries of the stochastic matrix

$$P = \begin{bmatrix} \frac{7}{10} & * & \frac{1}{5} \\ * & \frac{3}{10} & * \\ \frac{1}{10} & \frac{3}{5} & \frac{3}{10} \end{bmatrix}$$

and find its steady-state vector.

20. If P is an $n \times n$ stochastic matrix, and if M is a $1 \times n$ matrix whose entries are all 1's, then $MP = $ _____.

21. If P is a regular stochastic matrix with steady-state vector $\mathbf{q}$, what can you say about the sequence of products

$$P\mathbf{q}, \quad P^2\mathbf{q}, \quad P^3\mathbf{q}, \dots, \quad P^k\mathbf{q}, \dots$$

as $k \to \infty$?

22. (a) If P is a regular $n \times n$ stochastic matrix with steady-state vector $\mathbf{q}$, and if $\mathbf{e}_1, \mathbf{e}_2, \dots, \mathbf{e}_n$ are the standard unit vectors in column form, what can you say about the behavior of the sequence

$$P\mathbf{e}_i, \quad P^2\mathbf{e}_i, \quad P^3\mathbf{e}_i, \dots, \quad P^k\mathbf{e}_i, \dots$$

as $k \to \infty$ for each $i = 1, 2, \dots, n$?

(b) What does this tell you about the behavior of the column vectors of P^k as $k \to \infty$?

23. Prove that the product of two stochastic matrices is a stochastic matrix. [*Hint:* Write each column of the product as a linear combination of the columns of the first factor.]

24. Prove that if P is a stochastic matrix whose entries are all greater than or equal to ρ, then the entries of P^2 are greater than or equal to ρ.

True-False Exercises

In parts (a)–(e) determine whether the statement is true or false, and justify your answer.

(a) The vector $\begin{bmatrix} \frac{1}{3} \\ 0 \\ \frac{2}{3} \end{bmatrix}$ is a probability vector.

(b) The matrix $\begin{bmatrix} 0.2 & 1 \\ 0.8 & 0 \end{bmatrix}$ is a regular stochastic matrix.

(c) The column vectors of a transition matrix are probability vectors.

(d) A steady-state vector for a Markov chain with transition matrix P is any solution of the linear system $(I - P)\mathbf{q} = \mathbf{0}$.

(e) The square of every regular stochastic matrix is stochastic.

Chapter 4 Supplementary Exercises

1. Let V be the set of all ordered pairs of real numbers, and consider the following addition and scalar multiplication operations on $\mathbf{u} = (u_1, u_2, u_3)$ and $\mathbf{v} = (v_1, v_2, v_3)$:

$$\mathbf{u} + \mathbf{v} = (u_1 + v_1, u_2 + v_2, u_3 + v_3), \quad k\mathbf{u} = (ku_1, 0, 0)$$

(a) Compute $\mathbf{u} + \mathbf{v}$ and $k\mathbf{u}$ for $\mathbf{u} = (3, -2, 4)$, $\mathbf{v} = (1, 5, -2)$, and $k = -1$.

(b) In words, explain why V is closed under addition and scalar multiplication.

(c) Since the addition operation on V is the standard addition operation on R^3, certain vector space axioms hold for V because they are known to hold for R^3. Which axioms in Definition 1 of Section 4.1 are they?

(d) Show that Axioms 7, 8, and 9 hold.

(e) Show that Axiom 10 fails for the given operations.

2. In each part, the solution space of the system is a subspace of R^3 and so must be a line through the origin, a plane through the origin, all of R^3, or the origin only. For each system, determine which is the case. If the subspace is a plane, find an equation for it, and if it is a line, find parametric equations.

(a) $0x + 0y + 0z = 0$

(b) $2x - 3y + z = 0$
     $6x - 9y + 3z = 0$
     $-4x + 6y - 2z = 0$

(c) $x - 2y + 7z = 0$
     $-4x + 8y + 5z = 0$
     $2x - 4y + 3z = 0$

(d) $x + 4y + 8z = 0$
     $2x + 5y + 6z = 0$
     $3x + y - 4z = 0$

3. For what values of s is the solution space of

$$x_1 + x_2 + sx_3 = 0$$
$$x_1 + sx_2 + x_3 = 0$$
$$sx_1 + x_2 + x_3 = 0$$

the origin only, a line through the origin, a plane through the origin, or all of R^3?

4. (a) Express $(4a, a - b, a + 2b)$ as a linear combination of $(4, 1, 1)$ and $(0, -1, 2)$.

(b) Express $(3a + b + 3c, -a + 4b - c, 2a + b + 2c)$ as a linear combination of $(3, -1, 2)$ and $(1, 4, 1)$.

(c) Express $(2a - b + 4c, 3a - c, 4b + c)$ as a linear combination of three nonzero vectors.

5. Let W be the space spanned by $\mathbf{f} = \sin x$ and $\mathbf{g} = \cos x$.

(a) Show that for any value of θ, $\mathbf{f}_1 = \sin(x + \theta)$ and $\mathbf{g}_1 = \cos(x + \theta)$ are vectors in W.

(b) Show that $\mathbf{f}_1$ and $\mathbf{g}_1$ form a basis for W.

6. (a) Express $\mathbf{v} = (1, 1)$ as a linear combination of $\mathbf{v}_1 = (1, -1)$, $\mathbf{v}_2 = (3, 0)$, and $\mathbf{v}_3 = (2, 1)$ in two different ways.

(b) Explain why this does not violate Theorem 4.4.1.

7. Let A be an $n \times n$ matrix, and let $\mathbf{v}_1, \mathbf{v}_2, \ldots, \mathbf{v}_n$ be linearly independent vectors in R^n expressed as $n \times 1$ matrices. What must be true about A for $A\mathbf{v}_1, A\mathbf{v}_2, \ldots, A\mathbf{v}_n$ to be linearly independent?

8. Must a basis for P_n contain a polynomial of degree k for each $k = 0, 1, 2, \ldots, n$? Justify your answer.

9. For the purpose of this exercise, let us define a "checkerboard matrix" to be a square matrix $A = [a_{ij}]$ such that

$$a_{ij} = \begin{cases} 1 & \text{if } i + j \text{ is even} \\ 0 & \text{if } i + j \text{ is odd} \end{cases}$$

Find the rank and nullity of the following checkerboard matrices.

(a) The 3×3 checkerboard matrix.

(b) The 4×4 checkerboard matrix.

(c) The $n \times n$ checkerboard matrix.

10. For the purpose of this exercise, let us define an "X-matrix" to be a square matrix with an odd number of rows and columns that has 0's everywhere except on the two diagonals where it has 1's. Find the rank and nullity of the following X-matrices.

(a) $\begin{bmatrix} 1 & 0 & 1 \\ 0 & 1 & 0 \\ 1 & 0 & 1 \end{bmatrix}$

(b) $\begin{bmatrix} 1 & 0 & 0 & 0 & 1 \\ 0 & 1 & 0 & 1 & 0 \\ 0 & 0 & 1 & 0 & 0 \\ 0 & 1 & 0 & 1 & 0 \\ 1 & 0 & 0 & 0 & 1 \end{bmatrix}$

(c) the X-matrix of size $(2n + 1) \times (2n + 1)$

11. In each part, show that the stated set of polynomials is a subspace of P_n and find a basis for it.

(a) All polynomials in P_n such that $p(-x) = p(x)$.

(b) All polynomials in P_n such that $p(0) = 0$.

12. (**Calculus required**) Show that the set of all polynomials in P_n that have a horizontal tangent at $x = 0$ is a subspace of P_n. Find a basis for this subspace.

13. (a) Find a basis for the vector space of all 3×3 symmetric matrices.

(b) Find a basis for the vector space of all 3×3 skew-symmetric matrices.

14. Various advanced texts in linear algebra prove the following determinant criterion for rank: *The rank of a matrix A is r if and only if A has some $r \times r$ submatrix with a nonzero determinant, and all square submatrices of larger size have determinant zero.* [*Note:* A submatrix of A is any matrix obtained by deleting rows or columns of A. The matrix A itself is also considered to be a submatrix of A.] In each part, use this criterion to find the rank of the matrix.

(a) $\begin{bmatrix} 1 & 2 & 0 \\ 2 & 4 & -1 \end{bmatrix}$

(b) $\begin{bmatrix} 1 & 2 & 3 \\ 2 & 4 & 6 \end{bmatrix}$

(c) $\begin{bmatrix} 1 & 0 & 1 \\ 2 & -1 & 3 \\ 3 & -1 & 4 \end{bmatrix}$

(d) $\begin{bmatrix} 1 & -1 & 2 & 0 \\ 3 & 1 & 0 & 0 \\ -1 & 2 & 4 & 0 \end{bmatrix}$

15. Use the result in Exercise 14 above to find the possible ranks for matrices of the form

$$\begin{bmatrix} 0 & 0 & 0 & 0 & 0 & a_{16} \\ 0 & 0 & 0 & 0 & 0 & a_{26} \\ 0 & 0 & 0 & 0 & 0 & a_{36} \\ 0 & 0 & 0 & 0 & 0 & a_{46} \\ a_{51} & a_{52} & a_{53} & a_{54} & a_{55} & a_{56} \end{bmatrix}$$

16. Prove: If S is a basis for a vector space V, then for any vectors $\mathbf{u}$ and $\mathbf{v}$ in V and any scalar k, the following relationships hold.

(a) $(\mathbf{u} + \mathbf{v})_S = (\mathbf{u})_S + (\mathbf{v})_S$

(b) $(k\mathbf{u})_S = k(\mathbf{u})_S$

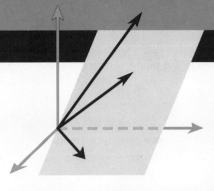

CHAPTER 5

Applications of Linear Algebra

INTRODUCTION This chapter consists of 8 applications of linear algebra. With one clearly marked exception, each application is in its own independent section, so sections can be deleted or permuted as desired. Each topic begins with a list of linear algebra prerequisites.

Because our primary objective in this chapter is to present applications of linear algebra, proofs are often omitted. Whenever results from other fields are needed, they are stated precisely, with motivation where possible, but usually without proof.

5.1 Constructing Curves and Surfaces Through Specified Points

In this section we describe a technique that uses determinants to construct lines, circles, and general conic sections through specified points in the plane. The procedure is also used to pass planes and spheres in 3-space through fixed points.

> **PREREQUISITES:** Linear Systems
> Determinants
> Analytic Geometry

The following theorem follows from Theorem 2.3.8.

> **THEOREM 5.1.1** *A homogeneous linear system with as many equations as unknowns has a nontrivial solution if and only if the determinant of the coefficient matrix is zero.*

We will now show how this result can be used to determine equations of various curves and surfaces through specified points.

A Line Through Two Points

Suppose that (x_1, y_1) and (x_2, y_2) are two distinct points in the plane. There exists a unique line

$$c_1 x + c_2 y + c_3 = 0 \qquad (1)$$

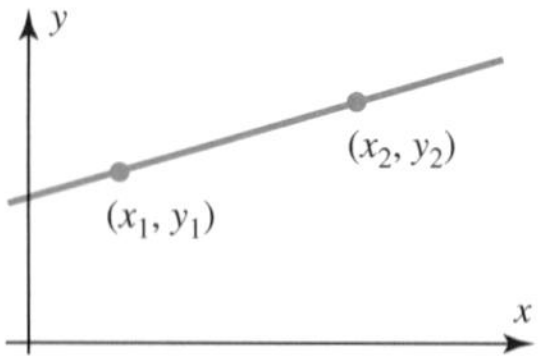

that passes through these two points (Figure 5.1.1). Note that c_1, c_2, and c_3 are not all zero and that these coefficients are unique only up to a multiplicative constant. Because (x_1, y_1) and (x_2, y_2) lie on the line, substituting them in (1) gives the two equations

$$c_1 x_1 + c_2 y_1 + c_3 = 0 \qquad (2)$$
$$c_1 x_2 + c_2 y_2 + c_3 = 0 \qquad (3)$$

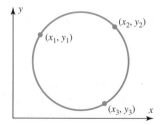

▲ Figure 5.1.1

The three equations, (1), (2), and (3), can be grouped together and rewritten as

$$x c_1 + y c_2 + c_3 = 0$$
$$x_1 c_1 + y_1 c_2 + c_3 = 0$$
$$x_2 c_1 + y_2 c_2 + c_3 = 0$$

which is a homogeneous linear system of three equations for c_1, c_2, and c_3. Because c_1, c_2, and c_3 are not all zero, this system has a nontrivial solution, so the determinant of the coefficient matrix of the system must be zero. That is,

$$\begin{vmatrix} x & y & 1 \\ x_1 & y_1 & 1 \\ x_2 & y_2 & 1 \end{vmatrix} = 0 \qquad (4)$$

Consequently, every point (x, y) on the line satisfies (4); conversely, it can be shown that every point (x, y) that satisfies (4) lies on the line.

▶ **EXAMPLE 1 Equation of a Line**

Find the equation of the line that passes through the two points $(2, 1)$ and $(3, 7)$.

Solution Substituting the coordinates of the two points into Equation (4) gives

$$\begin{vmatrix} x & y & 1 \\ 2 & 1 & 1 \\ 3 & 7 & 1 \end{vmatrix} = 0$$

The cofactor expansion of this determinant along the first row then gives

$$-6x + y + 11 = 0 \quad ◀$$

A Circle Through Three Points

Suppose that there are three distinct points in the plane, (x_1, y_1), (x_2, y_2), and (x_3, y_3), not all lying on a straight line. From analytic geometry we know that there is a unique circle, say,

$$c_1(x^2 + y^2) + c_2 x + c_3 y + c_4 = 0 \qquad (5)$$

that passes through them (Figure 5.1.2). Substituting the coordinates of the three points into this equation gives

$$c_1(x_1^2 + y_1^2) + c_2 x_1 + c_3 y_1 + c_4 = 0 \qquad (6)$$
$$c_1(x_2^2 + y_2^2) + c_2 x_2 + c_3 y_2 + c_4 = 0 \qquad (7)$$
$$c_1(x_3^2 + y_3^2) + c_2 x_3 + c_3 y_3 + c_4 = 0 \qquad (8)$$

▲ Figure 5.1.2

As before, Equations (5) through (8) form a homogeneous linear system with a nontrivial solution for c_1, c_2, c_3, and c_4. Thus the determinant of the coefficient matrix is zero:

$$\begin{vmatrix} x^2 + y^2 & x & y & 1 \\ x_1^2 + y_1^2 & x_1 & y_1 & 1 \\ x_2^2 + y_2^2 & x_2 & y_2 & 1 \\ x_3^2 + y_3^2 & x_3 & y_3 & 1 \end{vmatrix} = 0 \tag{9}$$

This is a determinant form for the equation of the circle.

▶ **EXAMPLE 2 Equation of a Circle**

Find the equation of the circle that passes through the three points $(1, 7)$, $(6, 2)$, and $(4, 6)$.

Solution Substituting the coordinates of the three points into Equation (9) gives

$$\begin{vmatrix} x^2 + y^2 & x & y & 1 \\ 50 & 1 & 7 & 1 \\ 40 & 6 & 2 & 1 \\ 52 & 4 & 6 & 1 \end{vmatrix} = 0$$

which reduces to

$$10(x^2 + y^2) - 20x - 40y - 200 = 0$$

In standard form this is

$$(x - 1)^2 + (y - 2)^2 = 5^2$$

Thus the circle has center $(1, 2)$ and radius 5. ◀

A General Conic Section Through Five Points

In his momumental work *Principia Mathematica*, Issac Newton posed and solved the following problem (Book I, Proposition 22, Problem 14): "To describe a conic that shall pass through five given points." Newton solved this problem geometrically, as shown in Figure 5.1.3, in which he passed an ellipse through the points A, B, D, P, C; however, the methods of this section can also be applied.

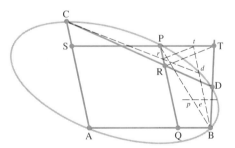

▶ Figure 5.1.3

The general equation of a conic section in the plane (a parabola, hyperbola, or ellipse, or degenerate forms of these curves) is given by

$$c_1 x^2 + c_2 xy + c_3 y^2 + c_4 x + c_5 y + c_6 = 0$$

This equation contains six coefficients, but we can reduce the number to five if we divide through by any one of them that is not zero. Thus only five coefficients must be determined, so five distinct points in the plane are sufficient to determine the equation of the conic section (Figure 5.1.4). As before, the equation can be put in determinant

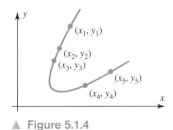

▲ Figure 5.1.4

form (see Exercise 7):

$$\begin{vmatrix} x^2 & xy & y^2 & x & y & 1 \\ x_1^2 & x_1 y_1 & y_1^2 & x_1 & y_1 & 1 \\ x_2^2 & x_2 y_2 & y_2^2 & x_2 & y_2 & 1 \\ x_3^2 & x_3 y_3 & y_3^2 & x_3 & y_3 & 1 \\ x_4^2 & x_4 y_4 & y_4^2 & x_4 & y_4 & 1 \\ x_5^2 & x_5 y_5 & y_5^2 & x_5 & y_5 & 1 \end{vmatrix} = 0 \tag{10}$$

▶ **EXAMPLE 3 Equation of an Orbit**

An astronomer who wants to determine the orbit of an asteroid about the Sun sets up a Cartesian coordinate system in the plane of the orbit with the Sun at the origin. Astronomical units of measurement are used along the axes (1 astronomical unit = mean distance of Earth to Sun = 93 million miles). By Kepler's first law, the orbit must be an ellipse, so the astronomer makes five observations of the asteroid at five different times and finds five points along the orbit to be

$$(8.025, 8.310), \ (10.170, 6.355), \ (11.202, 3.212), \ (10.736, 0.375), \ (9.092, -2.267)$$

Find the equation of the orbit.

Solution Substituting the coordinates of the five given points into (10) and rounding to three decimal places give

$$\begin{vmatrix} x^2 & xy & y^2 & x & y & 1 \\ 64.401 & 66.688 & 69.056 & 8.025 & 8.310 & 1 \\ 103.429 & 64.630 & 40.386 & 10.170 & 6.355 & 1 \\ 125.485 & 35.981 & 10.317 & 11.202 & 3.212 & 1 \\ 115.262 & 4.026 & 0.141 & 10.736 & 0.375 & 1 \\ 82.664 & -20.612 & 5.139 & 9.092 & -2.267 & 1 \end{vmatrix} = 0$$

The cofactor expansion of this determinant along the first row yields

$$386.802 x^2 - 102.895 xy + 446.029 y^2 - 2476.443 x - 1427.998 y - 17109.375 = 0$$

Figure 5.1.5 is an accurate diagram of the orbit, together with the five given points. ◀

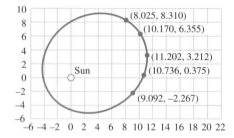

▶ **Figure 5.1.5**

A Plane Through Three Points

In Exercise 8 we ask you to show the following: The plane in 3-space with equation

$$c_1 x + c_2 y + c_3 z + c_4 = 0$$

that passes through three noncollinear points (x_1, y_1, z_1), (x_2, y_2, z_2), and (x_3, y_3, z_3) is given by the determinant equation

$$\begin{vmatrix} x & y & z & 1 \\ x_1 & y_1 & z_1 & 1 \\ x_2 & y_2 & z_2 & 1 \\ x_3 & y_3 & z_3 & 1 \end{vmatrix} = 0 \tag{11}$$

▶ **EXAMPLE 4** **Equation of a Plane**

The equation of the plane that passes through the three noncollinear points $(1, 1, 0)$, $(2, 0, -1)$, and $(2, 9, 2)$ is

$$\begin{vmatrix} x & y & z & 1 \\ 1 & 1 & 0 & 1 \\ 2 & 0 & -1 & 1 \\ 2 & 9 & 2 & 1 \end{vmatrix} = 0$$

which reduces to

$$2x - y + 3z - 1 = 0 \quad ◀$$

A Sphere Through Four *Points*

In Exercise 9 we ask you to show the following: The sphere in 3-space with equation

$$c_1(x^2 + y^2 + z^2) + c_2 x + c_3 y + c_4 z + c_5 = 0$$

that passes through four noncoplanar points (x_1, y_1, z_1), (x_2, y_2, z_2), (x_3, y_3, z_3), and (x_4, y_4, z_4) is given by the following determinant equation:

$$\begin{vmatrix} x^2 + y^2 + z^2 & x & y & z & 1 \\ x_1^2 + y_1^2 + z_1^2 & x_1 & y_1 & z_1 & 1 \\ x_2^2 + y_2^2 + z_2^2 & x_2 & y_2 & z_2 & 1 \\ x_3^2 + y_3^2 + z_3^2 & x_3 & y_3 & z_3 & 1 \\ x_4^2 + y_4^2 + z_4^2 & x_4 & y_4 & z_4 & 1 \end{vmatrix} = 0 \qquad (12)$$

▶ **EXAMPLE 5** **Equation of a Sphere**

The equation of the sphere that passes through the four points $(0, 3, 2)$, $(1, -1, 1)$, $(2, 1, 0)$, and $(5, 1, 3)$ is

$$\begin{vmatrix} x^2 + y^2 + z^2 & x & y & z & 1 \\ 13 & 0 & 3 & 2 & 1 \\ 3 & 1 & -1 & 1 & 1 \\ 5 & 2 & 1 & 0 & 1 \\ 35 & 5 & 1 & 3 & 1 \end{vmatrix} = 0$$

This reduces to

$$x^2 + y^2 + z^2 - 4x - 2y - 6z + 5 = 0$$

which in standard form is

$$(x - 2)^2 + (y - 1)^2 + (z - 3)^2 = 9 \quad ◀$$

Exercise Set 5.1

1. Find the equations of the lines that pass through the following points:

(a) $(1, -1)$, $(2, 2)$ (b) $(0, 1)$, $(1, -1)$

2. Find the equations of the circles that pass through the following points:

(a) $(2, 6)$, $(2, 0)$, $(5, 3)$ (b) $(2, -2)$, $(3, 5)$, $(-4, 6)$

3. Find the equation of the conic section that passes through the points $(0, 0)$, $(0, -1)$, $(2, 0)$, $(2, -5)$, and $(4, -1)$.

4. Find the equations of the planes in 3-space that pass through the following points:

(a) $(1, 1, -3)$, $(1, -1, 1)$, $(0, -1, 2)$

(b) $(2, 3, 1)$, $(2, -1, -1)$, $(1, 2, 1)$

5. (a) Alter Equation (11) so that it determines the plane that passes through the origin and is parallel to the plane that passes through three specified noncollinear points.

(b) Find the two planes described in part (a) corresponding to the triplets of points in Exercises 4(a) and 4(b).

6. Find the equations of the spheres in 3-space that pass through the following points:

(a) $(1, 2, 3)$, $(-1, 2, 1)$, $(1, 0, 1)$, $(1, 2, -1)$

(b) $(0, 1, -2)$, $(1, 3, 1)$, $(2, -1, 0)$, $(3, 1, -1)$

7. Show that Equation (10) is the equation of the conic section that passes through five given distinct points in the plane.

8. Show that Equation (11) is the equation of the plane in 3-space that passes through three given noncollinear points.

9. Show that Equation (12) is the equation of the sphere in 3-space that passes through four given noncoplanar points.

10. Find a determinant equation for the parabola of the form

$$c_1 y + c_2 x^2 + c_3 x + c_4 = 0$$

that passes through three given noncollinear points in the plane.

11. What does Equation (9) become if the three distinct points are collinear?

12. What does Equation (11) become if the three distinct points are collinear?

13. What does Equation (12) become if the four points are coplanar?

Section 5.1 Technology Exercises

The following exercises are designed to be solved using a technology utility. Typically, this will be MATLAB, *Mathematica*, Maple, Derive, or Mathcad, but it may also be some other type of linear algebra software or a scientific calculator with some linear algebra capabilities. For each exercise you will need to read the relevant documentation for the particular utility you are using. The goal of these exercises is to provide you with a basic proficiency with your technology utility. Once you have mastered the techniques in these exercises, you will be able to use your technology utility to solve many of the problems in the regular exercise sets.

T1. The general equation of a quadric surface is given by

$$a_1 x^2 + a_2 y^2 + a_3 z^2 + a_4 xy + a_5 xz$$
$$+ a_6 yz + a_7 x + a_8 y + a_9 z + a_{10} = 0$$

Given nine points on this surface, it may be possible to determine its equation.

(a) Show that if the nine points (x_i, y_i) for $i = 1, 2, 3, \ldots, 9$ lie on this surface, and if they determine uniquely the equation of this surface, then its equation can be written in determinant form as

$$\begin{vmatrix} x^2 & y^2 & z^2 & xy & xz & yz & x & y & z & 1 \\ x_1^2 & y_1^2 & z_1^2 & x_1 y_1 & x_1 z_1 & y_1 z_1 & x_1 & y_1 & z_1 & 1 \\ x_2^2 & y_2^2 & z_2^2 & x_2 y_2 & x_2 z_2 & y_2 z_2 & x_2 & y_2 & z_2 & 1 \\ x_3^2 & y_3^2 & z_3^2 & x_3 y_3 & x_3 z_3 & y_3 z_3 & x_3 & y_3 & z_3 & 1 \\ x_4^2 & y_4^2 & z_4^2 & x_4 y_4 & x_4 z_4 & y_4 z_4 & x_4 & y_4 & z_4 & 1 \\ x_5^2 & y_5^2 & z_5^2 & x_5 y_5 & x_5 z_5 & y_5 z_5 & x_5 & y_5 & z_5 & 1 \\ x_6^2 & y_6^2 & z_6^2 & x_6 y_6 & x_6 z_6 & y_6 z_6 & x_6 & y_6 & z_6 & 1 \\ x_7^2 & y_7^2 & z_7^2 & x_7 y_7 & x_7 z_7 & y_7 z_7 & x_7 & y_7 & z_7 & 1 \\ x_8^2 & y_8^2 & z_8^2 & x_8 y_8 & x_8 z_8 & y_8 z_8 & x_8 & y_8 & z_8 & 1 \\ x_9^2 & y_9^2 & z_9^2 & x_9 y_9 & x_9 z_9 & y_9 z_9 & x_9 & y_9 & z_9 & 1 \end{vmatrix} = 0$$

(b) Use the result in part (a) to determine the equation of the quadric surface that passes through the points $(1, 2, 3)$, $(2, 1, 7)$, $(0, 4, 6)$, $(3, -1, 4)$, $(3, 0, 11)$, $(-1, 5, 8)$, $(9, -8, 3)$, $(4, 5, 3)$, and $(-2, 6, 10)$.

T2. (a) A hyperplane in the n-dimensional Euclidean space R^n has an equation of the form

$$a_1 x_1 + a_2 x_2 + a_3 x_3 + \cdots + a_n x_n + a_{n+1} = 0$$

where a_i, $i = 1, 2, 3, \ldots, n + 1$, are constants, not all zero, and x_i, $i = 1, 2, 3, \ldots, n$, are variables for which

$$(x_1, x_2, x_3, \ldots, x_n) \in R^n$$

A point

$$(x_{10}, x_{20}, x_{30}, \ldots, x_{n0}) \in R^n$$

lies on this hyperplane if

$$a_1 x_{10} + a_2 x_{20} + a_3 x_{30} + \cdots + a_n x_{n0} + a_{n+1} = 0$$

Given that the n points $(x_{1i}, x_{2i}, x_{3i}, \ldots, x_{ni})$, $i = 1, 2, 3, \ldots, n$, lie on this hyperplane and that they uniquely determine the equation of the hyperplane, show that the equation of the hyperplane can be written in determinant form as

$$\begin{vmatrix} x_1 & x_2 & x_3 & \cdots & x_n & 1 \\ x_{11} & x_{21} & x_{31} & \cdots & x_{n1} & 1 \\ x_{12} & x_{22} & x_{32} & \cdots & x_{n2} & 1 \\ x_{13} & x_{23} & x_{33} & \cdots & x_{n3} & 1 \\ \vdots & \vdots & \vdots & \ddots & \vdots & \vdots \\ x_{1n} & x_{2n} & x_{3n} & \cdots & x_{nn} & 1 \end{vmatrix} = 0$$

(b) Determine the equation of the hyperplane in R^9 that goes through the following nine points:

$(1, 2, 3, 4, 5, 6, 7, 8, 9)$ $(2, 3, 4, 5, 6, 7, 8, 9, 1)$
$(3, 4, 5, 6, 7, 8, 9, 1, 2)$ $(4, 5, 6, 7, 8, 9, 1, 2, 3)$
$(5, 6, 7, 8, 9, 1, 2, 3, 4)$ $(6, 7, 8, 9, 1, 2, 3, 4, 5)$
$(7, 8, 9, 1, 2, 3, 4, 5, 6)$ $(8, 9, 1, 2, 3, 4, 5, 6, 7)$
$(9, 1, 2, 3, 4, 5, 6, 7, 8)$

5.2 Geometric Linear Programming

In this section we describe a geometric technique for maximizing or minimizing a linear expression in two variables subject to a set of linear constraints.

PREREQUISITES: Linear Systems
Linear Inequalities

Linear Programming   The study of linear programming theory has expanded greatly since the pioneering work of George Dantzig in the late 1940s. Today, linear programming is applied to a wide variety of problems in industry and science. In this section we present a geometric approach to the solution of simple linear programming problems. Let us begin with some examples.

▶ **EXAMPLE 1 Maximizing Sales Revenue**

A candy manufacturer has 130 pounds of chocolate-covered cherries and 170 pounds of chocolate-covered mints in stock. He decides to sell them in the form of two different mixtures. One mixture will contain half cherries and half mints by weight and will sell for $2.00 per pound. The other mixture will contain one-third cherries and two-thirds mints by weight and will sell for $1.25 per pound. How many pounds of each mixture should the candy manufacturer prepare in order to maximize his sales revenue?

Mathematical Formulation   Let the mixture of half cherries and half mints be called mix A, and let x_1 be the number of pounds of this mixture to be prepared. Let the mixture of one-third cherries and two-thirds mints be called mix B, and let x_2 be the number of pounds of this mixture to be prepared. Since mix A sells for $2.00 per pound and mix B sells for $1.25 per pound, the total sales z (in dollars) will be

$$z = 2.00x_1 + 1.25x_2$$

Since each pound of mix A contains $\frac{1}{2}$ pound of cherries and each pound of mix B contains $\frac{1}{3}$ pound of cherries, the total number of pounds of cherries used in both mixtures is

$$\tfrac{1}{2}x_1 + \tfrac{1}{3}x_2$$

Similarly, since each pound of mix A contains $\frac{1}{2}$ pound of mints and each pound of mix B contains $\frac{2}{3}$ pound of mints, the total number of pounds of mints used in both mixtures is

$$\tfrac{1}{2}x_1 + \tfrac{2}{3}x_2$$

Because the manufacturer can use at most 130 pounds of cherries and 170 pounds of mints, we must have

$$\tfrac{1}{2}x_1 + \tfrac{1}{3}x_2 \le 130$$
$$\tfrac{1}{2}x_1 + \tfrac{2}{3}x_2 \le 170$$

Furthermore, since x_1 and x_2 cannot be negative numbers, we must have

$$x_1 \ge 0 \quad \text{and} \quad x_2 \ge 0$$

The problem can therefore be formulated mathematically as follows: Find values of x_1 and x_2 that maximize

$$z = 2.00x_1 + 1.25x_2$$

subject to

$$\tfrac{1}{2}x_1 + \tfrac{1}{3}x_2 \le 130$$
$$\tfrac{1}{2}x_1 + \tfrac{2}{3}x_2 \le 170$$
$$x_1 \ge 0$$
$$x_2 \ge 0$$

Later in this section we will show how to solve this type of mathematical problem geometrically.

▶ **EXAMPLE 2 Maximizing Annual Yield**

A woman has up to $10,000 to invest. Her broker suggests investing in two bonds, A and B. Bond A is a rather risky bond with an annual yield of 10%, and bond B is a rather safe bond with an annual yield of 7%. After some consideration, she decides to invest at most $6000 in bond A, to invest at least $2000 in bond B, and to invest at least as much in bond A as in bond B. How should she invest her money in order to maximize her annual yield?

Mathematical Formulation Let x_1 be the number of dollars to be invested in bond A, and let x_2 be the number of dollars to be invested in bond B. Since each dollar invested in bond A earns $.10 per year and each dollar invested in bond B earns $.07 per year, the total dollar amount z earned each year by both bonds is

$$z = .10x_1 + .07x_2$$

The constraints imposed can be formulated mathematically as follows:

Invest no more than $10,000:	$x_1 + x_2 \le 10,000$
Invest at most $6000 in bond A:	$x_1 \le 6000$
Invest at least $2000 in bond B:	$x_2 \ge 2000$
Invest at least as much in bond A as in bond B:	$x_1 \ge x_2$

We also have the implicit assumption that x_1 and x_2 are nonnegative:

$$x_1 \ge 0 \quad \text{and} \quad x_2 \ge 0$$

Thus the complete mathematical formulation of the problem is as follows: Find values of x_1 and x_2 that maximize

$$z = .10x_1 + .07x_2$$

subject to

$$x_1 + x_2 \le 10,000$$
$$x_1 \le 6000$$
$$x_2 \ge 2000$$
$$x_1 - x_2 \ge 0$$
$$x_1 \ge 0$$
$$x_2 \ge 0$$

▶ **EXAMPLE 3 Minimizing Cost**

A student desires to design a breakfast of cornflakes and milk that is as economical as possible. On the basis of what he eats during his other meals, he decides that his breakfast should supply him with at least 9 grams of protein, at least $\frac{1}{3}$ the recommended daily allowance (RDA) of vitamin D, and at least $\frac{1}{4}$ the RDA of calcium. He finds the following nutrition and cost information on the milk and cornflakes containers:

	Milk ($\frac{1}{2}$ cup)	Cornflakes (1 ounce)
Cost	7.5 cents	5.0 cents
Protein	4 grams	2 grams
Vitamin D	$\frac{1}{8}$ of RDA	$\frac{1}{10}$ of RDA
Calcium	$\frac{1}{6}$ of RDA	None

In order not to have his mixture too soggy or too dry, the student decides to limit himself to mixtures that contain 1 to 3 ounces of cornflakes per cup of milk, inclusive. What quantities of milk and cornflakes should he use to minimize the cost of his breakfast?

Mathematical Formulation Let x_1 be the quantity of milk used (measured in $\frac{1}{2}$-cup units), and let x_2 be the quantity of cornflakes used (measured in 1-ounce units). Then if z is the cost of the breakfast in cents, we may write the following.

Cost of breakfast:	$z = 7.5x_1 + 5.0x_2$
At least 9 grams protein:	$4x_1 + 2x_2 \geq 9$
At least $\frac{1}{3}$ RDA vitamin D:	$\frac{1}{8}x_1 + \frac{1}{10}x_2 \geq \frac{1}{3}$
At least $\frac{1}{4}$ RDA calcium:	$\frac{1}{6}x_1 \geq \frac{1}{4}$
At least 1 ounce cornflakes per cup (two $\frac{1}{2}$-cups) of milk:	$\frac{x_2}{x_1} \geq \frac{1}{2}$ (or $x_1 - 2x_2 \leq 0$)
At most 3 ounces cornflakes per cup (two $\frac{1}{2}$-cups) of milk:	$\frac{x_2}{x_1} \leq \frac{3}{2}$ (or $3x_1 - 2x_2 \geq 0$)

As before, we also have the implicit assumption that $x_1 \geq 0$ and $x_2 \geq 0$. Thus the complete mathematical formulation of the problem is as follows: Find values of x_1 and x_2 that minimize

$$z = 7.5x_1 + 5.0x_2$$

subject to

$$4x_1 + 2x_2 \geq 9$$
$$\tfrac{1}{8}x_1 + \tfrac{1}{10}x_2 \geq \tfrac{1}{3}$$
$$\tfrac{1}{6}x_1 \geq \tfrac{1}{4}$$
$$x_1 - 2x_2 \leq 0$$
$$3x_1 - 2x_2 \geq 0$$
$$x_1 \geq 0$$
$$x_2 \geq 0 \quad \blacktriangleleft$$

Geometric Solution of Linear Programming Problems

Each of the preceding three examples is a special case of the following problem.

Problem Find values of x_1 and x_2 that either maximize or minimize

$$z = c_1x_1 + c_2x_2 \tag{1}$$

subject to

$$a_{11}x_1 + a_{12}x_2 \ (\leq)(\geq)(=) \ b_1$$
$$a_{21}x_1 + a_{22}x_2 \ (\leq)(\geq)(=) \ b_2$$
$$\vdots \qquad \vdots \qquad \qquad \vdots \tag{2}$$
$$a_{m1}x_1 + a_{m2}x_2 \ (\leq)(\geq)(=) \ b_m$$

and

$$x_1 \geq 0, \qquad x_2 \geq 0 \tag{3}$$

In each of the m conditions of (2), any one of the symbols $\leq$, $\geq$, and $=$ may be used.

The problem above is called the **general linear programming problem** in two variables. The linear function z in (1) is called the **objective function**. Equations (2) and (3) are called the **constraints**; in particular, the equations in (3) are called the **nonnegativity constraints** on the variables x_1 and x_2.

We will now show how to solve a linear programming problem in two variables graphically. A pair of values (x_1, x_2) that satisfy all of the constraints is called a **feasible solution**. The set of all feasible solutions determines a subset of the x_1x_2-plane called the **feasible region**. Our desire is to find a feasible solution that maximizes the objective function. Such a solution is called an **optimal solution**.

To examine the feasible region of a linear programming problem, let us note that each constraint of the form

$$a_{i1}x_1 + a_{i2}x_2 = b_i$$

defines a line in the x_1x_2-plane, whereas each constraint of the form

$$a_{i1}x_1 + a_{i2}x_2 \leq b_i \quad \text{or} \quad a_{i1}x_1 + a_{i2}x_2 \geq b_i$$

defines a half-plane that includes its boundary line

$$a_{i1}x_1 + a_{i2}x_2 = b_i$$

Thus the feasible region is always an intersection of finitely many lines and half-planes. For example, the four constraints

$$\tfrac{1}{2}x_1 + \tfrac{1}{3}x_2 \leq 130$$
$$\tfrac{1}{2}x_1 + \tfrac{2}{3}x_2 \leq 170$$
$$x_1 \geq 0$$
$$x_2 \geq 0$$

of Example 1 define the half-planes illustrated in parts (a), (b), (c), and (d) of Figure 5.2.1. The feasible region of this problem is thus the intersection of these four half-planes, which is illustrated in Figure 5.2.1e.

It can be shown that the feasible region of a linear programming problem has a boundary consisting of a finite number of straight line segments. If the feasible region can be enclosed in a sufficiently large circle, it is called **bounded** (Figure 5.2.1e); otherwise, it is called **unbounded** (see Figure 5.2.5). If the feasible region is *empty* (contains no points), then the constraints are inconsistent and the linear programming problem has no solution (see Figure 5.2.6).

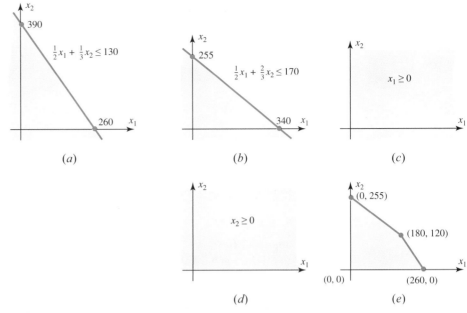

▲ Figure 5.2.1

Those boundary points of a feasible region that are intersections of two of the straight line boundary segments are called ***extreme points***. (They are also called *corner points* and *vertex points*.) For example, in Figure 5.2.1*e*, we see that the feasible region of Example 1 has four extreme points:

$$(0, 0), \quad (0, 255), \quad (180, 120), \quad (260, 0) \tag{4}$$

The importance of the extreme points of a feasible region is shown by the following theorem.

THEOREM 5.2.1 Maximum and Minimum Values

If the feasible region of a linear programming problem is nonempty and bounded, then the objective function attains both a maximum and a minimum value, and these occur at extreme points of the feasible region. If the feasible region is unbounded, then the objective function may or may not attain a maximum or minimum value; however, if it attains a maximum or minimum value, it does so at an extreme point.

Figure 5.2.2 suggests the idea behind the proof of this theorem. Since the objective function

$$z = c_1 x_1 + c_2 x_2$$

of a linear programming problem is a linear function of x_1 and x_2, its level curves (the curves along which z has constant values) are straight lines. As we move in a direction perpendicular to these level curves, the objective function either increases or decreases monotonically. Within a bounded feasible region, the maximum and minimum values of z must therefore occur at extreme points, as Figure 5.2.2 indicates.

In the next few examples we use Theorem 5.2.1 to solve several linear programming problems and illustrate the variations in the nature of the solutions that may occur.

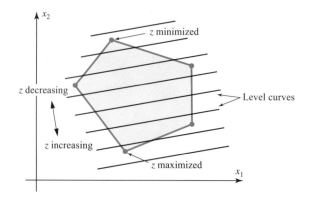

▶ Figure 5.2.2

▶ **EXAMPLE 4 Example 1 Revisited**

Figure 5.2.1*e* shows that the feasible region of Example 1 is bounded. Consequently, from Theorem 5.2.1 the objective function

$$z = 2.00x_1 + 1.25x_2$$

attains both its minimum and maximum values at extreme points. The four extreme points and the corresponding values of z are given in the following table.

Extreme Point (x_1, x_2)	Value of $z = 2.00x_1 + 1.25x_2$
(0, 0)	0
(0, 255)	318.75
(180, 120)	510.00
(260, 0)	520.00

We see that the largest value of z is 520.00 and the corresponding optimal solution is (260, 0). Thus the candy manufacturer attains maximum sales of \$520 when he produces 260 pounds of mixture A and none of mixture B.

▶ **EXAMPLE 5 Using Theorem 5.2.1**

Find values of x_1 and x_2 that maximize

$$z = x_1 + 3x_2$$

subject to

$$2x_1 + 3x_2 \leq 24$$
$$x_1 - x_2 \leq 7$$
$$x_2 \leq 6$$
$$x_1 \geq 0$$
$$x_2 \geq 0$$

Solution In Figure 5.2.3 we have drawn the feasible region of this problem. Since it is bounded, the maximum value of z is attained at one of the five extreme points. The values of the objective function at the five extreme points are given in the following table. From this table, the maximum value of z is 21, which is attained at $x_1 = 3$ and $x_2 = 6$.

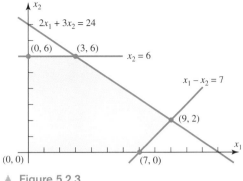

Extreme Point (x_1, x_2)	Value of $z = x_1 + 3x_2$
(0, 6)	18
(3, 6)	21
(9, 2)	15
(7, 0)	7
(0, 0)	0

▲ Figure 5.2.3

▶ **EXAMPLE 6 Using Theorem 5.2.1**

Find values of x_1 and x_2 that maximize

$$z = 4x_1 + 6x_2$$

subject to

$$2x_1 + 3x_2 \leq 24$$
$$x_1 - x_2 \leq 7$$
$$x_2 \leq 6$$
$$x_1 \geq 0$$
$$x_2 \geq 0$$

Solution The constraints in this problem are identical to the constraints in Example 5, so the feasible region of this problem is also given by Figure 5.2.3. The values of the objective function at the extreme points are given in the following table.

Extreme Point (x_1, x_2)	Value of $z = 4x_1 + 6x_2$
(0, 6)	36
(3, 6)	48
(9, 2)	48
(7, 0)	28
(0, 0)	0

We see that the objective function attains a maximum value of 48 at two adjacent extreme points, (3, 6) and (9, 2). This shows that an optimal solution to a linear programming problem need not be unique. As we ask you to show in Exercise 10, if the objective function has the same value at two adjacent extreme points, it has the same value at all points on the straight line boundary segment connecting the two extreme points. Thus, in this example the maximum value of z is attained at all points on the straight line segment connecting the extreme points (3, 6) and (9, 2).

▶ **EXAMPLE 7 The Feasible Region Is a Line Segment**

Find values of x_1 and x_2 that minimize

$$z = 2x_1 - x_2$$

subject to

$$2x_1 + 3x_2 = 12$$
$$2x_1 - 3x_2 \geq 0$$
$$x_1 \geq 0$$
$$x_2 \geq 0$$

Solution In Figure 5.2.4 we have drawn the feasible region of this problem. Because one of the constraints is an equality constraint, the feasible region is a straight line segment with two extreme points. The values of z at the two extreme points are given in the following table.

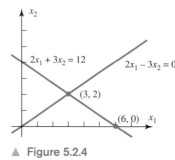

Extreme Point (x_1, x_2)	Value of $z = 2x_1 - x_2$
$(3, 2)$	4
$(6, 0)$	12

▲ Figure 5.2.4

The minimum value of z is thus 4 and is attained at $x_1 = 3$ and $x_2 = 2$.

▶ **EXAMPLE 8 Using Theorem 5.2.1**

Find values of x_1 and x_2 that maximize

$$z = 2x_1 + 5x_2$$

subject to

$$2x_1 + x_2 \geq 8$$
$$-4x_1 + x_2 \leq 2$$
$$2x_1 - 3x_2 \leq 0$$
$$x_1 \geq 0$$
$$x_2 \geq 0$$

Solution The feasible region of this linear programming problem is illustrated in Figure 5.2.5. Since it is unbounded, we are not assured by Theorem 5.2.1 that the objective function attains a maximum value. In fact, it is easily seen that since the feasible region contains points for which both x_1 and x_2 are arbitrarily large and positive, the objective function

$$z = 2x_1 + 5x_2$$

can be made arbitrarily large and positive. This problem has no optimal solution. Instead, we say the problem has an ***unbounded solution***.

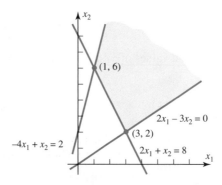

▶ Figure 5.2.5

▶ **EXAMPLE 9 Using Theorem 5.2.1**

Find values of x_1 and x_2 that maximize

$$z = -5x_1 + x_2$$

subject to

$$2x_1 + x_2 \geq 8$$
$$-4x_1 + x_2 \leq 2$$
$$2x_1 - 3x_2 \leq 0$$
$$x_1 \geq 0$$
$$x_2 \geq 0$$

Solution The above constraints are the same as those in Example 8, so the feasible region of this problem is also given by Figure 5.2.5. In Exercise 11 we ask you to show that the objective function of this problem attains a maximum within the feasible region. By Theorem 5.2.1, this maximum must be attained at an extreme point. The values of z at the two extreme points of the feasible region are given in the following table.

Extreme Point (x_1, x_2)	Value of $z = -5x_1 + x_2$
(1, 6)	1
(3, 2)	−13

The maximum value of z is thus 1 and is attained at the extreme point $x_1 = 1$, $x_2 = 6$.

▶ **EXAMPLE 10 Inconsistent Constraints**

Find values of x_1 and x_2 that minimize

$$z = 3x_1 - 8x_2$$

subject to

$$2x_1 - x_2 \leq 4$$
$$3x_1 + 11x_2 \leq 33$$
$$3x_1 + 4x_2 \geq 24$$
$$x_1 \geq 0$$
$$x_2 \geq 0$$

Solution As can be seen from Figure 5.2.6, the intersection of the five half-planes defined by the five constraints is empty. This linear programming problem has no feasible solutions since the constraints are inconsistent. ◀

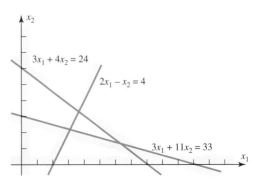

▶ Figure 5.2.6 There are no points common to all five shaded half-planes.

Exercise Set 5.2

1. Find values of x_1 and x_2 that maximize

$$z = 3x_1 + 2x_2$$

subject to

$$2x_1 + 3x_2 \leq 6$$
$$2x_1 - x_2 \geq 0$$
$$x_1 \leq 2$$
$$x_2 \leq 1$$
$$x_1 \geq 0$$
$$x_2 \geq 0$$

2. Find values of x_1 and x_2 that minimize

$$z = 3x_1 - 5x_2$$

subject to

$$2x_1 - x_2 \leq -2$$
$$4x_1 - x_2 \geq 0$$
$$x_2 \leq 3$$
$$x_1 \geq 0$$
$$x_2 \geq 0$$

3. Find values of x_1 and x_2 that minimize

$$z = -3x_1 + 2x_2$$

subject to

$$3x_1 - x_2 \geq -5$$
$$-x_1 + x_2 \geq 1$$
$$2x_1 + 4x_2 \geq 12$$
$$x_1 \geq 0$$
$$x_2 \geq 0$$

4. Solve the linear programming problem posed in Example 2.

5. Solve the linear programming problem posed in Example 3.

6. In Example 5 the constraint $x_1 - x_2 \leq 7$ is said to be *nonbinding* because it can be removed from the problem without affecting the solution. Likewise, the constraint $x_2 \leq 6$ is said to be *binding* because removing it will change the solution.

(a) Which of the remaining constraints are nonbinding and which are binding?

(b) For what values of the right-hand side of the nonbinding constraint $x_1 - x_2 \leq 7$ will this constraint become binding? For what values will the resulting feasible set be empty?

(c) For what values of the right-hand side of the binding constraints $x_2 \leq 6$ will this constraint become nonbinding? For what values will the resulting feasible set be empty?

7. A trucking firm ships the containers of two companies, A and B. Each container from company A weighs 40 pounds and is 2 cubic feet in volume. Each container from company B weighs 50 pounds and is 3 cubic feet in volume. The trucking firm charges company A $2.20 for each container shipped and charges company B $3.00 for each container shipped. If one of the firm's trucks cannot carry more than 37,000 pounds and cannot hold more than 2000 cubic feet, how many containers from companies A and B should a truck carry to maximize the shipping charges?

8. Repeat Exercise 7 if the trucking firm raises its price for shipping a container from company A to $2.50.

9. A manufacturer produces sacks of chicken feed from two ingredients, A and B. Each sack is to contain at least 10 ounces of nutrient N_1, at least 8 ounces of nutrient N_2, and at least 12 ounces of nutrient N_3. Each pound of ingredient A contains 2 ounces of nutrient N_1, 2 ounces of nutrient N_2, and 6 ounces of nutrient N_3. Each pound of ingredient B contains 5 ounces of nutrient N_1, 3 ounces of nutrient N_2, and 4 ounces of nutrient N_3. If ingredient A costs 8 cents per pound and ingredient B costs 9 cents per pound, how much of each ingredient should the manufacturer use in each sack of feed to minimize his costs?

10. If the objective function of a linear programming problem has the same value at two adjacent extreme points, show that it has the same value at all points on the straight line segment connecting the two extreme points. [*Hint:* If (x_1', x_2') and (x_1'', x_2'') are any two points in the plane, a point (x_1, x_2) lies on the straight line segment connecting them if

$$x_1 = tx_1' + (1 - t)x_1''$$

and

$$x_2 = tx_2' + (1 - t)x_2''$$

where t is a number in the interval $[0, 1]$.]

11. Show that the objective function in Example 9 attains a maximum value in the feasible set. [*Hint:* Examine the level curves of the objective function.]

Section 5.2 Technology Exercises

The following exercises are designed to be solved using a technology utility. Typically, this will be MATLAB, *Mathematica*, Maple, Derive, or Mathcad, but it may also be some other type of linear algebra software or a scientific calculator with some linear algebra capabilities. For each exercise you will need to read the relevant documentation for the particular utility you are using. The goal of these exercises is to provide you with a basic proficiency with your technology utility. Once you have mastered the techniques in these exercises, you will be able to use your technology utility to solve many of the problems in the regular exercise sets.

T1. Consider the feasible region consisting of $0 \leq x, 0 \leq y$ along with the set of inequalities

$$x \cos \left(\frac{(2k+1)\pi}{4n} \right) + y \sin \left(\frac{(2k+1)\pi}{4n} \right) \leq \cos \left(\frac{\pi}{4n} \right)$$

for $k = 0, 1, 2, \ldots, n-1$. Maximize the objective function

$$z = 3x + 4y$$

assuming that (a) $n = 1$, (b) $n = 2$, (c) $n = 3$, (d) $n = 4$, (e) $n = 5$, (f) $n = 6$, (g) $n = 7$, (h) $n = 8$, (i) $n = 9$, (j) $n = 10$, and (k) $n = 11$. (l) Next, maximize this objective function using the nonlinear feasible region, $0 \leq x, 0 \leq y$, and

$$x^2 + y^2 \leq 1$$

(m) Let the results of parts (a) through (k) begin a sequence of values for z_{max}. Do these values approach the value determined in part (l)? Explain.

T2. Repeat Exercise T1 using the objective function $z = x + y$.

5.3 The Simplex Method

In this section, we describe a procedure involving matrix algebra for maximizing a linear expression of two or more variables subject to a set of linear constraints.

> **PREREQUISITES:** Linear Systems

All linear optimizing (also known as linear programming) problems can be solved using an algorithm called the simplex method. This procedure essentially uses matrices and elementary row operations. Its applications in everyday life—transportation logistics, production planning to name just two—are as frequent as they are varied.

Standard Maximizing Problem

In this text, we shall focus our attention on the standard linear programming problem which consists of **maximizing** a linear function that is accompanied by constraints whose inequalities are of the "$\leq$" type. In addition, all variables must be assumed nonnegative.

Let us begin with an example, which we shall use to introduce the vocabulary and steps associated to the simplex algorithm.

▶ **EXAMPLE 1 Maximizing a Two-Variable Problem**

Find the values of variables x and y that maximize the objective function

$$P = 4x + 5y$$

subject to

$$x + 3y \leq 24$$
$$x - y \leq 4$$
$$x \leq 6$$
$$x \leq 0$$
$$y \geq 0$$

The following graph illustrates this problem's feasible region. We shall refer to it frequently to demonstrate the effect of the simplex method's steps.

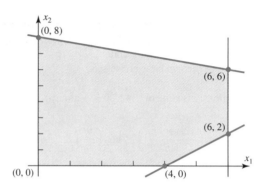

Solution

Step 1. **Converting the Problem into Standard Dorm**

Convert all constraints into equations by adding a "slack" variable to their left-hand side.

Rewrite the objective function with all variables on the left-hand side.

$$\begin{array}{llll}
\text{constraint 1:} & x + 3y \leq 24 & \Rightarrow & x + 3y + s_1 = 24 \\
\text{constraint 2:} & x - y \leq 4 & \Rightarrow & x - y + s_2 = 4 \\
\text{constraint 3:} & x \leq 6 & \Rightarrow & x + s_3 = 6 \\
\text{objective:} & P = 4x + 5y & \Rightarrow & -4x - 5y + P = 0
\end{array}$$

The new variables s_1, s_2, and s_3 are called **slack variables**. They are assumed to be nonnegative and are added to the left-hand side of each constraint to create equations rather than inequalities.

For instance, the coordinate point $(6, 2)$ of the feasible region satisfies the constraint $x + 3y \leq 24$ since $6 + 3(2) = 12 \leq 24$. There is, however, a difference of 12 between the two sides of the inequality. Point $(6, 2)$ would satisfy the equation $x + 3y + x_1 = 24$ were we to let the slack variable $s_1 = 12$.

Coordinate point $(6, 6)$ also satisfies the inequality $x + 3y \leq 24$ since $6 + 3(6) = 24 \leq 24$. This time, notice that there is no difference between the two sides of the inequality. Point $(6, 6)$ would therefore satisfy the equation $x + 3y + s_1 = 24$ were we to let the slack variable $s_1 = 0$.

Including slack variables seems artificial at first but keep in mind that we have never performed matrix operations involving inequalities. Matrices are designed to represent equations, which explains the necessity to "create" them ourselves in this case.

Step 2. **Set Up the Simplex Matrix**

Rewrite all equations (constraints and objective function) in their *standard* form into a matrix, placing the objective function in the last row. As always, the matrix will keep track only of each variable's coefficients.

$$\begin{array}{ccccccc}
x & y & s_1 & s_2 & s_3 & P & \\
\end{array}$$
$$\left[\begin{array}{cccccc|c}
1 & 3 & 1 & 0 & 0 & 0 & 24 \\
1 & -1 & 0 & 1 & 0 & 0 & 4 \\
1 & 0 & 0 & 0 & 1 & 0 & 6 \\
\hline
-4 & -5 & 0 & 0 & 0 & 1 & 0
\end{array} \right]$$

Let us now introduce some vocabulary.

We shall refer to as **basic** all variables whose columns are filled with a single 1 and with zeros everywhere else. One will notice these variables are somewhat comparable to "leading variables." All other variables are considered **nonbasic**.

From our previous simplex matrix, we can deduce that s_1, s_2, s_3, and P are all basic variables, leaving x and y as nonbasic variables.

$$
\begin{array}{cccccc}
x & y & s_1 & s_2 & s_3 & P \\
\end{array}
$$

$$
\left[
\begin{array}{cccccc|c}
1 & 3 & ① & 0 & 0 & 0 & 24 \\
1 & -1 & 0 & ① & 0 & 0 & 4 \\
1 & 0 & 0 & 0 & ① & 0 & 6 \\
\hline
-4 & -5 & 0 & 0 & 0 & ① & 0 \\
\end{array}
\right]
$$

By convention, we must assign to nonbasic variables the value 0. For instance, since the above matrix depicts x and y as being nonbasic variables, we must consider that $x = 0$, and that $y = 0$. Once this has been established, the other variables' values can be read directly from the matrix:

Row 1: $s_1 = 24$ Row 3: $s_3 = 6$

Row 2: $s_2 = 4$ Row 4: $P = 0$

Graphically, we are currently located at corner $(x = 0, y = 0)$ of the feasible region. It makes sense that $P = 0$ since the objective function is defined as $4x + 5y$. One can guess that this is certainly not the maximum value function P can attain.

So, how does one know if and when a maximum has been reached?
When the objective function has been maximized, no negative numbers should appear at the bottom row of the simplex matrix. One can observe that "-4" and "-5" appear in the initial matrix of our example. Clearly, the maximum solution has not yet been reached. To accomplish this, operations must be done to the matrix.

Step 3.   **Simplex Operations (Pivoting)**
Identify the column with the largest negative number in the bottom row.

Considering only strictly positive numbers, identify the entry in the chosen column whose ratio with the constants in the rightmost column is smallest. The ratio is the result of the constant in the rightmost column divided by the corresponding entry in the chosen column.

Transform this entry into a "pivoting 1"; use it to make all numbers above and below it zeros.

In our example, "-5" is the largest negative number in the bottom row and is found in the y-column. Only one number, that is the circled 3, in this column is neither zero nor negative. Its corresponding ratio with the right-hand column is 8 (no comparison need be done with other ratios since 3 is the only positive number in the y-column).

$$
\begin{array}{cccccc}
x & y & s_1 & s_2 & s_3 & P \\
\end{array}
$$

$$
\left[
\begin{array}{cccccc|c}
1 & ③ & 1 & 0 & 0 & 0 & 24 \\
1 & -1 & 0 & 1 & 0 & 0 & 4 \\
1 & 0 & 0 & 0 & 1 & 0 & 6 \\
\hline
-4 & -5 & 0 & 0 & 0 & 1 & 0 \\
\end{array}
\right]
\qquad \xleftarrow{\quad} \quad \frac{24}{3} = 8 \quad \textit{ratio}
$$

We must transform the 3 into a pivoting 1 by multiplying row 1 by $\frac{1}{3}$.

$$
\begin{array}{ccccccc}
x & y & s_1 & s_2 & s_3 & P & \\
\end{array}
$$

$$
\left[\begin{array}{cccccc|c}
\frac{1}{3} & 1 & \frac{1}{3} & 0 & 0 & 0 & 8 \\
1 & -1 & 0 & 1 & 0 & 0 & 4 \\
1 & 0 & 0 & 0 & 1 & 0 & 6 \\
\hline
-4 & -5 & 0 & 0 & 0 & 1 & 0
\end{array}\right]
$$

Row operations are then necessary to replace the remaining entries of column y by zeros. Note that these operations must involve using the new "pivoting 1." After having added row 1 to row 2, and 5 times row 1 to row 4, we obtain the following matrix:

$$
\begin{array}{cccccc}
x & y & s_1 & s_2 & s_3 & P \\
\end{array}
$$

$$
\left[\begin{array}{cccccc|c}
\frac{1}{3} & ① & \frac{1}{3} & 0 & 0 & 0 & 8 \\
\frac{4}{3} & 0 & \frac{1}{3} & ① & 0 & 0 & 12 \\
1 & 0 & 0 & 0 & ① & 0 & 6 \\
\hline
-\frac{7}{3} & 0 & \frac{5}{3} & 0 & 0 & ① & 40
\end{array}\right]
$$

Before we move on, one can observe that the basic variables are now y, s_2, s_3 and P. Variables x and s_1 are nonbasic. Recall that nonbasic variables are systematically attributed the value 0. As a result, the current value of each basic variable is given by

$$
\begin{array}{ll}
\text{Row 1:} \quad y = 8 & \text{Row 3:} \quad s_3 = 6 \\
\text{Row 2:} \quad s_2 = 12 & \text{Row 4:} \quad P = 40
\end{array}
$$

We are now located at corner $(x = 0, y = 8)$ of the feasible region.

Our objective, which is to maximize function P, has not yet been reached since a negative still appears at the bottom row. More operations will need to be done. We shall repeat step 3 until no more negatives appear in the last row.

Where will we now insert a new leading 1?

The column containing the largest (and only) negative number in its bottom row is the x-column. From this column, we then must identify the positive entry whose ratio with the rightmost constant is smallest.

$$
\begin{array}{cccccc}
x & y & s_1 & s_2 & s_3 & P \\
\end{array}
$$

$$
\left[\begin{array}{cccccc|c}
\frac{1}{3} & 1 & \frac{1}{3} & 0 & 0 & 0 & 8 \\
\frac{4}{3} & 0 & \frac{1}{3} & 1 & 0 & 0 & 12 \\
① & 0 & 0 & 0 & 1 & 0 & 6 \\
\hline
-\frac{7}{3} & 0 & \frac{5}{3} & 0 & 0 & 1 & 40
\end{array}\right]
$$

ratio

$8 \div \frac{1}{3} = 24$

$12 \div \frac{4}{3} = 9$

$6 \div 1 = 6$

Since the boxed entry is already a 1, we can immediately begin creating zeros above and below it. Adding appropriate multiples of row 3 to the others, we obtain

$$
\begin{array}{cccccc}
x & y & s_1 & s_2 & s_3 & P \\
\end{array}
$$

$$
\left[\begin{array}{cccccc|c}
0 & ① & \frac{1}{3} & 0 & -\frac{1}{3} & 0 & 6 \\
0 & 0 & \frac{1}{3} & ① & -\frac{4}{3} & 0 & 4 \\
① & 0 & 0 & 0 & 1 & 0 & 6 \\
\hline
0 & 0 & \frac{5}{3} & 0 & \frac{7}{3} & ① & 54
\end{array}\right]
$$

This time, no negatives appear in the last row. We have therefore reached the optimal solution.

What solution does this final matrix table represent?
Observe that x, y, s_2 and P are basic variables. The others, s_1 and s_3, are nonbasic and hence equal 0. By inspecting each row, one can identify the values of the basic variables:

$$\text{Row 1:} \quad y = 6 \qquad \text{Row 3:} \quad x = 6$$
$$\text{Row 2:} \quad s_2 = 4 \qquad \text{Row 4:} \quad P = 54$$

Hence, we are now located at corner $(x = 6, y = 6)$ of the feasible region. One will observe that, essentially, the effect of our simplex operations was to move from one corner of the feasible region to the next, until a maximum was finally reached.
    The solution we have obtained can be confirmed through a quick verification.
If $x = 6$ and $y = 6$, we should obtain

$$P = 4(6) + 5(6) = 24 + 30 = 54$$

Constraint 1: $\quad (6) + 3(6) = 24 \leq 24 \quad$ (no slack, so $s_1 = 0$)

Constraint 2: $\quad (6) - (6) = 0 \leq 4 \qquad$ (difference of 4, so $s_2 = 4$)

Constraint 3: $\quad (6) \leq 6 \qquad\qquad\qquad$ (no slack, so $s_3 = 0$) ◀

    The following example will underline one of the advantages of the simplex method; it can easily be generalized to problems containing more than two variables. This cannot be said about the graphical method for obvious reasons.

▶ **EXAMPLE 2 Maximizing a Three-Variable Problem**
Find the values of variables x, y and z that maximize the objective function

$$P = 9x + 6y + 2z$$

subject to

$$x + y + z \leq 100$$
$$3x + 2y + 5z \leq 120$$
$$2x + z \leq 60$$
$$(x, y, z \geq 0)$$

Solution First, we must convert the constraints and objective function into *standard* form

constraint 1:	$x + y + z \leq 100$	$\Rightarrow \quad x + y + z + s_1 = 100$
constraint 2:	$3x + 2y + 5z \leq 120$	$\Rightarrow \quad 3x + 2y + 5z + s_2 = 120$
constraint 3:	$2x + z \leq 60$	$\Rightarrow \quad 2x + z + s_3 = 60$
objective:	$P = 9x + 6y + 2z$	$\Rightarrow \quad -9x - 6y - 2z + P = 0$

The initial simplex matrix is obtained from the above equations.

$$
\begin{array}{ccccccc|c}
x & y & z & s_1 & s_2 & s_3 & P & \\
\hline
1 & 1 & 1 & 1 & 0 & 0 & 0 & 100 \\
3 & 2 & 5 & 0 & 1 & 0 & 0 & 120 \\
2 & 0 & 1 & 0 & 0 & 1 & 0 & 60 \\
\hline
-9 & -6 & -2 & 0 & 0 & 0 & 1 & 0 \\
\end{array}
$$

Because negative numbers appear in the bottom row, we know a maximum value has not yet been attained. We must therefore perform simplex operations in order to improve our current situation. The x-column has the largest negative number in its bottom row.

x	y	z	s_1	s_2	s_3	P		ratio
1	1	1	1	0	0	0	100	100
3	2	5	0	1	0	0	120	40
②	0	1	0	0	1	0	60	30
−9	−6	−2	0	0	0	1	0	

We have found that the circled 2 is to be turned into a pivoting 1, which can be accomplished by multiplying row 3 by $\frac{1}{2}$. Multiples of row 3 can be then be used to replace the remaining entries of the x-column by zeros. The result of these operations are shown in the following matrix:

x	y	z	s_1	s_2	s_3	P		ratio
0	1	$\frac{1}{2}$	1	0	$-\frac{1}{2}$	0	70	70
0	②	$\frac{7}{2}$	0	1	$-\frac{3}{2}$	0	30	15
1	0	$\frac{1}{2}$	0	0	$\frac{1}{2}$	0	30	
0	−6	$\frac{5}{2}$	0	0	$\frac{9}{2}$	1	270	

A negative number still appears at the bottom of the y-column, which means the objective function can still be increased. By comparing ratios, it is obvious we must now make the 2 appearing in the y-column our new pivoting 1. Elementary row operations involving this pivoting 1 are used to create zeros in the remaining entries of column y.

x	y	z	s_1	s_2	s_3	P	
0	0	$-\frac{5}{4}$	1	$-\frac{1}{2}$	$\frac{1}{4}$	0	55
0	1	$\frac{7}{4}$	0	$\frac{1}{2}$	$-\frac{3}{4}$	0	15
1	0	$\frac{1}{2}$	0	0	$\frac{1}{2}$	0	30
0	0	13	0	3	0	1	360

The fact no more negative values appear in the bottom row indicates function P has reached its maximum value. According to this final matrix, $z = 0$, $s_2 = 0$, and $s_3 = 0$ since they are nonbasic variables. The remaining variables' values are

$$x = 30, \quad y = 15, \quad s_1 = 55, \quad P = 360 \blacktriangleleft$$

Post-Optimal Analysis   Once a problem has been solved by using the simplex algorithm, it is of utmost importance and utility to be able to interpret the meaning of the final matrix. This table displays the optimal solution, but it can also be used to

- determine how this solution would vary if constraints were to change;
- identify which constraints are binding or nonbinding;
- detect the existence of other solutions.

Interpreting the Slack Variables' Columns

Let us take a another look at the final matrix obtained in Example 2.

$$
\begin{array}{ccccccc}
x & y & z & s_1 & s_2 & s_3 & P \\
\end{array}
$$

$$
\left[
\begin{array}{ccccccc|c}
0 & 0 & -\frac{5}{4} & 1 & -\frac{1}{2} & \frac{1}{4} & 0 & 55 \\
0 & 1 & \frac{7}{4} & 0 & \frac{1}{2} & -\frac{3}{4} & 0 & 15 \\
1 & 0 & \frac{1}{2} & 0 & 0 & \frac{1}{2} & 0 & 30 \\
\hline
0 & 0 & 13 & 0 & 3 & 0 & 1 & 360
\end{array}
\right]
$$

Useless in appearance, the entries in the slack variables' columns actually provide invaluable information. Every number indicates how the basic variable in the corresponding row will be affected when the right-hand side of that constraint is increased by one unit.

This means, for instance, that if the right-hand side of contraint 2 were to be increased by one unit:

- the basic variable in row 1 (s_1) would decrease by $\frac{1}{2}$;
- the basic variable in row 2 (y) would increase by $\frac{1}{2}$;
- the basic variable in row 3 (x) would be unaffected;
- the basic variable in row 4 (P) would increase by 3.

Unless nonnegativity restrictions are not observed, all nonbasic variables remain nonbasic.

▶ EXAMPLE 3 Effect of a Changing Constraint

Let us assume that the values of x, y, and z that maximize the objective function

$$
P = 9x + 6y + 2z
$$

subject to the constraints

$$
x + y + z \leq 100
$$
$$
3x + 2y + 5z \leq 120
$$
$$
2x + z \leq 60
$$
$$
(x, y, z \geq 0)
$$

are $x = 30$, $y = 15$, $z = 0$, and that the following is the final matrix obtained after using the simplex method:

$$
\begin{array}{ccccccc}
x & y & z & s_1 & s_2 & s_3 & P \\
\end{array}
$$

$$
\left[
\begin{array}{ccccccc|c}
0 & 0 & -\frac{5}{4} & 1 & -\frac{1}{2} & \frac{1}{4} & 0 & 55 \\
0 & 1 & \frac{7}{4} & 0 & \frac{1}{2} & -\frac{3}{4} & 0 & 15 \\
1 & 0 & \frac{1}{2} & 0 & 0 & \frac{1}{2} & 0 & 30 \\
\hline
0 & 0 & 13 & 0 & 3 & 0 & 1 & 360
\end{array}
\right]
$$

Find the new optimal solution if constraint 2 were changed from $3x + 2y + 5z \leq 120$ to $3x + 2y + 5z \leq 124$.

Solution We shall denote the change to the right-hand side of constraint 2 by $\Delta_2 = 4$. Recall that the entries in column s_2 indicate how each basic variable in the corresponding row will be affected when constraint 2 is increased by one unit. Of course, the effect will be 4 times greater in our current situation since the right hand of the latter has undergone an increase of 4 units.

Rather than looking at each basic variable's change individually, it is convenient to use vector forms, as shown below, to determine the new optimal solution.

$$
\underbrace{\begin{bmatrix} s_1 \\ y \\ x \\ P \end{bmatrix} = \begin{bmatrix} 55 \\ 15 \\ 30 \\ 360 \end{bmatrix}}_{\substack{\text{original} \\ \text{solution}}} + \Delta_2 \underbrace{\begin{bmatrix} -\frac{1}{2} \\ \frac{1}{2} \\ 0 \\ 3 \end{bmatrix}}_{\substack{\text{column} \\ s_2}} = \begin{bmatrix} 55 \\ 15 \\ 30 \\ 360 \end{bmatrix} + 4 \begin{bmatrix} -\frac{1}{2} \\ \frac{1}{2} \\ 0 \\ 3 \end{bmatrix} = \underbrace{\begin{bmatrix} 53 \\ 17 \\ 30 \\ 372 \end{bmatrix}}_{\substack{\text{new} \\ \text{optimal} \\ \text{solution}}}
$$

All original basic variables have remained nonnegative. All variables that were previously nonbasic remain as such (thus $z = 0$). ◀

If more than one constraint is modified, the effects of each change are added. Changing many constraints therefore requires practically the same amount of work as changing only one.

▶ **EXAMPLE 4** **Effect of Changing Multiple Constraints**

Consider once again the optimizing problem discussed in Example 2. What would be the new optimal solution if all three constraints were changed as follows:

$$
\begin{aligned}
x + y + z &\le 100 && \xrightarrow{\Delta_1 = -10} & x + y + z &\le 90 \\
3x + 2y + 5z &\le 120 && \xrightarrow{\Delta_2 = 8} & 3x + 2y + 5z &\le 128 \\
2x + z &\le 60 && \xrightarrow{\Delta_3 = -4} & 2x + z &\le 56
\end{aligned}
$$

Solution The s_1, s_2, and s_3 columns of the final simplex table will be needed in this case since all their corresponding constraints are undergoing changes. The total variation to the original solution will be obtained by adding the impact caused by each constraint's change, that is

$$
\begin{bmatrix} s_1 \\ y \\ x \\ P \end{bmatrix} = \begin{bmatrix} 55 \\ 15 \\ 30 \\ 360 \end{bmatrix} + \Delta_1 \begin{bmatrix} 1 \\ 0 \\ 0 \\ 0 \end{bmatrix} + \Delta_2 \begin{bmatrix} -\frac{1}{2} \\ \frac{1}{2} \\ 0 \\ 3 \end{bmatrix} + \Delta_3 \begin{bmatrix} \frac{1}{4} \\ -\frac{3}{4} \\ \frac{1}{2} \\ 0 \end{bmatrix}
$$

Replacing the Δ's by their respective values, we obtain

$$
\begin{bmatrix} s_1 \\ y \\ x \\ P \end{bmatrix} = \begin{bmatrix} 55 \\ 15 \\ 30 \\ 360 \end{bmatrix} + (-10) \begin{bmatrix} 1 \\ 0 \\ 0 \\ 0 \end{bmatrix} + (8) \begin{bmatrix} -\frac{1}{2} \\ \frac{1}{2} \\ 0 \\ 3 \end{bmatrix} + (-4) \begin{bmatrix} \frac{1}{4} \\ -\frac{3}{4} \\ \frac{1}{2} \\ 0 \end{bmatrix} = \begin{bmatrix} 40 \\ 22 \\ 28 \\ 384 \end{bmatrix}
$$

Note that all nonbasic variables (z, s_2, and s_3) remain nonbasic. ◀

Identifying Binding Constraints A constraint is binding when its limiting value is attained at the optimal solution. In other words, no slack remains between the left- and right-hand side of this particular constraint.

▶ **EXAMPLE 5** **Finding Binding Constraints**

Let us consider the final matrix obtained from Example 2.

x	y	z	s_1	s_2	s_3	P	
0	0	$-\frac{5}{4}$	1	$-\frac{1}{2}$	$\frac{1}{4}$	0	55
0	1	$\frac{7}{4}$	0	$\frac{1}{2}$	$-\frac{3}{4}$	0	15
1	0	$\frac{1}{2}$	0	0	$\frac{1}{2}$	0	30
0	0	13	0	3	0	1	360

The optimal solution depicts s_2 and s_3 as nonbasic variables. This means

$$s_2 = 0 \quad \rightarrow \quad \text{constraint 2 is binding}$$
$$s_3 = 0 \quad \rightarrow \quad \text{constraint 3 is binding}$$

However, one will notice that s_1 is basic. More specifically, $s_1 = 55$. This means the optimal solution does not occur at the boundary value of constraint 1. It is therefore a nonbinding constraint. In an economic context, this would mean it is not advantageous to use all the resources represented by constraint 1 in order to maximize the objective function (profit, for example). ◀

Existence of Multiple Solutions   One can detect the existence of multiple solutions of a linear optimizing problem if the objective function reaches its maximum value at different points of the feasible region. In terms of the simplex matrix, this means that basic variables can be changed without altering the objective's maximum value. The following example will illustrate how one can detect the existence of multiple solutions, and how the latter can be found.

▶ **EXAMPLE 6**

According to the final table of Example 2, shown below, the objective function (P) is maximized when $x = 30$, $y = 15$, $z = 0$.

x	y	z	s_1	s_2	s_3	P	
0	0	$-\frac{5}{4}$	1	$-\frac{1}{2}$	$\frac{1}{4}$	0	55
0	1	$\frac{7}{4}$	0	$\frac{1}{2}$	$-\frac{3}{4}$	0	15
1	0	$\frac{1}{2}$	0	0	$\frac{1}{2}$	0	30
0	0	13	0	3	0	1	360

Explain why there are multiple optimal solutions to this linear programming problem, and find one alternative solution.

Solution According to the final table's last row, $0x + 0y + 13z + 0s_1 + 3s_2 + 0s_3 + 1P = 360$.

Isolating P from the previous equation, we obtain

$$P = 360 - 0x - 0y - 13z - 0s_1 - 3s_2 - 0s_3$$

Variable z is nonbasic, and for good reason. If z were increased by 1 unit, the value of P would be reduced by 13. Since our objective is to maximize function P, having z as a basic variable would not be wise. Variable s_2 is also nonbasic. Function P would be reduced by 3 for every unit increase of s_2. It is therefore not profitable to make s_2 a basic variable. One will quickly conclude that nothing can be done to make the value of P any greater than it already is (360). However, although s_3 is nonbasic, the 0 coefficient at the bottom of its column indicates there will be no effect to the maximum if s_3's value were to be changed. In other words, s_3 could be made basic without diminishing the current value of P.

In short, the existence of multiple optimal solutions can be detected when one of the nonbasic variables has a "0" in its bottom row. An alternative optimal solution is obtained by making this variable basic through a regular pivot operation. We shall treat this 0 the same way we treated negative numbers in the bottom row earlier. By identifying the lowest ratio, one can identify the location of a new pivoting 1:

$$
\begin{array}{cccccccc}
x & y & z & s_1 & s_2 & s_3 & P & \\
\end{array}
$$

$$
\left[\begin{array}{ccccccc|c}
0 & 0 & -\frac{5}{4} & 1 & -\frac{1}{2} & \frac{1}{4} & 0 & 55 \\
0 & 1 & \frac{7}{4} & 0 & \frac{1}{2} & -\frac{3}{4} & 0 & 15 \\
1 & 0 & \frac{1}{2} & 0 & 0 & \left(\frac{1}{2}\right) & 0 & 30 \\
\hline
0 & 0 & 13 & 0 & 3 & 0 & 1 & 360
\end{array}\right]
\qquad
\begin{array}{c}
\textit{ratio} \\
220 \\
\\
60 \\
\\
\end{array}
$$

Multiplying row 3 by 2, the $\frac{1}{2}$ is transformed into a pivoting 1.

$$
\begin{array}{cccccccc}
x & y & z & s_1 & s_2 & s_3 & P & \\
\end{array}
$$

$$
\left[\begin{array}{ccccccc|c}
0 & 0 & -\frac{5}{4} & 1 & -\frac{1}{2} & \frac{1}{4} & 0 & 55 \\
0 & 1 & \frac{7}{4} & 0 & \frac{1}{2} & -\frac{3}{4} & 0 & 15 \\
2 & 0 & 1 & 0 & 0 & ① & 0 & 60 \\
\hline
0 & 0 & 13 & 0 & 3 & 0 & 1 & 360
\end{array}\right]
$$

The reader can verify that appropriate row operations involving the new pivoting 1 reveals the following alternative final table:

$$
\begin{array}{cccccccc}
x & y & z & s_1 & s_2 & s_3 & P & \\
\end{array}
$$

$$
\left[\begin{array}{ccccccc|c}
-\frac{1}{2} & 0 & -\frac{3}{2} & 1 & -\frac{1}{2} & 0 & 0 & 40 \\
\frac{3}{2} & 1 & \frac{5}{2} & 0 & \frac{1}{2} & 0 & 0 & 60 \\
2 & 0 & 1 & 0 & 0 & 1 & 0 & 60 \\
\hline
0 & 0 & 13 & 0 & 3 & 0 & 1 & 360
\end{array}\right]
$$

According to the above table, the objective's maximum value of 360 is obtained when $x = 0$, $y = 60$, $z = 0$.

One can show that the objective function remains equal to 360 at any point of the segment joining the original solution ($x = 30$, $y = 15$, $z = 0$) and the one we have just obtained. In addition, all constraints are respected along this segment. ◄

Exercise Set 5.3

▶ In Exercises 1–5, solve the given standard linear programming problem using the simplex method.

1. Maximize

$$P = 5x + 3y$$

Subject to the constraints

$$x + 2y \le 20$$
$$x - y \le 5$$
$$x \le 8$$
$$(x, y \ge 0)$$

2. Maximize

$$f = 3x + 2y$$

Subject to the constraints

$$x + 2y \le 16$$
$$x + y \le 10$$
$$2x + y \le 16$$
$$(x, y \ge 0)$$

3. Maximize
$$f = 5x + 2y$$

Subject to the constraints
$$x + 4y \leq 36$$
$$x - y \leq 6$$
$$x \leq 8$$
$$(x, y \geq 0)$$

4. Maximize
$$P = 7x + 5y + 6z$$

Subject to the constraints
$$x + y - z \leq 3$$
$$x + 2y + z \leq 8$$
$$x + y \leq 5$$
$$(x, y, z \geq 0)$$

5. Maximize
$$P = x + 2y + 5z + t$$

Subject to the constraints
$$x + 2y + z - t \leq 20$$
$$-x + y + z + t \leq 12$$
$$2x + y + z - t \leq 30$$
$$(x, y, z, t \geq 0)$$

6. Consider the linear programming problem described in Exercise 2.

(a) Identify all constraints that are binding?

(b) How many solutions does this problem have?

(c) What are the values of x, y, and P that produce the optimal solution if the second constraint were changed to $x + y \leq 8$?

7. RiverFlo produces two types of River Kayaks. The *Coolit* model is designed for the "drifters," that is for leisurely promenades on the water. The *MoveOver* model is designed for the more adventurous kayakers. There are constraints to be observed during the production process. The company orders 500 m^2 of carbon fibre and 200 m^2 of fibreglass every week. The company disposes of a maximum of 240 man-hours (for example, 6 people who work 40 hours each) per week to complete its production. The *MoveOver* model requires more production time because of handcrafted elements.

The following table describes the material composition as well as the required production time for each model.

Model	Carbon Fibre (m^2)	Fibreglass (m^2)	Production Time (hours)
Coolit	15	10	4
MoveOver	20	5	6

The cost of every square metre of carbon fibre is $6, whereas $4 is the cost for every square metre of fibreglass. Suppose all employees are paid at an hourly wage of $20. A *Coolit* kayak is sold at a price of $255. The *Moveover* kayak is sold for $300.

The objective is to determine how many kayaks of each model should be produced in order to maximize profits.

(a) Define the variables for this problem.

(b) Set up the objective function and the constraints in terms of these variables.

(c) Solve the problem using the simplex method. What is the maximum profit the company can obtain?

(d) Identify all binding constraints.

8. The following is the optimal table obtained after solving a linear programming problem through the simplex method.

x	y	z	s_1	s_2	s_3	P	
1	0	2	0	1	1.5	0	24
0	1	1	0	-1	1.5	0	16
0	0	1	1	0.5	2	0	6
0	0	0	0	2	5	1	100

(a) If constraint 3 were increased by 2 units, how will the optimal solution be affected?

(b) Explain why a new solution cannot be obtained by post-optimal analysis if constraint 3 were reduced by 4.

(c) Explain how one knows there are many optimal solutions to the linear programming problem.

(d) If constraint 2 were increased by 6 and constraint 3 were decreased by 2 units, what values would x, y, z, and P take at the new optimal solution?

(e) The z variable is currently nonbasic, which means we have chosen to produce no units of product z. Find two other solutions with the same optimal profit but where z is basic.

9. Consider the following linear programming problem:

Maximize $P = 2x + 5y$

Subject to $x + 2y \leq 22$

                $2x + y \leq 20$

                $x \geq 2$

                $y \geq 3$

The above is not a standard maximizing problem due to the form of the last two constraints. Standardize and solve the problem by replacing x by $u + 2$ and y by $v + 3$.

10. A small Montreal-based airline, Queeject, is providing flights to Canada's other major cities: Toronto, Calgary, and Vancouver. Daily profits are generated by computing the difference between the company's revenues (from ticket sales) and its costs (fuel, employees on board, meals provided, rent paid to airport, etc.). The company's daily profit is described as follows.

$$P = 400t + 360c + 450v$$

where the variables t, c, and v are

t: number of planes used for Montreal-Toronto round-trips

c: number of planes used for Montreal-Calgary round-trips

v: number of planes used for Montreal-Vancouver round-trips

The following table shows the staffing requirements for each destination, as well as the demand to be fulfilled.

	Pilots	On-board staff	Minimum number of flights per day
Toronto	2	4	3
Calgary	3	6	2
Vancouver	3	8	1

Queeject employs 56 pilots, has a total personnel of 120 attendants to provide service during the flights, and has a fleet of 18 airplanes at its disposal.

(a) How many planes should Queeject send to each city if it wishes to maximize its profits?

(b) Are there many solutions to this optimizing problem?

(c) The company wants to expand and considers buying 2 more planes and hiring 12 extra on-board attendants. How will the optimal solution be affected?

(d) According to the Queeject Pilots Union, the pilots are already overworked. In their opinion, the changes proposed in part (c) will worsen their working conditions. Is the union exaggerating?

Section 5.3   Technology Exercises

The following exercises are designed to be solved using a technology utility. Typically, this will be MATLAB, *Mathematica*, Maple, Derive, or Mathcad, but it may also be some other type of linear algebra software or a scientific calculator with some linear algebra capabilities. For each exercise you will need to read the relevant documentation for the particular utility you are using. The goal of these exercises is to provide you with a basic proficiency with your technology utility. Once you have mastered the techniques in these exercises, you will be able to use your technology utility to solve many of the problems in the regular exercise sets.

T1. A financial advisor must invest all of the $500,000 submitted to him by a client. There are 4 major investment sectors from which he may choose, all of which are presented in the table below along with their average yearly return.

Investment Sector	Average Return (μ)
Government bonds	4%
New technologies	15%
Natural resources	6%
International investing	8%

Based on return alone, one would invest everything in new technologies. However, it is an internal policy of the advisor's firm to place at least 30% of the total investment in the safer sectors, bonds, and natural resources. International investing is a tempting sector, but legislation imposes that it represent no more than 10% of one's portfolio.

Let x_{GB}, x_{NT}, x_{NR}, x_{II} represent the amounts that will be invested in each sector. How should the financial advisor distribute the sum to maximize the profit function

$$P = 0.04x_{GB} + 0.15x_{NT} + 0.06x_{NR} + 0.08x_{II}$$

while respecting all constraints?

T2. In Exercise T1, the profit function depended only on the average return of each investment sector, without consideration for their possible variations. Recent experience has shown that certain sectors are much riskier than others, but can also be very rewarding. The table below shows each sector's standard deviation.

Investment Sector	Standard Deviation (σ)
Government bonds	0%
New technologies	18%
Natural resources	3%
International investing	6%

One will notice that the return on government bonds does not vary. In contrast, it would not be unusual for investments in new technologies to vary by 18% from their average return. An aggressive investor may be interested in this sector because of its high-reward possibility. A conservative investor may perceive this sector as very risky. In taking into account the different perceptions of clients, financial planners have modeled the following *utility* function

$$U = 0.04x_{GB} + 0.15x_{NT} + 0.06x_{NR} + 0.08x_{II}$$
$$+ k\left[(0x_{GB})^2 + (0.18x_{NT})^2 + (0.03x_{NR})^2 + (0.06x_{II})^2\right]$$

The value of k depends on the investor's tolerance to risk. The aggressive investor is attributed the value $k = 1$, whereas this pa-

rameter's value is -1 for the conservative investor. We use $k = 0$ for investors who are ambivalent to risk (this is the case you have treated in T1).

For aggressive and conservative investors, how should the financial advisor distribute the sum to maximize the utility function, while respecting all constraints?

5.4 Markov Chains

In this section we describe a general model of a system that changes from state to state. We then apply the model to several concrete problems.

> **PREREQUISITES:** Linear Systems
> Matrices
> Intuitive Understanding of Limits

A Markov Process

Suppose a physical or mathematical system undergoes a process of change such that at any moment it can occupy one of a finite number of states. For example, the weather in a certain city could be in one of three possible states: sunny, cloudy, or rainy. Or an individual could be in one of four possible emotional states: happy, sad, angry, or apprehensive. Suppose that such a system changes with time from one state to another and at scheduled times the state of the system is observed. If the state of the system at any observation cannot be predicted with certainty, but the probability that a given state occurs can be predicted by just knowing the state of the system at the preceding observation, then the process of change is called a ***Markov chain*** or ***Markov process***.

> **DEFINITION 1**   If a Markov chain has k possible states, which we label as $1, 2, \ldots, k$, then the probability that the system is in state i at any observation after it was in state j at the preceding observation is denoted by p_{ij} and is called the ***transition probability*** from state j to state i. The matrix $P = [p_{ij}]$ is called the ***transition matrix of the Markov chain***.

For example, in a three-state Markov chain, the transition matrix has the form

$$\begin{array}{c} \text{Preceding State} \\ \begin{array}{ccc} 1 & 2 & 3 \end{array} \\ \begin{bmatrix} p_{11} & p_{12} & p_{13} \\ p_{21} & p_{22} & p_{23} \\ p_{31} & p_{32} & p_{33} \end{bmatrix} \begin{array}{l} 1 \\ 2 \\ 3 \end{array} \end{array} \quad \text{New State}$$

In this matrix, p_{32} is the probability that the system will change from state 2 to state 3, p_{11} is the probability that the system will still be in state 1 if it was previously in state 1, and so forth.

▶ **EXAMPLE 1 Transition Matrix of the Markov Chain**

A car rental agency has three rental locations, denoted by 1, 2, and 3. A customer may rent a car from any of the three locations and return the car to any of the three locations.

The manager finds that customers return the cars to the various locations according to the following probabilities:

Rented from Location

$$
\begin{array}{c}
 \quad 1 \quad\ \ 2 \quad\ \ 3 \\
\begin{bmatrix}
.8 & .3 & .2 \\
.1 & .2 & .6 \\
.1 & .5 & .2
\end{bmatrix}
\begin{array}{l}
1 \quad \text{Returned} \\
2 \quad\ \ \text{to} \\
3 \quad \text{Location}
\end{array}
\end{array}
$$

This matrix is the transition matrix of the system considered as a Markov chain. From this matrix, the probability is .6 that a car rented from location 3 will be returned to location 2, the probability is .8 that a car rented from location 1 will be returned to location 1, and so forth.

▶ **EXAMPLE 2 Transition Matrix of the Markov Chain**

By reviewing its donation records, the alumni office of a college finds that 80% of its alumni who contribute to the annual fund one year will also contribute the next year, and 30% of those who do not contribute one year will contribute the next. This can be viewed as a Markov chain with two states: state 1 corresponds to an alumnus giving a donation in any one year, and state 2 corresponds to the alumnus not giving a donation in that year. The transition matrix is

$$
P = \begin{bmatrix} .8 & .3 \\ .2 & .7 \end{bmatrix} \quad \blacktriangleleft
$$

In the examples above, the transition matrices of the Markov chains have the property that the entries in any column sum to 1. This is not accidental. If $P = [p_{ij}]$ is the transition matrix of any Markov chain with k states, then for each j we must have

$$
p_{1j} + p_{2j} + \cdots + p_{kj} = 1 \tag{1}
$$

because if the system is in state j at one observation, it is certain to be in one of the k possible states at the next observation.

A matrix with property (1) is called a ***stochastic matrix***, a ***probability matrix***, or a ***Markov matrix***. From the preceding discussion, it follows that the transition matrix for a Markov chain must be a stochastic matrix.

In a Markov chain, the state of the system at any observation time cannot generally be determined with certainty. The best one can usually do is specify probabilities for each of the possible states. For example, in a Markov chain with three states, we might describe the possible state of the system at some observation time by a column vector

$$
\mathbf{x} = \begin{bmatrix} x_1 \\ x_2 \\ x_3 \end{bmatrix}
$$

in which x_1 is the probability that the system is in state 1, x_2 the probability that it is in state 2, and x_3 the probability that it is in state 3. In general we make the following definition.

DEFINITION 2 The ***state vector*** for an observation of a Markov chain with k states is a column vector $\mathbf{x}$ whose ith component x_i is the probability that the system is in the ith state at that time.

Observe that the entries in any state vector for a Markov chain are nonnegative and have a sum of 1. (Why?) A column vector that has this property is called a ***probability vector***.

Let us suppose now that we know the state vector $\mathbf{x}^{(0)}$ for a Markov chain at some initial observation. The following theorem will enable us to determine the state vectors

$$\mathbf{x}^{(1)}, \mathbf{x}^{(2)}, \ldots, \mathbf{x}^{(n)}, \ldots$$

at the subsequent observation times.

> **THEOREM 5.4.1** *If P is the transition matrix of a Markov chain and* $\mathbf{x}^{(n)}$ *is the state vector at the nth observation, then* $\mathbf{x}^{(n+1)} = P\mathbf{x}^{(n)}$.

The proof of this theorem involves ideas from probability theory and will not be given here. From this theorem, it follows that

$$\mathbf{x}^{(1)} = P\mathbf{x}^{(0)}$$
$$\mathbf{x}^{(2)} = P\mathbf{x}^{(1)} = P^2\mathbf{x}^{(0)}$$
$$\mathbf{x}^{(3)} = P\mathbf{x}^{(2)} = P^3\mathbf{x}^{(0)}$$
$$\vdots$$
$$\mathbf{x}^{(n)} = P\mathbf{x}^{(n-1)} = P^n\mathbf{x}^{(0)}$$

In this way, the initial state vector $\mathbf{x}^{(0)}$ and the transition matrix P determine $\mathbf{x}^{(n)}$ for $n = 1, 2, \ldots$.

▶ **EXAMPLE 3 Example 2 Revisited**

The transition matrix in Example 2 was

$$P = \begin{bmatrix} .8 & .3 \\ .2 & .7 \end{bmatrix}$$

We now construct the probable future donation record of a new graduate who did not give a donation in the initial year after graduation. For such a graduate the system is initially in state 2 with certainty, so the initial state vector is

$$\mathbf{x}^{(0)} = \begin{bmatrix} 0 \\ 1 \end{bmatrix}$$

From Theorem 5.4.1 we then have

$$\mathbf{x}^{(1)} = P\mathbf{x}^{(0)} = \begin{bmatrix} .8 & .3 \\ .2 & .7 \end{bmatrix}\begin{bmatrix} 0 \\ 1 \end{bmatrix} = \begin{bmatrix} .3 \\ .7 \end{bmatrix}$$

$$\mathbf{x}^{(2)} = P\mathbf{x}^{(1)} = \begin{bmatrix} .8 & .3 \\ .2 & .7 \end{bmatrix}\begin{bmatrix} .3 \\ .7 \end{bmatrix} = \begin{bmatrix} .45 \\ .55 \end{bmatrix}$$

$$\mathbf{x}^{(3)} = P\mathbf{x}^{(2)} = \begin{bmatrix} .8 & .3 \\ .2 & .7 \end{bmatrix}\begin{bmatrix} .45 \\ .55 \end{bmatrix} = \begin{bmatrix} .525 \\ .475 \end{bmatrix}$$

Thus, after three years the alumnus can be expected to make a donation with probability .525. Beyond three years, we find the following state vectors (to three decimal places):

$$\mathbf{x}^{(4)} = \begin{bmatrix} .563 \\ .438 \end{bmatrix}, \quad \mathbf{x}^{(5)} = \begin{bmatrix} .581 \\ .419 \end{bmatrix}, \quad \mathbf{x}^{(6)} = \begin{bmatrix} .591 \\ .409 \end{bmatrix}, \quad \mathbf{x}^{(7)} = \begin{bmatrix} .595 \\ .405 \end{bmatrix}$$

$$\mathbf{x}^{(8)} = \begin{bmatrix} .598 \\ .402 \end{bmatrix}, \quad \mathbf{x}^{(9)} = \begin{bmatrix} .599 \\ .401 \end{bmatrix}, \quad \mathbf{x}^{(10)} = \begin{bmatrix} .599 \\ .401 \end{bmatrix}, \quad \mathbf{x}^{(11)} = \begin{bmatrix} .600 \\ .400 \end{bmatrix}$$

For all n beyond 11, we have

$$\mathbf{x}^{(n)} = \begin{bmatrix} .600 \\ .400 \end{bmatrix}$$

to three decimal places. In other words, the state vectors converge to a fixed vector as the number of observations increases. (We will discuss this further below.)

▶ **EXAMPLE 4 Example 1 Revisited**

The transition matrix in Example 1 was

$$\begin{bmatrix} .8 & .3 & .2 \\ .1 & .2 & .6 \\ .1 & .5 & .2 \end{bmatrix}$$

If a car is rented initially from location 2, then the initial state vector is

$$\mathbf{x}^{(0)} = \begin{bmatrix} 0 \\ 1 \\ 0 \end{bmatrix}$$

Using this vector and Theorem 5.4.1, one obtains the later state vectors listed in Table 1. For all values of n greater than 11, all state vectors are equal to $\mathbf{x}^{(11)}$ to three decimal places.

Table 1

$x^{(n)}$ \ n	0	1	2	3	4	5	6	7	8	9	10	11
$x_1^{(n)}$	0	.300	.400	.477	.511	.533	.544	.550	.553	.555	.556	.557
$x_2^{(n)}$	1	.200	.370	.252	.261	.240	.238	.233	.232	.231	.230	.230
$x_3^{(n)}$	0	.500	.230	.271	.228	.227	.219	.217	.215	.214	.214	.213

Two things should be observed in this example. First, it was not necessary to know how long a customer kept the car. That is, in a Markov process the time period between observations need not be regular. Second, the state vectors approach a fixed vector as n increases, just as in the first example. ◀

▶ **EXAMPLE 5 Using Theorem 5.4.1**

A traffic officer is assigned to control the traffic at the eight intersections indicated in Figure 5.4.1. She is instructed to remain at each intersection for an hour and then to either remain at the same intersection or move to a neighboring intersection. To avoid establishing a pattern, she is told to choose her new intersection on a random basis, with each possible choice equally likely. For example, if she is at intersection 5, her next intersection can be 2, 4, 5, or 8, each with probability $\frac{1}{4}$. Every day she starts at the

▲ Figure 5.4.1

location where she stopped the day before. The transition matrix for this Markov chain is

Old Intersection

$$
\begin{array}{c}
\begin{array}{cccccccc} 1 & 2 & 3 & 4 & 5 & 6 & 7 & 8 \end{array} \\
\begin{bmatrix}
\frac{1}{3} & \frac{1}{3} & 0 & \frac{1}{5} & 0 & 0 & 0 & 0 \\
\frac{1}{3} & \frac{1}{3} & 0 & 0 & \frac{1}{4} & 0 & 0 & 0 \\
0 & 0 & \frac{1}{3} & \frac{1}{5} & 0 & \frac{1}{3} & 0 & 0 \\
\frac{1}{3} & 0 & \frac{1}{3} & \frac{1}{5} & \frac{1}{4} & 0 & \frac{1}{4} & 0 \\
0 & \frac{1}{3} & 0 & \frac{1}{5} & \frac{1}{4} & 0 & 0 & \frac{1}{3} \\
0 & 0 & \frac{1}{3} & 0 & 0 & \frac{1}{3} & \frac{1}{4} & 0 \\
0 & 0 & 0 & \frac{1}{5} & 0 & \frac{1}{3} & \frac{1}{4} & \frac{1}{3} \\
0 & 0 & 0 & 0 & \frac{1}{4} & 0 & \frac{1}{4} & \frac{1}{3}
\end{bmatrix}
\begin{array}{l} 1 \\ 2 \\ 3 \\ 4 \\ 5 \\ 6 \\ 7 \\ 8 \end{array}
\end{array}
$$

New Intersection

If the traffic officer begins at intersection 5, her probable locations, hour by hour, are given by the state vectors given in Table 2. For all values of n greater than 22, all state vectors are equal to $\mathbf{x}^{(22)}$ to three decimal places. Thus, as with the first two examples, the state vectors approach a fixed vector as n increases. ◄

Table 2

$x^{(n)}$ \ n	0	1	2	3	4	5	10	15	20	22
$x_1^{(n)}$	0	.000	.116	.130	.123	.113	.109	.108	.107	.107
$x_2^{(n)}$	0	.250	.146	.163	.140	.138	.115	.109	.108	.107
$x_3^{(n)}$	0	.000	.050	.039	.067	.073	.100	.106	.107	.107
$x_4^{(n)}$	0	.250	.113	.187	.162	.178	.178	.179	.179	.179
$x_5^{(n)}$	1	.250	.279	.190	.190	.168	.149	.144	.143	.143
$x_6^{(n)}$	0	.000	.000	.050	.056	.074	.099	.105	.107	.107
$x_7^{(n)}$	0	.000	.133	.104	.131	.125	.138	.142	.143	.143
$x_8^{(n)}$	0	.250	.146	.152	.124	.121	.108	.107	.107	.107

Limiting Behavior of the State Vectors

In our examples we saw that the state vectors approached some fixed vector as the number of observations increased. We now ask whether the state vectors always approach a fixed vector in a Markov chain. A simple example shows that this is not the case.

► **EXAMPLE 6 System Oscillates Between Two State Vectors**

Let

$$
P = \begin{bmatrix} 0 & 1 \\ 1 & 0 \end{bmatrix} \quad \text{and} \quad \mathbf{x}^{(0)} = \begin{bmatrix} 1 \\ 0 \end{bmatrix}
$$

Then, because $P^2 = I$ and $P^3 = P$, we have that

$$
\mathbf{x}^{(0)} = \mathbf{x}^{(2)} = \mathbf{x}^{(4)} = \cdots = \begin{bmatrix} 1 \\ 0 \end{bmatrix}
$$

and

$$
\mathbf{x}^{(1)} = \mathbf{x}^{(3)} = \mathbf{x}^{(5)} = \cdots = \begin{bmatrix} 0 \\ 1 \end{bmatrix}
$$

This system oscillates indefinitely between the two state vectors $\begin{bmatrix} 1 \\ 0 \end{bmatrix}$ and $\begin{bmatrix} 0 \\ 1 \end{bmatrix}$, so it does not approach any fixed vector. ◀

However, if we impose a mild condition on the transition matrix, we can show that a fixed limiting state vector is approached. This condition is described by the following definition.

> **DEFINITION 3** A transition matrix is **regular** if some integer power of it has all positive entries.

Thus, for a regular transition matrix P, there is some positive integer m such that all entries of P^m are positive. This is the case with the transition matrices of Examples 1 and 2 for $m = 1$. In Example 5 it turns out that P^4 has all positive entries. Consequently, in all three examples the transition matrices are regular.

A Markov chain that is governed by a regular transition matrix is called a **regular Markov chain**. We will see that every regular Markov chain has a fixed state vector $\mathbf{q}$ such that $P^n \mathbf{x}^{(0)}$ approaches $\mathbf{q}$ as n increases for any choice of $\mathbf{x}^{(0)}$. This result is of major importance in the theory of Markov chains. It is based on the following theorem.

> **THEOREM 5.4.2** **Behavior of P^n as $n \to \infty$**
>
> *If P is a regular transition matrix, then as $n \to \infty$,*
>
> $$P^n \to \begin{bmatrix} q_1 & q_1 & \cdots & q_1 \\ q_2 & q_2 & \cdots & q_2 \\ \vdots & \vdots & & \vdots \\ q_k & q_k & \cdots & q_k \end{bmatrix}$$
>
> *where the q_i are positive numbers such that $q_1 + q_2 + \cdots + q_k = 1$.*

We will not prove this theorem here. We refer you to a more specialized text, such as J. Kemeny and J. Snell, *Finite Markov Chains* (New York: Springer-Verlag, 1976).

Let us set

$$Q = \begin{bmatrix} q_1 & q_1 & \cdots & q_1 \\ q_2 & q_2 & \cdots & q_2 \\ \vdots & \vdots & & \vdots \\ q_k & q_k & \cdots & q_k \end{bmatrix} \quad \text{and} \quad \mathbf{q} = \begin{bmatrix} q_1 \\ q_2 \\ \vdots \\ q_k \end{bmatrix}$$

Thus, Q is a transition matrix, all of whose columns are equal to the probability vector $\mathbf{q}$. Q has the property that if $\mathbf{x}$ is any probability vector, then

$$Q\mathbf{x} = \begin{bmatrix} q_1 & q_1 & \cdots & q_1 \\ q_2 & q_2 & \cdots & q_2 \\ \vdots & \vdots & & \vdots \\ q_k & q_k & \cdots & q_k \end{bmatrix} \begin{bmatrix} x_1 \\ x_2 \\ \vdots \\ x_k \end{bmatrix} = \begin{bmatrix} q_1 x_1 + q_1 x_2 + \cdots + q_1 x_k \\ q_2 x_1 + q_2 x_2 + \cdots + q_2 x_k \\ \vdots & \vdots & \vdots \\ q_k x_1 + q_k x_2 + \cdots + q_k x_k \end{bmatrix}$$

$$= (x_1 + x_2 + \cdots + x_k) \begin{bmatrix} q_1 \\ q_2 \\ \vdots \\ q_k \end{bmatrix} = (1)\mathbf{q} = \mathbf{q}$$

That is, Q transforms any probability vector $\mathbf{x}$ into the fixed probability vector $\mathbf{q}$. This result leads to the following theorem.

THEOREM 5.4.3 Behavior of $P^n\mathbf{x}$ as $n \to \infty$

If P is a regular transition matrix and $\mathbf{x}$ is any probability vector, then as $n \to \infty$,

$$P^n\mathbf{x} \to \begin{bmatrix} q_1 \\ q_2 \\ \vdots \\ q_k \end{bmatrix} = \mathbf{q}$$

where $\mathbf{q}$ is a fixed probability vector, independent of n, all of whose entries are positive.

This result holds since Theorem 5.4.2 implies that $P^n \to Q$ as $n \to \infty$. This in turn implies that $P^n\mathbf{x} \to Q\mathbf{x} = \mathbf{q}$ as $n \to \infty$. Thus, for a regular Markov chain, the system eventually approaches a fixed state vector $\mathbf{q}$. The vector $\mathbf{q}$ is called the **steady-state vector** of the regular Markov chain.

For systems with many states, usually the most efficient technique of computing the steady-state vector $\mathbf{q}$ is simply to calculate $P^n\mathbf{x}$ for some large n. Our examples illustrate this procedure. Each is a regular Markov process, so that convergence to a steady-state vector is ensured. Another way of computing the steady-state vector is to make use of the following theorem.

THEOREM 5.4.4 Steady-State Vector

The steady-state vector $\mathbf{q}$ of a regular transition matrix P is the unique probability vector that satisfies the equation $P\mathbf{q} = \mathbf{q}$.

To see this, consider the matrix identity $PP^n = P^{n+1}$. By Theorem 5.4.2, both P^n and P^{n+1} approach Q as $n \to \infty$. Thus, we have $PQ = Q$. Any one column of this matrix equation gives $P\mathbf{q} = \mathbf{q}$. To show that $\mathbf{q}$ is the only probability vector that satisfies this equation, suppose $\mathbf{r}$ is another probability vector such that $P\mathbf{r} = \mathbf{r}$. Then also $P^n\mathbf{r} = \mathbf{r}$ for $n = 1, 2, \ldots$. When we let $n \to \infty$, Theorem 5.4.3 leads to $\mathbf{q} = \mathbf{r}$.

Theorem 5.4.4 can also be expressed by the statement that the homogeneous linear system

$$(I - P)\mathbf{q} = \mathbf{0}$$

has a unique solution vector $\mathbf{q}$ with nonnegative entries that satisfy the condition $q_1 + q_2 + \cdots + q_k = 1$. We can apply this technique to the computation of the steady-state vectors for our examples.

▶ **EXAMPLE 7** Example 2 Revisited

In Example 2 the transition matrix was

$$P = \begin{bmatrix} .8 & .3 \\ .2 & .7 \end{bmatrix}$$

so the linear system $(I - P)\mathbf{q} = \mathbf{0}$ is

$$\begin{bmatrix} .2 & -.3 \\ -.2 & .3 \end{bmatrix} \begin{bmatrix} q_1 \\ q_2 \end{bmatrix} = \begin{bmatrix} 0 \\ 0 \end{bmatrix} \tag{2}$$

This leads to the single independent equation

$$.2q_1 - .3q_2 = 0$$

or

$$q_1 = 1.5q_2$$

Thus, when we set $q_2 = s$, any solution of (2) is of the form

$$\mathbf{q} = s \begin{bmatrix} 1.5 \\ 1 \end{bmatrix}$$

where s is an arbitrary constant. To make the vector $\mathbf{q}$ a probability vector, we set $s = 1/(1.5 + 1) = .4$. Consequently,

$$\mathbf{q} = \begin{bmatrix} .6 \\ .4 \end{bmatrix}$$

is the steady-state vector of this regular Markov chain. This means that over the long run, 60% of the alumni will give a donation in any one year, and 40% will not. Observe that this agrees with the result obtained numerically in Example 3.

▶ **EXAMPLE 8 Example 1 Revisited**

In Example 1 the transition matrix was

$$P = \begin{bmatrix} .8 & .3 & .2 \\ .1 & .2 & .6 \\ .1 & .5 & .2 \end{bmatrix}$$

so the linear system $(I - P)\mathbf{q} = \mathbf{0}$ is

$$\begin{bmatrix} .2 & -.3 & -.2 \\ -.1 & .8 & -.6 \\ -.1 & -.5 & .8 \end{bmatrix} \begin{bmatrix} q_1 \\ q_2 \\ q_3 \end{bmatrix} = \begin{bmatrix} 0 \\ 0 \\ 0 \end{bmatrix}$$

The reduced row echelon form of the coefficient matrix is (verify)

$$\begin{bmatrix} 1 & 0 & -\frac{34}{13} \\ 0 & 1 & -\frac{14}{13} \\ 0 & 0 & 0 \end{bmatrix}$$

so the original linear system is equivalent to the system

$$q_1 = \left(\tfrac{34}{13}\right)q_3$$
$$q_2 = \left(\tfrac{14}{13}\right)q_3$$

When we set $q_3 = s$, any solution of the linear system is of the form

$$\mathbf{q} = s \begin{bmatrix} \frac{34}{13} \\ \frac{14}{13} \\ 1 \end{bmatrix}$$

To make this a probability vector, we set

$$s = \frac{1}{\frac{34}{13} + \frac{14}{13} + 1} = \frac{13}{61}$$

Thus, the steady-state vector of the system is

$$\mathbf{q} = \begin{bmatrix} \frac{34}{61} \\ \frac{14}{61} \\ \frac{13}{61} \end{bmatrix} = \begin{bmatrix} .5573\ldots \\ .2295\ldots \\ .2131\ldots \end{bmatrix}$$

This agrees with the result obtained numerically in Table 1. The entries of $\mathbf{q}$ give the long-run probabilities that any one car will be returned to location 1, 2, or 3, respectively. If the car rental agency has a fleet of 1000 cars, it should design its facilities so that there are at least 558 spaces at location 1, at least 230 spaces at location 2, and at least 214 spaces at location 3.

▶ **EXAMPLE 9 Example 5 Revisited**

We will not give the details of the calculations but simply state that the unique probability vector solution of the linear system $(I - P)\mathbf{q} = \mathbf{0}$ is

$$\mathbf{q} = \begin{bmatrix} \frac{3}{28} \\ \frac{3}{28} \\ \frac{3}{28} \\ \frac{5}{28} \\ \frac{4}{28} \\ \frac{3}{28} \\ \frac{4}{28} \\ \frac{3}{28} \end{bmatrix} = \begin{bmatrix} .1071\ldots \\ .1071\ldots \\ .1071\ldots \\ .1785\ldots \\ .1428\ldots \\ .1071\ldots \\ .1428\ldots \\ .1071\ldots \end{bmatrix}$$

The entries in this vector indicate the proportion of time the traffic officer spends at each intersection over the long term. Thus, if the objective is for her to spend the same proportion of time at each intersection, then the strategy of random movement with equal probabilities from one intersection to another is not a good one. (See Exercise 5.) ◀

Exercise Set 5.4

1. Consider the transition matrix

$$P = \begin{bmatrix} .4 & .5 \\ .6 & .5 \end{bmatrix}$$

(a) Calculate $\mathbf{x}^{(n)}$ for $n = 1, 2, 3, 4, 5$ if $\mathbf{x}^{(0)} = \begin{bmatrix} 1 \\ 0 \end{bmatrix}$.

(b) State why P is regular and find its steady-state vector.

2. Consider the transition matrix

$$P = \begin{bmatrix} .2 & .1 & .7 \\ .6 & .4 & .2 \\ .2 & .5 & .1 \end{bmatrix}$$

(a) Calculate $\mathbf{x}^{(1)}$, $\mathbf{x}^{(2)}$, and $\mathbf{x}^{(3)}$ to three decimal places if

$$\mathbf{x}^{(0)} = \begin{bmatrix} 0 \\ 0 \\ 1 \end{bmatrix}$$

(b) State why P is regular and find its steady-state vector.

3. Find the steady-state vectors of the following regular transition matrices:

(a) $\begin{bmatrix} \frac{1}{3} & \frac{3}{4} \\ \frac{2}{3} & \frac{1}{4} \end{bmatrix}$ (b) $\begin{bmatrix} .81 & .26 \\ .19 & .74 \end{bmatrix}$ (c) $\begin{bmatrix} \frac{1}{3} & \frac{1}{2} & 0 \\ \frac{1}{3} & 0 & \frac{1}{4} \\ \frac{1}{3} & \frac{1}{2} & \frac{3}{4} \end{bmatrix}$

4. Let P be the transition matrix

$$\begin{bmatrix} \frac{1}{2} & 0 \\ \frac{1}{2} & 1 \end{bmatrix}$$

(a) Show that P is not regular.

(b) Show that as n increases, $P^n \mathbf{x}^{(0)}$ approaches $\begin{bmatrix} 0 \\ 1 \end{bmatrix}$ for any initial state vector $\mathbf{x}^{(0)}$.

(c) What conclusion of Theorem 5.4.3 is not valid for the steady state of this transition matrix?

5. Verify that if P is a $k \times k$ regular transition matrix all of whose row sums are equal to 1, then the entries of its steady-state vector are all equal to $1/k$.

6. Show that the transition matrix

$$P = \begin{bmatrix} 0 & \frac{1}{2} & \frac{1}{2} \\ \frac{1}{2} & \frac{1}{2} & 0 \\ \frac{1}{2} & 0 & \frac{1}{2} \end{bmatrix}$$

is regular, and use Exercise 5 to find its steady-state vector.

7. John is either happy or sad. If he is happy one day, then he is happy the next day four times out of five. If he is sad one day, then he is sad the next day one time out of three. Over the long term, what are the chances that John is happy on any given day?

8. A country is divided into three demographic regions. It is found that each year 5% of the residents of region 1 move to region 2, and 5% move to region 3. Of the residents of region 2, 15% move to region 1 and 10% move to region 3. And of the residents of region 3, 10% move to region 1 and 5% move to region 2. What percentage of the population resides in each of the three regions after a long period of time?

Section 5.4 Technology Exercises

The following exercises are designed to be solved using a technology utility. Typically, this will be MATLAB, *Mathematica*, Maple, Derive, or Mathcad, but it may also be some other type of linear algebra software or a scientific calculator with some linear algebra capabilities. For each exercise you will need to read the relevant documentation for the particular utility you are using. The goal of these exercises is to provide you with a basic proficiency with your technology utility. Once you have mastered the techniques in these exercises, you will be able to use your technology utility to solve many of the problems in the regular exercise sets.

T1. Consider the sequence of transition matrices

$$\{P_2, P_3, P_4, \ldots\}$$

with

$$P_2 = \begin{bmatrix} 0 & \frac{1}{2} \\ 1 & \frac{1}{2} \end{bmatrix}, \qquad P_3 = \begin{bmatrix} 0 & 0 & \frac{1}{3} \\ 0 & \frac{1}{2} & \frac{1}{3} \\ 1 & \frac{1}{2} & \frac{1}{3} \end{bmatrix},$$

$$P_4 = \begin{bmatrix} 0 & 0 & 0 & \frac{1}{4} \\ 0 & 0 & \frac{1}{3} & \frac{1}{4} \\ 0 & \frac{1}{2} & \frac{1}{3} & \frac{1}{4} \\ 1 & \frac{1}{2} & \frac{1}{3} & \frac{1}{4} \end{bmatrix}, \qquad P_5 = \begin{bmatrix} 0 & 0 & 0 & 0 & \frac{1}{5} \\ 0 & 0 & 0 & \frac{1}{4} & \frac{1}{5} \\ 0 & 0 & \frac{1}{3} & \frac{1}{4} & \frac{1}{5} \\ 0 & \frac{1}{2} & \frac{1}{3} & \frac{1}{4} & \frac{1}{5} \\ 1 & \frac{1}{2} & \frac{1}{3} & \frac{1}{4} & \frac{1}{5} \end{bmatrix},$$

and so on.

(a) Use a computer to show that each of these four matrices is regular by computing their squares.

(b) Verify Theorem 5.4.2 by computing the 100th power of P_k for $k = 2, 3, 4, 5$. Then make a conjecture as to the limiting value of P_k^n as $n \to \infty$ for all $k = 2, 3, 4, \ldots$.

(c) Verify that the common column $\mathbf{q}_k$ of the limiting matrix you found in part (b) satisfies the equation $P_k \mathbf{q}_k = \mathbf{q}_k$, as required by Theorem 5.4.4.

T2. A mouse is placed in a box with nine rooms as shown in the accompanying figure. Assume that it is equally likely that the mouse goes through any door in the room or stays in the room.

(a) Construct the 9×9 transition matrix for this problem and show that it is regular.

(b) Determine the steady-state vector for the matrix.

(c) Use a symmetry argument to show that this problem may be solved using only a 3×3 matrix.

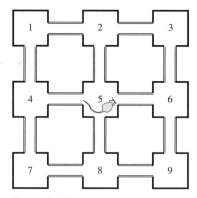

Figure Ex-T2

5.5 Graph Theory

In this section we introduce matrix representations of relations among members of a set. We use matrix arithmetic to analyze these relationships.

PREREQUISITES: Matrix Addition and Multiplication

Relations Among Members of a Set

There are countless examples of sets with finitely many members in which some relation exists among members of the set. For example, the set could consist of a collection of people, animals, countries, companies, sports teams, or cities; and the relation between two members, A and B, of such a set could be that person A dominates person B, animal A feeds on animal B, country A militarily supports country B, company A sells its product to company B, sports team A consistently beats sports team B, or city A has a direct airline flight to city B.

We will now show how the theory of *directed graphs* can be used to mathematically model relations such as those in the preceding examples.

Directed Graphs

A ***directed graph*** is a finite set of elements, $\{P_1, P_2, \ldots, P_n\}$, together with a finite collection of ordered pairs (P_i, P_j) of distinct elements of this set, with no ordered pair being repeated. The elements of the set are called ***vertices***, and the ordered pairs are called ***directed edges***, of the directed graph. We use the notation $P_i \rightarrow P_j$ (which is read "P_i is connected to P_j") to indicate that the directed edge (P_i, P_j) belongs to the directed graph. Geometrically, we can visualize a directed graph (Figure 5.5.1) by representing the vertices as points in the plane and representing the directed edge $P_i \rightarrow P_j$ by drawing a line or arc from vertex P_i to vertex P_j, with an arrow pointing from P_i to P_j. If both $P_i \rightarrow P_j$ and $P_j \rightarrow P_i$ hold (denoted $P_i \leftrightarrow P_j$), we draw a single line between P_i and P_j with two oppositely pointing arrows (as with P_2 and P_3 in the figure).

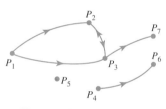

▲ Figure 5.5.1

As in Figure 5.5.1, for example, a directed graph may have separate "components" of vertices that are connected only among themselves; and some vertices, such as P_5, may not be connected with any other vertex. Also, because $P_i \rightarrow P_i$ is not permitted in a directed graph, a vertex cannot be connected with itself by a single arc that does not pass through any other vertex.

Figure 5.5.2 shows diagrams representing three more examples of directed graphs. With a directed graph having n vertices, we may associate an $n \times n$ matrix $M = [m_{ij}]$, called the ***vertex matrix*** of the directed graph. Its elements are defined by

$$m_{ij} = \begin{cases} 1, & \text{if } P_i \rightarrow P_j \\ 0, & \text{otherwise} \end{cases}$$

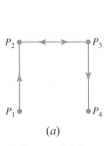
(*a*)

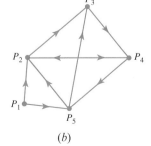
(*b*)

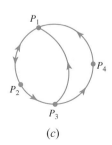
(*c*)

▲ Figure 5.5.2

for $i, j = 1, 2, \ldots, n$. For the three directed graphs in Figure 5.5.2, the corresponding vertex matrices are

Figure 5.5.2a: $\quad M = \begin{bmatrix} 0 & 1 & 0 & 0 \\ 0 & 0 & 1 & 0 \\ 0 & 1 & 0 & 1 \\ 0 & 0 & 0 & 0 \end{bmatrix}$

Figure 5.5.2b: $\quad M = \begin{bmatrix} 0 & 1 & 0 & 0 & 1 \\ 0 & 0 & 1 & 1 & 0 \\ 0 & 0 & 0 & 1 & 0 \\ 0 & 1 & 0 & 0 & 1 \\ 0 & 1 & 1 & 0 & 0 \end{bmatrix}$

Figure 5.5.2c: $\quad M = \begin{bmatrix} 0 & 1 & 0 & 0 \\ 1 & 0 & 1 & 0 \\ 1 & 0 & 0 & 1 \\ 1 & 0 & 0 & 0 \end{bmatrix}$

By their definition, vertex matrices have the following two properties:

(i)   All entries are either 0 or 1.

(ii) All diagonal entries are 0.

Conversely, any matrix with these two properties determines a unique directed graph having the given matrix as its vertex matrix. For example, the matrix

$$M = \begin{bmatrix} 0 & 1 & 1 & 0 \\ 0 & 0 & 1 & 0 \\ 1 & 0 & 0 & 1 \\ 0 & 0 & 0 & 0 \end{bmatrix}$$

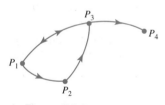

▲ Figure 5.5.3

determines the directed graph in Figure 5.5.3.

▶ **EXAMPLE 1 Influences Within a Family**

A certain family consists of a mother, father, daughter, and two sons. The family members have influence, or power, over each other in the following ways: the mother can influence the daughter and the oldest son; the father can influence the two sons; the daughter can influence the father; the oldest son can influence the youngest son; and the youngest son can influence the mother. We may model this family influence pattern with a directed graph whose vertices are the five family members. If family member A influences family member B, we write $A \rightarrow B$. Figure 5.5.4 is the resulting directed graph, where we have used obvious letter designations for the five family members. The vertex matrix of this directed graph is

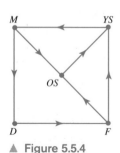

▲ Figure 5.5.4

$$\begin{array}{c} \\ M \\ F \\ D \\ OS \\ YS \end{array} \begin{array}{ccccc} M & F & D & OS & YS \end{array} \\ \begin{bmatrix} 0 & 0 & 1 & 1 & 0 \\ 0 & 0 & 0 & 1 & 1 \\ 0 & 1 & 0 & 0 & 0 \\ 0 & 0 & 0 & 0 & 1 \\ 1 & 0 & 0 & 0 & 0 \end{bmatrix}$$

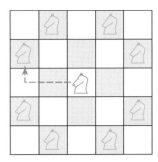

▲ Figure 5.5.5

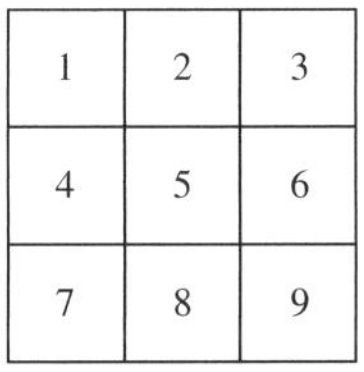

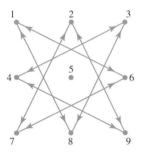

▲ Figure 5.5.6

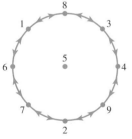

▲ Figure 5.5.7

▲ Figure 5.5.8

▶ **EXAMPLE 2** Vertex Matrix: Moves on a Chessboard

In chess the knight moves in an "L"-shaped pattern about the chessboard. For the board in Figure 5.5.5 it may move horizontally two squares and then vertically one square, or it may move vertically two squares and then horizontally one square. Thus, from the center square in the figure, the knight may move to any of the eight marked shaded squares. Suppose that the knight is restricted to the nine numbered squares in Figure 5.5.6. If by $i \rightarrow j$ we mean that the knight may move from square i to square j, the directed graph in Figure 5.5.7 illustrates all possible moves that the knight may make among these nine squares. In Figure 5.5.8 we have "unraveled" Figure 5.5.7 to make the pattern of possible moves clearer.

The vertex matrix of this directed graph is given by

$$M = \begin{bmatrix} 0 & 0 & 0 & 0 & 0 & 1 & 0 & 1 & 0 \\ 0 & 0 & 0 & 0 & 0 & 0 & 1 & 0 & 1 \\ 0 & 0 & 0 & 1 & 0 & 0 & 0 & 1 & 0 \\ 0 & 0 & 1 & 0 & 0 & 0 & 0 & 0 & 1 \\ 0 & 0 & 0 & 0 & 0 & 0 & 0 & 0 & 0 \\ 1 & 0 & 0 & 0 & 0 & 0 & 1 & 0 & 0 \\ 0 & 1 & 0 & 0 & 0 & 1 & 0 & 0 & 0 \\ 1 & 0 & 1 & 0 & 0 & 0 & 0 & 0 & 0 \\ 0 & 1 & 0 & 1 & 0 & 0 & 0 & 0 & 0 \end{bmatrix} \quad ◀$$

In Example 1 the father cannot directly influence the mother; that is, $F \rightarrow M$ is not true. But he can influence the youngest son, who can then influence the mother. We write this as $F \rightarrow YS \rightarrow M$ and call it a **2-step connection** from F to M. Analogously, we call $M \rightarrow D$ a **1-step connection**, $F \rightarrow OS \rightarrow YS \rightarrow M$ a **3-step connection**, and so forth. Let us now consider a technique for finding the number of all possible r-step connections $(r = 1, 2, \ldots)$ from one vertex P_i to another vertex P_j of an arbitrary directed graph. (This will include the case when P_i and P_j are the same vertex.) The number of 1-step connections from P_i to P_j is simply m_{ij}. That is, there is either zero or one 1-step connection from P_i to P_j, depending on whether m_{ij} is zero or one. For the number of 2-step connections, we consider the square of the vertex matrix. If we let $m_{ij}^{(2)}$ be the (i, j)-th element of M^2, we have

$$m_{ij}^{(2)} = m_{i1}m_{1j} + m_{i2}m_{2j} + \cdots + m_{in}m_{nj} \tag{1}$$

Now, if $m_{i1} = m_{1j} = 1$, there is a 2-step connection $P_i \rightarrow P_1 \rightarrow P_j$ from P_i to P_j. But if either m_{i1} or m_{1j} is zero, such a 2-step connection is not possible. Thus $P_i \rightarrow P_1 \rightarrow P_j$ is a 2-step connection if and only if $m_{i1}m_{1j} = 1$. Similarly, for any $k = 1, 2, \ldots, n$, $P_i \rightarrow P_k \rightarrow P_j$ is a 2-step connection from P_i to P_j if and only if the term $m_{ik}m_{kj}$ on the right side of (1) is one; otherwise, the term is zero. Thus, the right side of (1) is the total number of two 2-step connections from P_i to P_j.

A similar argument will work for finding the number of 3-, 4-, ..., r-step connections from P_i to P_j. In general, we have the following result.

THEOREM 5.5.1 *Let M be the vertex matrix of a directed graph and let $m_{ij}^{(r)}$ be the (i, j)-th element of M^r. Then $m_{ij}^{(r)}$ is equal to the number of r-step connections from P_i to P_j.*

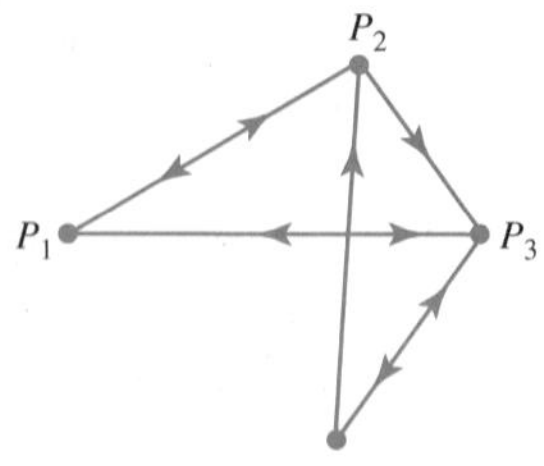

▶ **EXAMPLE 3** Using Theorem 5.5.1

Figure 5.5.9 is the route map of a small airline that services the four cities P_1, P_2, P_3, P_4. As a directed graph, its vertex matrix is

$$M = \begin{bmatrix} 0 & 1 & 1 & 0 \\ 1 & 0 & 1 & 0 \\ 1 & 0 & 0 & 1 \\ 0 & 1 & 1 & 0 \end{bmatrix}$$

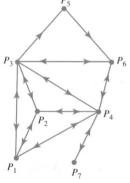

▲ Figure 5.5.9

We have that

$$M^2 = \begin{bmatrix} 2 & 0 & 1 & 1 \\ 1 & 1 & 1 & 1 \\ 0 & 2 & 2 & 0 \\ 2 & 0 & 1 & 1 \end{bmatrix} \quad \text{and} \quad M^3 = \begin{bmatrix} 1 & 3 & 3 & 1 \\ 2 & 2 & 3 & 1 \\ 4 & 0 & 2 & 2 \\ 1 & 3 & 3 & 1 \end{bmatrix}$$

If we are interested in connections from city P_4 to city P_3, we may use Theorem 5.5.1 to find their number. Because $m_{43} = 1$, there is one 1-step connection; because $m_{43}^{(2)} = 1$, there is one 2-step connection; and because $m_{43}^{(3)} = 3$, there are three 3-step connections. To verify this, from Figure 5.5.9 we find

1-step connections from P_4 to P_3: $P_4 \rightarrow P_3$

2-step connections from P_4 to P_3: $P_4 \rightarrow P_2 \rightarrow P_3$

3-step connections from P_4 to P_3: $P_4 \rightarrow P_3 \rightarrow P_4 \rightarrow P_3$

$P_4 \rightarrow P_2 \rightarrow P_1 \rightarrow P_3$

$P_4 \rightarrow P_3 \rightarrow P_1 \rightarrow P_3$ ◀

Cliques   In everyday language a "clique" is a closely knit group of people (usually three or more) that tends to communicate within itself and has no place for outsiders. In graph theory this concept is given a more precise meaning.

DEFINITION 1 A subset of a directed graph is called a ***clique*** if it satisfies the following three conditions:

   (i)   The subset contains at least three vertices.

   (ii) For each pair of vertices P_i and P_j in the subset, both $P_i \rightarrow P_j$ and $P_j \rightarrow P_i$ are true.

   (iii) The subset is as large as possible; that is, it is not possible to add another vertex to the subset and still satisfy condition (ii).

This definition suggests that cliques are maximal subsets that are in perfect "communication" with each other. For example, if the vertices represent cities, and $P_i \rightarrow P_j$ means that there is a direct airline flight from city P_i to city P_j, then there is a direct flight between any two cities within a clique in either direction.

▶ **EXAMPLE 4** A Directed Graph with Two Cliques

The directed graph illustrated in Figure 5.5.10 (which might represent the route map of an airline) has two cliques:

$$\{P_1, P_2, P_3, P_4\} \quad \text{and} \quad \{P_3, P_4, P_6\}$$

▲ Figure 5.5.10

This example shows that a directed graph may contain several cliques and that a vertex may simultaneously belong to more than one clique. ◀

For simple directed graphs, cliques can be found by inspection. But for large directed graphs, it would be desirable to have a systematic procedure for detecting cliques. For this purpose, it will be helpful to define a matrix $S = [s_{ij}]$ related to a given directed graph as follows:

$$s_{ij} = \begin{cases} 1, & \text{if } P_i \leftrightarrow P_j \\ 0, & \text{otherwise} \end{cases}$$

The matrix S determines a directed graph that is the same as the given directed graph, with the exception that the directed edges with only one arrow are deleted. For example, if the original directed graph is given by Figure 5.5.11a, the directed graph that has S as its vertex matrix is given in Figure 5.5.11b. The matrix S may be obtained from the vertex matrix M of the original directed graph by setting $s_{ij} = 1$ if $m_{ij} = m_{ji} = 1$ and setting $s_{ij} = 0$ otherwise.

The following theorem, which uses the matrix S, is helpful for identifying cliques.

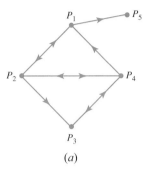

(a)

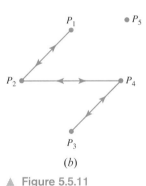

(b)

▲ Figure 5.5.11

THEOREM 5.5.2 Identifying Cliques

Let $s_{ij}^{(3)}$ be the (i, j)-th element of S^3. Then a vertex P_i belongs to some clique if and only if $s_{ii}^{(3)} \neq 0$.

Proof If $s_{ii}^{(3)} \neq 0$, then there is at least one 3-step connection from P_i to itself in the modified directed graph determined by S. Suppose it is $P_i \to P_j \to P_k \to P_i$. In the modified directed graph, all directed relations are two-way, so we also have the connections $P_i \leftrightarrow P_j \leftrightarrow P_k \leftrightarrow P_i$. But this means that $\{P_i, P_j, P_k\}$ is either a clique or a subset of a clique. In either case, P_i must belong to some clique. The converse statement, "if P_i belongs to a clique, then $s_{ii}^{(3)} \neq 0$," follows in a similar manner. ◀

▶ **EXAMPLE 5 Using Theorem 5.5.2**

Suppose that a directed graph has as its vertex matrix

$$M = \begin{bmatrix} 0 & 1 & 1 & 1 \\ 1 & 0 & 1 & 0 \\ 0 & 1 & 0 & 1 \\ 1 & 0 & 0 & 0 \end{bmatrix}$$

Then

$$S = \begin{bmatrix} 0 & 1 & 0 & 1 \\ 1 & 0 & 1 & 0 \\ 0 & 1 & 0 & 0 \\ 1 & 0 & 0 & 0 \end{bmatrix} \quad \text{and} \quad S^3 = \begin{bmatrix} 0 & 3 & 0 & 2 \\ 3 & 0 & 2 & 0 \\ 0 & 2 & 0 & 1 \\ 2 & 0 & 1 & 0 \end{bmatrix}$$

Because all diagonal entries of S^3 are zero, it follows from Theorem 5.5.2 that the directed graph has no cliques.

▶ **EXAMPLE 6** Using Theorem 5.5.2

Suppose that a directed graph has as its vertex matrix

$$M = \begin{bmatrix} 0 & 1 & 0 & 1 & 1 \\ 1 & 0 & 0 & 1 & 0 \\ 1 & 1 & 0 & 1 & 0 \\ 1 & 1 & 0 & 0 & 0 \\ 1 & 0 & 0 & 1 & 0 \end{bmatrix}$$

Then

$$S = \begin{bmatrix} 0 & 1 & 0 & 1 & 1 \\ 1 & 0 & 0 & 1 & 0 \\ 0 & 0 & 0 & 0 & 0 \\ 1 & 1 & 0 & 0 & 0 \\ 1 & 0 & 0 & 0 & 0 \end{bmatrix} \quad \text{and} \quad S^3 = \begin{bmatrix} 2 & 4 & 0 & 4 & 3 \\ 4 & 2 & 0 & 3 & 1 \\ 0 & 0 & 0 & 0 & 0 \\ 4 & 3 & 0 & 2 & 1 \\ 3 & 1 & 0 & 1 & 0 \end{bmatrix}$$

The nonzero diagonal entries of S^3 are $s_{11}^{(3)}$, $s_{22}^{(3)}$, and $s_{44}^{(3)}$. Consequently, in the given directed graph, P_1, P_2, and P_4 belong to cliques. Because a clique must contain at least three vertices, the directed graph has only one clique, $\{P_1, P_2, P_4\}$. ◀

Dominance-Directed Graphs

In many groups of individuals or animals, there is a definite "pecking order" or dominance relation between any two members of the group. That is, given any two individuals A and B, either A dominates B or B dominates A, but not both. In terms of a directed graph in which $P_i \to P_j$ means P_i dominates P_j, this means that for all distinct pairs, either $P_i \to P_j$ or $P_j \to P_i$, but not both. In general, we have the following definition.

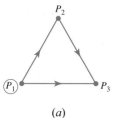

DEFINITION 2 A ***dominance-directed graph*** is a directed graph such that for any distinct pair of vertices P_i and P_j, either $P_i \to P_j$ or $P_j \to P_i$, but not both.

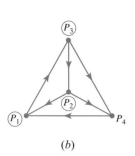

(b)

An example of a directed graph satisfying this definition is a league of n sports teams that play each other exactly one time, as in one round of a round-robin tournament in which no ties are allowed. If $P_i \to P_j$ means that team P_i beat team P_j in their single match, it is easy to see that the definition of a dominance-directed group is satisfied. For this reason, dominance-directed graphs are sometimes called ***tournaments***.

Figure 5.5.12 illustrates some dominance-directed graphs with three, four, and five vertices, respectively. In these three graphs, the circled vertices have the following interesting property: from each one there is either a 1-step or a 2-step connection to any other vertex in its graph. In a sports tournament, these vertices would correspond to the most "powerful" teams in the sense that these teams either beat any given team or beat some other team that beat the given team. We can now state and prove a theorem that guarantees that any dominance-directed graph has at least one vertex with this property.

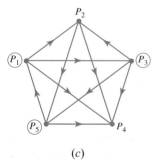

(c)

▲ Figure 5.5.12

THEOREM 5.5.3 Connections in Dominance-Directed Graphs

In any dominance-directed graph, there is at least one vertex from which there is a 1-step or 2-step connection to any other vertex.

Proof Consider a vertex (there may be several) with the largest total number of 1-step and 2-step connections to other vertices in the graph. By renumbering the vertices, we may assume that P_1 is such a vertex. Suppose there is some vertex P_i such that there is no 1-step or 2-step connection from P_1 to P_i. Then, in particular, $P_1 \to P_i$ is not true, so that by definition of a dominance-directed graph, it must be that $P_i \rightarrow P_1$. Next, let P_k be any vertex such that $P_1 \rightarrow P_k$ is true. Then we cannot have $P_k \rightarrow P_i$, as then $P_1 \rightarrow P_k \rightarrow P_i$ would be a 2-step connection from P_1 to P_i. Thus, it must be that $P_i \rightarrow P_k$. That is, P_i has 1-step connections to all the vertices to which P_1 has 1-step connections. The vertex P_i must then also have 2-step connections to all the vertices to which P_1 has 2-step connections. But because, in addition, we have that $P_i \rightarrow P_1$, this means that P_i has more 1-step and 2-step connections to other vertices than does P_1. However, this contradicts the way in which P_1 was chosen. Hence, there can be no vertex P_i to which P_1 has no 1-step or 2-step connection. ◄

This proof shows that a vertex with the largest total number of 1-step and 2-step connections to other vertices has the property stated in the theorem. There is a simple way of finding such vertices using the vertex matrix M and its square M^2. The sum of the entries in the ith row of M is the total number of 1-step connections from P_i to other vertices, and the sum of the entries of the ith row of M^2 is the total number of 2-step connections from P_i to other vertices. Consequently, the sum of the entries of the ith row of the matrix $A = M + M^2$ is the total number of 1-step and 2-step connections from P_i to other vertices. In other words, a row of $A = M + M^2$ with the largest row sum identifies a vertex having the property stated in Theorem 5.5.3.

▶ **EXAMPLE 7** **Using Theorem 5.5.3**

Suppose that five baseball teams play each other exactly once, and the results are as indicated in the dominance-directed graph of Figure 5.5.13. The vertex matrix of the graph is

$$M = \begin{bmatrix} 0 & 0 & 1 & 1 & 0 \\ 1 & 0 & 1 & 0 & 1 \\ 0 & 0 & 0 & 1 & 0 \\ 0 & 1 & 0 & 0 & 0 \\ 1 & 0 & 1 & 1 & 0 \end{bmatrix}$$

so

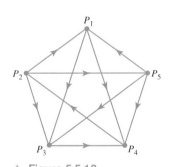

▲ Figure 5.5.13

$$A = M + M^2 = \begin{bmatrix} 0 & 0 & 1 & 1 & 0 \\ 1 & 0 & 1 & 0 & 1 \\ 0 & 0 & 0 & 1 & 0 \\ 0 & 1 & 0 & 0 & 0 \\ 1 & 0 & 1 & 1 & 0 \end{bmatrix} + \begin{bmatrix} 0 & 1 & 0 & 1 & 0 \\ 1 & 0 & 2 & 3 & 0 \\ 0 & 1 & 0 & 0 & 0 \\ 1 & 0 & 1 & 0 & 1 \\ 0 & 1 & 1 & 2 & 0 \end{bmatrix} = \begin{bmatrix} 0 & 1 & 1 & 2 & 0 \\ 2 & 0 & 3 & 3 & 1 \\ 0 & 1 & 0 & 1 & 0 \\ 1 & 1 & 1 & 0 & 1 \\ 1 & 1 & 2 & 3 & 0 \end{bmatrix}$$

The row sums of A are

$$\text{1st row sum} = 4$$
$$\text{2nd row sum} = 9$$
$$\text{3rd row sum} = 2$$
$$\text{4th row sum} = 4$$
$$\text{5th row sum} = 7$$

Because the second row has the largest row sum, the vertex P_2 must have a 1-step or 2-step connection to any other vertex. This is easily verified from Figure 5.5.13. ◄

We have informally suggested that a vertex with the largest number of 1-step and 2-step connections to other vertices is a "powerful" vertex. We can formalize this concept with the following definition.

DEFINITION 3 The ***power*** of a vertex of a dominance-directed graph is the total number of 1-step and 2-step connections from it to other vertices. Alternatively, the power of a vertex P_i is the sum of the entries of the ith row of the matrix $A = M + M^2$, where M is the vertex matrix of the directed graph.

▶ **EXAMPLE 8 Example 7 Revisited**

Let us rank the five baseball teams in Example 7 according to their powers. From the calculations for the row sums in that example, we have

$$\text{Power of team } P_1 = 4$$
$$\text{Power of team } P_2 = 9$$
$$\text{Power of team } P_3 = 2$$
$$\text{Power of team } P_4 = 4$$
$$\text{Power of team } P_5 = 7$$

Hence, the ranking of the teams according to their powers would be

$$P_2 \text{ (first)}, \quad P_5 \text{ (second)}, \quad P_1 \text{ and } P_4 \text{ (tied for third)}, \quad P_3 \text{ (last)} \blacktriangleleft$$

Exercise Set 5.5

1. Construct the vertex matrix for each of the directed graphs illustrated in Figure Ex-1.

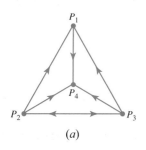

(a)

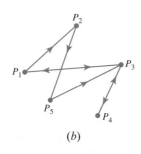

(b)

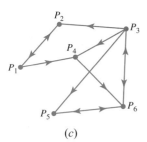

(c)

▲ **Figure Ex-1**

2. Draw a diagram of the directed graph corresponding to each of the following vertex matrices.

(a)
$$\begin{bmatrix} 0 & 1 & 1 & 0 \\ 1 & 0 & 0 & 0 \\ 0 & 0 & 0 & 1 \\ 1 & 0 & 1 & 0 \end{bmatrix}$$

(b)
$$\begin{bmatrix} 0 & 0 & 1 & 0 & 0 \\ 1 & 0 & 0 & 0 & 1 \\ 0 & 1 & 0 & 1 & 1 \\ 0 & 0 & 0 & 0 & 0 \\ 1 & 1 & 1 & 0 & 0 \end{bmatrix}$$

(c)
$$\begin{bmatrix} 0 & 1 & 0 & 1 & 0 & 1 \\ 1 & 0 & 0 & 0 & 1 & 0 \\ 0 & 0 & 0 & 0 & 0 & 0 \\ 1 & 1 & 0 & 0 & 1 & 0 \\ 0 & 0 & 0 & 1 & 0 & 1 \\ 0 & 1 & 0 & 0 & 1 & 0 \end{bmatrix}$$

3. Let M be the following vertex matrix of a directed graph:

$$\begin{bmatrix} 0 & 1 & 1 & 1 \\ 1 & 0 & 0 & 0 \\ 0 & 1 & 0 & 1 \\ 0 & 1 & 1 & 0 \end{bmatrix}$$

(a) Draw a diagram of the directed graph.

(b) Use Theorem 5.5.1 to find the number of 1-, 2-, and 3-step connections from the vertex P_1 to the vertex P_2. Verify your answer by listing the various connections as in Example 3.

(c) Repeat part (b) for the 1-, 2-, and 3-step connections from P_1 to P_4.

4. (a) Compute the matrix product $M^T M$ for the vertex matrix M in Example 1.

(b) Verify that the kth diagonal entry of $M^T M$ is the number of family members who influence the kth family member. Why is this true?

(c) Find a similar interpretation for the values of the nondiagonal entries of $M^T M$.

5. By inspection, locate all cliques in each of the directed graphs illustrated in Figure Ex-5.

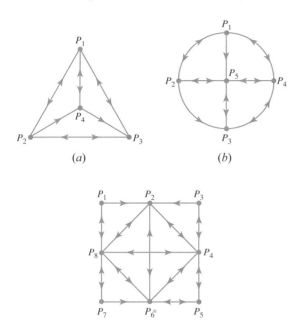

(a)

(b)

(c)

▲ Figure Ex-5

6. For each of the following vertex matrices, use Theorem 5.5.2 to find all cliques in the corresponding directed graphs.

(a) $\begin{bmatrix} 0 & 1 & 0 & 1 & 0 \\ 1 & 0 & 1 & 0 & 1 \\ 0 & 1 & 0 & 1 & 1 \\ 1 & 0 & 0 & 0 & 1 \\ 1 & 0 & 1 & 1 & 0 \end{bmatrix}$ (b) $\begin{bmatrix} 0 & 1 & 0 & 1 & 1 & 0 \\ 1 & 0 & 1 & 0 & 1 & 1 \\ 0 & 1 & 0 & 1 & 0 & 1 \\ 1 & 0 & 1 & 0 & 1 & 1 \\ 0 & 1 & 0 & 1 & 0 & 0 \\ 0 & 0 & 1 & 1 & 1 & 0 \end{bmatrix}$

7. For the dominance-directed graph illustrated in Figure Ex-7 construct the vertex matrix and find the power of each vertex.

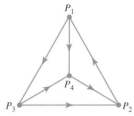

◀ Figure Ex-7

8. Five baseball teams play each other one time with the following results:

$$A \text{ beats } B, C, D$$
$$B \text{ beats } C, E$$
$$C \text{ beats } D, E$$
$$D \text{ beats } B$$
$$E \text{ beats } A, D$$

Rank the five baseball teams in accordance with the powers of the vertices they correspond to in the dominance-directed graph representing the outcomes of the games.

Section 5.5   Technology Exercises

The following exercises are designed to be solved using a technology utility. Typically, this will be MATLAB, *Mathematica*, Maple, Derive, or Mathcad, but it may also be some other type of linear algebra software or a scientific calculator with some linear algebra capabilities. For each exercise you will need to read the relevant documentation for the particular utility you are using. The goal of these exercises is to provide you with a basic proficiency with your technology utility. Once you have mastered the techniques in these exercises, you will be able to use your technology utility to solve many of the problems in the regular exercise sets.

T1. A graph having n vertices such that every vertex is connected to every other vertex has a vertex matrix given by

$$M_n = \begin{bmatrix} 0 & 1 & 1 & 1 & 1 & \cdots & 1 \\ 1 & 0 & 1 & 1 & 1 & \cdots & 1 \\ 1 & 1 & 0 & 1 & 1 & \cdots & 1 \\ 1 & 1 & 1 & 0 & 1 & \cdots & 1 \\ 1 & 1 & 1 & 1 & 0 & \cdots & 1 \\ \vdots & \vdots & \vdots & \vdots & \vdots & \ddots & \vdots \\ 1 & 1 & 1 & 1 & 1 & \cdots & 0 \end{bmatrix}$$

In this problem we develop a formula for M_n^k whose (i, j)-th entry equals the number of k-step connections from P_i to P_j.

(a) Use a computer to compute the eight matrices M_n^k for $n = 2, 3$ and for $k = 2, 3, 4, 5$.

(b) Use the results in part (a) and symmetry arguments to show that M_n^k can be written as

$$M_n^k = \begin{bmatrix} 0 & 1 & 1 & 1 & 1 & \cdots & 1 \\ 1 & 0 & 1 & 1 & 1 & \cdots & 1 \\ 1 & 1 & 0 & 1 & 1 & \cdots & 1 \\ 1 & 1 & 1 & 0 & 1 & \cdots & 1 \\ 1 & 1 & 1 & 1 & 0 & \cdots & 1 \\ \vdots & \vdots & \vdots & \vdots & \vdots & \ddots & \vdots \\ 1 & 1 & 1 & 1 & 1 & \cdots & 0 \end{bmatrix}^k$$

$$= \begin{bmatrix} \alpha_k & \beta_k & \beta_k & \beta_k & \beta_k & \cdots & \beta_k \\ \beta_k & \alpha_k & \beta_k & \beta_k & \beta_k & \cdots & \beta_k \\ \beta_k & \beta_k & \alpha_k & \beta_k & \beta_k & \cdots & \beta_k \\ \beta_k & \beta_k & \beta_k & \alpha_k & \beta_k & \cdots & \beta_k \\ \beta_k & \beta_k & \beta_k & \beta_k & \alpha_k & \cdots & \beta_k \\ \vdots & \vdots & \vdots & \vdots & \vdots & \ddots & \vdots \\ \beta_k & \beta_k & \beta_k & \beta_k & \beta_k & \cdots & \alpha_k \end{bmatrix}$$

(c) Using the fact that $M_n^k = M_n M_n^{k-1}$, show that

$$\begin{bmatrix} \alpha_k \\ \beta_k \end{bmatrix} = \begin{bmatrix} 0 & n-1 \\ 1 & n-2 \end{bmatrix} \begin{bmatrix} \alpha_{k-1} \\ \beta_{k-1} \end{bmatrix}$$

with

$$\begin{bmatrix} \alpha_1 \\ \beta_1 \end{bmatrix} = \begin{bmatrix} 0 \\ 1 \end{bmatrix}$$

(d) Using part (c), show that

$$\begin{bmatrix} \alpha_k \\ \beta_k \end{bmatrix} = \begin{bmatrix} 0 & n-1 \\ 1 & n-2 \end{bmatrix}^{k-1} \begin{bmatrix} 0 \\ 1 \end{bmatrix}$$

(e) Use the methods of Section 5.2 to compute

$$\begin{bmatrix} 0 & n-1 \\ 1 & n-2 \end{bmatrix}^{k-1}$$

and thereby obtain expressions for α_k and β_k, and eventually show that

$$M_n^k = \left(\frac{(n-1)^k - (-1)^k}{n} \right) U_n + (-1)^k I_n$$

where U_n is the $n \times n$ matrix all of whose entries are ones and I_n is the $n \times n$ identity matrix.

(f) Show that for $n > 2$, all vertices for these directed graphs belong to cliques.

T2. Consider a round-robin tournament among n players (labeled $a_1, a_2, a_3, \ldots, a_n$) where a_1 beats a_2, a_2 beats a_3, a_3 beats $a_4, \ldots,$ a_{n-1} beats a_n, and a_n beats a_1. Compute the "power" of each player, showing that they all have the same power; then determine that common power. [*Hint:* Use a computer to study the cases $n = 3, 4, 5, 6$; then make a conjecture and prove your conjecture to be true.]

5.6 Leontief Economic Models

In this section we discuss two linear models for economic systems. Some results about nonnegative matrices are applied to determine equilibrium price structures and outputs necessary to satisfy demand.

> **PREREQUISITES:** Linear Systems
> Matrices

Economic Systems

Matrix theory has been very successful in describing the interrelations among prices, outputs, and demands in economic systems. In this section we discuss some simple models based on the ideas of Nobel laureate Wassily Leontief. We examine two different but related models: the closed or input-output model, and the open or production model. In each, we are given certain economic parameters that describe the interrelations between the "industries" in the economy under consideration. Using matrix theory, we then evaluate certain other parameters, such as prices or output levels, in order to satisfy a desired economic objective. We begin with the closed model.

Leontief Closed (Input-Output) Model

First we present a simple example; then we proceed to the general theory of the model.

▶ **EXAMPLE 1 An Input-Output Model**

Three homeowners—a carpenter, an electrician, and a plumber—agree to make repairs in their three homes. They agree to work a total of 10 days each according to the following schedule:

	Work Performed by		
	Carpenter	**Electrician**	**Plumber**
Days of Work in Home of Carpenter	2	1	6
Days of Work in Home of Electrician	4	5	1
Days of Work in Home of Plumber	4	4	3

For tax purposes, they must report and pay each other a reasonable daily wage, even for the work each does on his or her own home. Their normal daily wages are about $100, but they agree to adjust their respective daily wages so that each homeowner will come out even—that is, so that the total amount paid out by each is the same as the total amount each receives. We can set

$$p_1 = \text{daily wage of carpenter}$$
$$p_2 = \text{daily wage of electrician}$$
$$p_3 = \text{daily wage of plumber}$$

To satisfy the "equilibrium" condition that each homeowner comes out even, we require that

$$\text{total expenditures} = \text{total income}$$

for each of the homeowners for the 10-day period. For example, the carpenter pays a total of $2p_1 + p_2 + 6p_3$ for the repairs in his own home and receives a total income of $10p_1$ for the repairs that he performs on all three homes. Equating these two expressions then gives the first of the following three equations:

$$2p_1 + p_2 + 6p_3 = 10p_1$$
$$4p_1 + 5p_2 + p_3 = 10p_2$$
$$4p_1 + 4p_2 + 3p_3 = 10p_3$$

The remaining two equations are the equilibrium equations for the electrician and the plumber. Dividing these equations by 10 and rewriting them in matrix form yields

$$\begin{bmatrix} .2 & .1 & .6 \\ .4 & .5 & .1 \\ .4 & .4 & .3 \end{bmatrix} \begin{bmatrix} p_1 \\ p_2 \\ p_3 \end{bmatrix} = \begin{bmatrix} p_1 \\ p_2 \\ p_3 \end{bmatrix} \tag{1}$$

Equation (1) can be rewritten as a homogeneous system by subtracting the left side from the right side to obtain

$$\begin{bmatrix} .8 & -.1 & -.6 \\ -.4 & .5 & -.1 \\ -.4 & -.4 & .7 \end{bmatrix} \begin{bmatrix} p_1 \\ p_2 \\ p_3 \end{bmatrix} = \begin{bmatrix} 0 \\ 0 \\ 0 \end{bmatrix}$$

The solution of this homogeneous system is found to be (verify)

$$\begin{bmatrix} p_1 \\ p_2 \\ p_3 \end{bmatrix} = s \begin{bmatrix} 31 \\ 32 \\ 36 \end{bmatrix}$$

where s is an arbitrary constant. This constant is a scale factor, which the homeowners may choose for their convenience. For example, they can set $s = 3$ so that the corresponding daily wages—$93, $96, and $108—are about $100. ◀

This example illustrates the salient features of the Leontief input-output model of a closed economy. In the basic Equation (1), each column sum of the coefficient matrix is 1, corresponding to the fact that each of the homeowners' "output" of labor is completely distributed among these same homeowners in the proportions given by the entries in the column. Our problem is to determine suitable "prices" for these outputs so as to put the system in equilibrium—that is, so that each homeowner's total expenditures equal his or her total income.

In the general model we have an economic system consisting of a finite number of "industries," which we number as industries $1, 2, \ldots, k$. Over some fixed period of time, each industry produces an "output" of some good or service that is completely utilized in a predetermined manner by the k industries. An important problem is to find suitable "prices" to be charged for these k outputs so that for each industry, total expenditures equal total income. Such a price structure represents an equilibrium position for the economy.

For the fixed time period in question, let us set

p_i = price charged by the ith industry for its total output

e_{ij} = fraction of the total output of the jth industry purchased by the ith industry

for $i, j = 1, 2, \ldots, k$. By definition, we have

(i) $p_i \geq 0$, $i = 1, 2, \ldots, k$

(ii) $e_{ij} \geq 0$, $i, j = 1, 2, \ldots, k$

(iii) $e_{1j} + e_{2j} + \cdots + e_{kj} = 1$, $j = 1, 2, \ldots, k$

With these quantities, we form the *price vector*

$$\mathbf{p} = \begin{bmatrix} p_1 \\ p_2 \\ \vdots \\ p_k \end{bmatrix}$$

and the *exchange matrix* or *input-output matrix*

$$E = \begin{bmatrix} e_{11} & e_{12} & \cdots & e_{1k} \\ e_{21} & e_{22} & \cdots & e_{2k} \\ \vdots & \vdots & & \vdots \\ e_{k1} & e_{k2} & \cdots & e_{kk} \end{bmatrix}$$

Condition (iii) expresses the fact that all the column sums of the exchange matrix are 1.

As in the example, in order that the expenditures of each industry be equal to its income, the following matrix equation must be satisfied [see (1)]:

$$E\mathbf{p} = \mathbf{p} \tag{2}$$

or

$$(I - E)\mathbf{p} = \mathbf{0} \tag{3}$$

Equation (3) is a homogeneous linear system for the price vector $\mathbf{p}$. It will have a nontrivial solution if and only if the determinant of its coefficient matrix $I - E$ is zero. In Exercise 7 we ask you to show that this is the case for any exchange matrix E. Thus, (3) always has nontrivial solutions for the price vector $\mathbf{p}$.

Actually, for our economic model to make sense, we need more than just the fact that (3) has nontrivial solutions for $\mathbf{p}$. We also need the prices p_i of the k outputs to be nonnegative numbers. We express this condition as $\mathbf{p} \geq 0$. (In general, if A is any vector or matrix, the notation $A \geq 0$ means that every entry of A is nonnegative, and the notation $A > 0$ means that every entry of A is positive. Similarly, $A \geq B$ means $A - B \geq 0$, and $A > B$ means $A - B > 0$.) To show that (3) has a nontrivial solution for which $\mathbf{p} \geq 0$ is a bit more difficult than showing merely that some nontrivial solution exists. But it is true, and we state this fact without proof in the following theorem.

THEOREM 5.6.1 *If E is an exchange matrix, then $E\mathbf{p} = \mathbf{p}$ always has a nontrivial solution $\mathbf{p}$ whose entries are nonnegative.*

Let us consider a few simple examples of this theorem.

▶ **EXAMPLE 2** **Using Theorem 5.6.1**

Let

$$E = \begin{bmatrix} \frac{1}{2} & 0 \\ \frac{1}{2} & 1 \end{bmatrix}$$

Then $(I - E)\mathbf{p} = \mathbf{0}$ is

$$\begin{bmatrix} \frac{1}{2} & 0 \\ -\frac{1}{2} & 0 \end{bmatrix} \begin{bmatrix} p_1 \\ p_2 \end{bmatrix} = \begin{bmatrix} 0 \\ 0 \end{bmatrix}$$

which has the general solution

$$\mathbf{p} = s \begin{bmatrix} 0 \\ 1 \end{bmatrix}$$

where s is an arbitrary constant. We then have nontrivial solutions $\mathbf{p} \geq 0$ for any $s > 0$.

▶ **EXAMPLE 3** **Using Theorem 5.6.1**

Let

$$E = \begin{bmatrix} 1 & 0 \\ 0 & 1 \end{bmatrix}$$

Then $(I - E)\mathbf{p} = \mathbf{0}$ has the general solution

$$\mathbf{p} = s \begin{bmatrix} 1 \\ 0 \end{bmatrix} + t \begin{bmatrix} 0 \\ 1 \end{bmatrix}$$

where s and t are independent arbitrary constants. Nontrivial solutions $\mathbf{p} \geq 0$ then result from any $s \geq 0$ and $t \geq 0$, not both zero. ◀

Example 2 indicates that in some situations one of the prices must be zero in order to satisfy the equilibrium condition. Example 3 indicates that there may be several linearly independent price structures available. Neither of these situations describes a truly interdependent economic structure. The following theorem gives sufficient conditions for both cases to be excluded.

THEOREM 5.6.2 *Let E be an exchange matrix such that for some positive integer m all the entries of E^m are positive. Then there is exactly one linearly independent solution of $(I - E)\mathbf{p} = \mathbf{0}$, and it may be chosen so that all its entries are positive.*

We will not give a proof of this theorem. If you have read Section 5.4 on Markov chains, observe that this theorem is essentially the same as Theorem 5.4.4. What we are calling exchange matrices in this section were called stochastic or Markov matrices in Section 5.4.

▶ **EXAMPLE 4** **Using Theorem 5.6.2**

The exchange matrix in Example 1 was

$$E = \begin{bmatrix} .2 & .1 & .6 \\ .4 & .5 & .1 \\ .4 & .4 & .3 \end{bmatrix}$$

Because $E > 0$, the condition $E^m > 0$ in Theorem 5.6.2 is satisfied for $m = 1$. Consequently, we are guaranteed that there is exactly one linearly independent solution of $(I - E)\mathbf{p} = \mathbf{0}$, and it can be chosen so that $\mathbf{p} > 0$. In that example, we found that

$$\mathbf{p} = \begin{bmatrix} 31 \\ 32 \\ 36 \end{bmatrix}$$

is such a solution. ◄

Leontief Open (Production) Model

In contrast with the closed model, in which the outputs of k industries are distributed only among themselves, the open model attempts to satisfy an outside demand for the outputs. Portions of these outputs can still be distributed among the industries themselves, to keep them operating, but there is to be some excess, some net production, with which to satisfy the outside demand. In the closed model the outputs of the industries are fixed, and our objective is to determine prices for these outputs so that the equilibrium condition, that expenditures equal incomes, is satisfied. In the open model it is the prices that are fixed, and our objective is to determine levels of the outputs of the industries needed to satisfy the outside demand. We will measure the levels of the outputs in terms of their economic values using the fixed prices. To be precise, over some fixed period of time, let

x_i = monetary value of the total output of the ith industry

d_i = monetary value of the output of the ith industry needed to satisfy the outside demand

c_{ij} = monetary value of the output of the ith industry needed by the jth industry to produce one unit of monetary value of its own output

With these quantities, we define the **production vector**

$$\mathbf{x} = \begin{bmatrix} x_1 \\ x_2 \\ \vdots \\ x_k \end{bmatrix}$$

the **demand vector**

$$\mathbf{d} = \begin{bmatrix} d_1 \\ d_2 \\ \vdots \\ d_k \end{bmatrix}$$

and the **consumption matrix**

$$C = \begin{bmatrix} c_{11} & c_{12} & \cdots & c_{1k} \\ c_{21} & c_{22} & \cdots & c_{2k} \\ \vdots & \vdots & & \vdots \\ c_{k1} & c_{k2} & \cdots & c_{kk} \end{bmatrix}$$

By their nature, we have that

$$\mathbf{x} \geq 0, \quad \mathbf{d} \geq 0, \quad \text{and} \quad C \geq 0$$

From the definition of c_{ij} and x_j, it can be seen that the quantity

$$c_{i1}x_1 + c_{i2}x_2 + \cdots + c_{ik}x_k$$

is the value of the output of the ith industry needed by all k industries to produce a total output specified by the production vector $\mathbf{x}$. Because this quantity is simply the ith entry of the column vector $C\mathbf{x}$, we can say further that the ith entry of the column vector

$$\mathbf{x} - C\mathbf{x}$$

is the value of the excess output of the ith industry available to satisfy the outside demand. The value of the outside demand for the output of the ith industry is the ith entry of the demand vector $\mathbf{d}$. Consequently, we are led to the following equation

$$\mathbf{x} - C\mathbf{x} = \mathbf{d}$$

or

$$(I - C)\mathbf{x} = \mathbf{d} \tag{4}$$

for the demand to be exactly met, without any surpluses or shortages. Thus, given C and $\mathbf{d}$, our objective is to find a production vector $\mathbf{x} \geq 0$ that satisfies Equation (4).

▶ **EXAMPLE 5 Production Vector for a Town**

A town has three main industries: a coal-mining operation, an electric power-generating plant, and a local railroad. To mine $1 of coal, the mining operation must purchase $.25 of electricity to run its equipment and $.25 of transportation for its shipping needs. To produce $1 of electricity, the generating plant requires $.65 of coal for fuel, $.05 of its own electricity to run auxiliary equipment, and $.05 of transportation. To provide $1 of transportation, the railroad requires $.55 of coal for fuel and $.10 of electricity for its auxiliary equipment. In a certain week the coal-mining operation receives orders for $50,000 of coal from outside the town, and the generating plant receives orders for $25,000 of electricity from outside. There is no outside demand for the local railroad. How much must each of the three industries produce in that week to exactly satisfy their own demand and the outside demand?

Solution For the one-week period let

$$x_1 = \text{value of total output of coal-mining operation}$$
$$x_2 = \text{value of total output of power-generating plant}$$
$$x_3 = \text{value of total output of local railroad}$$

From the information supplied, the consumption matrix of the system is

$$C = \begin{bmatrix} 0 & .65 & .55 \\ .25 & .05 & .10 \\ .25 & .05 & 0 \end{bmatrix}$$

The linear system $(I - C)\mathbf{x} = \mathbf{d}$ is then

$$\begin{bmatrix} 1.00 & -.65 & -.55 \\ -.25 & .95 & -.10 \\ -.25 & -.05 & 1.00 \end{bmatrix} \begin{bmatrix} x_1 \\ x_2 \\ x_3 \end{bmatrix} = \begin{bmatrix} 50,000 \\ 25,000 \\ 0 \end{bmatrix}$$

The coefficient matrix on the left is invertible, and the solution is given by

$$\mathbf{x} = (I - C)^{-1}\mathbf{d} = \frac{1}{503} \begin{bmatrix} 756 & 542 & 470 \\ 220 & 690 & 190 \\ 200 & 170 & 630 \end{bmatrix} \begin{bmatrix} 50,000 \\ 25,000 \\ 0 \end{bmatrix} = \begin{bmatrix} 102,087 \\ 56,163 \\ 28,330 \end{bmatrix}$$

Thus, the total output of the coal-mining operation should be $102,087, the total output of the power-generating plant should be $56,163, and the total output of the railroad should be $28,330. ◀

Let us reconsider Equation (4):

$$(I - C)\mathbf{x} = \mathbf{d}$$

If the square matrix $I - C$ is invertible, we can write

$$\mathbf{x} = (I - C)^{-1}\mathbf{d} \tag{5}$$

In addition, if the matrix $(I - C)^{-1}$ has only nonnegative entries, then we are guaranteed that for any $\mathbf{d} \geq 0$, Equation (5) has a unique nonnegative solution for $\mathbf{x}$. This is a particularly desirable situation, as it means that any outside demand can be met. The terminology used to describe this case is given in the following definition.

DEFINITION 1 A consumption matrix C is said to be **productive** if $(I - C)^{-1}$ exists and
$$(I - C)^{-1} \geq 0$$

We will now consider some simple criteria that guarantee that a consumption matrix is productive. The first is given in the following theorem.

THEOREM 5.6.3 Productive Consumption Matrix

A consumption matrix C is productive if and only if there is some production vector $\mathbf{x} \geq 0$ such that $\mathbf{x} > C\mathbf{x}$.

(The proof is outlined in Exercise 9.) The condition $\mathbf{x} > C\mathbf{x}$ means that there is some production schedule possible such that each industry produces more than it consumes.

Theorem 5.6.3 has two interesting corollaries. Suppose that all the row sums of C are less than 1. If

$$\mathbf{x} = \begin{bmatrix} 1 \\ 1 \\ \vdots \\ 1 \end{bmatrix}$$

then $C\mathbf{x}$ is a column vector whose entries are these row sums. Therefore, $\mathbf{x} > C\mathbf{x}$, and the condition of Theorem 5.6.3 is satisfied. Thus, we arrive at the following corollary:

COROLLARY 5.6.4 *A consumption matrix is productive if each of its row sums is less than 1.*

As we ask you to show in Exercise 8, this corollary leads to the following:

COROLLARY 5.6.5 *A consumption matrix is productive if each of its column sums is less than 1.*

Recalling the definition of the entries of the consumption matrix C, we see that the jth column sum of C is the total value of the outputs of all k industries needed to produce one unit of value of output of the jth industry. The jth industry is thus said to be **profitable** if that jth column sum is less than 1. In other words, Corollary 5.6.5 says that a consumption matrix is productive if all k industries in the economic system are profitable.

▶ **EXAMPLE 6 Using Corollary 5.6.5**

The consumption matrix in Example 5 was

$$
C = \begin{bmatrix} 0 & .65 & .55 \\ .25 & .05 & .10 \\ .25 & .05 & 0 \end{bmatrix}
$$

All three column sums in this matrix are less than 1, so all three industries are profitable. Consequently, by Corollary 5.6.5, the consumption matrix C is productive. This can also be seen in the calculations in Example 5, as $(I - C)^{-1}$ is nonnegative. ◀

Exercise Set 5.6

1. For the following exchange matrices, find nonnegative price vectors that satisfy the equilibrium condition (3).

   (a) $\begin{bmatrix} \frac{1}{2} & \frac{1}{3} \\ \frac{1}{2} & \frac{2}{3} \end{bmatrix}$

   (b) $\begin{bmatrix} \frac{1}{2} & 0 & \frac{1}{2} \\ \frac{1}{3} & 0 & \frac{1}{2} \\ \frac{1}{6} & 1 & 0 \end{bmatrix}$

   (c) $\begin{bmatrix} .35 & .50 & .30 \\ .25 & .20 & .30 \\ .40 & .30 & .40 \end{bmatrix}$

2. Using Theorem 5.6.3 and its corollaries, show that each of the following consumption matrices is productive.

   (a) $\begin{bmatrix} .8 & .1 \\ .3 & .6 \end{bmatrix}$

   (b) $\begin{bmatrix} .70 & .30 & .25 \\ .20 & .40 & .25 \\ .05 & .15 & .25 \end{bmatrix}$

   (c) $\begin{bmatrix} .7 & .3 & .2 \\ .1 & .4 & .3 \\ .2 & .4 & .1 \end{bmatrix}$

3. Using Theorem 5.6.2, show that there is only one linearly independent price vector for the closed economic system with exchange matrix

   $$
   E = \begin{bmatrix} 0 & .2 & .5 \\ 1 & .2 & .5 \\ 0 & .6 & 0 \end{bmatrix}
   $$

4. Three neighbors have backyard vegetable gardens. Neighbor A grows tomatoes, neighbor B grows corn, and neighbor C grows lettuce. They agree to divide their crops among themselves as follows: A gets $\frac{1}{2}$ of the tomatoes, $\frac{1}{3}$ of the corn, and $\frac{1}{4}$ of the lettuce. B gets $\frac{1}{3}$ of the tomatoes, $\frac{1}{3}$ of the corn, and $\frac{1}{4}$ of the lettuce. C gets $\frac{1}{6}$ of the tomatoes, $\frac{1}{3}$ of the corn, $\frac{1}{2}$ of the lettuce. What prices should the neighbors assign to their respective crops if the equilibrium condition of a closed economy is to be satisfied, and if the lowest-priced crop is to have a price of $100?

5. Three engineers—a civil engineer (CE), an electrical engineer (EE), and a mechanical engineer (ME)—each have a consulting firm. The consulting they do is of a multidisciplinary nature, so they buy a portion of each others' services. For each $1

they buy a portion of each others' services. For each $1 of consulting the CE does, she buys $.10 of the EE's services and $.30 of the ME's services. For each $1 of consulting the EE does, she buys $.20 of the CE's services and $.40 of the ME's services. And for each $1 of consulting the ME does, she buys $.30 of the CE's services and $.40 of the EE's services. In a certain week the CE receives outside consulting orders of $500, the EE receives outside consulting orders of $700, and the ME receives outside consulting orders of $600. What dollar amount of consulting does each engineer perform in that week?

6. (a) Suppose that the demand d_i for the output of the ith industry increases by one unit. Explain why the ith column of the matrix $(I - C)^{-1}$ is the increase that must be made to the production vector $\mathbf{x}$ to satisfy this additional demand.

   (b) Referring to Example 5, use the result in part (a) to determine the increase in the value of the output of the coal-mining operation needed to satisfy a demand of one additional unit in the value of the output of the power-generating plant.

7. Using the fact that the column sums of an exchange matrix E are all 1, show that the column sums of $I - E$ are zero. From this, show that $I - E$ has zero determinant, and so $(I - E)\mathbf{p} = \mathbf{0}$ has nontrivial solutions for $\mathbf{p}$.

8. Show that Corollary 5.6.5 follows from Corollary 5.6.4. [*Hint:* Use the fact that $(A^T)^{-1} = (A^{-1})^T$ for any invertible matrix A.]

9. (*Calculus required*) Prove Theorem 5.6.3 as follows:

   (a) Prove the "only if" part of the theorem; that is, show that if C is a productive consumption matrix, then there is a vector $\mathbf{x} \geq 0$ such that $\mathbf{x} > C\mathbf{x}$.

   (b) Prove the "if" part of the theorem as follows:

   Step 1. Show that if there is a vector $\mathbf{x}^* \geq 0$ such that $C\mathbf{x}^* < \mathbf{x}^*$, then $\mathbf{x}^* > 0$.

   Step 2. Show that there is a number λ such that $0 < \lambda < 1$ and $C\mathbf{x}^* < \lambda\mathbf{x}^*$.

   Step 3. Show that $C^n\mathbf{x}^* < \lambda^n\mathbf{x}^*$ for $n = 1, 2, \ldots$.

   Step 4. Show that $C^n \to 0$ as $n \to \infty$.

   Step 5. By multiplying out, show that

   $$(I - C)(I + C + C^2 + \cdots + C^{n-1}) = I - C^n$$

   for $n = 1, 2, \ldots$.

Step 6. By letting $n \to \infty$ in Step 5, show that the matrix infinite sum

$$S = I + C + C^2 + \cdots$$

exists and that $(I - C)S = I$.

Step 7. Show that $S \geq 0$ and that $S = (I - C)^{-1}$.

Step 8. Show that C is a productive consumption matrix.

Section 5.6 Technology Exercises

The following exercises are designed to be solved using a technology utility. Typically, this will be MATLAB, *Mathematica*, Maple, Derive, or Mathcad, but it may also be some other type of linear algebra software or a scientific calculator with some linear algebra capabilities. For each exercise you will need to read the relevant documentation for the particular utility you are using. The goal of these exercises is to provide you with a basic proficiency with your technology utility. Once you have mastered the techniques in these exercises, you will be able to use your technology utility to solve many of the problems in the regular exercise sets.

T1. Consider a sequence of exchange matrices $\{E_2, E_3, E_4, E_5, \ldots, E_n\}$, where

$$E_2 = \begin{bmatrix} 0 & \frac{1}{2} \\ 1 & \frac{1}{2} \end{bmatrix}, \qquad E_3 = \begin{bmatrix} 0 & \frac{1}{2} & \frac{1}{3} \\ 1 & 0 & \frac{1}{3} \\ 0 & \frac{1}{2} & \frac{1}{3} \end{bmatrix},$$

$$E_4 = \begin{bmatrix} 0 & \frac{1}{2} & \frac{1}{3} & \frac{1}{4} \\ 1 & 0 & \frac{1}{3} & \frac{1}{4} \\ 0 & \frac{1}{2} & 0 & \frac{1}{4} \\ 0 & 0 & \frac{1}{3} & \frac{1}{4} \end{bmatrix}, \qquad E_5 = \begin{bmatrix} 0 & \frac{1}{2} & \frac{1}{3} & \frac{1}{4} & \frac{1}{5} \\ 1 & 0 & \frac{1}{3} & \frac{1}{4} & \frac{1}{5} \\ 0 & \frac{1}{2} & 0 & \frac{1}{4} & \frac{1}{5} \\ 0 & 0 & \frac{1}{3} & 0 & \frac{1}{5} \\ 0 & 0 & 0 & \frac{1}{4} & \frac{1}{5} \end{bmatrix}$$

and so on. Use a computer to show that $E_2^2 > 0_2$, $E_3^3 > 0_3$, $E_4^4 > 0_4$, $E_5^5 > 0_5$, and make the conjecture that although $E_n^n > 0_n$ is true, $E_n^k > 0_n$ is not true for $k = 1, 2, 3, \ldots, n-1$. Next, use a computer to determine the vectors $\mathbf{p}_n$ such that $E_n \mathbf{p}_n = \mathbf{p}_n$ (for $n = 2, 3, 4, 5, 6$), and then see if you can discover a pattern that would allow you to compute $\mathbf{p}_{n+1}$ easily from $\mathbf{p}_n$. Test your discovery by first constructing $\mathbf{p}_8$ from

$$\mathbf{p}_7 = \begin{bmatrix} 2520 \\ 3360 \\ 1890 \\ 672 \\ 175 \\ 36 \\ 7 \end{bmatrix}$$

and then checking to see whether $E_8 \mathbf{p}_8 = \mathbf{p}_8$.

T2. Consider an open production model having n industries with $n > 1$. In order to produce \$1 of its own output, the jth industry must spend \$$(1/n)$ for the output of the ith industry (for all $i \neq j$), but the jth industry (for all $j = 1, 2, 3, \ldots, n$) spends nothing for its own output. Construct the consumption matrix C_n, show that it is productive, and determine an expression for $(I_n - C_n)^{-1}$. In determining an expression for $(I_n - C_n)^{-1}$, use a computer to study the cases when $n = 2, 3, 4$, and 5; then make a conjecture and prove your conjecture to be true. [*Hint:* If $F_n = [1]_{n \times n}$ (i.e., the $n \times n$ matrix with every entry equal to 1), first show that

$$F_n^2 = n F_n$$

and then express your value of $(I_n - C_n)^{-1}$ in terms of n, I_n, and F_n.]

5.7 Computer Graphics

In this section we assume that a view of a three-dimensional object is displayed on a video screen and show how matrix algebra can be used to obtain new views of the object by rotation, translation, and scaling.

> **PREREQUISITES:** Matrix Algebra
> Analytic Geometry

Visualization of a Three-Dimensional Object

Suppose that we want to visualize a three-dimensional object by displaying various views of it on a video screen. The object we have in mind to display is to be determined by a finite number of straight line segments. As an example, consider the truncated right pyramid with hexagonal base illustrated in Figure 5.7.1. We first introduce an xyz-coordinate system in which to embed the object. As in Figure 5.7.1, we orient the coordinate system so that its origin is at the center of the video screen and the xy-plane coincides with the plane of the screen. Consequently, an observer will see only the projection of the view of the three-dimensional object onto the two-dimensional xy-plane.

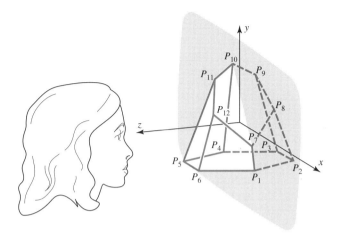

▷ **Figure 5.7.1**

In the xyz-coordinate system, the endpoints $P_1, P_2, \ldots, P_n$ of the straight line segments that determine the view of the object will have certain coordinates—say,

$$(x_1, y_1, z_1), \quad (x_2, y_2, z_2), \ldots, \quad (x_n, y_n, z_n)$$

These coordinates, together with a specification of which pairs are to be connected by straight line segments, are to be stored in the memory of the video display system. For example, assume that the 12 vertices of the truncated pyramid in Figure 5.7.1 have the following coordinates (the screen is 4 units wide by 3 units high):

$$P_1: (1.000, -.800, .000), \qquad P_2: (.500, -.800, -.866),$$
$$P_3: (-.500, -.800, -.866), \qquad P_4: (-1.000, -.800, .000),$$
$$P_5: (-.500, -.800, .866), \qquad P_6: (.500, -.800, .866),$$
$$P_7: (.840, -.400, .000), \qquad P_8: (.315, .125, -.546),$$
$$P_9: (-.210, .650, -.364), \qquad P_{10}: (-.360, .800, .000),$$
$$P_{11}: (-.210, .650, .364), \qquad P_{12}: (.315, .125, .546)$$

These 12 vertices are connected pairwise by 18 straight line segments as follows, where $P_i \leftrightarrow P_j$ denotes that point P_i is connected to point P_j:

$$P_1 \leftrightarrow P_2, \quad P_2 \leftrightarrow P_3, \quad P_3 \leftrightarrow P_4, \quad P_4 \leftrightarrow P_5, \quad P_5 \leftrightarrow P_6, \quad P_6 \leftrightarrow P_1,$$
$$P_7 \leftrightarrow P_8, \quad P_8 \leftrightarrow P_9, \quad P_9 \leftrightarrow P_{10}, \quad P_{10} \leftrightarrow P_{11}, \quad P_{11} \leftrightarrow P_{12}, \quad P_{12} \leftrightarrow P_7,$$
$$P_1 \leftrightarrow P_7, \quad P_2 \leftrightarrow P_8, \quad P_3 \leftrightarrow P_9, \quad P_4 \leftrightarrow P_{10}, \quad P_5 \leftrightarrow P_{11}, \quad P_6 \leftrightarrow P_{12}$$

In View 1 these 18 straight line segments are shown as they would appear on the video screen. It should be noticed that only the x- and y-coordinates of the vertices are needed by the video display system to draw the view, because only the projection of the object onto the xy-plane is displayed. However, we must keep track of the z-coordinates to carry out certain transformations discussed later.

We now show how to form new views of the object by scaling, translating, or rotating the initial view. We first construct a $3 \times n$ matrix P, referred to as the *coordinate matrix*

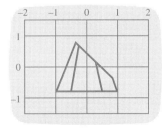

▷ **View 1**

of the view, whose columns are the coordinates of the n points of a view:

$$P = \begin{bmatrix} x_1 & x_2 & \cdots & x_n \\ y_1 & y_2 & \cdots & y_n \\ z_1 & z_2 & \cdots & z_n \end{bmatrix}$$

For example, the coordinate matrix P corresponding to View 1 is the 3×12 matrix

$$\begin{bmatrix} 1.000 & .500 & -.500 & -1.000 & -.500 & .500 & .840 & .315 & -.210 & -.360 & -.210 & .315 \\ -.800 & -.800 & -.800 & -.800 & -.800 & -.800 & -.400 & .125 & .650 & .800 & .650 & .125 \\ .000 & -.866 & -.866 & .000 & .866 & .866 & .000 & -.546 & -.364 & .000 & .364 & .546 \end{bmatrix}$$

We will show below how to transform the coordinate matrix P of a view to a new coordinate matrix P' corresponding to a new view of the object. The straight line segments connecting the various points move with the points as they are transformed. In this way, each view is uniquely determined by its coordinate matrix once we have specified which pairs of points in the original view are to be connected by straight lines.

Scaling   The first type of transformation we consider consists of scaling a view along the x, y, and z directions by factors of α, β, and γ, respectively. By this we mean that if a point P_i has coordinates (x_i, y_i, z_i) in the original view, it is to move to a new point P'_i with coordinates $(\alpha x_i, \beta y_i, \gamma z_i)$ in the new view. This has the effect of transforming a unit cube in the original view to a rectangular parallelepiped of dimensions $\alpha \times \beta \times \gamma$ (Figure 5.7.2). Mathematically, this may be accomplished with matrix multiplication as follows. Define a 3×3 diagonal matrix

$$S = \begin{bmatrix} \alpha & 0 & 0 \\ 0 & \beta & 0 \\ 0 & 0 & \gamma \end{bmatrix}$$

Then, if a point P_i in the original view is represented by the column vector

$$\begin{bmatrix} x_i \\ y_i \\ z_i \end{bmatrix}$$

then the transformed point P'_i is represented by the column vector

$$\begin{bmatrix} x'_i \\ y'_i \\ z'_i \end{bmatrix} = \begin{bmatrix} \alpha & 0 & 0 \\ 0 & \beta & 0 \\ 0 & 0 & \gamma \end{bmatrix} \begin{bmatrix} x_i \\ y_i \\ z_i \end{bmatrix}$$

Using the coordinate matrix P, which contains the coordinates of all n points of the original view as its columns, we can transform these n points simultaneously to produce the coordinate matrix P' of the scaled view, as follows:

$$SP = \begin{bmatrix} \alpha & 0 & 0 \\ 0 & \beta & 0 \\ 0 & 0 & \gamma \end{bmatrix} \begin{bmatrix} x_1 & x_2 & \cdots & x_n \\ y_1 & y_2 & \cdots & y_n \\ z_1 & z_2 & \cdots & z_n \end{bmatrix}$$

$$= \begin{bmatrix} \alpha x_1 & \alpha x_2 & \cdots & \alpha x_n \\ \beta y_1 & \beta y_2 & \cdots & \beta y_n \\ \gamma z_1 & \gamma z_2 & \cdots & \gamma z_n \end{bmatrix} = P'$$

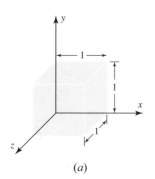

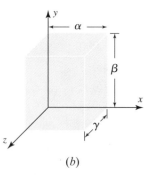

▲ Figure 5.7.2

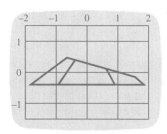

▲ **View 2** View 1 scaled by $\alpha = 1.8$, $\beta = 0.5$, $\gamma = 3.0$.

The new coordinate matrix can then be entered into the video display system to produce the new view of the object. As an example, View 2 is View 1 scaled by setting $\alpha = 1.8$,

$\beta = 0.5$, and $\gamma = 3.0$. Note that the scaling $\gamma = 3.0$ along the z-axis is not visible in View 2, since we see only the projection of the object onto the xy-plane.

Translation

We next consider the transformation of translating or displacing an object to a new position on the screen. Referring to Figure 5.7.3, suppose we desire to change an existing view so that each point P_i with coordinates (x_i, y_i, z_i) moves to a new point P_i' with coordinates $(x_i + x_0, y_i + y_0, z_i + z_0)$. The vector

$$\begin{bmatrix} x_0 \\ y_0 \\ z_0 \end{bmatrix}$$

is called the **translation vector** of the transformation. By defining a $3 \times n$ matrix T as

$$T = \begin{bmatrix} x_0 & x_0 & \cdots & x_0 \\ y_0 & y_0 & \cdots & y_0 \\ z_0 & z_0 & \cdots & z_0 \end{bmatrix}$$

we can translate all n points of the view determined by the coordinate matrix P by matrix addition via the equation

$$P' = P + T$$

The coordinate matrix P' then specifies the new coordinates of the n points. For example, if we wish to translate View 1 according to the translation vector

$$\begin{bmatrix} 1.2 \\ 0.4 \\ 1.7 \end{bmatrix}$$

the result is View 3. Note, again, that the translation $z_0 = 1.7$ along the z-axis does not show up explicitly in View 3.

In Exercise 7, a technique of performing translations by matrix multiplication rather than by matrix addition is explained.

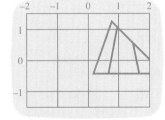

▲ **View 3** View 1 translated by $x_0 = 1.2$, $y_0 = 0.4$, $z_0 = 1.7$.

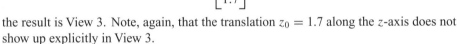

▶ **Figure 5.7.3**

Rotation

A more complicated type of transformation is a rotation of a view about one of the three coordinate axes. We begin with a rotation about the z-axis (the axis perpendicular to the screen) through an angle θ. Given a point P_i in the original view with coordinates (x_i, y_i, z_i), we wish to compute the new coordinates (x_i', y_i', z_i') of the rotated point P_i'. Referring to Figure 5.7.4 and using a little trigonometry, you should be able to derive the following:

$$x_i' = \rho \cos(\phi + \theta) = \rho \cos\phi \cos\theta - \rho \sin\phi \sin\theta = x_i \cos\theta - y_i \sin\theta$$
$$y_i' = \rho \sin(\phi + \theta) = \rho \cos\phi \sin\theta + \rho \sin\phi \cos\theta = x_i \sin\theta + y_i \cos\theta$$
$$z_i' = z_i$$

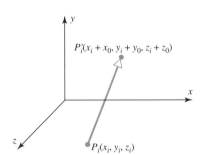

▲ Figure 5.7.4

These equations can be written in matrix form as

$$
\begin{bmatrix} x_i' \\ y_i' \\ z_i' \end{bmatrix} = \begin{bmatrix} \cos\theta & -\sin\theta & 0 \\ \sin\theta & \cos\theta & 0 \\ 0 & 0 & 1 \end{bmatrix} \begin{bmatrix} x_i \\ y_i \\ z_i \end{bmatrix}
$$

If we let R denote the 3×3 matrix in this equation, all n points can be rotated by the matrix product

$$P' = RP$$

to yield the coordinate matrix P' of the rotated view.

Rotations about the x- and y-axes can be accomplished analogously, and the resulting rotation matrices are given with Views 4, 5, and 6. These three new views of the truncated pyramid correspond to rotations of View 1 about the x-, y-, and z-axes, respectively, each through an angle of $90°$.

Rotation about the x-axis

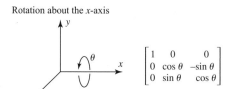

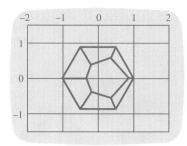

▲ **View 4** View 1 rotated $90°$ about the x-axis.

Rotation about the y-axis

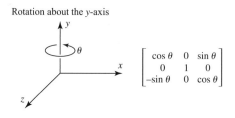

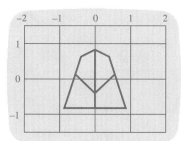

▲ **View 5** View 1 rotated $90°$ about the y-axis.

Rotation about the z-axis

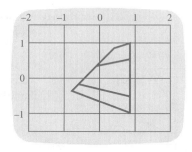

▲ **View 6** View 1 rotated $90°$ about the z-axis.

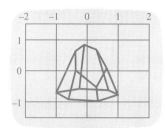

▲ **View 7** Oblique view of truncated pyramid.

Rotations about three coordinate axes may be combined to give oblique views of an object. For example, View 7 is View 1 rotated first about the x-axis through 30°, then about the y-axis through $-70°$, and finally about the z-axis through $-27°$. Mathematically, these three successive rotations can be embodied in the single transformation equation $P' = RP$, where R is the product of three individual rotation matrices:

$$R_1 = \begin{bmatrix} 1 & 0 & 0 \\ 0 & \cos(30°) & -\sin(30°) \\ 0 & \sin(30°) & \cos(30°) \end{bmatrix}$$

$$R_2 = \begin{bmatrix} \cos(-70°) & 0 & \sin(-70°) \\ 0 & 1 & 0 \\ -\sin(-70°) & 0 & \cos(-70°) \end{bmatrix}$$

$$R_3 = \begin{bmatrix} \cos(-27°) & -\sin(-27°) & 0 \\ \sin(-27°) & \cos(-27°) & 0 \\ 0 & 0 & 1 \end{bmatrix}$$

in the order

$$R = R_3 R_2 R_1 = \begin{bmatrix} .305 & -.025 & -.952 \\ -.155 & .985 & -.076 \\ .940 & .171 & .296 \end{bmatrix}$$

As a final illustration, in View 8 we have two separate views of the truncated pyramid, which constitute a stereoscopic pair. They were produced by first rotating View 7 about the y-axis through an angle of $-3°$ and translating it to the right, then rotating the same View 7 about the y-axis through an angle of $+3°$ and translating it to the left. The translation distances were chosen so that the stereoscopic views are about $2\frac{1}{2}$ inches apart—the approximate distance between a pair of eyes.

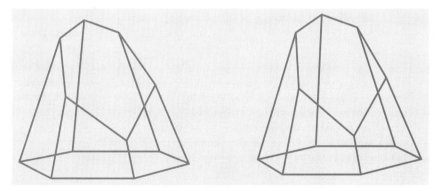

▲ **View 8** Stereoscopic figure of truncated pyramid. The three-dimensionality of the diagram can be seen by holding the book about one foot away and focusing on a distant object. Then by shifting your gaze to View 8 without refocusing, you can make the two views of the stereoscopic pair merge together and produce the desired effect.

Exercise Set 5.7

1. View 9 is a view of a square with vertices $(0, 0, 0)$, $(1, 0, 0)$, $(1, 1, 0)$, and $(0, 1, 0)$.

(a) What is the coordinate matrix of View 9?

(b) What is the coordinate matrix of View 9 after it is scaled by a factor $1\frac{1}{2}$ in the x-direction and $\frac{1}{2}$ in the y-direction? Draw a sketch of the scaled view.

(c) What is the coordinate matrix of View 9 after it is translated by the following vector?

$$\begin{bmatrix} -2 \\ -1 \\ 3 \end{bmatrix}$$

Draw a sketch of the translated view.

(d) What is the coordinate matrix of View 9 after it is rotated through an angle of $-30°$ about the z-axis? Draw a sketch of the rotated view.

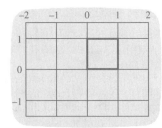

▲ **View 9**   Square with vertices $(0, 0, 0)$, $(1, 0, 0)$, $(1, 1, 0)$, and $(0, 1, 0)$ (Exercises 1 and 2).

2. (a) If the coordinate matrix of View 9 is multiplied by the matrix

$$\begin{bmatrix} 1 & \frac{1}{2} & 0 \\ 0 & 1 & 0 \\ 0 & 0 & 1 \end{bmatrix}$$

the result is the coordinate matrix of View 10. Such a transformation is called a *shear in the x-direction with factor $\frac{1}{2}$ with respect to the y-coordinate*. Show that under such a transformation, a point with coordinates (x_i, y_i, z_i) has new coordinates $(x_i + \frac{1}{2}y_i, y_i, z_i)$.

(b) What are the coordinates of the four vertices of the shear square in View 10?

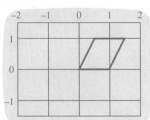

▲ **View 10**   View 9 sheared along the x-axis by $\frac{1}{2}$ with respect to the y-coordinate (Exercise 2).

(c) The matrix

$$\begin{bmatrix} 1 & 0 & 0 \\ .6 & 1 & 0 \\ 0 & 0 & 1 \end{bmatrix}$$

determines a *shear in the y-direction with factor .6 with respect to the x-coordinate* (an example appears in View 11). Sketch a view of the square in View 9 after such a shearing transformation, and find the new coordinates of its four vertices.

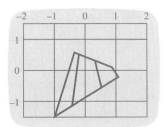

▲ **View 11**   View 1 sheared along the y-axis by .6 with respect to the x-coordinate (Exercise 2).

3. (a) The *reflection about the xz-plane* is defined as the transformation that takes a point (x_i, y_i, z_i) to the point $(x_i, -y_i, z_i)$ (e.g., View 12). If P and P' are the coordinate matrices of a view and its reflection about the xz-plane, respectively, find a matrix M such that $P' = MP$.

(b) Analogous to part (a), define the *reflection about the yz-plane* and construct the corresponding transformation matrix. Draw a sketch of View 1 reflected about the yz-plane.

(c) Analogous to part (a), define the *reflection about the xy-plane* and construct the corresponding transformation matrix. Draw a sketch of View 1 reflected about the xy-plane.

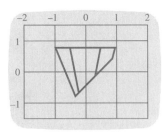

▲ **View 12**   View 1 reflected about the xz-plane (Exercise 3).

4. (a) View 13 is View 1 subject to the following five transformations:

1.   Scale by a factor of $\frac{1}{2}$ in the x-direction, 2 in the y-direction, and $\frac{1}{3}$ in the z-direction.

2.   Translate $\frac{1}{2}$ unit in the x-direction.

3. Rotate 20° about the x-axis.

4. Rotate −45° about the y-axis.

5. Rotate 90° about the z-axis.

Construct the five matrices M_1, M_2, M_3, M_4, and M_5 associated with these five transformations.

(b) If P is the coordinate matrix of View 1 and P' is the coordinate matrix of View 13, express P' in terms of M_1, M_2, M_3, M_4, M_5, and P.

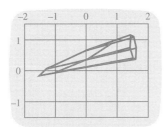

▲ **View 13** View 1 scaled, translated, and rotated (Exercise 4).

5. (a) View 14 is View 1 subject to the following seven transformations:

1. Scale by a factor of .3 in the x-direction and by a factor of .5 in the y-direction.

2. Rotate 45° about the x-axis.

3. Translate 1 unit in the x-direction.

4. Rotate 35° about the y-axis.

5. Rotate −45° about the z-axis.

6. Translate 1 unit in the z-direction.

7. Scale by a factor of 2 in the x-direction.

Construct the matrices M_1, M_2, ..., M_7 associated with these seven transformations.

(b) If P is the coordinate matrix of View 1 and P' is the coordinate matrix of View 14, express P' in terms of M_1, M_2, ..., M_7, and P.

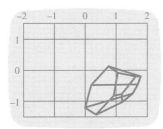

▲ **View 14** View 1 scaled, translated, and rotated (Exercise 5).

6. Suppose that a view with coordinate matrix P is to be rotated through an angle θ about an axis through the origin and specified by two angles α and β (see Figure Ex-6). If P' is the

coordinate matrix of the rotated view, find rotation matrices R_1, R_2, R_3, R_4, and R_5 such that

$$P' = R_5 R_4 R_3 R_2 R_1 P$$

[*Hint:* The desired rotation can be accomplished in the following five steps:

1. Rotate through an angle of β about the y-axis.

2. Rotate through an angle of α about the z-axis.

3. Rotate through an angle of θ about the y-axis.

4. Rotate through an angle of $-\alpha$ about the z-axis.

5. Rotate through an angle of $-\beta$ about the y-axis.]

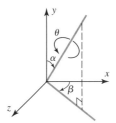

◀ **Figure Ex-6**

7. This exercise illustrates a technique for translating a point with coordinates (x_i, y_i, z_i) to a point with coordinates $(x_i + x_0, y_i + y_0, z_i + z_0)$ by matrix multiplication rather than matrix addition.

(a) Let the point (x_i, y_i, z_i) be associated with the column vector

$$\mathbf{v}_i = \begin{bmatrix} x_i \\ y_i \\ z_i \\ 1 \end{bmatrix}$$

and let the point $(x_i + x_0, y_i + y_0, z_i + z_0)$ be associated with the column vector

$$\mathbf{v}'_i = \begin{bmatrix} x_i + x_0 \\ y_i + y_0 \\ z_i + z_0 \\ 1 \end{bmatrix}$$

Find a 4×4 matrix M such that $\mathbf{v}'_i = M\mathbf{v}_i$.

(b) Find the specific 4×4 matrix of the above form that will effect the translation of the point $(4, -2, 3)$ to the point $(-1, 7, 0)$.

8. For the three rotation matrices given with Views 4, 5, and 6, show that

$$R^{-1} = R^T$$

(A matrix with this property is called an ***orthogonal matrix***. See Section 7.1.)

Section 5.7   Technology Exercises

The following exercises are designed to be solved using a technology utility. Typically, this will be MATLAB, *Mathematica*, Maple, Derive, or Mathcad, but it may also be some other type of linear algebra software or a scientific calculator with some linear algebra capabilities. For each exercise you will need to read the relevant documentation for the particular utility you are using. The goal of these exercises is to provide you with a basic proficiency with your technology utility. Once you have mastered the techniques in these exercises, you will be able to use your technology utility to solve many of the problems in the regular exercise sets.

T1. Let (a, b, c) be a unit vector normal to the plane $ax + by + cz = 0$, and let $\mathbf{r} = (x, y, z)$ be a vector. It can be shown that the mirror image of the vector $\mathbf{r}$ through the above plane has coordinates $\mathbf{r}_m = (x_m, y_m, z_m)$, where

$$\begin{bmatrix} x_m \\ y_m \\ z_m \end{bmatrix} = M \begin{bmatrix} x \\ y \\ z \end{bmatrix}$$

with

$$M = I - 2\mathbf{n}\mathbf{n}^T = \begin{bmatrix} 1 & 0 & 0 \\ 0 & 1 & 0 \\ 0 & 0 & 1 \end{bmatrix} - 2 \begin{bmatrix} a \\ b \\ c \end{bmatrix} \begin{bmatrix} a & b & c \end{bmatrix}$$

(a) Show that $M^2 = I$ and give a physical reason why this must be so. [*Hint:* Use the fact that (a, b, c) is a unit vector to show that $\mathbf{n}^T\mathbf{n} = 1$.]

(b) Use a computer to show that $\det(M) = -1$.

(c) The eigenvectors of M satisfy the equation

$$\begin{bmatrix} x_m \\ y_m \\ z_m \end{bmatrix} = M \begin{bmatrix} x \\ y \\ z \end{bmatrix} = \lambda \begin{bmatrix} x \\ y \\ z \end{bmatrix}$$

and therefore correspond to those vectors whose direction is not affected by a reflection through the plane. Use a computer to determine the eigenvectors and eigenvalues of M, and then give a physical argument to support your answer.

T2. A vector $\mathbf{v} = (x, y, z)$ is rotated by an angle θ about an axis having unit vector (a, b, c), thereby forming the rotated vector $\mathbf{v}_R = (x_R, y_R, z_R)$. It can be shown that

$$\begin{bmatrix} x_R \\ y_R \\ z_R \end{bmatrix} = R(\theta) \begin{bmatrix} x \\ y \\ z \end{bmatrix}$$

with

$$R(\theta) = \cos(\theta) \begin{bmatrix} 1 & 0 & 0 \\ 0 & 1 & 0 \\ 0 & 0 & 1 \end{bmatrix} + (1 - \cos(\theta)) \begin{bmatrix} a \\ b \\ c \end{bmatrix} \begin{bmatrix} a & b & c \end{bmatrix}$$

$$+ \sin(\theta) \begin{bmatrix} 0 & -c & b \\ c & 0 & -a \\ -b & a & 0 \end{bmatrix}$$

(a) Use a computer to show that $R(\theta)R(\varphi) = R(\theta + \varphi)$, and then give a physical reason why this must be so. Depending on the sophistication of the computer you are using, you may have to experiment using different values of a, b, and

$$c = \sqrt{1 - a^2 - b^2}$$

(b) Show also that $R^{-1}(\theta) = R(-\theta)$ and give a physical reason why this must be so.

(c) Use a computer to show that $\det(R(\theta)) = +1$.

5.8   Cryptography

In this section we present a method of encoding and decoding messages. We also examine modular arithmetic and show how Gaussian elimination can sometimes be used to break an opponent's code.

> **PREREQUISITES:** Matrices
> Gaussian Elimination
> Matrix Operations
> Linear Independence
> Linear Transformations (Section 4.9)

Ciphers   The study of encoding and decoding secret messages is called *cryptography*. Although secret codes date to the earliest days of written communication, there has been a recent surge of interest in the subject because of the need to maintain the privacy of information transmitted over public lines of communication. In the language of cryptography, codes are called **ciphers**, uncoded messages are called **plaintext**, and coded messages are called **ciphertext**. The process of converting from plaintext to ciphertext is called **enciphering**, and the reverse process of converting from ciphertext to plaintext is called **deciphering**.

The simplest ciphers, called **substitution ciphers**, are those that replace each letter of the alphabet by a different letter. For example, in the substitution cipher

Plain   *A B C D E F G H I J K L M N O P Q R S T U V W X Y Z*
Cipher *D E F G H I J K L M N O P Q R S T U V W X Y Z A B C*

the plaintext letter *A* is replaced by *D*, the plaintext letter *B* by *E*, and so forth. With this cipher the plaintext message

ROME WAS NOT BUILT IN A DAY

becomes

URPH ZDV QRW EXLOW LQ D GDB

Hill Ciphers   A disadvantage of substitution ciphers is that they preserve the frequencies of individual letters, making it relatively easy to break the code by statistical methods. One way to overcome this problem is to divide the plaintext into groups of letters and encipher the plaintext group by group, rather than one letter at a time. A system of cryptography in which the plaintext is divided into sets of *n* letters, each of which is replaced by a set of *n* cipher letters, is called a **polygraphic system**. In this section we will study a class of polygraphic systems based on matrix transformations. [The ciphers that we will discuss are called **Hill ciphers** after Lester S. Hill, who introduced them in two papers: "Cryptography in an Algebraic Alphabet," *American Mathematical Monthly, 36* (June–July 1929), pp. 306–312; and "Concerning Certain Linear Transformation Apparatus of Cryptography," *American Mathematical Monthly, 38* (March 1931), pp. 135–154.]

In the discussion to follow, we assume that each plaintext and ciphertext letter except *Z* is assigned the numerical value that specifies its position in the standard alphabet (Table 1). For reasons that will become clear later, *Z* is assigned a value of zero.

Table 1

A	*B*	*C*	*D*	*E*	*F*	*G*	*H*	*I*	*J*	*K*	*L*	*M*	*N*	*O*	*P*	*Q*	*R*	*S*	*T*	*U*	*V*	*W*	*X*	*Y*	*Z*
1	2	3	4	5	6	7	8	9	10	11	12	13	14	15	16	17	18	19	20	21	22	23	24	25	0

In the simplest Hill ciphers, successive *pairs* of plaintext are transformed into ciphertext by the following procedure:

Step 1.   Choose a 2×2 matrix with integer entries

$$A = \begin{bmatrix} a_{11} & a_{12} \\ a_{21} & a_{22} \end{bmatrix}$$

to perform the encoding. Certain additional conditions on *A* will be imposed later.

Step 2.   Group successive plaintext letters into pairs, adding an arbitrary "dummy" letter to fill out the last pair if the plaintext has an odd number of letters, and replace each plaintext letter by its numerical value.

Step 3.   Successively convert each plaintext pair $p_1 p_2$ into a column vector

$$\mathbf{p} = \begin{bmatrix} p_1 \\ p_2 \end{bmatrix}$$

and form the product $A\mathbf{p}$. We will call $\mathbf{p}$ a **plaintext vector** and $A\mathbf{p}$ the corresponding **ciphertext vector**.

Step 4.   Convert each ciphertext vector into its alphabetic equivalent.

▶ **EXAMPLE 1** **Hill Cipher of a Message**

Use the matrix

$$\begin{bmatrix} 1 & 2 \\ 0 & 3 \end{bmatrix}$$

to obtain the Hill cipher for the plaintext message

I AM HIDING

Solution If we group the plaintext into pairs and add the dummy letter *G* to fill out the last pair, we obtain

    IA   *MH*   *ID*   *IN*   *G G*

or, equivalently, from Table 1,

      9 1   13 8   9 4   9 14   7 7

To encipher the pair *IA*, we form the matrix product

$$\begin{bmatrix} 1 & 2 \\ 0 & 3 \end{bmatrix} \begin{bmatrix} 9 \\ 1 \end{bmatrix} = \begin{bmatrix} 11 \\ 3 \end{bmatrix}$$

which, from Table 1, yields the ciphertext *KC*.

To encipher the pair *MH*, we form the product

$$\begin{bmatrix} 1 & 2 \\ 0 & 3 \end{bmatrix} \begin{bmatrix} 13 \\ 8 \end{bmatrix} = \begin{bmatrix} 29 \\ 24 \end{bmatrix} \tag{1}$$

However, there is a problem here, because the number 29 has no alphabet equivalent (Table 1). To resolve this problem, we make the following agreement:

> *Whenever an integer greater than* 25 *occurs, it will be replaced by the remainder that results when this integer is divided by* 26.

Because the remainder after division by 26 is one of the integers $0, 1, 2, \ldots, 25$, this procedure will always yield an integer with an alphabet equivalent.

Thus, in (1) we replace 29 by 3, which is the remainder after dividing 29 by 26. It now follows from Table 1 that the ciphertext for the pair *MH* is *CX*.

The computations for the remaining ciphertext vectors are

$$\begin{bmatrix} 1 & 2 \\ 0 & 3 \end{bmatrix} \begin{bmatrix} 9 \\ 4 \end{bmatrix} = \begin{bmatrix} 17 \\ 12 \end{bmatrix}$$

$$\begin{bmatrix} 1 & 2 \\ 0 & 3 \end{bmatrix} \begin{bmatrix} 9 \\ 14 \end{bmatrix} = \begin{bmatrix} 37 \\ 42 \end{bmatrix} \quad \text{or} \quad \begin{bmatrix} 11 \\ 16 \end{bmatrix}$$

$$\begin{bmatrix} 1 & 2 \\ 0 & 3 \end{bmatrix} \begin{bmatrix} 7 \\ 7 \end{bmatrix} = \begin{bmatrix} 21 \\ 21 \end{bmatrix}$$

These correspond to the ciphertext pairs *QL*, *KP*, and *UU*, respectively. In summary, the entire ciphertext message is

    KC   *CX*   *QL*   *KP*   *UU*

which would usually be transmitted as a single string without spaces:

KCCXQLKPUU ◀

Because the plaintext was grouped in pairs and enciphered by a 2×2 matrix, the Hill cipher in Example 1 is referred to as a **Hill 2-cipher**. It is obviously also possible to group the plaintext in triples and encipher by a 3×3 matrix with integer entries; this is called a *Hill 3-cipher*. In general, for a *Hill n-cipher*, plaintext is grouped into sets of n letters and enciphered by an $n \times n$ matrix with integer entries.

Modular Arithmetic
In Example 1, integers greater than 25 were replaced by their remainders after division by 26. This technique of working with remainders is at the core of a body of mathematics called *modular arithmetic*. Because of its importance in cryptography, we will digress for a moment to touch on some of the main ideas in this area.

In modular arithmetic we are given a positive integer m, called the *modulus*, and any two integers whose difference is an integer multiple of the modulus are regarded as "equal" or "equivalent" with respect to the modulus. More precisely, we make the following definition.

DEFINITION 1 If m is a positive integer and a and b are any integers, then we say that a is *equivalent* to b modulo m, written

$$a = b \quad (\text{mod } m)$$

if $a - b$ is an integer multiple of m.

▶ **EXAMPLE 2 Various Equivalences**

$$7 = 2 \quad (\text{mod } 5)$$
$$19 = 3 \quad (\text{mod } 2)$$
$$-1 = 25 \quad (\text{mod } 26)$$
$$12 = 0 \quad (\text{mod } 4) \quad \blacktriangleleft$$

For any modulus m it can be proved that every integer a is equivalent, modulo m, to exactly one of the integers

$$0, 1, 2, \ldots, m - 1$$

We call this integer the *residue* of a modulo m, and we write

$$Z_m = \{0, 1, 2, \ldots, m - 1\}$$

to denote the set of residues modulo m.

If a is a *nonnegative* integer, then its residue modulo m is simply the remainder that results when a is divided by m. For an arbitrary integer a, the residue can be found using the following theorem.

THEOREM 5.8.1 *For any integer a and modulus m, let*

$$R = \text{remainder of } \frac{|a|}{m}$$

Then the residue r of a modulo m is given by

$$r = \begin{cases} R & \text{if } a \geq 0 \\ m - R & \text{if } a < 0 \quad \text{and} \quad R \neq 0 \\ 0 & \text{if } a < 0 \quad \text{and} \quad R = 0 \end{cases}$$

▶ **EXAMPLE 3 Residues mod 26**

Find the residue modulo 26 of (a) 87, (b) −38, and (c) −26.

Solution (a) Dividing $|87| = 87$ by 26 yields a remainder of $R = 9$, so $r = 9$. Thus,

$$87 = 9 \quad (\text{mod } 26)$$

Solution (b) Dividing $|-38| = 38$ by 26 yields a remainder of $R = 12$, so $r = 26 - 12 = 14$. Thus,

$$-38 = 14 \quad (\text{mod } 26)$$

Solution (c) Dividing $|-26| = 26$ by 26 yields a remainder of $R = 0$. Thus,

$$-26 = 0 \quad (\text{mod } 26) \quad ◀$$

In ordinary arithmetic every nonzero number a has a *reciprocal* or *multiplicative inverse*, denoted by a^{-1}, such that

$$aa^{-1} = a^{-1}a = 1$$

In modular arithmetic we have the following corresponding concept:

DEFINITION 2 If a is a number in Z_m, then a number a^{-1} in Z_m is called a **reciprocal** or **multiplicative inverse** of a modulo m if $aa^{-1} = a^{-1}a = 1 \ (\text{mod } m)$.

It can be proved that if a and m have no common prime factors, then a has a unique reciprocal modulo m; conversely, if a and m have a common prime factor, then a has no reciprocal modulo m.

▶ **EXAMPLE 4 Reciprocal of 3 mod 26**

The number 3 has a reciprocal modulo 26 because 3 and 26 have no common prime factors. This reciprocal can be obtained by finding the number x in Z_{26} that satisfies the modular equation

$$3x = 1 \quad (\text{mod } 26)$$

Although there are general methods for solving such modular equations, it would take us too far afield to study them. However, because 26 is relatively small, this equation can be solved by trying the possible solutions, 0 to 25, one at a time. With this approach we find that $x = 9$ is the solution, because

$$3 \cdot 9 = 27 = 1 \quad (\text{mod } 26)$$

Thus,

$$3^{-1} = 9 \quad (\text{mod } 26)$$

▶ **EXAMPLE 5 A Number with No Reciprocal mod 26**

The number 4 has no reciprocal modulo 26, because 4 and 26 have 2 as a common prime factor (see Exercise 8). ◀

For future reference, in Table 2 we provide the following reciprocals modulo 26:

Table 2   Reciprocals Modulo 26

a	1	3	5	7	9	11	15	17	19	21	23	25
a^{-1}	1	9	21	15	3	19	7	23	11	5	17	25

Deciphering Every useful cipher must have a procedure for decipherment. In the case of a Hill cipher, decipherment uses the inverse (mod 26) of the enciphering matrix. To be precise, if m is a positive integer, then a square matrix A with entries in Z_m is said to be ***invertible modulo m*** if there is a matrix B with entries in Z_m such that

$$AB = BA = I \quad (\bmod \ m)$$

Suppose now that

$$A = \begin{bmatrix} a_{11} & a_{12} \\ a_{21} & a_{22} \end{bmatrix}$$

is invertible modulo 26 and this matrix is used in a Hill 2-cipher. If

$$\mathbf{p} = \begin{bmatrix} p_1 \\ p_2 \end{bmatrix}$$

is a plaintext vector, then

$$\mathbf{c} = A\mathbf{p} \quad (\bmod \ 26)$$

is the corresponding ciphertext vector and

$$\mathbf{p} = A^{-1}\mathbf{c} \quad (\bmod \ 26)$$

Thus, each plaintext vector can be recovered from the corresponding ciphertext vector by multiplying it on the left by A^{-1} (mod 26).

In cryptography it is important to know which matrices are invertible modulo 26 and how to obtain their inverses. We now investigate these questions.

In ordinary arithmetic, a square matrix A is invertible if and only if $\det(A) \neq 0$, or, equivalently, if and only if $\det(A)$ has a reciprocal. The following theorem is the analog of this result in modular arithmetic.

THEOREM 5.8.2 *A square matrix A with entries in Z_m is invertible modulo m if and only if the residue of $\det(A)$ modulo m has a reciprocal modulo m.*

Because the residue of $\det(A)$ modulo m will have a reciprocal modulo m if and only if this residue and m have no common prime factors, we have the following corollary.

COROLLARY 5.8.3 *A square matrix A with entries in Z_m is invertible modulo m if and only if m and the residue of $\det(A)$ modulo m have no common prime factors.*

Because the only prime factors of $m = 26$ are 2 and 13, we have the following corollary, which is useful in cryptography.

COROLLARY 5.8.4 *A square matrix A with entries in Z_{26} is invertible modulo 26 if and only if the residue of $\det(A)$ modulo 26 is not divisible by 2 or 13.*

We leave it for you to verify that if

$$A = \begin{bmatrix} a & b \\ c & d \end{bmatrix}$$

has entries in Z_{26} and the residue of $\det(A) = ad - bc$ modulo 26 is not divisible by 2 or 13, then the inverse of A (mod 26) is given by

$$A^{-1} = (ad - bc)^{-1} \begin{bmatrix} d & -b \\ -c & a \end{bmatrix} \quad (\bmod \ 26) \tag{2}$$

where $(ad - bc)^{-1}$ is the reciprocal of the residue of $ad - bc$ (mod 26).

▶ **EXAMPLE 6** **Inverse of a Matrix mod 26**

Find the inverse of

$$A = \begin{bmatrix} 5 & 6 \\ 2 & 3 \end{bmatrix}$$

modulo 26.

Solution

$$\det(A) = ad - bc = 5 \cdot 3 - 6 \cdot 2 = 3$$

so from Table 2,

$$(ad - bc)^{-1} = 3^{-1} = 9 \quad (\text{mod } 26)$$

Thus, from (2),

$$A^{-1} = 9 \begin{bmatrix} 3 & -6 \\ -2 & 5 \end{bmatrix} = \begin{bmatrix} 27 & -54 \\ -18 & 45 \end{bmatrix} = \begin{bmatrix} 1 & 24 \\ 8 & 19 \end{bmatrix} \quad (\text{mod } 26)$$

As a check,

$$AA^{-1} = \begin{bmatrix} 5 & 6 \\ 2 & 3 \end{bmatrix} \begin{bmatrix} 1 & 24 \\ 8 & 19 \end{bmatrix} = \begin{bmatrix} 53 & 234 \\ 26 & 105 \end{bmatrix} = \begin{bmatrix} 1 & 0 \\ 0 & 1 \end{bmatrix} \quad (\text{mod } 26)$$

Similarly, $A^{-1}A = I$.

▶ **EXAMPLE 7** **Decoding a Hill 2-Cipher**

Decode the following Hill 2-cipher, which was enciphered by the matrix in Example 6:

$$GTNKGKDUSK$$

Solution From Table 1 the numerical equivalent of this ciphertext is

$$7 \ 20 \qquad 14 \ 11 \qquad 7 \ 11 \qquad 4 \ 21 \qquad 19 \ 11$$

To obtain the plaintext pairs, we multiply each ciphertext vector by the inverse of A (obtained in Example 6):

$$\begin{bmatrix} 1 & 24 \\ 8 & 19 \end{bmatrix} \begin{bmatrix} 7 \\ 20 \end{bmatrix} = \begin{bmatrix} 487 \\ 436 \end{bmatrix} = \begin{bmatrix} 19 \\ 20 \end{bmatrix} \quad (\text{mod } 26)$$

$$\begin{bmatrix} 1 & 24 \\ 8 & 19 \end{bmatrix} \begin{bmatrix} 14 \\ 11 \end{bmatrix} = \begin{bmatrix} 278 \\ 321 \end{bmatrix} = \begin{bmatrix} 18 \\ 9 \end{bmatrix} \quad (\text{mod } 26)$$

$$\begin{bmatrix} 1 & 24 \\ 8 & 19 \end{bmatrix} \begin{bmatrix} 7 \\ 11 \end{bmatrix} = \begin{bmatrix} 271 \\ 265 \end{bmatrix} = \begin{bmatrix} 11 \\ 5 \end{bmatrix} \quad (\text{mod } 26)$$

$$\begin{bmatrix} 1 & 24 \\ 8 & 19 \end{bmatrix} \begin{bmatrix} 4 \\ 21 \end{bmatrix} = \begin{bmatrix} 508 \\ 431 \end{bmatrix} = \begin{bmatrix} 14 \\ 15 \end{bmatrix} \quad (\text{mod } 26)$$

$$\begin{bmatrix} 1 & 24 \\ 8 & 19 \end{bmatrix} \begin{bmatrix} 19 \\ 11 \end{bmatrix} = \begin{bmatrix} 283 \\ 361 \end{bmatrix} = \begin{bmatrix} 23 \\ 23 \end{bmatrix} \quad (\text{mod } 26)$$

From Table 1, the alphabet equivalents of these vectors are

$$ST \quad RI \quad KE \quad NO \quad WW$$

which yields the message

$$STRIKE \quad NOW \quad ◀$$

Breaking a Hill Cipher   Because the purpose of enciphering messages and information is to prevent "opponents" from learning their contents, cryptographers are concerned with the *security* of their ciphers—that is, how readily they can be broken (deciphered by their opponents). We will conclude this section by discussing one technique for breaking Hill ciphers.

Suppose that you are able to obtain some corresponding plaintext and ciphertext from an opponent's message. For example, on examining some intercepted ciphertext, you may be able to deduce that the message is a letter that begins *DEAR SIR*. We will show that with a small amount of such data, it may be possible to determine the deciphering matrix of a Hill code and consequently obtain access to the rest of the message.

It is a basic result in linear algebra that a linear transformation is completely determined by its values at a basis. This principle suggests that if we have a Hill *n*-cipher, and if

$$\mathbf{p}_1, \mathbf{p}_2, \ldots, \mathbf{p}_n$$

are linearly independent plaintext vectors whose corresponding ciphertext vectors

$$A\mathbf{p}_1, A\mathbf{p}_2, \ldots, A\mathbf{p}_n$$

are known, then there is enough information available to determine the matrix A and hence A^{-1} (mod m).

The following theorem, whose proof is discussed in the exercises, provides a way to do this.

THEOREM 5.8.5 Determining the Deciphering Matrix

Let $\mathbf{p}_1, \mathbf{p}_2, \ldots, \mathbf{p}_n$ *be linearly independent plaintext vectors, and let* $\mathbf{c}_1, \mathbf{c}_2, \ldots, \mathbf{c}_n$ *be the corresponding ciphertext vectors in a Hill n-cipher. If*

$$P = \begin{bmatrix} \mathbf{p}_1^T \\ \mathbf{p}_2^T \\ \vdots \\ \mathbf{p}_n^T \end{bmatrix}$$

is the n × n matrix with row vectors $\mathbf{p}_1^T, \mathbf{p}_2^T, \ldots, \mathbf{p}_n^T$ *and if*

$$C = \begin{bmatrix} \mathbf{c}_1^T \\ \mathbf{c}_2^T \\ \vdots \\ \mathbf{c}_n^T \end{bmatrix}$$

is the n × n matrix with row vectors $\mathbf{c}_1^T, \mathbf{c}_2^T, \ldots, \mathbf{c}_n^T$, *then the sequence of elementary row operations that reduces C to I transforms P to* $(A^{-1})^T$.

This theorem tells us that to find the transpose of the deciphering matrix A^{-1}, we must find a sequence of row operations that reduces C to I and then perform this same sequence of operations on P. The following example illustrates a simple algorithm for doing this.

▶ **EXAMPLE 8 Using Theorem 5.8.5**

The following Hill 2-cipher is intercepted:

$$IOSBTGXESPXHOPDE$$

Decipher the message, given that it starts with the word *DEAR*.

Solution From Table 1, the numerical equivalent of the known plaintext is

$$
\begin{array}{cc}
DE & AR \\
4\ 5 & 1\ 18
\end{array}
$$

and the numerical equivalent of the corresponding ciphertext is

$$
\begin{array}{cc}
IO & SB \\
9\ 15 & 19\ 2
\end{array}
$$

so the corresponding plaintext and ciphertext vectors are

$$
\mathbf{p}_1 = \begin{bmatrix} 4 \\ 5 \end{bmatrix} \leftrightarrow \mathbf{c}_1 = \begin{bmatrix} 9 \\ 15 \end{bmatrix}
$$

$$
\mathbf{p}_2 = \begin{bmatrix} 1 \\ 18 \end{bmatrix} \leftrightarrow \mathbf{c}_2 = \begin{bmatrix} 19 \\ 2 \end{bmatrix}
$$

We want to reduce

$$
C = \begin{bmatrix} \mathbf{c}_1^T \\ \mathbf{c}_2^T \end{bmatrix} = \begin{bmatrix} 9 & 15 \\ 19 & 2 \end{bmatrix}
$$

to I by elementary row operations and simultaneously apply these operations to

$$
P = \begin{bmatrix} \mathbf{p}_1^T \\ \mathbf{p}_2^T \end{bmatrix} = \begin{bmatrix} 4 & 5 \\ 1 & 18 \end{bmatrix}
$$

to obtain $(A^{-1})^T$ (the transpose of the deciphering matrix). This can be accomplished by adjoining P to the right of C and applying row operations to the resulting matrix $[C \mid P]$ until the left side is reduced to I. The final matrix will then have the form $[I \mid (A^{-1})^T]$. The computations can be carried out as follows:

$$
\begin{bmatrix} 9 & 15 & | & 4 & 5 \\ 19 & 2 & | & 1 & 18 \end{bmatrix}
$$

⟵ We formed the matrix $[C \mid P]$.

$$
\begin{bmatrix} 1 & 45 & | & 12 & 15 \\ 19 & 2 & | & 1 & 18 \end{bmatrix}
$$

⟵ We multiplied the first row by $9^{-1} = 3$.

$$
\begin{bmatrix} 1 & 19 & | & 12 & 15 \\ 19 & 2 & | & 1 & 18 \end{bmatrix}
$$

⟵ We replaced 45 by its residue modulo 26.

$$
\begin{bmatrix} 1 & 19 & | & 12 & 15 \\ 0 & -359 & | & -227 & -267 \end{bmatrix}
$$

⟵ We added -19 times the first row to the second.

$$
\begin{bmatrix} 1 & 19 & | & 12 & 15 \\ 0 & 5 & | & 7 & 19 \end{bmatrix}
$$

⟵ We replaced the entries in the second row by their residues modulo 26.

$$
\begin{bmatrix} 1 & 19 & | & 12 & 15 \\ 0 & 1 & | & 147 & 399 \end{bmatrix}
$$

⟵ We multiplied the second row by $5^{-1} = 21$.

$$
\begin{bmatrix} 1 & 19 & | & 12 & 15 \\ 0 & 1 & | & 17 & 9 \end{bmatrix}
$$

⟵ We replaced the entries in the second row by their residues modulo 26.

$$
\begin{bmatrix} 1 & 0 & | & -311 & -156 \\ 0 & 1 & | & 17 & 9 \end{bmatrix}
$$

⟵ We added -19 times the second row to the first.

$$
\begin{bmatrix} 1 & 0 & | & 1 & 0 \\ 0 & 1 & | & 17 & 9 \end{bmatrix}
$$

⟵ We replaced the entries in the first row by their residues modulo 26.

Thus,

$$(A^{-1})^T = \begin{bmatrix} 1 & 0 \\ 17 & 9 \end{bmatrix}$$

so the deciphering matrix is

$$A^{-1} = \begin{bmatrix} 1 & 17 \\ 0 & 9 \end{bmatrix}$$

To decipher the message, we first group the ciphertext into pairs and find the numerical equivalent of each letter:

IO	*SB*	*TG*	*XE*	*SP*	*XH*	*OP*	*DE*
9 15	19 2	20 7	24 5	19 16	24 8	15 16	4 5

Next, we multiply successive ciphertext vectors on the left by A^{-1} and find the alphabet equivalents of the resulting plaintext pairs:

$$\begin{bmatrix} 1 & 17 \\ 0 & 9 \end{bmatrix}\begin{bmatrix} 9 \\ 15 \end{bmatrix} = \begin{bmatrix} 4 \\ 5 \end{bmatrix} \quad \begin{matrix} D \\ E \end{matrix}$$

$$\begin{bmatrix} 1 & 17 \\ 0 & 9 \end{bmatrix}\begin{bmatrix} 19 \\ 2 \end{bmatrix} = \begin{bmatrix} 1 \\ 18 \end{bmatrix} \quad \begin{matrix} A \\ R \end{matrix}$$

$$\begin{bmatrix} 1 & 17 \\ 0 & 9 \end{bmatrix}\begin{bmatrix} 20 \\ 7 \end{bmatrix} = \begin{bmatrix} 9 \\ 11 \end{bmatrix} \quad \begin{matrix} I \\ K \end{matrix}$$

$$\begin{bmatrix} 1 & 17 \\ 0 & 9 \end{bmatrix}\begin{bmatrix} 24 \\ 5 \end{bmatrix} = \begin{bmatrix} 5 \\ 19 \end{bmatrix} \quad \begin{matrix} E \\ S \end{matrix} \qquad \text{(mod 26)}$$

$$\begin{bmatrix} 1 & 17 \\ 0 & 9 \end{bmatrix}\begin{bmatrix} 19 \\ 16 \end{bmatrix} = \begin{bmatrix} 5 \\ 14 \end{bmatrix} \quad \begin{matrix} E \\ N \end{matrix}$$

$$\begin{bmatrix} 1 & 17 \\ 0 & 9 \end{bmatrix}\begin{bmatrix} 24 \\ 8 \end{bmatrix} = \begin{bmatrix} 4 \\ 20 \end{bmatrix} \quad \begin{matrix} D \\ T \end{matrix}$$

$$\begin{bmatrix} 1 & 17 \\ 0 & 9 \end{bmatrix}\begin{bmatrix} 15 \\ 16 \end{bmatrix} = \begin{bmatrix} 1 \\ 14 \end{bmatrix} \quad \begin{matrix} A \\ N \end{matrix}$$

$$\begin{bmatrix} 1 & 17 \\ 0 & 9 \end{bmatrix}\begin{bmatrix} 4 \\ 5 \end{bmatrix} = \begin{bmatrix} 11 \\ 19 \end{bmatrix} \quad \begin{matrix} K \\ S \end{matrix}$$

Finally, we construct the message from the plaintext pairs:

DE   AR   IK   ES   EN   DT   AN   KS

DEAR IKE SEND TANKS ◄

FURTHER READINGS

Readers interested in learning more about mathematical cryptography are referred to the following books, the first of which is elementary and the second more advanced.

1. ABRAHAM SINKOV, *Elementary Cryptanalysis, a Mathematical Approach* (Mathematical Association of America, 2009).

2. ALAN G. KONHEIM, *Cryptography, a Primer* (New York: Wiley-Interscience, 1981).

Exercise Set 5.8

1. Obtain the Hill cipher of the message

DARK NIGHT

for each of the following enciphering matrices:

(a) $\begin{bmatrix} 1 & 3 \\ 2 & 1 \end{bmatrix}$
(b) $\begin{bmatrix} 4 & 3 \\ 1 & 2 \end{bmatrix}$

2. In each part determine whether the matrix is invertible modulo 26. If so, find its inverse modulo 26 and check your work by verifying that $AA^{-1} = A^{-1}A = I$ (mod 26).

(a) $A = \begin{bmatrix} 9 & 1 \\ 7 & 2 \end{bmatrix}$
(b) $A = \begin{bmatrix} 3 & 1 \\ 5 & 3 \end{bmatrix}$
(c) $A = \begin{bmatrix} 8 & 11 \\ 1 & 9 \end{bmatrix}$

(d) $A = \begin{bmatrix} 2 & 1 \\ 1 & 7 \end{bmatrix}$
(e) $A = \begin{bmatrix} 3 & 1 \\ 6 & 2 \end{bmatrix}$
(f) $A = \begin{bmatrix} 1 & 8 \\ 1 & 3 \end{bmatrix}$

3. Decode the message

SAKNOXAOJX

given that it is a Hill cipher with enciphering matrix

$$\begin{bmatrix} 4 & 1 \\ 3 & 2 \end{bmatrix}$$

4. A Hill 2-cipher is intercepted that starts with the pairs

SL   HK

Find the deciphering and enciphering matrices, given that the plaintext is known to start with the word *ARMY*.

5. Decode the following Hill 2-cipher if the last four plaintext letters are known to be *ATOM*.

LNGIHGYBVRENJYQO

6. Decode the following Hill 3-cipher if the first nine plaintext letters are *IHAVECOME*:

HPAFQGGDUGDDHPGODYNOR

7. All of the results of this section can be generalized to the case where the plaintext is a binary message; that is, it is a sequence of 0's and 1's. In this case we do all of our modular arithmetic using modulus 2 rather than modulus 26. Thus, for example,

$1 + 1 = 0$ (mod 2). Suppose we want to encrypt the message 110101111. Let us first break it into triplets to form the three

vectors $\begin{bmatrix} 1 \\ 1 \\ 0 \end{bmatrix}$, $\begin{bmatrix} 1 \\ 0 \\ 1 \end{bmatrix}$, $\begin{bmatrix} 1 \\ 1 \\ 1 \end{bmatrix}$, and let us take $\begin{bmatrix} 1 & 1 & 0 \\ 0 & 1 & 1 \\ 1 & 1 & 1 \end{bmatrix}$ as our

enciphering matrix.

(a) Find the encoded message.

(b) Find the inverse modulo 2 of the enciphering matrix, and verify that it decodes your encoded message.

8. If, in addition to the standard alphabet, a period, comma, and question mark were allowed, then 29 plaintext and ciphertext symbols would be available and all matrix arithmetic would be done modulo 29. Under what conditions would a matrix with entries in Z_{29} be invertible modulo 29?

9. Show that the modular equation $4x = 1$ (mod 26) has no solution in Z_{26} by successively substituting the values $x = 0, 1, 2, \ldots, 25$.

10. (a) Let P and C be the matrices in Theorem 5.8.5. Show that
$P = C(A^{-1})^T$.

(b) To prove Theorem 5.8.5, let $E_1, E_2, \ldots, E_n$ be the elementary matrices that correspond to the row operations that reduce C to I, so

$$E_n \cdots E_2 E_1 C = I$$

Show that

$$E_n \cdots E_2 E_1 P = (A^{-1})^T$$

from which it follows that the same sequence of row operations that reduces C to I converts P to $(A^{-1})^T$.

11. (a) If A is the enciphering matrix of a Hill n-cipher, show that
$$A^{-1} = (C^{-1}P)^T \quad (\text{mod } 26)$$
where C and P are the matrices defined in Theorem 5.8.5.

(b) Instead of using Theorem 5.8.5 as in the text, find the deciphering matrix A^{-1} of Example 8 by using the result in part (a) and Equation (2) to compute C^{-1}. [*Note:* Although this method is practical for Hill 2-ciphers, Theorem 5.8.5 is more efficient for Hill n-ciphers with $n > 2$.]

Section 5.8   Technology Exercises

The following exercises are designed to be solved using a technology utility. Typically, this will be MATLAB, *Mathematica*, Maple, Derive, or Mathcad, but it may also be some other type of linear algebra software or a scientific calculator with some linear algebra capabilities. For each exercise you will need to read the relevant documentation for the particular utility you are using. The goal of these exercises is to provide you with a basic proficiency with your technology utility. Once you have mastered the techniques in these exercises, you will be able to use your technology utility to solve many of the problems in the regular exercise sets.

T1. Two integers that have no common factors (except 1) are said to be relatively prime. Given a positive integer n, let $S_n = \{a_1, a_2, a_3, \ldots, a_m\}$, where $a_1 < a_2 < a_3 < \cdots < a_m$, be the set of all positive integers less than n and relatively prime to n. For example, if $n = 9$, then

$$S_9 = \{a_1, a_2, a_3, \ldots, a_6\} = \{1, 2, 4, 5, 7, 8\}$$

(a) Construct a table consisting of n and S_n for $n = 2, 3, \ldots, 15$, and then compute

$$\sum_{k=1}^{m} a_k \quad \text{and} \quad \left(\sum_{k=1}^{m} a_k\right) \pmod{n}$$

in each case. Draw a conjecture for $n > 15$ and prove your conjecture to be true. [*Hint:* Use the fact that if a is relatively prime to n, then $n - a$ is also relatively prime to n.]

(b) Given a positive integer n and the set S_n, let P_n be the $m \times m$ matrix

$$P_n = \begin{bmatrix} a_1 & a_2 & a_3 & \cdots & a_{m-1} & a_m \\ a_2 & a_3 & a_4 & \cdots & a_m & a_1 \\ a_3 & a_4 & a_5 & \cdots & a_1 & a_2 \\ \vdots & \vdots & \vdots & \ddots & \vdots & \vdots \\ a_{m-1} & a_m & a_1 & \cdots & a_{m-3} & a_{m-2} \\ a_m & a_1 & a_2 & \cdots & a_{m-2} & a_{m-1} \end{bmatrix}$$

so that, for example,

$$P_9 = \begin{bmatrix} 1 & 2 & 4 & 5 & 7 & 8 \\ 2 & 4 & 5 & 7 & 8 & 1 \\ 4 & 5 & 7 & 8 & 1 & 2 \\ 5 & 7 & 8 & 1 & 2 & 4 \\ 7 & 8 & 1 & 2 & 4 & 5 \\ 8 & 1 & 2 & 4 & 5 & 7 \end{bmatrix}$$

Use a computer to compute $\det(P_n)$ and $\det(P_n) \pmod{n}$ for $n = 2, 3, \ldots, 15$, and then use these results to construct a conjecture.

(c) Use the results of part (a) to prove your conjecture to be true. [*Hint:* Add the first $m - 1$ rows of P_n to its last row and then use Theorem 2.2.3.] What do these results imply about the inverse of $P_n \pmod{n}$?

T2. Given a positive integer n greater than 1, the number of positive integers less than n and relatively prime to n is called the **Euler phi function** of n and is denoted by $\varphi(n)$. For example, $\varphi(6) = 2$ since only two positive integers (1 and 5) are less than 6 and have no common factor with 6.

(a) Using a computer, for each value of $n = 2, 3, \ldots, 25$ compute and print out all positive integers that are less than n and relatively prime to n. Then use these integers to determine the values of $\varphi(n)$ for $n = 2, 3, \ldots, 25$. Can you discover a pattern in the results?

(b) It can be shown that if $\{p_1, p_2, p_3, \ldots, p_m\}$ are all the distinct prime factors of n, then

$$\varphi(n) = n \left(1 - \frac{1}{p_1}\right)\left(1 - \frac{1}{p_2}\right)\left(1 - \frac{1}{p_3}\right)\cdots\left(1 - \frac{1}{p_m}\right)$$

For example, since $\{2, 3\}$ are the distinct prime factors of 12, we have

$$\varphi(12) = 12 \left(1 - \frac{1}{2}\right)\left(1 - \frac{1}{3}\right) = 4$$

which agrees with the fact that $\{1, 5, 7, 11\}$ are the only positive integers less than 12 and relatively prime to 12. Using a computer, print out all the prime factors of n for $n = 2, 3, \ldots, 25$. Then compute $\varphi(n)$ using the formula above and compare it to your results in part (a).

Exercise Set 1.1 (page 9)

1. (a), (c), and (f) are linear equations; (b), (d) and (e) are not linear equations

3. (a) and (d) are linear systems; (b) and (c) are not linear systems   5. (a) and (d) are both consistent

7. (a), (d), and (e) are solutions; (b) and (c) are not solutions   9. (a) $x = \frac{5}{7}t + \frac{3}{7}$ (b) $x_1 = \frac{1}{4}r - \frac{5}{8}s + \frac{3}{4}t - \frac{1}{8}$
$y = t$ $x_2 = r$
$x_3 = s$
$x_4 = t$

11. (a) $2x_1 \qquad = 0$ (b) $3x_1 \qquad - \quad 2x_3 = 5$ (c) $7x_1 + 2x_2 + x_3 - 3x_4 = 5$
$3x_1 - 4x_2 \quad = 0$ $7x_1 + \quad x_2 + 4x_3 = -3$ $x_1 + 2x_2 + 4x_3 \qquad = 1$
$x_2 \quad = 1$ $-2x_2 + x_3 = 7$

(d) $x_1 \qquad = 7$
$x_2 \qquad = -2$
$x_3 \qquad = 3$
$x_4 = 4$

13. (a) $\begin{bmatrix} -2 & 6 \\ 3 & 8 \\ 9 & -3 \end{bmatrix}$ (b) $\begin{bmatrix} 6 & -1 & 3 & 4 \\ 0 & 5 & -1 & 1 \end{bmatrix}$ (c) $\begin{bmatrix} 0 & 2 & 0 & -3 & 1 & 0 \\ -3 & -1 & 1 & 0 & 0 & -1 \\ 6 & 2 & -1 & 2 & -3 & 6 \end{bmatrix}$

(d) $\begin{bmatrix} 1 & 0 & 0 & 0 & -1 & 7 \end{bmatrix}$

True/False 1.1

(a) True   (b) False   (c) True   (d) True   (e) False   (f) False   (g) True   (h) False

Exercise Set 1.2 (page 22)

1. (a) Both   (b) Both   (c) Both   (d) Both   (e) Both   (f) Both   (g) Row echelon

3. (a) $x_1 = -37, \ x_2 = -8, \ x_3 = 5$ (b) $x_1 = 13t - 10, \ x_2 = 13t - 5, \ x_3 = -t + 2, \ x_4 = t$
(c) $x_1 = -7s + 2t - 11, \ x_2 = s, \ x_3 = -3t - 4, \ x_4 = -3t + 9, \ x_5 = t$ (d) Inconsistent

5. $x_1 = 3, \ x_2 = 1, \ x_3 = 2$ 7. $x = t - 1, y = 2s, z = s, w = t$ 9. $x_1 = 3, \ x_2 = 1, \ x_3 = 2$

11. $x = t - 1, y = 2s, z = s, w = t$ 13. Has nontrivial solutions   15. Has nontrivial solutions   17. $x_1 = 0, \ x_2 = 0, \ x_3 = 0$

19. $x_1 = -s, \ x_2 = -t - s, \ x_3 = 4s, \ x_4 = t$ 21. $w = t, x = -t, y = t, z = 0$ 23. $I_1 = -1, \ I_2 = 0, \ I_3 = 1, \ I_4 = 2$

25. If $a = 4$, there are infinitely many solutions; if $a = -4$, there are no solutions; if $a \neq \pm 4$, there is exactly one solution.

27. If $a = 3$, there are infinitely many solutions; if $a = -3$, there are no solutions; if $a \neq \pm 3$, there is exactly one solution.

29. $x = \frac{2a}{3} - \frac{b}{9}, \ y = -\frac{a}{3} + \frac{2b}{9}$ 31. $\begin{bmatrix} 1 & 3 \\ 0 & 1 \end{bmatrix}$ and $\begin{bmatrix} 1 & 0 \\ 0 & 1 \end{bmatrix}$ are possible answers.   35. $x = \pm 1, \ y = \pm\sqrt{3}, \ z = \pm\sqrt{2}$

37. $a = 1, b = -6, c = 2, d = 10$ 39. The nonhomogeneous system will have exactly one solution.

True/False 1.2

(a) True   (b) False   (c) False   (d) True   (e) True   (f) False   (g) True   (h) False   (i) False

Exercise Set 1.3 (page 35)

1. (a) Undefined   (b) 4×2 (c) Undefined   (d) Undefined   (e) 5×5 (f) 5×2 (g) Undefined   (h) 5×2

3. (a) $\begin{bmatrix} 7 & 6 & 5 \\ -2 & 1 & 3 \\ 7 & 3 & 7 \end{bmatrix}$ **(b)** $\begin{bmatrix} -5 & 4 & -1 \\ 0 & -1 & -1 \\ -1 & 1 & 1 \end{bmatrix}$ **(c)** $\begin{bmatrix} 15 & 0 \\ -5 & 10 \\ 5 & 5 \end{bmatrix}$ **(d)** $\begin{bmatrix} -7 & -28 & -14 \\ -21 & -7 & -35 \end{bmatrix}$ **(e)** Undefined

(f) $\begin{bmatrix} 22 & -6 & 8 \\ -2 & 4 & 6 \\ 10 & 0 & 4 \end{bmatrix}$ **(g)** $\begin{bmatrix} -39 & -21 & -24 \\ 9 & -6 & -15 \\ -33 & -12 & -30 \end{bmatrix}$ **(h)** $\begin{bmatrix} 0 & 0 \\ 0 & 0 \\ 0 & 0 \end{bmatrix}$ **(i)** 5   **(j)** -25 **(k)** 168   **(l)** Undefined

5. (a) $\begin{bmatrix} 12 & -3 \\ -4 & 5 \\ 4 & 1 \end{bmatrix}$ **(b)** Undefined **(c)** $\begin{bmatrix} 42 & 108 & 75 \\ 12 & -3 & 21 \\ 36 & 78 & 63 \end{bmatrix}$ **(d)** $\begin{bmatrix} 3 & 45 & 9 \\ 11 & -11 & 17 \\ 7 & 17 & 13 \end{bmatrix}$ **(e)** $\begin{bmatrix} 3 & 45 & 9 \\ 11 & -11 & 17 \\ 7 & 17 & 13 \end{bmatrix}$

(f) $\begin{bmatrix} 21 & 17 \\ 17 & 35 \end{bmatrix}$ **(g)** $\begin{bmatrix} 0 & -2 & 11 \\ 12 & 1 & 8 \end{bmatrix}$ **(h)** $\begin{bmatrix} 12 & 6 & 9 \\ 48 & -20 & 14 \\ 24 & 8 & 16 \end{bmatrix}$ **(i)** 61   **(j)** 35   **(k)** 28   **(l)** 99

7. (a) $[67 \; 41 \; 41]$ **(b)** $[63 \; 67 \; 57]$ **(c)** $\begin{bmatrix} 41 \\ 21 \\ 67 \end{bmatrix}$ **(d)** $\begin{bmatrix} 6 \\ 6 \\ 63 \end{bmatrix}$ **(e)** $[24 \; 56 \; 97]$ **(f)** $\begin{bmatrix} 76 \\ 98 \\ 97 \end{bmatrix}$

9. (a) $\begin{bmatrix} -3 \\ 48 \\ 24 \end{bmatrix} = 3\begin{bmatrix} 3 \\ 6 \\ 0 \end{bmatrix} + 6\begin{bmatrix} -2 \\ 5 \\ 4 \end{bmatrix}$; $\begin{bmatrix} 12 \\ 29 \\ 56 \end{bmatrix} = -2\begin{bmatrix} 3 \\ 6 \\ 0 \end{bmatrix} + 5\begin{bmatrix} -2 \\ 5 \\ 4 \end{bmatrix} + 4\begin{bmatrix} 7 \\ 4 \\ 9 \end{bmatrix}$;

$\begin{bmatrix} 76 \\ 98 \\ 97 \end{bmatrix} = 7\begin{bmatrix} 3 \\ 6 \\ 0 \end{bmatrix} + 4\begin{bmatrix} -2 \\ 5 \\ 4 \end{bmatrix} + 9\begin{bmatrix} 7 \\ 4 \\ 9 \end{bmatrix}$

(b) $\begin{bmatrix} 64 \\ 21 \\ 77 \end{bmatrix} = 6\begin{bmatrix} 6 \\ 0 \\ 7 \end{bmatrix} + 7\begin{bmatrix} 4 \\ 3 \\ 5 \end{bmatrix}$; $\begin{bmatrix} 14 \\ 22 \\ 28 \end{bmatrix} = -2\begin{bmatrix} 6 \\ 0 \\ 7 \end{bmatrix} + \begin{bmatrix} -2 \\ 1 \\ 7 \end{bmatrix} + 7\begin{bmatrix} 4 \\ 3 \\ 5 \end{bmatrix}$;

$\begin{bmatrix} 38 \\ 18 \\ 74 \end{bmatrix} = 4\begin{bmatrix} 6 \\ 0 \\ 7 \end{bmatrix} + 3\begin{bmatrix} -2 \\ 1 \\ 7 \end{bmatrix} + 5\begin{bmatrix} 4 \\ 3 \\ 5 \end{bmatrix}$

11. (a) $\begin{bmatrix} 2 & -3 & 5 \\ 9 & -1 & 1 \\ 1 & 5 & 4 \end{bmatrix}\begin{bmatrix} x_1 \\ x_2 \\ x_3 \end{bmatrix} = \begin{bmatrix} 7 \\ -1 \\ 0 \end{bmatrix}$ **(b)** $\begin{bmatrix} 4 & 0 & -3 & 1 \\ 5 & 1 & 0 & -8 \\ 2 & -5 & 9 & -1 \\ 0 & 3 & -1 & 7 \end{bmatrix}\begin{bmatrix} x_1 \\ x_2 \\ x_3 \\ x_4 \end{bmatrix} = \begin{bmatrix} 1 \\ 3 \\ 0 \\ 2 \end{bmatrix}$

13. (a) $\begin{aligned} 5x_1 &+ 6x_2 &- 7x_3 &= 2 \\ -x_1 &- 2x_2 &+ 3x_3 &= 0 \\ &\quad 4x_2 &- x_3 &= 3 \end{aligned}$ **(b)** $\begin{aligned} x_1 &+ x_2 &+ x_3 &= 2 \\ 2x_1 &+ 3x_2 & &= 2 \\ 5x_1 &- 3x_2 &- 6x_3 &= -9 \end{aligned}$

15. -1 **17.** $a = 4, b = -6, c = -1, d = 1$

23. (a) $\begin{bmatrix} a_{11} & 0 & 0 & 0 & 0 & 0 \\ 0 & a_{22} & 0 & 0 & 0 & 0 \\ 0 & 0 & a_{33} & 0 & 0 & 0 \\ 0 & 0 & 0 & a_{44} & 0 & 0 \\ 0 & 0 & 0 & 0 & a_{55} & 0 \\ 0 & 0 & 0 & 0 & 0 & a_{66} \end{bmatrix}$ **(b)** $\begin{bmatrix} a_{11} & a_{12} & a_{13} & a_{14} & a_{15} & a_{16} \\ 0 & a_{22} & a_{23} & a_{24} & a_{25} & a_{26} \\ 0 & 0 & a_{33} & a_{34} & a_{35} & a_{36} \\ 0 & 0 & 0 & a_{44} & a_{45} & a_{46} \\ 0 & 0 & 0 & 0 & a_{55} & a_{56} \\ 0 & 0 & 0 & 0 & 0 & a_{66} \end{bmatrix}$

(c) $\begin{bmatrix} a_{11} & 0 & 0 & 0 & 0 & 0 \\ a_{21} & a_{22} & 0 & 0 & 0 & 0 \\ a_{31} & a_{32} & a_{33} & 0 & 0 & 0 \\ a_{41} & a_{42} & a_{43} & a_{44} & 0 & 0 \\ a_{51} & a_{52} & a_{53} & a_{54} & a_{55} & 0 \\ a_{61} & a_{62} & a_{63} & a_{64} & a_{65} & a_{66} \end{bmatrix}$ **(d)** $\begin{bmatrix} a_{11} & a_{12} & 0 & 0 & 0 & 0 \\ a_{21} & a_{22} & a_{23} & 0 & 0 & 0 \\ 0 & a_{32} & a_{33} & a_{34} & 0 & 0 \\ 0 & 0 & a_{43} & a_{44} & a_{45} & 0 \\ 0 & 0 & 0 & a_{54} & a_{55} & a_{56} \\ 0 & 0 & 0 & 0 & a_{65} & a_{66} \end{bmatrix}$

25. $f\begin{pmatrix} x_1 \\ x_2 \end{pmatrix} = \begin{pmatrix} x_1 + x_2 \\ x_2 \end{pmatrix}$

  (a) $f\begin{pmatrix} 1 \\ 1 \end{pmatrix} = \begin{pmatrix} 2 \\ 1 \end{pmatrix}$ **(b)** $f\begin{pmatrix} 2 \\ 0 \end{pmatrix} = \begin{pmatrix} 2 \\ 0 \end{pmatrix}$ **(c)** $f\begin{pmatrix} 4 \\ 3 \end{pmatrix} = \begin{pmatrix} 7 \\ 3 \end{pmatrix}$ **(d)** $f\begin{pmatrix} 2 \\ -2 \end{pmatrix} = \begin{pmatrix} 0 \\ -2 \end{pmatrix}$

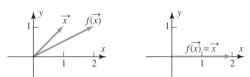 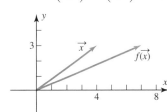 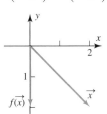

27. One; namely, $A = \begin{bmatrix} 1 & 1 & 0 \\ 1 & -1 & 0 \\ 0 & 0 & 0 \end{bmatrix}$

29. (a) $\begin{bmatrix} 1 & 1 \\ 1 & 1 \end{bmatrix}$ and $\begin{bmatrix} -1 & -1 \\ -1 & -1 \end{bmatrix}$ **(b)** Four; $\begin{bmatrix} \sqrt{5} & 0 \\ 0 & 3 \end{bmatrix}, \begin{bmatrix} -\sqrt{5} & 0 \\ 0 & 3 \end{bmatrix}, \begin{bmatrix} \sqrt{5} & 0 \\ 0 & -3 \end{bmatrix}, \begin{bmatrix} -\sqrt{5} & 0 \\ 0 & -3 \end{bmatrix}$

True/False 1.3

(a) True   **(b)** False   **(c)** False   **(d)** False   **(e)** True   **(f)** False   **(g)** False   **(h)** True   **(i)** True   **(j)** True

(k) True   **(l)** False   **(m)** True   **(n)** True   **(o)** False

Exercise Set 1.4   (page 49)

5. $B^{-1} = \begin{bmatrix} \frac{1}{5} & \frac{3}{20} \\ -\frac{1}{5} & \frac{1}{10} \end{bmatrix}$ **7.** $D^{-1} = \begin{bmatrix} \frac{1}{2} & 0 \\ 0 & \frac{1}{3} \end{bmatrix}$ **9.** $\begin{bmatrix} \frac{1}{2}(e^x + e^{-x}) & -\frac{1}{2}(e^x - e^{-x}) \\ -\frac{1}{2}(e^x - e^{-x}) & \frac{1}{2}(e^x + e^{-x}) \end{bmatrix}$

15. $A = \begin{bmatrix} \frac{2}{7} & 1 \\ \frac{1}{7} & \frac{3}{7} \end{bmatrix}$ **17.** $\begin{bmatrix} -\frac{9}{13} & \frac{1}{13} \\ \frac{2}{13} & -\frac{6}{13} \end{bmatrix}$

19. (a) $\begin{bmatrix} 41 & 15 \\ 30 & 11 \end{bmatrix}$ **(b)** $\begin{bmatrix} 11 & -15 \\ -30 & 41 \end{bmatrix}$ **(c)** $\begin{bmatrix} 6 & 2 \\ 4 & 2 \end{bmatrix}$ **(d)** $\begin{bmatrix} 1 & 1 \\ 2 & -1 \end{bmatrix}$ **(e)** $\begin{bmatrix} 20 & 7 \\ 14 & 6 \end{bmatrix}$ **(f)** $\begin{bmatrix} 39 & 13 \\ 26 & 13 \end{bmatrix}$

21. (a) $\begin{bmatrix} 27 & 0 & 0 \\ 0 & 26 & -18 \\ 0 & 18 & 26 \end{bmatrix}$ **(b)** $\begin{bmatrix} \frac{1}{27} & 0 & 0 \\ 0 & 0.026 & 0.018 \\ 0 & -0.018 & 0.026 \end{bmatrix}$ **(c)** $\begin{bmatrix} 4 & 0 & 0 \\ 0 & -5 & -12 \\ 0 & 12 & -5 \end{bmatrix}$ **(d)** $\begin{bmatrix} 1 & 0 & 0 \\ 0 & -3 & 3 \\ 0 & -3 & -3 \end{bmatrix}$

   (e) $\begin{bmatrix} 16 & 0 & 0 \\ 0 & -14 & -15 \\ 0 & 15 & -14 \end{bmatrix}$ **(f)** $\begin{bmatrix} 25 & 0 & 0 \\ 0 & 32 & -24 \\ 0 & 24 & 32 \end{bmatrix}$

27. $\begin{bmatrix} \frac{1}{a_{11}} & 0 & \cdots & 0 \\ 0 & \frac{1}{a_{22}} & \cdots & 0 \\ \vdots & \vdots & & \vdots \\ 0 & 0 & \cdots & \frac{1}{a_{nn}} \end{bmatrix}$ **31.** $D = CA^{-1}B^{-1}A^{-2}BC^2(B^T)^{-1}A^2$ **33.** B^{-1} **35.** $A^{-1} = \begin{bmatrix} \frac{1}{2} & \frac{1}{2} & -\frac{1}{2} \\ -\frac{1}{2} & \frac{1}{2} & \frac{1}{2} \\ \frac{1}{2} & -\frac{1}{2} & \frac{1}{2} \end{bmatrix}$

37. $A^{-1} = \begin{bmatrix} \frac{1}{2} & \frac{1}{2} & -\frac{1}{2} \\ -\frac{1}{2} & \frac{1}{2} & \frac{1}{2} \\ 1 & 0 & 0 \end{bmatrix}$ **39.** $x_1 = \frac{1}{23}, x_2 = \frac{13}{23}$ **41.** $x_1 = -\frac{1}{11}, x_2 = \frac{6}{11}$

True/False 1.4

(a) False   **(b)** False   **(c)** False   **(d)** False   **(e)** False   **(f)** True   **(g)** True   **(h)** True   **(i)** False   **(j)** True

(k) False

Exercise Set 1.5 (page 58)

1. **(a)** Elementary **(b)** Not elementary **(c)** Not elementary **(d)** Not elementary

3. **(a)** Add 3 times row 2 to row 1: $\begin{bmatrix} 1 & 3 \\ 0 & 1 \end{bmatrix}$ **(b)** Multiply row 1 by $-\frac{1}{7}$: $\begin{bmatrix} -\frac{1}{7} & 0 & 0 \\ 0 & 1 & 0 \\ 0 & 0 & 1 \end{bmatrix}$

(c) Add 5 times row 1 to row 3: $\begin{bmatrix} 1 & 0 & 0 \\ 0 & 1 & 0 \\ 5 & 0 & 1 \end{bmatrix}$ **(d)** Swap rows 1 and 3: $\begin{bmatrix} 0 & 0 & 1 & 0 \\ 0 & 1 & 0 & 0 \\ 1 & 0 & 0 & 0 \\ 0 & 0 & 0 & 1 \end{bmatrix}$

5. **(a)** Swap rows 1 and 2: $EA = \begin{bmatrix} 3 & -6 & -6 & -6 \\ -1 & -2 & 5 & -1 \end{bmatrix}$

(b) Add -3 times row 2 to row 3: $EA = \begin{bmatrix} 2 & -1 & 0 & -4 & -4 \\ 1 & -3 & -1 & 5 & 3 \\ -1 & 9 & 4 & -12 & -10 \end{bmatrix}$

(c) Add 4 times row 3 to row 1: $EA = \begin{bmatrix} 13 & 28 \\ 2 & 5 \\ 3 & 6 \end{bmatrix}$

7. **(a)** $\begin{bmatrix} 0 & 0 & 1 \\ 0 & 1 & 0 \\ 1 & 0 & 0 \end{bmatrix}$ **(b)** $\begin{bmatrix} 0 & 0 & 1 \\ 0 & 1 & 0 \\ 1 & 0 & 0 \end{bmatrix}$ **(c)** $\begin{bmatrix} 1 & 0 & 0 \\ 0 & 1 & 0 \\ -2 & 0 & 1 \end{bmatrix}$ **(d)** $\begin{bmatrix} 1 & 0 & 0 \\ 0 & 1 & 0 \\ 2 & 0 & 1 \end{bmatrix}$

9. $\begin{bmatrix} -7 & 4 \\ 2 & -1 \end{bmatrix}$ 11. $\begin{bmatrix} \frac{2}{7} & \frac{3}{7} \\ \frac{3}{7} & \frac{1}{7} \end{bmatrix}$ 13. $\begin{bmatrix} \frac{3}{2} & -\frac{11}{10} & -\frac{6}{5} \\ -1 & 1 & 1 \\ -\frac{1}{2} & \frac{7}{10} & \frac{2}{5} \end{bmatrix}$ 15. No inverse 17. $\begin{bmatrix} \frac{1}{2} & -\frac{1}{2} & \frac{1}{2} \\ -\frac{1}{2} & \frac{1}{2} & \frac{1}{2} \\ \frac{1}{2} & \frac{1}{2} & -\frac{1}{2} \end{bmatrix}$

19. $\begin{bmatrix} \frac{7}{2} & 0 & -3 \\ -1 & 1 & 0 \\ 0 & -1 & 1 \end{bmatrix}$ 21. $\begin{bmatrix} \frac{1}{4} & \frac{1}{2} & -3 & 0 \\ -\frac{1}{8} & \frac{1}{4} & -\frac{3}{2} & 0 \\ 0 & 0 & \frac{1}{2} & 0 \\ \frac{1}{40} & -\frac{1}{20} & -\frac{1}{10} & -\frac{1}{5} \end{bmatrix}$ 23. $\begin{bmatrix} -\frac{7}{12} & \frac{5}{24} & \frac{5}{8} & -\frac{1}{4} \\ \frac{5}{6} & \frac{5}{12} & \frac{1}{4} & -\frac{1}{2} \\ \frac{5}{12} & \frac{5}{24} & \frac{5}{8} & -\frac{1}{4} \\ -\frac{1}{12} & -\frac{1}{24} & -\frac{1}{8} & \frac{1}{4} \end{bmatrix}$

25. **(a)** $\begin{bmatrix} \frac{1}{k_1} & 0 & 0 & 0 \\ 0 & \frac{1}{k_2} & 0 & 0 \\ 0 & 0 & \frac{1}{k_3} & 0 \\ 0 & 0 & 0 & \frac{1}{k_4} \end{bmatrix}$ **(b)** $\begin{bmatrix} \frac{1}{k} & -\frac{1}{k} & 0 & 0 \\ 0 & 1 & 0 & 0 \\ 0 & 0 & \frac{1}{k} & -\frac{1}{k} \\ 0 & 0 & 0 & 1 \end{bmatrix}$ 27. $c \neq 0, 1$

29. $\begin{bmatrix} -3 & 1 \\ 2 & 2 \end{bmatrix} = \begin{bmatrix} 1 & 0 \\ 0 & 2 \end{bmatrix}\begin{bmatrix} 1 & 1 \\ 0 & 1 \end{bmatrix}\begin{bmatrix} -4 & 0 \\ 0 & 1 \end{bmatrix}\begin{bmatrix} 1 & 0 \\ 1 & 1 \end{bmatrix}$

31. $\begin{bmatrix} 1 & 0 & -2 \\ 0 & 4 & 3 \\ 0 & 0 & 1 \end{bmatrix} = \begin{bmatrix} 1 & 0 & -2 \\ 0 & 1 & 0 \\ 0 & 0 & 1 \end{bmatrix}\begin{bmatrix} 1 & 0 & 0 \\ 0 & 1 & 3 \\ 0 & 0 & 1 \end{bmatrix}\begin{bmatrix} 1 & 0 & 0 \\ 0 & 4 & 0 \\ 0 & 0 & 1 \end{bmatrix}$

33. $\begin{bmatrix} -\frac{1}{4} & \frac{1}{8} \\ \frac{1}{4} & \frac{3}{8} \end{bmatrix} = \begin{bmatrix} 1 & 0 \\ -1 & 1 \end{bmatrix}\begin{bmatrix} -\frac{1}{4} & 0 \\ 0 & 1 \end{bmatrix}\begin{bmatrix} 1 & -1 \\ 0 & 1 \end{bmatrix}\begin{bmatrix} 1 & 0 \\ 0 & \frac{1}{2} \end{bmatrix}$

35. $\begin{bmatrix} 1 & 0 & 2 \\ 0 & \frac{1}{4} & -\frac{3}{4} \\ 0 & 0 & 1 \end{bmatrix} = \begin{bmatrix} 1 & 0 & 0 \\ 0 & \frac{1}{4} & 0 \\ 0 & 0 & 1 \end{bmatrix}\begin{bmatrix} 1 & 0 & 0 \\ 0 & 1 & -3 \\ 0 & 0 & 1 \end{bmatrix}\begin{bmatrix} 1 & 0 & 2 \\ 0 & 1 & 0 \\ 0 & 0 & 1 \end{bmatrix}$

37. Add -1 times the first row to the second row. Add -1 times the first row to the third row. Add -1 times the second row to the first row. Add the second row to the third row.

True/False 1.5

(a) False   (b) True   (c) True   (d) True   (e) True   (f) True   (g) False

Exercise Set 1.6   (page 65)

1. $x_1 = 3$, $x_2 = -1$ 3. $x_1 = -1$, $x_2 = 4$, $x_3 = -7$ 5. $x = 1$, $y = 5$, $z = -1$ 7. $x_1 = 2b_1 - 5b_2$, $x_2 = -b_1 + 3b_2$

9. **(i)** $x_1 = \frac{22}{17}$, $x_2 = \frac{1}{17}$ **(ii)** $x_1 = \frac{21}{17}$, $x_2 = \frac{11}{17}$

11. **(i)** $x_1 = \frac{7}{15}$, $x_2 = \frac{4}{15}$ **(ii)** $x_1 = \frac{34}{15}$, $x_2 = \frac{28}{15}$ **(iii)** $x_1 = \frac{19}{15}$, $x_2 = \frac{13}{15}$ **(iv)** $x_1 = -\frac{1}{5}$, $x_2 = \frac{3}{5}$

13. No conditions on b_1 and b_2 15. $b_3 = b_1 - b_2$ 17. $b_1 = b_3 + b_4$, $b_2 = 2b_3 + b_4$

19. $X = \begin{bmatrix} 11 & 12 & -3 & 27 & 26 \\ -6 & -8 & 1 & -18 & -17 \\ -15 & -21 & 9 & -38 & -35 \end{bmatrix}$

True/False 1.6

(a) True   (b) True   (c) True   (d) True   (e) True   (f) True   (g) True

Exercise Set 1.7   (page 71)

1. $\begin{bmatrix} \frac{1}{2} & 0 \\ 0 & -\frac{1}{5} \end{bmatrix}$ 3. $\begin{bmatrix} -1 & 0 & 0 \\ 0 & \frac{1}{2} & 0 \\ 0 & 0 & 3 \end{bmatrix}$ 5. $\begin{bmatrix} 6 & 3 \\ 4 & -1 \\ 4 & 10 \end{bmatrix}$ 7. $\begin{bmatrix} -15 & 10 & 0 & 20 & -20 \\ 2 & -10 & 6 & 0 & 6 \\ 18 & -6 & -6 & -6 & -6 \end{bmatrix}$

9. $A^2 = \begin{bmatrix} 1 & 0 \\ 0 & 4 \end{bmatrix}$, $A^{-2} = \begin{bmatrix} 1 & 0 \\ 0 & \frac{1}{4} \end{bmatrix}$, $A^{-k} = \begin{bmatrix} 1 & 0 \\ 0 & 1/(-2)^k \end{bmatrix}$

11. $A^2 = \begin{bmatrix} \frac{1}{4} & 0 & 0 \\ 0 & \frac{1}{9} & 0 \\ 0 & 0 & \frac{1}{16} \end{bmatrix}$, $A^{-2} = \begin{bmatrix} 4 & 0 & 0 \\ 0 & 9 & 0 \\ 0 & 0 & 16 \end{bmatrix}$, $A^{-k} = \begin{bmatrix} 2^k & 0 & 0 \\ 0 & 3^k & 0 \\ 0 & 0 & 4^k \end{bmatrix}$

13. Not symmetric   15. Symmetric   17. Not symmetric   19. Not symmetric   21. Not invertible   23. $a = -8$

25. $x \neq 1$, -2, 4 27. $\begin{bmatrix} 1 & 0 & 0 \\ 0 & -1 & 0 \\ 0 & 0 & -1 \end{bmatrix}$ 35. **(a)** Yes   **(b)** No (unless $n = 1$)   **(c)** Yes   **(d)** No (unless $n = 1$)

39. $\begin{bmatrix} 0 & 0 & -8 \\ 0 & 0 & -4 \\ 8 & 4 & 0 \end{bmatrix}$ 43. $A = \begin{bmatrix} 1 & 10 \\ 0 & -2 \end{bmatrix}$

True/False 1.7

(a) True   (b) False   (c) False   (d) True   (e) True   (f) False   (g) False   (h) True   (i) True   (j) False
(k) False   (l) False   (m) True

Exercise Set 1.8 (page 84)

1.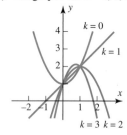

3. **(a)** $x_3 - x_4 = -500$, $-x_1 + x_4 = 100$, $x_1 - x_2 = 300$, $x_2 - x_3 = 100$
   (b) $x_1 = -100 + t$, $x_2 = -400 + t$, $x_3 = -500 + t$, $x_4 = t$
   (c) For all rates to be nonnegative, we need $t = 500$ cars per hour, so $x_1 = 400$, $x_2 = 100$, $x_3 = 0$, $x_4 = 500$

5. $I_1 = \frac{13}{5}$ A, $I_2 = -\frac{2}{5}$ A, $I_3 = \frac{11}{5}$ A   7. $I_1 = I_4 = I_5 = I_6 = \frac{1}{2}$ A, $I_2 = I_3 = 0$ A

9. $x_1 = 1$, $x_2 = 5$, $x_3 = 3$, and $x_4 = 4$; the balanced equation is $C_3H_8 + 5O_2 \rightarrow 3CO_2 + 4H_2O$

11. $x_1 = x_2 = x_3 = x_4 = t$; the balanced equation is $CH_3COF + H_2O \rightarrow CH_3COOH + HF$

13. $p(x) = x^2 - 2x + 2$ 15. $p(x) = 1 + \frac{13}{6}x - \frac{1}{6}x^3$

17. **(a)** Using $a_1 = k$ as a parameter, $p(x) = 1 + kx + (1 - k)x^2$ where $-\infty < k < \infty$.
    (b) The graphs for k = 0, 1, 2, and 3 are shown.

True/False 1.8

(a) True   **(b)** False   **(c)** True   **(d)** False   **(e)** False

Exercise Set 1.9 (page 90)

1. **(a)** $\begin{bmatrix} 0.50 & 0.25 \\ 0.25 & 0.10 \end{bmatrix}$ **(b)** $\begin{bmatrix} \$25,290 \\ \$22,581 \end{bmatrix}$ 3. **(a)** $\begin{bmatrix} 0.1 & 0.6 & 0.4 \\ 0.3 & 0.2 & 0.3 \\ 0.4 & 0.1 & 0.2 \end{bmatrix}$ **(b)** $\begin{bmatrix} \$31,500 \\ \$26,500 \\ \$26,300 \end{bmatrix}$ 5. $\begin{bmatrix} 123.08 \\ 202.56 \end{bmatrix}$

True/False 1.9

(a) False   **(b)** True   **(c)** False   **(d)** True   **(e)** True

Chapter 1 Supplementary Exercises (page 91)

1. $\begin{aligned} 3x_1 &- x_2 &&+ x_4 &= 1 \\ 2x_1 && + 3x_3 &+ 3x_4 &= -1 \end{aligned}$
   $x_1 = -\frac{3}{2}s - \frac{3}{2}t - \frac{1}{2}$, $x_2 = -\frac{9}{2}s - \frac{1}{2}t - \frac{5}{2}$, $x_3 = s$, $x_4 = t$

3. $\begin{aligned} 2x_1 &- 4x_2 &+ x_3 &= 6 \\ -4x_1 && + 3x_3 &= -1 \\ x_2 &- x_3 &&= 3 \end{aligned}$
   $x_1 = -\frac{17}{2}$, $x_2 = -\frac{26}{3}$, $x_3 = -\frac{35}{3}$

5. $x' = \frac{3}{5}x + \frac{4}{5}y$, $y' = -\frac{4}{5}x + \frac{3}{5}y$ 7. $x = 4$, $y = 2$, $z = 3$

9. **(a)** $a \neq 0$, $b \neq 2$ **(b)** $a \neq 0$, $b = 2$ **(c)** $a = 0$, $b = 2$ **(d)** $a = 0$, $b \neq 2$

11. $K = \begin{bmatrix} 0 & 2 \\ 1 & 1 \end{bmatrix}$ 13. **(a)** $X = \begin{bmatrix} -1 & 3 & -1 \\ 6 & 0 & 1 \end{bmatrix}$ **(b)** $X = \begin{bmatrix} 1 & -2 \\ 3 & 1 \end{bmatrix}$ **(c)** $X = \begin{bmatrix} -\frac{113}{37} & -\frac{160}{37} \\ -\frac{20}{37} & -\frac{46}{37} \end{bmatrix}$

15. $a = 1$, $b = -2$, $c = 3$

Exercise Set 2.1 (page 98)

1. $M_{11} = 29,\ C_{11} = 29$
  $M_{12} = 21,\ C_{12} = -21$
  $M_{13} = 27,\ C_{13} = 27$
  $M_{21} = -11,\ C_{21} = 11$
  $M_{22} = 13,\ C_{22} = 13$
  $M_{23} = -5,\ C_{23} = 5$
  $M_{31} = -19,\ C_{31} = -19$
  $M_{32} = -19,\ C_{32} = 19$
  $M_{33} = 19,\ C_{33} = 19$

3. **(a)** $M_{13} = 0,\ C_{13} = 0$
  (b) $M_{23} = -96,\ C_{23} = 96$
  (c) $M_{22} = -48,\ C_{22} = -48$
  (d) $M_{21} = 72,\ C_{21} = -72$

5. $22;\ \begin{bmatrix} \frac{2}{11} & -\frac{5}{22} \\ \frac{1}{11} & \frac{3}{22} \end{bmatrix}$ **7.** $59;\ \begin{bmatrix} -\frac{2}{59} & -\frac{7}{59} \\ \frac{7}{59} & -\frac{5}{59} \end{bmatrix}$

9. $a^2 - 5a + 21$ **11.** -65 **13.** -123 **15.** $\lambda = 1$ or -3 **17.** $\lambda = 1$ or -1 **19.** (all parts) -123 **21.** -40

23. 0 **25.** -240 **27.** -1 **29.** 0 **31.** 6 **33.** The determinant is $\sin^2\theta + \cos^2\theta = 1$.   **35.** $d_2 = d_1 + \lambda$

True/False 2.1

(a) False   **(b)** False   **(c)** True   **(d)** True   **(e)** True   **(f)** True   **(g)** False   **(h)** False   **(i)** False   **(j)** True

Exercise Set 2.2 (page 105)

5. -5 **7.** -1 **9.** 1 **11.** 5 **13.** 33 **15.** 6 **17.** -2

19. Exercise 14: 39; Exercise 15: 6; Exercise 16: $-\frac{1}{6}$; Exercise 17: -2 **21.** -6 **23.** 72 **25.** -6 **27.** 18

True/False 2.2

(a) True   **(b)** True   **(c)** False   **(d)** False   **(e)** True   **(f)** True

Exercise Set 2.3 (page 115)

7. Invertible   **9.** Invertible   **11.** Not invertible   **13.** Invertible   **15.** $k \neq \frac{5 \pm \sqrt{17}}{2}$ **17.** $k \neq -1$

19. $A^{-1} = \begin{bmatrix} 3 & -5 & -5 \\ -3 & 4 & 5 \\ 2 & -2 & -3 \end{bmatrix}$ **21.** $A^{-1} = \begin{bmatrix} \frac{1}{2} & \frac{3}{2} & 1 \\ 0 & 1 & \frac{3}{2} \\ 0 & 0 & \frac{1}{2} \end{bmatrix}$ **23.** $A^{-1} = \begin{bmatrix} -4 & 3 & 0 & -1 \\ 2 & -1 & 0 & 0 \\ -7 & 0 & -1 & 8 \\ 6 & 0 & 1 & -7 \end{bmatrix}$

25. $x = \frac{3}{11},\ y = \frac{2}{11},\ z = -\frac{1}{11}$ **27.** $x_1 = -\frac{30}{11},\ x_2 = -\frac{38}{11},\ x_3 = -\frac{40}{11}$ **29.** Cramer's rule does not apply.   **31.** $y = 0$

35. **(a)** -189 **(b)** $-\frac{1}{7}$ **(c)** $-\frac{8}{7}$ **(d)** $-\frac{1}{56}$ **(e)** 7 **37.** **(a)** 189 **(b)** $\frac{1}{7}$ **(c)** $\frac{8}{7}$ **(d)** $\frac{1}{56}$

True/False 2.3

(a) False   **(b)** False   **(c)** True   **(d)** False   **(e)** True   **(f)** True   **(g)** True   **(h)** True   **(i)** True   **(j)** True

(k) True   **(l)** False

Chapter 2 Supplementary Exercises (page 117)

1. -18 **3.** 24 **5.** -10 **7.** 329 **9.** Exercise 3: 24; Exercise 4: 0; Exercise 5: -10; Exercise 6: -48

11. The matrices in Exercises 1–3 are invertible, the matrix in Exercise 4 is not.

13. $-b^2 + 5b - 21$ **15.** -120 **17.** $\begin{bmatrix} -\frac{1}{6} & \frac{1}{9} \\ \frac{1}{6} & \frac{2}{9} \end{bmatrix}$ **19.** $\begin{bmatrix} \frac{1}{8} & -\frac{1}{8} & -\frac{3}{8} \\ \frac{1}{8} & \frac{5}{24} & -\frac{1}{24} \\ \frac{1}{4} & -\frac{7}{12} & -\frac{1}{12} \end{bmatrix}$ **21.** $\begin{bmatrix} \frac{1}{5} & \frac{2}{5} & -\frac{1}{10} \\ \frac{1}{5} & -\frac{3}{5} & \frac{2}{5} \\ -\frac{2}{5} & \frac{6}{5} & -\frac{3}{10} \end{bmatrix}$

23.
$$\begin{bmatrix} \frac{10}{329} & -\frac{2}{329} & \frac{52}{329} & -\frac{27}{329} \\ \frac{55}{329} & -\frac{11}{329} & -\frac{43}{329} & \frac{16}{329} \\ -\frac{3}{47} & \frac{10}{47} & -\frac{25}{47} & -\frac{6}{47} \\ -\frac{31}{329} & \frac{72}{329} & \frac{102}{329} & -\frac{15}{329} \end{bmatrix}$$

25. $x' = \frac{3}{5}x + \frac{4}{5}y, \; y' = -\frac{4}{5}x + \frac{3}{5}y$

29. (b) $\cos\beta = \frac{c^2+a^2-b^2}{2ac}, \; \cos\gamma = \frac{a^2+b^2-c^2}{2ab}$

Exercise Set 3.1 (page 128)

1. (a) 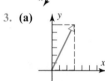 **(b)** $z\,(-3,4,5)$ **(c)** $(3,-4,5)$ **(d)** $(3,4,-5)$ **(e)** $(-3,-4,5)$ **(f)** $(-3,4,-5)$

3. (a) 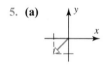 **(b)** 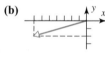 **(c)**   **(d)** 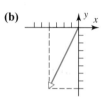 **(e)**   **(f)**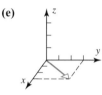

5. (a)   **(b)** 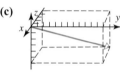 **(c)**

7. (a) $\overrightarrow{P_1P_2} = (-1, \; 3)$ **(b)** $\overrightarrow{P_1P_2} = (-3, \; 6, \; 1)$

9. (a) The terminal point is $B(2,3)$. **(b)** The initial point is $A(-2,-2,-1)$.

11. (a) $\mathbf{u} = (-1, 2, -4)$ is one possible answer. **(b)** $\mathbf{u} = (7, -2, -6)$ is one possible answer.

13. (a) $\mathbf{u}+\mathbf{w} = (1,-4)$ **(b)** $\mathbf{v}-3\mathbf{u} = (-12,8)$ **(c)** $2(\mathbf{u}-5\mathbf{w}) = (38,28)$ **(d)** $3\mathbf{v}-2(\mathbf{u}+2\mathbf{w}) = (4,29)$
(e) $-3(\mathbf{w}-2\mathbf{u}+\mathbf{v}) = (33,-12)$ **(f)** $(-2\mathbf{u}-\mathbf{v})-5(\mathbf{v}+3\mathbf{w}) = (37,17)$

15. (a) $(-1,9,-11,1)$ **(b)** $(22,53,-19,14)$ **(c)** $(-13,13,-36,-2)$ **(d)** $(-90,-114,60,-36)$ **(e)** $(-9,-5,-5,-3)$
(f) $(27,29,-27,9)$

17. (a) $\mathbf{w}-\mathbf{u} = (-9,3,-3,-8,5)$ **(b)** $2\mathbf{v}+3\mathbf{u} = (13,-5,14,13,-9)$ **(c)** $-\mathbf{w}+3(\mathbf{v}-\mathbf{u}) = (-14,-2,24,2,7)$
(d) $5(-\mathbf{v}+4\mathbf{u}-\mathbf{w}) = (125,-25,-20,75,-70)$ **(e)** $-2(3\mathbf{w}+\mathbf{v})+(2\mathbf{u}+\mathbf{w}) = (32,-10,1,27,-16)$
(f) $\frac{1}{2}(\mathbf{w}-5\mathbf{v}+2\mathbf{u})+\mathbf{v} = (\frac{9}{2}, \frac{3}{2}, -12, -\frac{5}{2}, -2)$

19. (a) $\mathbf{v}-\mathbf{w} = (-2,1,-4,-2,7)$ **(b)** $6\mathbf{u}+2\mathbf{v} = (-10,6,-4,26,28)$ **(c)** $(2\mathbf{u}-7\mathbf{w})-(8\mathbf{v}+\mathbf{u}) = (-77,8,94,-25,23)$

21. $\mathbf{x} = \left(-\frac{8}{3}, \frac{1}{2}, \frac{8}{3}, \frac{2}{3}, \frac{11}{6}\right)$ **23. (a)** Not parallel **(b)** Parallel **(c)** Parallel **25.** $a = 3, b = -1$

27. $c_1 = 2, \; c_2 = -1, \; c_3 = 5$ **29.** $c_1 = 1, \; c_2 = 1, \; c_3 = -1, \; c_4 = 1$ **33. (a)** $\left(\frac{9}{2}, -\frac{1}{2}, -\frac{1}{2}\right)$ **(b)** $\left(\frac{23}{4}, -\frac{9}{4}, \frac{1}{4}\right)$

True/False 3.1

(a) False **(b)** False **(c)** False **(d)** True **(e)** True **(f)** False **(g)** False **(h)** True **(i)** False
(j) True **(k)** False

Exercise Set 3.2 (page 141)

1. (a) $\|\mathbf{v}\| = 5, \; \frac{\mathbf{v}}{\|\mathbf{v}\|} = \left(\frac{4}{5}, -\frac{3}{5}\right), \; -\frac{\mathbf{v}}{\|\mathbf{v}\|} = \left(-\frac{4}{5}, \frac{3}{5}\right)$ **(b)** $\|\mathbf{v}\| = 2\sqrt{3}, \; \frac{\mathbf{v}}{\|\mathbf{v}\|} = \left(\frac{1}{\sqrt{3}}, \frac{1}{\sqrt{3}}, \frac{1}{\sqrt{3}}\right), \; -\frac{\mathbf{v}}{\|\mathbf{v}\|} = \left(-\frac{1}{\sqrt{3}}, -\frac{1}{\sqrt{3}}, -\frac{1}{\sqrt{3}}\right)$
(c) $\|\mathbf{v}\| = \sqrt{15}, \; \frac{\mathbf{v}}{\|\mathbf{v}\|} = \frac{1}{\sqrt{15}}(1, 0, 2, 1, 3), \; -\frac{\mathbf{v}}{\|\mathbf{v}\|} = -\frac{1}{\sqrt{15}}(1, 0, 2, 1, 3)$

3. (a) $\|\mathbf{u}+\mathbf{v}\| = \sqrt{83}$ **(b)** $\|\mathbf{u}\| + \|\mathbf{v}\| = \sqrt{17} + \sqrt{26}$ **(c)** $\|-2\mathbf{u}+2\mathbf{v}\| = 2\sqrt{3}$ **(d)** $\|3\mathbf{u}-5\mathbf{v}+\mathbf{w}\| = \sqrt{466}$

5. (a) $\|3\mathbf{u}-5\mathbf{v}+\mathbf{w}\| = \sqrt{2570}$ **(b)** $\|3\mathbf{u}\| - 5\|\mathbf{v}\| + \|\mathbf{w}\| = 3\sqrt{46} - 10\sqrt{21} + \sqrt{42}$ **(c)** $\|-\|\mathbf{u}\|\mathbf{v}\| = 2\sqrt{966}$

7. $k = \frac{5}{7}$, $k = -\frac{5}{7}$ 9. **(a)** $\mathbf{u} \cdot \mathbf{v} = -8$, $\mathbf{u} \cdot \mathbf{u} = 26$, $\mathbf{v} \cdot \mathbf{v} = 24$ **(b)** $\mathbf{u} \cdot \mathbf{v} = 0$, $\mathbf{u} \cdot \mathbf{u} = 54$, $\mathbf{v} \cdot \mathbf{v} = 21$

11. **(a)** $\|\mathbf{u} - \mathbf{v}\| = \sqrt{14}$ **(b)** $\|\mathbf{u} - \mathbf{v}\| = \sqrt{59}$ **(c)** $\|\mathbf{u} - \mathbf{v}\| = \sqrt{677}$

13. **(a)** $\cos\theta = \frac{15}{\sqrt{27}\sqrt{17}}$; θ is acute   **(b)** $\cos\theta = -\frac{4}{\sqrt{6}\sqrt{45}}$; θ is obtuse   **(c)** $\cos\theta = -\frac{136}{\sqrt{225}\sqrt{180}}$; θ is obtuse

15. $\mathbf{a} \cdot \mathbf{b} = 45\frac{\sqrt{3}}{2}$

17. **(a)** $\mathbf{u} \cdot (\mathbf{v} \cdot \mathbf{w})$ does not make sense because $\mathbf{v} \cdot \mathbf{w}$ is a scalar.   **(b)** $\mathbf{u} \cdot (\mathbf{v} + \mathbf{w})$ makes sense.
   (c) $\|\mathbf{u} \cdot \mathbf{v}\|$ does not make sense because the quantity inside the norm is a scalar.
   (d) $(\mathbf{u} \cdot \mathbf{v}) - \|\mathbf{u}\|$ makes sense since the terms are both scalars.

19. **(a)** $\left(-\frac{4}{5}, -\frac{3}{5}\right)$ **(b)** $\left(\frac{1}{5\sqrt{2}}, \frac{7}{5\sqrt{2}}\right)$ **(c)** $\left(-\frac{3}{4}, \frac{1}{2}, \frac{\sqrt{3}}{4}\right)$ **(d)** $\left(\frac{1}{\sqrt{55}}, \frac{2}{\sqrt{55}}, \frac{3}{\sqrt{55}}, \frac{4}{\sqrt{55}}, \frac{5}{\sqrt{55}}\right)$

23. **(a)** $\cos\theta = -\frac{11}{\sqrt{962}}$ **(b)** $\cos\theta = -\frac{3}{\sqrt{10}}$ **(c)** $\cos\theta = 0$ **(d)** $\cos\theta = 0$

25. **(a)** $|\mathbf{u} \cdot \mathbf{v}| = 10$, $\|\mathbf{u}\|\,\|\mathbf{v}\| = \sqrt{13}\sqrt{17} \approx 14.866$ **(b)** $|\mathbf{u} \cdot \mathbf{v}| = 7$, $\|\mathbf{u}\|\,\|\mathbf{v}\| = \sqrt{10}\sqrt{14} \approx 11.832$
   (c) $|\mathbf{u} \cdot \mathbf{v}| = 5$, $\|\mathbf{u}\|\,\|\mathbf{v}\| = (3)(2) = 6$

27. A sphere of radius 1 centered at $(x_0,\ y_0,\ z_0)$.

True/False 3.2

(a) True   **(b)** True   **(c)** False   **(d)** True   **(e)** True   **(f)** False   **(g)** False   **(h)** False   **(i)** True   **(j)** True

Exercise Set 3.3   (page 150)

1. **(a)** Orthogonal   **(b)** Not orthogonal   **(c)** Not orthogonal   **(d)** Not orthogonal

3. **(a)** Not an orthogonal set   **(b)** Orthogonal set   **(c)** Orthogonal set   **(d)** Not an orthogonal set

5. $\pm\left(\frac{1}{\sqrt{3}}, \frac{1}{\sqrt{3}}, -\frac{1}{\sqrt{3}}\right)$ 7. Yes   9. $-2(x+1) + (y-3) - (z+2) = 0$ 11. $2z = 0$ 13. Not parallel   15. Parallel

17. Not perpendicular   19. **(a)** $\frac{2}{5}$ **(b)** $\frac{18}{\sqrt{22}}$ 21. $(0,0)$, $(6,2)$ 23. $\left(-\frac{16}{13}, 0, -\frac{80}{13}\right)$, $\left(\frac{55}{13}, 1, -\frac{11}{13}\right)$

25. $\left(0, \frac{2}{5}, -\frac{1}{5}\right)$, $\left(1, \frac{3}{5}, \frac{6}{5}\right)$ 27. $\left(\frac{1}{5}, -\frac{1}{5}, \frac{1}{10}, -\frac{1}{10}\right)$, $\left(\frac{9}{5}, \frac{6}{5}, \frac{9}{10}, \frac{21}{10}\right)$ 29. 1   31. $\frac{1}{\sqrt{17}}$ 33. $\frac{5}{3}$ 35. $\frac{1}{\sqrt{29}}$

37. $\frac{11}{\sqrt{6}}$ 39. 0 (The planes coincide.)   41. **(b)** $\cos\beta = \frac{b}{\|\mathbf{v}\|}$, $\cos\gamma = \frac{c}{\|\mathbf{v}\|}$

True/False 3.3

(a) True   **(b)** True   **(c)** True   **(d)** True   **(e)** True   **(f)** False   **(g)** False

Exercise Set 3.4   (page 159)

1. Vector equation: $(x, y) = (-4, 1) + t(0, -8)$;
   parametric equations: $x = -4$, $y = 1 - 8t$

3. Vector equation: $(x, y, z) = t(-3, 0, 1)$;
   parametric equations: $x = -3t$, $y = 0$, $z = t$

5. Point: $(3, -6)$; parallel vector: $(-5, -1)$ 7. Point: $(4, 6)$; parallel vector: $(-6, -6)$

9. Vector equation: $(x, y, z) = (-3, 1, 0) + t_1(0, -3, 6) + t_2(-5, 1, 2)$;
   parametric equations: $x = -3 - 5t_2$, $y = 1 - 3t_1 + t_2$, $z = 6t_1 + 2t_2$

11. Vector equation: $(x, y, z) = (-1, 1, 4) + t_1(6, -1, 0) + t_2(-1, 3, 1)$;
   parametric equations: $x = -1 + 6t_1 - t_2$, $y = 1 - t_1 + 3t_2$, $z = 4 + t_2$

13. A possible answer is vector equation: $(x, y) = t(3, 2)$;
   parametric equations: $x = 3t$, $y = 2t$

15. A possible answer is vector equation: $(x, y, z) = t_1(0, 1, 0) + t_2(5, 0, 4)$;
parametric equations: $x = 5t_2$, $y = t_1$, $z = 4t_2$

17. $x_1 = -s - t$, $x_2 = s$, $x_3 = t$

19. $x_1 = \frac{3}{7}r - \frac{19}{4}s - \frac{8}{7}t$, $x_2 = -\frac{2}{7}r + \frac{1}{7}s + \frac{3}{7}t$, $x_3 = r$, $x_4 = s$, $x_5 = t$

21. (a) $(1, 0, 0) + s(-1, 1, 0) + t(-1, 0, 1)$ **(b)** a plane in R^3 passing through $P(1, 0, 0)$ and parallel to $(-1, 1, 0)$ and $(-1, 0, 1)$

23. (a)
$$\begin{array}{ccccccc} x & + & y & + & z & = & 0 \\ -2x & + & 3y & & & = & 0 \end{array}$$
(b) a line through the origin in R^3 **(c)** $x = -\frac{3}{5}t$, $y = -\frac{2}{5}t$, $z = t$

25. (a) $x_1 = -\frac{2}{3}s + \frac{1}{3}t$, $x_2 = s$, $x_3 = t$ **(c)** $x_1 = 1 - \frac{2}{3}s + \frac{1}{3}t$, $x_2 = s$, $x_3 = 1 + t$

27. $x_1 = \frac{1}{3} - \frac{4}{3}s - \frac{1}{3}t$, $x_2 = s$, $x_3 = t$, $x_4 = 1$; The general solution of the associated homogeneous system is $x_1 = -\frac{4}{3}s - \frac{1}{3}t$,
$x_2 = s$, $x_3 = t$, $x_4 = 0$. A particular solution of the given system is $x_1 = \frac{1}{3}$, $x_2 = 0$, $x_3 = 0$, $x_4 = 1$.

True/False 3.4

(a) True **(b)** False **(c)** True **(d)** True **(e)** False **(f)** True

Exercise Set 3.5 (page 168)

1. (a) $(32, -6, -4)$ **(b)** $(-14, -20, -82)$ **(c)** $(27, 40, -42)$ **3.** $(18, 36, -18)$ **5.** $(-3, 9, -3)$ **7.** $\sqrt{59}$ **9.** $\sqrt{101}$

11. 3 **13.** 7 **15.** $\frac{\sqrt{374}}{2}$ **17.** 16 **19.** The vectors do not lie in the same plane. **21.** -92 **23.** abc

25. (a) -3 **(b)** 3 **(c)** 3 **27. (a)** $\frac{\sqrt{26}}{2}$ **(b)** $\frac{\sqrt{26}}{3}$ **29.** $2(\mathbf{v} \times \mathbf{u})$ **37. (a)** $\frac{17}{6}$ **(b)** $\frac{1}{2}$

True/False 3.5

(a) True **(b)** True **(c)** False **(d)** True **(e)** False **(f)** False

Chapter 3 Supplementary Exercises (page 170)

1. (a) $3\mathbf{v} - 2\mathbf{u} = (13, -3, 10)$ **(b)** $\|\mathbf{u} + \mathbf{v} + \mathbf{w}\| = \sqrt{70}$ **(c)** $\sqrt{774}$ **(d)** $\text{proj}_{\mathbf{w}}\mathbf{u} = -\frac{12}{27}(2, -5, -5)$
(e) $\mathbf{u} \cdot (\mathbf{v} \times \mathbf{w}) = -122$ **(f)** $(-5\mathbf{v} + \mathbf{w}) \times ((\mathbf{u} \cdot \mathbf{v})\mathbf{w}) = (-3150, -2430, 1170)$

3. (a) $3\mathbf{v} - 2\mathbf{u} = (-5, -12, 20, -2)$ **(b)** $\|\mathbf{u} + \mathbf{v} + \mathbf{w}\| = \sqrt{106}$ **(c)** $\sqrt{2810}$ **(d)** $\text{proj}_{\mathbf{w}}\mathbf{u} = -\frac{15}{77}(9, 1, -6, -6)$

5. Not an orthogonal set

7. (a) A line through the origin, perpendicular to the given vector.
(b) A plane through the origin, perpendicular to the given vector. **(c)** $\{0\}$ (the origin)
(d) A line through the origin, perpendicular to the plane containing the two noncollinear vectors.

9. True **11.** $S(-1, -1, 5)$ **13.** $\sqrt{\frac{14}{17}}$ **15.** $\frac{11}{\sqrt{35}}$

17. Vector equation: $(x, y, z) = (-2, 1, 3) + t_1(1, -2, -2) + t_2(5, -1, -5)$;
parametric equations: $x = -2 + t_1 + 5t_2$, $y = 1 - 2t_1 - t_2$, $z = 3 - 2t_1 - 5t_2$

19. Vector equation: $(x, y) = (0, -3) + t(8, -1)$;
parametric equations: $x = 8t$, $y = -3 - t$

21. A possible answer is vector equation: $(x, y) = (0, -5) + t(1, 3)$; parametric equations: $x = t$, $y = -5 + 3t$

23. $3(x + 1) + 6(y - 5) + 2(z - 6) = 0$ **25.** $-18(x - 9) - 51y - 24(z - 4) = 0$ **29.** A plane

Exercise Set 4.1 (page 178)

1. (a) $\mathbf{u} + \mathbf{v} = (2, 6)$, $3\mathbf{u} = (0, 6)$ **(c)** Axioms 1–5 **3.** The set is a vector space with the given operations.

5. Not a vector space, Axioms 5 and 6 fail. **7.** Not a vector space. Axiom 8 fails.

9. The set is a vector space with the given operations. **11.** The set is a vector space with the given operations.

(a) False **(b)** False **(c)** True **(d)** False **(e)** False

Exercise Set 4.2 (page 188)

1. (a), (c), (e) **3.** (a), (b), (d) **5.** (a), (c), (d) **7.** (a), (b), (d) **9.** (a), (b), (c)

11. (a) The vectors span **(b)** The vectors do not span **(c)** The vectors do not span **(d)** The vectors span

13. The polynomials do not span

15. (a) Line; $x = -\frac{1}{2}t$, $y = -\frac{3}{2}t$, $z = t$ **(b)** Line; $x = 2t$, $y = t$, $z = 0$ **(c)** Origin **(d)** Origin
(e) Line; $x = -3t$, $y = -2t$, $z = t$ **(f)** Plane; $x - 3y + z = 0$

(a) True **(b)** True **(c)** False **(d)** False **(e)** False **(f)** True **(g)** True **(h)** False **(i)** False
(j) True **(k)** False

Exercise Set 4.3 (page 199)

1. (a) $\mathbf{u}_2$ is a scalar multiple of $\mathbf{u}_1$. **(b)** The vectors are linearly dependent by Theorem 4.3.3.
(c) $\mathbf{p}_2$ is a scalar multiple of $\mathbf{p}_1$. **(d)** B is a scalar multiple of A.

3. None

5. (a) They do not lie in a plane. **(b)** They do lie in a plane. **7. (b)** $\mathbf{v}_1 = \frac{2}{7}\mathbf{v}_2 - \frac{3}{7}\mathbf{v}_3$, $\mathbf{v}_2 = \frac{7}{2}\mathbf{v}_1 + \frac{3}{2}\mathbf{v}_3$, $\mathbf{v}_3 = -\frac{7}{3}\mathbf{v}_1 + \frac{2}{3}\mathbf{v}_2$

9. $\lambda = -\frac{1}{2}$, $\lambda = 1$

19. (a) They are linearly independent since $\mathbf{v}_1$, $\mathbf{v}_2$, and $\mathbf{v}_3$ do not lie in the same plane when they are placed with their initial points at the origin.
(b) They are not linearly independent since $\mathbf{v}_1$, $\mathbf{v}_2$, and $\mathbf{v}_3$ line in the same plane when they are placed with their initial points at the origin.

21. $W(x) = -x \sin x - \cos x \neq 0$ for some x. **23. (a)** $W(x) = e^x \neq 0$ **(b)** $W(x) = 2 \neq 0$
25. $W(x) = 2 \sin x \neq 0$ for some x.

(a) False **(b)** True **(c)** False **(d)** True **(e)** True **(f)** False **(g)** True **(h)** False

Exercise Set 4.4 (page 207)

1. (a) A basis for R^2 has two linearly independent vectors. **(b)** A basis for R^3 has three linearly independent vectors.
(c) A basis for P_2 has three linearly independent vectors. **(d)** A basis for M_{22} has four linearly independent vectors.

3. (a), (b) **7. (a)** $(\mathbf{w})_S = (3, -7)$ **(b)** $(\mathbf{w})_S = \left(\frac{5}{28}, \frac{3}{14}\right)$ **(c)** $(\mathbf{w})_S = \left(a, \frac{b-a}{2}\right)$

9. (a) $(\mathbf{v})_S = (3, -2, 1)$ **(b)** $(\mathbf{v})_S = (-2, 0, 1)$ **11.** $(A)_S = (-1, 1, -1, 3)$ **13.** $A = A_1 - A_2 + A_3 - A_4$

15. $\mathbf{p} = 7\mathbf{p}_1 - 8\mathbf{p}_2 + 3\mathbf{p}_3$ **17. (a)** $(2, 0)$ **(b)** $\left(\frac{2}{\sqrt{3}}, -\frac{1}{\sqrt{3}}\right)$ **(c)** $(0, 1)$ **(d)** $\left(\frac{2}{\sqrt{3}}a, b - \frac{a}{\sqrt{3}}\right)$

(a) False **(b)** False **(c)** True **(d)** True **(e)** False

Exercise Set 4.5 (page 216)

1. Basis: $(1, 0, 1)$; dimension $= 1$ **3.** Basis: $(4, 1, 0, 0)$, $(-3, 0, 1, 0)$, $(1, 0, 0, 1)$; dimension $= 3$ **5.** No basis; dimension $= 0$

7. (a) $\left(\frac{2}{3}, 1, 0\right)$, $\left(-\frac{5}{3}, 0, 1\right)$ **(b)** $(1, 1, 0)$, $(0, 0, 1)$ **(c)** $(2, -1, 4)$ **(d)** $(1, 1, 0)$, $(0, 1, 1)$

9. (a) n **(b)** $\frac{n(n+1)}{2}$ **(c)** $\frac{n(n+1)}{2}$ **13.** Any two of $(0, 1, 0, 0)$, $(0, 0, 1, 0)$, and $(0, 0, 0, 1)$ can be used.

15. $\mathbf{v}_3 = (a, b, c)$ with $9a - 3b - 5c \neq 0$

Exercise Set 4.6 (page 222)

1. (a) $[\mathbf{w}]_S = \begin{bmatrix} 3 \\ -7 \end{bmatrix}$ **(b)** $[\mathbf{w}]_S = \begin{bmatrix} \frac{5}{28} \\ \frac{3}{14} \end{bmatrix}$ **(c)** $[\mathbf{w}]_S = \begin{bmatrix} a \\ \frac{b-a}{2} \end{bmatrix}$

3. (a) $(\mathbf{p})_S = (4, -3, 1)$, $[\mathbf{p}]_S = \begin{bmatrix} 4 \\ -3 \\ 1 \end{bmatrix}$ **(b)** $(\mathbf{p})_S = (0, 2, -1)$, $[\mathbf{p}]_S = \begin{bmatrix} 0 \\ 2 \\ -1 \end{bmatrix}$

5. (a) $\mathbf{w} = (16, 10, 12)$ **(b)** $\mathbf{q} = 3 + 4x^2$ **(c)** $B = \begin{bmatrix} 15 & -1 \\ 6 & 3 \end{bmatrix}$

7. (a) $\begin{bmatrix} \frac{13}{10} & -\frac{1}{2} \\ -\frac{2}{5} & 0 \end{bmatrix}$ **(b)** $\begin{bmatrix} 0 & -\frac{5}{2} \\ -2 & -\frac{13}{2} \end{bmatrix}$ **(c)** $[\mathbf{w}]_B = \begin{bmatrix} -\frac{17}{10} \\ \frac{8}{5} \end{bmatrix}$, $[\mathbf{w}]_{B'} = \begin{bmatrix} -4 \\ -7 \end{bmatrix}$

9. (a) $\begin{bmatrix} 3 & 2 & \frac{5}{2} \\ -2 & -3 & -\frac{1}{2} \\ 5 & 1 & 6 \end{bmatrix}$ **(b)** $[\mathbf{w}]_B = \begin{bmatrix} 9 \\ -9 \\ 5 \end{bmatrix}$, $[\mathbf{w}]_{B'} = \begin{bmatrix} -\frac{7}{2} \\ \frac{23}{2} \\ 6 \end{bmatrix}$

11. (b) $\begin{bmatrix} 2 & 0 \\ 1 & 3 \end{bmatrix}$ **(c)** $\begin{bmatrix} \frac{1}{2} & 0 \\ -\frac{1}{6} & \frac{1}{3} \end{bmatrix}$ **(d)** $[\mathbf{h}]_B = \begin{bmatrix} 2 \\ -5 \end{bmatrix}$, $[\mathbf{h}]_{B'} = \begin{bmatrix} 1 \\ -2 \end{bmatrix}$

13. (a) $\begin{bmatrix} 1 & 2 & 3 \\ 2 & 5 & 3 \\ 1 & 0 & 8 \end{bmatrix}$ **(b)** $\begin{bmatrix} -40 & 16 & 9 \\ 13 & -5 & -3 \\ 5 & -2 & -1 \end{bmatrix}$ **(d)** $[\mathbf{w}]_B = \begin{bmatrix} -239 \\ 77 \\ 30 \end{bmatrix}$, $[\mathbf{w}]_S = \begin{bmatrix} 5 \\ -3 \\ 1 \end{bmatrix}$

(e) $[\mathbf{w}]_S = \begin{bmatrix} 3 \\ -5 \\ 0 \end{bmatrix}$, $[\mathbf{w}]_B = \begin{bmatrix} -200 \\ 64 \\ 25 \end{bmatrix}$

15. (a) $\begin{bmatrix} 3 & 5 \\ -1 & -2 \end{bmatrix}$ **(b)** $\begin{bmatrix} 2 & 5 \\ -1 & -3 \end{bmatrix}$ **(d)** $[\mathbf{w}]_{B_1} = \begin{bmatrix} 2 \\ -1 \end{bmatrix}$, $[\mathbf{w}]_{B_2} = \begin{bmatrix} -1 \\ 1 \end{bmatrix}$ **(e)** $[\mathbf{w}]_{B_2} = \begin{bmatrix} 3 \\ -1 \end{bmatrix}$, $[\mathbf{w}]_{B_1} = \begin{bmatrix} 4 \\ -1 \end{bmatrix}$

17. (a) $\begin{bmatrix} 3 & 2 & \frac{5}{2} \\ -2 & -3 & -\frac{1}{2} \\ 5 & 1 & 6 \end{bmatrix}$ **(b)** $[\mathbf{w}]_{B_1} = \begin{bmatrix} 9 \\ -9 \\ -5 \end{bmatrix}$, $[\mathbf{w}]_{B_2} = \begin{bmatrix} -\frac{7}{2} \\ \frac{23}{2} \\ 6 \end{bmatrix}$ **19. (a)** $\begin{bmatrix} \cos 2\theta & \sin 2\theta \\ \sin 2\theta & -\cos 2\theta \end{bmatrix}$

23. (a) $B = \{(1, 1, 0), (1, 0, 2), (0, 2, 1)\}$ **(b)** $B = \left\{\left(\frac{4}{5}, \frac{1}{5}, -\frac{2}{5}\right), \left(\frac{1}{5}, -\frac{1}{5}, \frac{2}{5}\right), \left(-\frac{2}{5}, \frac{2}{5}, \frac{1}{5}\right)\right\}$

Exercise Set 4.7 (page 235)

1. $r_1 = (2, -1, 0, 1)$, $r_2 = (3, 5, 7, -1)$, $r_3 = (1, 4, 2, 7)$;

$$c_1 = \begin{bmatrix} 2 \\ 3 \\ 1 \end{bmatrix}, \quad c_2 = \begin{bmatrix} -1 \\ 5 \\ 4 \end{bmatrix}, \quad c_3 = \begin{bmatrix} 0 \\ 7 \\ 2 \end{bmatrix}, \quad c_4 = \begin{bmatrix} 1 \\ -1 \\ 7 \end{bmatrix}$$

3. (a) $\begin{bmatrix} -2 \\ 10 \end{bmatrix} = \begin{bmatrix} 1 \\ 4 \end{bmatrix} - \begin{bmatrix} 3 \\ -6 \end{bmatrix}$ (b) **b** is not in the column space of A.   (c) $\begin{bmatrix} 1 \\ 9 \\ 1 \end{bmatrix} - 3\begin{bmatrix} -1 \\ 3 \\ 1 \end{bmatrix} + \begin{bmatrix} 1 \\ 1 \\ 1 \end{bmatrix} = \begin{bmatrix} 5 \\ 1 \\ -1 \end{bmatrix}$

(d) $\begin{bmatrix} 2 \\ 0 \\ 0 \end{bmatrix} = \begin{bmatrix} 1 \\ 1 \\ -1 \end{bmatrix} + (t-1)\begin{bmatrix} -1 \\ 1 \\ -1 \end{bmatrix} + t\begin{bmatrix} 1 \\ -1 \\ 1 \end{bmatrix}$ (e) $\begin{bmatrix} 4 \\ 3 \\ 5 \\ 7 \end{bmatrix} = -26\begin{bmatrix} 1 \\ 0 \\ 1 \\ 0 \end{bmatrix} + 13\begin{bmatrix} 2 \\ 1 \\ 2 \\ 1 \end{bmatrix} - 7\begin{bmatrix} 0 \\ 2 \\ 1 \\ 2 \end{bmatrix} + 4\begin{bmatrix} 1 \\ 1 \\ 3 \\ 2 \end{bmatrix}$

5. (a) $\begin{bmatrix} 1 \\ 0 \end{bmatrix} + t\begin{bmatrix} 3 \\ 1 \end{bmatrix}$; $t\begin{bmatrix} 3 \\ 1 \end{bmatrix}$ (b) $\begin{bmatrix} -2 \\ 7 \\ 0 \end{bmatrix} + t\begin{bmatrix} -1 \\ -1 \\ 1 \end{bmatrix}$; $t\begin{bmatrix} -1 \\ -1 \\ 1 \end{bmatrix}$

(c) $\begin{bmatrix} -1 \\ 0 \\ 0 \\ 0 \end{bmatrix} + r\begin{bmatrix} 2 \\ 1 \\ 0 \\ 0 \end{bmatrix} + s\begin{bmatrix} -1 \\ 0 \\ 1 \\ 0 \end{bmatrix} + t\begin{bmatrix} -2 \\ 0 \\ 0 \\ 1 \end{bmatrix}$; $r\begin{bmatrix} 2 \\ 1 \\ 0 \\ 0 \end{bmatrix} + s\begin{bmatrix} -1 \\ 0 \\ 1 \\ 0 \end{bmatrix} + t\begin{bmatrix} -2 \\ 0 \\ 0 \\ 1 \end{bmatrix}$

(d) $\begin{bmatrix} \frac{6}{5} \\ \frac{7}{5} \\ 0 \\ 0 \end{bmatrix} + s\begin{bmatrix} \frac{7}{5} \\ \frac{4}{5} \\ 1 \\ 0 \end{bmatrix} + t\begin{bmatrix} \frac{1}{5} \\ -\frac{3}{5} \\ 0 \\ 1 \end{bmatrix}$; $s\begin{bmatrix} \frac{7}{5} \\ \frac{4}{5} \\ 1 \\ 0 \end{bmatrix} + t\begin{bmatrix} \frac{1}{5} \\ -\frac{3}{5} \\ 0 \\ 1 \end{bmatrix}$

7. (a) $r_1 = [1\ 0\ 2]$, $r_2 = [0\ 0\ 1]$, $c_1 = \begin{bmatrix} 1 \\ 0 \\ 0 \end{bmatrix}$, $c_2 = \begin{bmatrix} 2 \\ 1 \\ 0 \end{bmatrix}$

(b) $r_1 = [1\ -3\ 0\ 0]$, $r_2 = [0\ 1\ 0\ 0]$, $c_1 = \begin{bmatrix} 1 \\ 0 \\ 0 \\ 0 \end{bmatrix}$, $c_2 = \begin{bmatrix} -3 \\ 1 \\ 0 \\ 0 \end{bmatrix}$

(c) $r_1 = [1\ 2\ 4\ 5]$, $r_2 = [0\ 1\ -3\ 0]$, $r_3 = [0\ 0\ 1\ -3]$, $r_4 = [0\ 0\ 0\ 1]$,

$$c_1 = \begin{bmatrix} 1 \\ 0 \\ 0 \\ 0 \\ 0 \end{bmatrix}, \quad c_2 = \begin{bmatrix} 2 \\ 1 \\ 0 \\ 0 \\ 0 \end{bmatrix}, \quad c_3 = \begin{bmatrix} 4 \\ -3 \\ 1 \\ 0 \\ 0 \end{bmatrix}, \quad c_4 = \begin{bmatrix} 5 \\ 0 \\ -3 \\ 1 \\ 0 \end{bmatrix}$$

(d) $r_1 = [1\ 2\ -1\ 5]$, $r_2 = [0\ 1\ 4\ 3]$, $r_3 = [0\ 0\ 1\ -7]$, $r_4 = [0\ 0\ 0\ 1]$

$$c_1 = \begin{bmatrix} 1 \\ 0 \\ 0 \\ 0 \end{bmatrix}, \quad c_2 = \begin{bmatrix} 2 \\ 1 \\ 0 \\ 0 \end{bmatrix}, \quad c_3 = \begin{bmatrix} -1 \\ 4 \\ 1 \\ 0 \end{bmatrix}, \quad c_4 = \begin{bmatrix} 5 \\ 3 \\ -7 \\ 1 \end{bmatrix}$$

9. (a) $r_1 = [1\ 0\ 2]$; $r_2 = [0\ 0\ 1]$; $c_1 = \begin{bmatrix} 1 \\ 0 \\ 0 \end{bmatrix}$; $c_2 = \begin{bmatrix} 2 \\ 1 \\ 0 \end{bmatrix}$

(b) $r_1 = [1\ -3\ 0\ 0]$; $r_2 = [0\ 1\ 0\ 0]$; $c_1 = \begin{bmatrix} 1 \\ 0 \\ 0 \\ 0 \end{bmatrix}$; $c_2 = \begin{bmatrix} -3 \\ 1 \\ 0 \\ 0 \end{bmatrix}$

(c) $\mathbf{r}_1 = \begin{bmatrix} 1 & 2 & 4 & 5 \end{bmatrix}$; $\mathbf{r}_2 = \begin{bmatrix} 0 & 1 & -3 & 0 \end{bmatrix}$; $\mathbf{r}_3 = \begin{bmatrix} 0 & 0 & 1 & -3 \end{bmatrix}$;

$$\mathbf{r}_4 = \begin{bmatrix} 0 & 0 & 0 & 1 \end{bmatrix}; \; \mathbf{c}_1 = \begin{bmatrix} 1 \\ 0 \\ 0 \\ 0 \\ 0 \end{bmatrix}; \; \mathbf{c}_2 = \begin{bmatrix} 2 \\ 1 \\ 0 \\ 0 \\ 0 \end{bmatrix}; \; \mathbf{c}_3 = \begin{bmatrix} 4 \\ -3 \\ 1 \\ 0 \\ 0 \end{bmatrix}; \; \mathbf{c}_4 = \begin{bmatrix} 5 \\ 0 \\ -3 \\ 1 \\ 0 \end{bmatrix}$$

(d) $\mathbf{r}_1 = \begin{bmatrix} 1 & 2 & -1 & 5 \end{bmatrix}$; $\mathbf{r}_2 = \begin{bmatrix} 0 & 1 & 4 & 3 \end{bmatrix}$; $\mathbf{r}_3 = \begin{bmatrix} 0 & 0 & 1 & -7 \end{bmatrix}$;

$$\mathbf{r}_4 = \begin{bmatrix} 0 & 0 & 0 & 1 \end{bmatrix}; \; \mathbf{c}_1 = \begin{bmatrix} 1 \\ 0 \\ 0 \\ 0 \end{bmatrix}; \; \mathbf{c}_2 = \begin{bmatrix} 2 \\ 1 \\ 0 \\ 0 \end{bmatrix}; \; \mathbf{c}_3 = \begin{bmatrix} -1 \\ 4 \\ 1 \\ 0 \end{bmatrix}; \; \mathbf{c}_4 = \begin{bmatrix} 5 \\ 3 \\ -7 \\ 1 \end{bmatrix}$$

11. **(a)** $(1, 1, -4 -3)$, $(0, 1, -5, -2)$, $\left(0, 0, 1, -\frac{1}{2}\right)$ **(b)** $(1, -1, 2, 0)$, $(0, 1, 0, 0)$, $\left(0, 0, 1, -\frac{1}{6}\right)$
   (c) $(1, 1, 0, 0)$, $(0, 1, 1, 1)$, $(0, 0, 1, 1)$, $(0, 0, 0, 1)$

15. **(b)** $\begin{bmatrix} 0 & 0 & 0 \\ 0 & 1 & 0 \\ 0 & 0 & 1 \end{bmatrix}$

17. **(a)** $\begin{bmatrix} 3a & -5a \\ 3b & -5b \end{bmatrix}$ for all real numbers a, b not both 0.

(b) Since A and B are invertible, their null spaces are the origin. The null space of C is the line $3x + y = 0$. The null space of D is the entire xy-plane.

True/False 4.7

(a) True   **(b)** False   **(c)** False   **(d)** False   **(e)** False   **(f)** True   **(g)** True   **(h)** False   **(i)** True   **(j)** False

Exercise Set 4.8   (page 246)

1. $\text{Rank}(A) = \text{Rank}(A^T) = 2$ 3. **(a)** 2; 1   **(b)** 1; 2   **(c)** 2; 2   **(d)** 2; 3   **(e)** 3; 2
5. **(a)** Rank $= 4$, nullity $= 0$ **(b)** Rank $= 3$, nullity $= 2$ **(c)** Rank $= 3$, nullity $= 0$
7. **(a)** Yes, 0   **(b)** No   **(c)** Yes, 2   **(d)** Yes, 7   **(e)** No   **(f)** Yes, 4   **(g)** Yes, 0
9. $b_1 = r$, $b_2 = s$, $b_3 = 4s - 3r$, $b_4 = 2r - s$, $b_5 = 8s - 7r$ 11. No   13. Rank is 2 if $r = 2$ and $s = 1$; the rank is never 1.
17. **(a)** 3   **(b)** 5   **(c)** 3   **(d)** 3   19. $A = \begin{bmatrix} 0 & 1 \\ 0 & 0 \end{bmatrix}$; $B = \begin{bmatrix} 1 & 2 \\ 2 & 4 \end{bmatrix}$

True/False 4.8

(a) False   **(b)** True   **(c)** False   **(d)** False   **(e)** True   **(f)** False   **(g)** False   **(h)** False   **(i)** True   **(j)** False

Exercise Set 4.9   (page 260)

1. **(a)** Domain: R^2; codomain: R^3 **(b)** Domain: R^3; codomain: R^2 **(c)** Domain: R^3; codomain: R^3
   (d) Domain: R^6; codomain: R^1

3. R^2, R^3, $(-1, 2, 3)$

5. **(a)** Linear; $R^3 \to R^2$ **(b)** Nonlinear; $R^2 \to R^3$ **(c)** Linear; $R^3 \to R^3$ **(d)** Nonlinear; $R^4 \to R^2$

7. (a) and (c) are matrix transformations; (b), (d), and (e) are not matrix transformations.

9. $\begin{bmatrix} 3 & 5 & -1 \\ 4 & -1 & 1 \\ 3 & 2 & -1 \end{bmatrix}$; $T(-1, 2, 4) = (3, -2, -3)$

11. (a) $\begin{bmatrix} 0 & 1 \\ -1 & 0 \\ 1 & 3 \\ 1 & -1 \end{bmatrix}$ **(b)** $\begin{bmatrix} 7 & 2 & -1 & 1 \\ 0 & 1 & 1 & 0 \\ -1 & 0 & 0 & 0 \end{bmatrix}$ **(c)** $\begin{bmatrix} 0 & 0 & 0 \\ 0 & 0 & 0 \\ 0 & 0 & 0 \\ 0 & 0 & 0 \\ 0 & 0 & 0 \end{bmatrix}$ **(d)** $\begin{bmatrix} 0 & 0 & 0 & 1 \\ 1 & 0 & 0 & 0 \\ 0 & 0 & 1 & 0 \\ 0 & 1 & 0 & 0 \\ 1 & 0 & -1 & 0 \end{bmatrix}$

13. (a) $T(-1, 4) = (5, 4)$ **(b)** $T(2, 1, -3) = (0, -2, 0)$ **15. (a)** $(2, -5, -3)$ **(b)** $(2, 5, 3)$ **(c)** $(-2, -5, 3)$

17. (a) $(-2, 1, 0)$ **(b)** $(-2, 0, 3)$ **(c)** $(0, 1, 3)$ **19. (a)** $\left(-2, \frac{\sqrt{3}-2}{2}, \frac{1+2\sqrt{3}}{2}\right)$ **(b)** $\left(0, 1, 2\sqrt{2}\right)$ **(c)** $(-1, -2, 2)$

21. (a) $\left(-2, \frac{\sqrt{3}+2}{2}, \frac{-1+2\sqrt{3}}{2}\right)$ **(b)** $\left(-2\sqrt{2}, 1, 0\right)$ **(c)** $(1, 2, 2)$

25. $\begin{bmatrix} -\frac{1}{9} & \frac{8}{9} & \frac{4}{9} \\ \frac{8}{9} & -\frac{1}{9} & \frac{4}{9} \\ \frac{4}{9} & \frac{4}{9} & -\frac{7}{9} \end{bmatrix}$ **29. (a)** Twice the orthogonal projection on the x-axis. **(b)** Twice the reflection about the x-axis.

31. Rotation through the angle 2θ.

33. Rotation through the angle θ and translation by $\mathbf{x}_0$; not a matrix transformation since $\mathbf{x}_0$ is nonzero. **35.** A line in R^n.

True/False 4.9

(a) False **(b)** False **(c)** False **(d)** True **(e)** False **(f)** True **(g)** False **(h)** False **(i)** True

Exercise Set 4.10 (page 271)

1. $T_B \circ T_A = \begin{bmatrix} 5 & -1 & 21 \\ 10 & -8 & 4 \\ 45 & 3 & 25 \end{bmatrix}$, $T_A \circ T_B = \begin{bmatrix} -8 & -3 & 1 \\ -5 & -15 & -8 \\ 44 & -11 & 45 \end{bmatrix}$

3. (a) $T_1 = \begin{bmatrix} 1 & 1 \\ 1 & -1 \end{bmatrix}$, $T_2 = \begin{bmatrix} 3 & 0 \\ 2 & 4 \end{bmatrix}$ **(b)** $T_2 \circ T_1 = \begin{bmatrix} 3 & 3 \\ 6 & -2 \end{bmatrix}$, $T_1 \circ T_2 = \begin{bmatrix} 5 & 4 \\ 1 & -4 \end{bmatrix}$

(c) $T_2(T_1(x_1, x_2)) = (3x_1 + 3x_2, 6x_1 - 2x_2)$,
$T_1(T_2(x_1, x_2)) = (5x_1 + 4x_2, x_1 - 4x_2)$

5. (a) $\begin{bmatrix} 1 & 0 \\ 0 & -1 \end{bmatrix}$ **(b)** $\begin{bmatrix} 0 & 0 \\ 0 & \frac{1}{2} \end{bmatrix}$ **(c)** $\begin{bmatrix} 3 & 0 \\ 0 & -3 \end{bmatrix}$

7. (a) $\begin{bmatrix} -1 & 0 & 0 \\ 0 & 0 & 0 \\ 0 & 0 & 1 \end{bmatrix}$ **(b)** $\begin{bmatrix} 1 & 0 & 1 \\ 0 & \sqrt{2} & 0 \\ -1 & 0 & 1 \end{bmatrix}$ **(c)** $\begin{bmatrix} -1 & 0 & 0 \\ 0 & 1 & 0 \\ 0 & 0 & 0 \end{bmatrix}$

9. (a) $T_1 \circ T_2 = T_2 \circ T_1$ **(b)** $T_1 \circ T_2 = T_2 \circ T_1$ **(c)** $T_1 \circ T_2 \neq T_2 \circ T_1$

11. (a) Not one-to-one **(b)** One-to-one **(c)** One-to-one **(d)** One-to-one **(e)** One-to-one **(f)** One-to-one
(g) One-to-one

13. (a) One-to-one; $\begin{bmatrix} \frac{1}{3} & -\frac{2}{3} \\ \frac{1}{3} & \frac{1}{3} \end{bmatrix}$; $T^{-1}(w_1, w_2) = \left(\frac{1}{3}w_1 - \frac{2}{3}w_2, \frac{1}{3}w_1 + \frac{1}{3}w_2\right)$

(b) Not one-to-one **(c)** One-to-one; $\begin{bmatrix} 0 & -1 \\ -1 & 0 \end{bmatrix}$; $T^{-1}(w_1, w_2) = (-w_2, -w_1)$

(d) Not one-to-one

15. (a) Reflection about the x-axis **(b)** Rotation through the angle $-\frac{\pi}{4}$ **(c)** Contraction by a factor of $\frac{1}{3}$
(d) Reflection about the yz-plane **(e)** Dilation by a factor of 5

17. (a) Matrix operator **(b)** Not a matrix operator **(c)** Matrix operator **(d)** Not a matrix operator

19. (a) Matrix transformation **(b)** Matrix transformation **21. (a)** $\begin{bmatrix} -1 & 0 \\ 0 & 0 \end{bmatrix}$ **(b)** $\begin{bmatrix} 0 & 1 \\ -1 & 0 \end{bmatrix}$ **(c)** $\begin{bmatrix} 0 & 0 \\ 3 & 0 \end{bmatrix}$

23. (a) $T_A(\mathbf{e}_1) = (-1, 2, 4)$, $T_A(\mathbf{e}_2) = (3, 1, 5)$, $T_A(\mathbf{e}_3) = (0, 2, -3)$ **(b)** $T_A(\mathbf{e}_1 + \mathbf{e}_2 + \mathbf{e}_3) = (2, 5, 6)$
(c) $T_A(7\mathbf{e}_3) = (0, 14, -21)$

25. **(a)** Yes   **(b)** Yes   **27.** **(b)** $T(x_1,\ x_2) = \left(x_1^2 + x_2^2,\ x_1 x_2\right)$

29. **(a)** The range of T is a proper subset of R^n.   **(b)** T must map infinitely many vectors to $\mathbf{0}$.

True/False 4.10

(a) False   **(b)** True   **(c)** True   **(d)** False   **(e)** False   **(f)** False

Exercise Set 4.11   (page 280)

1. (a) $\begin{bmatrix} 0 & -1 \\ -1 & 0 \end{bmatrix}$ **(b)** $\begin{bmatrix} -1 & 0 \\ 0 & -1 \end{bmatrix}$ **(c)** $\begin{bmatrix} 1 & 0 \\ 0 & 0 \end{bmatrix}$ **(d)** $\begin{bmatrix} 0 & 0 \\ 0 & 1 \end{bmatrix}$

3. (a) $\begin{bmatrix} 1 & 0 & 0 \\ 0 & 1 & 0 \\ 0 & 0 & -1 \end{bmatrix}$ **(b)** $\begin{bmatrix} 1 & 0 & 0 \\ 0 & -1 & 0 \\ 0 & 0 & 1 \end{bmatrix}$ **(c)** $\begin{bmatrix} -1 & 0 & 0 \\ 0 & 1 & 0 \\ 0 & 0 & 1 \end{bmatrix}$

5. (a) $\begin{bmatrix} 0 & -1 & 0 \\ 1 & 0 & 0 \\ 0 & 0 & 1 \end{bmatrix}$ **(b)** $\begin{bmatrix} 1 & 0 & 0 \\ 0 & 0 & -1 \\ 0 & 1 & 0 \end{bmatrix}$ **(c)** $\begin{bmatrix} 0 & 0 & 1 \\ 0 & 1 & 0 \\ -1 & 0 & 0 \end{bmatrix}$

7. Rectangle with vertices at $(0, 0)$, $(-3, 0)$, $(0, 1)$, $(-3, 1)$ **9. (a)** $\begin{bmatrix} 1 & 0 \\ 4 & 1 \end{bmatrix}$ **(b)** $\begin{bmatrix} 1 & -2 \\ 0 & 1 \end{bmatrix}$

11. (a) Expansion by a factor of 3 in the x-direction

   (b) Expansion by a factor of 5 in the y-direction and reflection about the x-axis

   (c) Shearing by a factor of 4 in the x-direction

13. (a) $\begin{bmatrix} \frac{1}{2} & 0 \\ 0 & 5 \end{bmatrix}$ **(b)** $\begin{bmatrix} 1 & 0 \\ 2 & 5 \end{bmatrix}$ **(c)** $\begin{bmatrix} 0 & -1 \\ -1 & 0 \end{bmatrix}$

17. (a) $y = \frac{2}{7}x$ **(b)** $y = x$ **(c)** $y = \frac{1}{2}x$ **(d)** $y = -2x$ **(e)** $y = -\frac{8+5\sqrt{3}}{11}x$ **19. (b)** No

23. (a) $\begin{bmatrix} 1 & 0 & k \\ 0 & 1 & k \\ 0 & 0 & 1 \end{bmatrix}$

   (b) Shear in the xz-direction with

   factor k maps (x, y, z) to $(x + ky, y, z + ky)$: $\begin{bmatrix} 1 & k & 0 \\ 0 & 1 & 0 \\ 0 & k & 1 \end{bmatrix}$.

   Shear in the yz-direction with factor k maps (x, y, z) to $(x, y + kx, z + kx)$: $\begin{bmatrix} 1 & 0 & 0 \\ k & 1 & 0 \\ k & 0 & 1 \end{bmatrix}$.

True/False 4.11

(a) False   **(b)** True   **(c)** True   **(d)** True   **(e)** False   **(f)** False   **(g)** True

Exercise Set 4.12   (page 290)

1. (a) Stochastic   **(b)** Not stochastic   **(c)** Stochastic   **(d)** Not stochastic   **3.** $\begin{bmatrix} 0.54545 \\ 0.45455 \end{bmatrix}$

5. (a) Regular   **(b)** Not regular   **(c)** Regular   **7.** $\begin{bmatrix} \frac{8}{17} \\ \frac{9}{17} \end{bmatrix}$ **9.** $\begin{bmatrix} \frac{4}{11} \\ \frac{4}{11} \\ \frac{3}{11} \end{bmatrix}$

11. (a) Probability that something in state 1 stays in state 1   **(b)** Probability that something in state 2 moves to state 1

   (c) 0.8   **(d)** 0.85

13. (a) $\begin{bmatrix} 0.95 & 0.55 \\ 0.05 & 0.45 \end{bmatrix}$ (b) 0.93 (c) 0.142 (d) 0.63

15. (a)

Year	1	2	3	4	5
City	95,750	91,840	88,243	84,933	81,889
Suburbs	29,250	33,160	36,757	40,067	43,111

(b)

City	46,875
Suburbs	78,125

17. (a) $\frac{23}{100}$ (b) $\begin{bmatrix} \frac{46}{159} \\ \frac{22}{53} \\ \frac{47}{159} \end{bmatrix}$ (c) 35, 50, 35 19. $P = \begin{bmatrix} \frac{7}{10} & \frac{1}{10} & \frac{1}{5} \\ \frac{1}{5} & \frac{3}{10} & \frac{1}{2} \\ \frac{1}{10} & \frac{3}{5} & \frac{3}{10} \end{bmatrix}$; $\mathbf{q} = \begin{bmatrix} \frac{1}{3} \\ \frac{1}{3} \\ \frac{1}{3} \end{bmatrix}$

21. $P^k \mathbf{q} = \mathbf{q}$ for every positive integer k

True/False 4.12

(a) True (b) True (c) True (d) False (e) True

Chapter 4 Supplementary Exercises (page 292)

1. (a) $\mathbf{u} + \mathbf{v} = (4, 3, 2)$, $-\mathbf{u} = (-3, 0, 0)$ (c) Axioms 1–5

3. If $s \neq 1$, -2, the solution space is the origin. If $s = 1$, the solution space is a plane through the origin. If $s = -2$, the solution space is a line through the origin.

7. A must be invertible

9. (a) Rank $= 2$, nullity $= 1$ (b) Rank $= 2$, nullity $= 2$ (c) Rank $= 2$, nullity $= n - 2$

11. (a) $\{1, x^2, x^4, \ldots, x^{2m}\}$ where $2m = n$ if n is even and $2m = n - 1$ if n is odd. (b) $\{x, x^2, x^3, \ldots, x^n\}$

13. (a) $\left\{ \begin{bmatrix} 1 & 0 & 0 \\ 0 & 0 & 0 \\ 0 & 0 & 0 \end{bmatrix}, \begin{bmatrix} 0 & 1 & 0 \\ 1 & 0 & 0 \\ 0 & 0 & 0 \end{bmatrix}, \begin{bmatrix} 0 & 0 & 1 \\ 0 & 0 & 0 \\ 1 & 0 & 0 \end{bmatrix}, \begin{bmatrix} 0 & 0 & 0 \\ 0 & 1 & 0 \\ 0 & 0 & 0 \end{bmatrix}, \begin{bmatrix} 0 & 0 & 0 \\ 0 & 0 & 1 \\ 0 & 1 & 0 \end{bmatrix}, \begin{bmatrix} 0 & 0 & 0 \\ 0 & 0 & 0 \\ 0 & 0 & 1 \end{bmatrix} \right\}$

(b) $\left\{ \begin{bmatrix} 0 & 1 & 0 \\ -1 & 0 & 0 \\ 0 & 0 & 0 \end{bmatrix}, \begin{bmatrix} 0 & 0 & 1 \\ 0 & 0 & 0 \\ -1 & 0 & 0 \end{bmatrix}, \begin{bmatrix} 0 & 0 & 0 \\ 0 & 0 & 1 \\ 0 & -1 & 0 \end{bmatrix} \right\}$

15. Possible ranks are 2, 1, and 0.

Exercise Set 5.1 (page 295)

1. (a) $y = 3x - 4$ (b) $y = -2x + 1$

2. (a) $x^2 + y^2 - 4x - 6y + 4 = 0$ or $(x - 2)^2 + (y - 3)^2 = 9$ (b) $x^2 + y^2 + 2x - 4y - 20 = 0$ or $(x + 1)^2 + (y - 2)^2 = 25$

3. $x^2 + 2xy + y^2 - 2x + y = 0$ (a parabola) 4. (a) $x + 2y + z = 0$ (b) $-x + y - 2z + 1 = 0$

5. (a) $\begin{vmatrix} x & y & z & 0 \\ x_1 & y_1 & z_1 & 1 \\ x_2 & y_2 & z_2 & 1 \\ x_3 & y_3 & z_3 & 1 \end{vmatrix} = 0$ (b) $x + 2y + z = 0$; $-x + y - 2z = 0$

6. (a) $x^2 + y^2 + z^2 - 2x - 4y - 2z = -2$ or $(x - 1)^2 + (y - 2)^2 + (z - 1)^2 = 4$
(b) $x^2 + y^2 + z^2 - 2x - 2y = 3$ or $(x - 1)^2 + (y - 1)^2 + z^2 = 5$

10. $\begin{vmatrix} y & x^2 & x & 1 \\ y_1 & x_1^2 & x_1 & 1 \\ y_2 & x_2^2 & x_2 & 1 \\ y_3 & x_3^2 & x_3 & 1 \end{vmatrix} = 0$ 11. The equation of the line through the three collinear points 12. $0 = 0$

13. The equation of the plane through the four coplanar points

Exercise Set 5.2 (page 301)

1. $x_1 = 2$, $x_2 = \frac{2}{3}$; maximum value of $z = \frac{22}{3}$ 2. No feasible solutions   3. Unbounded solution

4. Invest $6000 in bond A and $4000 in bond B; the annual yield is $880.

5. $\frac{7}{9}$ cup of milk, $\frac{25}{18}$ ounces of corn flakes; minimum cost $= \frac{335}{18} = 18.6\cent$

6. **(a)** $x_1 \geq 0$ and $x_2 \geq 0$ are nonbinding; $2x_1 + 3x_2 \leq 24$ is binding
   (b) $x_1 - x_2 \leq v$ for $v < -3$ is binding and for $v < -6$ yields the empty set.
   (c) $x_2 \leq v$ for $v < 8$ is nonbinding and for $v < 0$ yields the empty set.

7. 550 containers from company A and 300 containers from company B; maximum shipping charges $= \$2110$

8. 925 containers from company A and no containers from company B; maximum shipping charges $= \$2312.50$

9. 0.4 pound of ingredient A and 2.4 pounds of ingredient B; minimum cost $= 24.8\cent$

Exercise Set 5.3 (page 311)

1. **(a)** $\mathbf{u}_2$ is a scalar multiple of $\mathbf{u}_1$.
   (b) The vectors are linearly dependent by Theorem 5.3.3.
   (c) $\mathbf{p}_2$ is a scalar multiple of $\mathbf{p}_1$.
   (d) B is a scalar multiple of A.

3. None   5. **(a)** They do not lie in a plane.   **(b)** They do lie in a plane.

7. **(b)** $\mathbf{v}_1 = \frac{2}{7}\mathbf{v}_2 - \frac{3}{7}\mathbf{v}_3$, $\mathbf{v}_2 = \frac{7}{2}\mathbf{v}_1 + \frac{3}{2}\mathbf{v}_3$, $\mathbf{v}_3 = -\frac{7}{3}\mathbf{v}_1 + \frac{2}{3}\mathbf{v}_2$ 9. $\lambda = -\frac{1}{2}$, $\lambda = 1$

18. If and only if the vector is not zero

19. **(a)** They are linearly independent since $\mathbf{v}_1$, $\mathbf{v}_2$, and $\mathbf{v}_3$ do not lie in the same plane when they are placed with their initial points at the origin.
   (b) They are not linearly independent since $\mathbf{v}_1$, $\mathbf{v}_2$, and $\mathbf{v}_3$ lie in the same plane when they are placed with their initial points at the origin.

20. (a), (d), (e), (f)

24. **(a)** False   **(b)** False   **(c)** True   **(d)** False   27. **(a)** Yes

Exercise Set 5.4 (page 323)

1. **(a)** $\mathbf{x}^{(1)} = \begin{bmatrix} .4 \\ .6 \end{bmatrix}$, $\mathbf{x}^{(2)} = \begin{bmatrix} .46 \\ .54 \end{bmatrix}$, $\mathbf{x}^{(3)} = \begin{bmatrix} .454 \\ .546 \end{bmatrix}$, $\mathbf{x}^{(4)} = \begin{bmatrix} .4546 \\ .5454 \end{bmatrix}$, $\mathbf{x}^{(5)} = \begin{bmatrix} .45454 \\ .54546 \end{bmatrix}$

   (b) P is regular since all entries of P are positive; $\mathbf{q} = \begin{bmatrix} \frac{5}{11} \\ \frac{6}{11} \end{bmatrix}$

2. **(a)** $\mathbf{x}^{(1)} = \begin{bmatrix} .7 \\ .2 \\ .1 \end{bmatrix}$, $\mathbf{x}^{(2)} = \begin{bmatrix} .23 \\ .52 \\ .25 \end{bmatrix}$, $\mathbf{x}^{(3)} = \begin{bmatrix} .273 \\ .396 \\ .331 \end{bmatrix}$ **(b)** P is regular, since all entries of P are positive: $\mathbf{q} = \begin{bmatrix} \frac{22}{72} \\ \frac{29}{72} \\ \frac{21}{72} \end{bmatrix}$

3. **(a)** $\begin{bmatrix} \frac{9}{17} \\ \frac{8}{17} \end{bmatrix}$ **(b)** $\begin{bmatrix} \frac{26}{45} \\ \frac{19}{45} \end{bmatrix}$ **(c)** $\begin{bmatrix} \frac{3}{19} \\ \frac{4}{19} \\ \frac{12}{19} \end{bmatrix}$

4. **(a)** $P^n = \begin{bmatrix} \left(\frac{1}{2}\right)^n & 0 \\ 1 - \left(\frac{1}{2}\right)^n & 1 \end{bmatrix}$, $n = 1, 2, \dots$.   Thus, no integer power of P has all positive entries.

   (b) $P^n \to \begin{bmatrix} 0 & 0 \\ 1 & 1 \end{bmatrix}$ as n increases, so $P^n \mathbf{x}^{(0)} \to \begin{bmatrix} 0 \\ 1 \end{bmatrix}$ for any $\mathbf{x}^{(0)}$ as n increases.

   (c) The entries of the limiting vector $\begin{bmatrix} 0 \\ 1 \end{bmatrix}$ are not all positive.

6. $P^2 = \begin{bmatrix} \frac{1}{2} & \frac{1}{4} & \frac{1}{4} \\ \frac{1}{4} & \frac{1}{2} & \frac{1}{4} \\ \frac{1}{4} & \frac{1}{4} & \frac{1}{2} \end{bmatrix}$ has all positive entries; $\mathbf{q} = \begin{bmatrix} \frac{1}{3} \\ \frac{1}{3} \\ \frac{1}{3} \end{bmatrix}$ 7. $\frac{10}{13}$

8. $54\frac{1}{6}\%$ in region 1, $16\frac{2}{3}\%$ in region 2, and $29\frac{1}{6}\%$ in region 3

Exercise Set 5.5   (page 333)

1. **(a)** $\begin{bmatrix} 0 & 0 & 0 & 1 \\ 1 & 0 & 1 & 1 \\ 1 & 1 & 0 & 1 \\ 0 & 0 & 0 & 0 \end{bmatrix}$ **(b)** $\begin{bmatrix} 0 & 1 & 1 & 0 & 0 \\ 0 & 0 & 0 & 0 & 1 \\ 1 & 0 & 0 & 1 & 0 \\ 0 & 0 & 1 & 0 & 0 \\ 0 & 0 & 1 & 0 & 0 \end{bmatrix}$ **(c)** $\begin{bmatrix} 0 & 1 & 0 & 1 & 0 & 0 \\ 1 & 0 & 0 & 0 & 0 & 0 \\ 0 & 1 & 0 & 1 & 1 & 1 \\ 0 & 0 & 0 & 0 & 0 & 1 \\ 0 & 0 & 0 & 0 & 0 & 1 \\ 0 & 0 & 1 & 0 & 1 & 0 \end{bmatrix}$

2. **(a)** **(b)** **(c)**

3. **(a)** 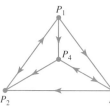   **(b)**   1-step: $P_1 \to P_2$ **(c)**   1-step: $P_1 \to P_4$

   2-step: $P_1 \to P_4 \to P_2$ 2-step: $P_1 \to P_3 \to P_4$

   $P_1 \to P_3 \to P_2$ 3-step: $P_1 \to P_2 \to P_1 \to P_4$

   3-step: $P_1 \to P_2 \to P_1 \to P_2$ $P_1 \to P_4 \to P_3 \to P_4$

   $P_1 \to P_3 \to P_4 \to P_2$

   $P_1 \to P_4 \to P_3 \to P_2$

4. **(a)** $\begin{bmatrix} 1 & 0 & 0 & 0 & 0 \\ 0 & 1 & 0 & 0 & 0 \\ 0 & 0 & 1 & 1 & 0 \\ 0 & 0 & 1 & 2 & 1 \\ 0 & 0 & 0 & 1 & 2 \end{bmatrix}$

   (c) The ijth entry is the number of family members who influence both the ith and jth family members.

5. **(a)** $\{P_1, P_2, P_3\}$ **(b)** $\{P_3, P_4, P_5\}$ **(c)** $\{P_2, P_4, P_6, P_8\}$ and $\{P_4, P_5, P_6\}$ 6. **(a)** None   **(b)** $\{P_3, P_4, P_6\}$

7. $\begin{bmatrix} 0 & 0 & 1 & 1 \\ 1 & 0 & 0 & 0 \\ 0 & 1 & 0 & 1 \\ 0 & 1 & 0 & 0 \end{bmatrix}$ Power of $P_1 = 5$

   Power of $P_2 = 3$

   Power of $P_3 = 4$

   Power of $P_4 = 2$

   8. First, A; second, B and E (tie); fourth, C; fifth, D

Exercise Set 5.6 (page 342)

1. **(a)** $\begin{bmatrix} 2 \\ 3 \end{bmatrix}$ **(b)** $\begin{bmatrix} 6 \\ 5 \\ 6 \end{bmatrix}$ **(c)** $\begin{bmatrix} 78 \\ 54 \\ 79 \end{bmatrix}$

2. **(a)** Use Corollary 10.8.4; all row sums are less than one.
   (b) Use Corollary 10.8.5; all column sums are less than one.

   (c) Use Theorem 10.8.3, with $\mathbf{x} = \begin{bmatrix} 2 \\ 1 \\ 1 \end{bmatrix} > C\mathbf{x} = \begin{bmatrix} 1.9 \\ .9 \\ .9 \end{bmatrix}$.

3. E^2 has all positive entries.   4. Price of tomatoes, \$120.00; price of corn, \$100.00; price of lettuce, \$106.67

5. \$1256 for the CE, \$1448 for the EE, \$1556 for the ME   6. **(b)** $\frac{542}{503}$

Exercise Set 5.7 (page 350)

1. **(a)** $\begin{bmatrix} 0 & 1 & 1 & 0 \\ 0 & 0 & 1 & 1 \\ 0 & 0 & 0 & 0 \end{bmatrix}$ **(b)** $\begin{bmatrix} 0 & \frac{3}{2} & \frac{3}{2} & 0 \\ 0 & 0 & \frac{1}{2} & \frac{1}{2} \\ 0 & 0 & 0 & 0 \end{bmatrix}$ 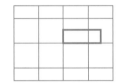 **(c)** $\begin{bmatrix} -2 & -1 & -1 & -2 \\ -1 & -1 & 0 & 0 \\ 3 & 3 & 3 & 3 \end{bmatrix}$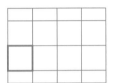

   (d) $\begin{bmatrix} 0 & .866 & 1.366 & .500 \\ 0 & -.500 & .366 & .866 \\ 0 & 0 & 0 & 0 \end{bmatrix}$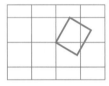

2. **(b)** $(0,0,0)$, $(1,0,0)$, $\left(1\frac{1}{2}, 1, 0\right)$, and $\left(\frac{1}{2}, 1, 0\right)$
   (c) $(0,0,0)$, $(1,.6,0)$, $(1,1.6,0)$, $(0,1,0)$

3. **(a)** $\begin{bmatrix} 1 & 0 & 0 \\ 0 & -1 & 0 \\ 0 & 0 & 1 \end{bmatrix}$ **(b)** $\begin{bmatrix} -1 & 0 & 0 \\ 0 & 1 & 0 \\ 0 & 0 & 1 \end{bmatrix}$ 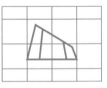 **(c)** $\begin{bmatrix} 1 & 0 & 0 \\ 0 & 1 & 0 \\ 0 & 0 & -1 \end{bmatrix}$

4. **(a)** $M_1 = \begin{bmatrix} \frac{1}{2} & 0 & 0 \\ 0 & 2 & 0 \\ 0 & 0 & \frac{1}{3} \end{bmatrix}$, $M_2 = \begin{bmatrix} \frac{1}{2} & \frac{1}{2} & \cdots & \frac{1}{2} \\ 0 & 0 & \cdots & 0 \\ 0 & 0 & \cdots & 0 \end{bmatrix}$, $M_3 = \begin{bmatrix} 1 & 0 & 0 \\ 0 & \cos 20° & -\sin 20° \\ 0 & \sin 20° & \cos 20° \end{bmatrix}$,

   $M_4 = \begin{bmatrix} \cos(-45°) & 0 & \sin(-45°) \\ 0 & 1 & 0 \\ -\sin(-45°) & 0 & \cos(-45°) \end{bmatrix}$, $M_5 = \begin{bmatrix} 0 & -1 & 0 \\ 1 & 0 & 0 \\ 0 & 0 & 1 \end{bmatrix}$
   (b) $P' = M_5 M_4 M_3 (M_1 P + M_2)$

5. (a) $M_1 = \begin{bmatrix} .3 & 0 & 0 \\ 0 & .5 & 0 \\ 0 & 0 & 1 \end{bmatrix}$, $M_2 = \begin{bmatrix} 1 & 0 & 0 \\ 0 & \cos 45° & -\sin 45° \\ 0 & \sin 45° & \cos 45° \end{bmatrix}$, $M_3 = \begin{bmatrix} 1 & 1 & \cdots & 1 \\ 0 & 0 & \cdots & 0 \\ 0 & 0 & \cdots & 0 \end{bmatrix}$,

$M_4 = \begin{bmatrix} \cos 35° & 0 & \sin 35° \\ 0 & 1 & 0 \\ -\sin 35° & 0 & \cos 35° \end{bmatrix}$, $M_5 = \begin{bmatrix} \cos(-45°) & -\sin(-45°) & 0 \\ \sin(-45°) & \cos(-45°) & 0 \\ 0 & 0 & 1 \end{bmatrix}$,

$M_6 = \begin{bmatrix} 0 & 0 & \cdots & 0 \\ 0 & 0 & \cdots & 0 \\ 1 & 1 & \cdots & 1 \end{bmatrix}$, $M_7 = \begin{bmatrix} 2 & 0 & 0 \\ 0 & 1 & 0 \\ 0 & 0 & 1 \end{bmatrix}$

(b) $P' = M_7(M_5 M_4(M_2 M_1 P + M_3) + M_6)$

6. $R_1 = \begin{bmatrix} \cos\beta & 0 & \sin\beta \\ 0 & 1 & 0 \\ -\sin\beta & 0 & \cos\beta \end{bmatrix}$, $R_2 = \begin{bmatrix} \cos\alpha & -\sin\alpha & 0 \\ \sin\alpha & \cos\alpha & 0 \\ 0 & 0 & 1 \end{bmatrix}$,   **7. (a)** $M = \begin{bmatrix} 1 & 0 & 0 & x_0 \\ 0 & 1 & 0 & y_0 \\ 0 & 0 & 1 & z_0 \\ 0 & 0 & 0 & 1 \end{bmatrix}$ **(b)** $\begin{bmatrix} 1 & 0 & 0 & -5 \\ 0 & 1 & 0 & 9 \\ 0 & 0 & 1 & -3 \\ 0 & 0 & 0 & 1 \end{bmatrix}$

$R_3 = \begin{bmatrix} \cos\theta & 0 & \sin\theta \\ 0 & 1 & 0 \\ -\sin\theta & 0 & \cos\theta \end{bmatrix}$, $R_4 = \begin{bmatrix} \cos\alpha & \sin\alpha & 0 \\ -\sin\alpha & \cos\alpha & 0 \\ 0 & 0 & 1 \end{bmatrix}$,

$R_5 = \begin{bmatrix} \cos\beta & 0 & -\sin\beta \\ 0 & 1 & 0 \\ \sin\beta & 0 & \cos\beta \end{bmatrix}$

Exercise Set 5.8   (page 358)

1. (a) *GIYUOKEVBH*   **(b)** *SFANEFZWJH*

2. (a) $A^{-1} = \begin{bmatrix} 12 & 7 \\ 23 & 15 \end{bmatrix}$ **(b)** Not invertible   **(c)** $A^{-1} = \begin{bmatrix} 1 & 19 \\ 23 & 24 \end{bmatrix}$ **(d)** Not invertible   **(e)** Not invertible

(f) $A^{-1} = \begin{bmatrix} 15 & 12 \\ 21 & 5 \end{bmatrix}$

3. *WE LOVE MATH*   **4.** Deciphering matrix $= \begin{bmatrix} 7 & 15 \\ 6 & 5 \end{bmatrix}$; enciphering matrix $= \begin{bmatrix} 7 & 5 \\ 2 & 15 \end{bmatrix}$

5. *THEY SPLIT THE ATOM*   **6.** *I HAVE COME TO BURY CAESAR*   **7. (a)** 010110001   **(b)** $\begin{bmatrix} 0 & 1 & 1 \\ 1 & 1 & 1 \\ 1 & 0 & 1 \end{bmatrix}$

8. A is invertible modulo 29 if and only if $\det(A) \neq 0 \pmod{29}$.

INDEX

A

Absolute value, 523
  of determinant, 166
Addition
  associative law for, 38, 122
  by scalars, 172
  of vectors in R^2 and R^3, 120, 122
  of vectors in R^n, 126
Additivity property
  of matrix transformation, 249, 269, 270
Adjoint, of a matrix, 110–111
Aeronautics, yaw, pitch, and roll, 256
Algebraic operations, using vector
  components, 126–127
Algebraic properties of matrices, 38–48
Algebraic properties of vectors, dot
  product, 135–136
Alleles, 292
Amps (unit), 76
Angle
  in R^n, 137, 143
  between vectors, 134–135, 137
Angle of rotation, 254
Approximate integration, 83
Area
  of parallelogram, 165
  of triangle, 154
Arithmetic operations
  matrices, 27–34, 38–42
  vectors in R^2 and R^3, 120–122
  vectors in R^n, 125
Associative law for addition, 38, 122
Associative law for matrix multiplication,
  38, 39–40
Astronautics, yaw, pitch, and roll, 256
Augmented matrices, 6–7, 11, 12, 18, 25,
  33
Axis of rotation, 254

B

Back-substitution, 19–20
Backward phase, 15
Balancing (of chemical equation), 79
Basis, 209–211
  change of, 217–222
  coordinate system for vector space,
    201–202
  finite basis, 201
  by inspection, 212
  linear combinations and, 233–234
  number of vectors in, 209, 211
  ordered basis, 205
  for row and column spaces, 229–230,
    231

by row reduction, 230, 231
  for row space of a matrix, 232–233
  standard basis, 202–203, 205–206
  transition matrix, 219
  uniqueness of basis representation, 204
  for vector space using row operations,
    232
Basis vectors, 201
Battery, 76
Bôcher, Maxime, 7, 184
Books, ISBN number of, 141
Bounded feasible regions, 304, 305
Branches (network), 73
Brightness, graphical images, 124
Bunyakovsky, Viktor Yakovlevich, 137

C

Calculus of variations, 162
Cancellation law, 41
Carroll, Lewis, 96
Cauchy, Augustin, 109, 137, 172
Cauchy-Schwarz inequality, 137
Cayley, Arthur, 29, 34, 43
Change-of-basis problem, 217–218
Chemical equations, balancing with linear
  systems, 78–80
Chemical formulas, 78
Circle, through three points, 296
Cliques, directed graphs, 336–338
Clockwise closed-loop convention, 76
Closed economies, 86
Closed Leontief model, 342–346
Closure under addition, 172
Closure under scalar, 172
Codomain, 247
Coefficient matrices, 33
Coefficients
  of linear combination of matrices, 31
  of linear combination of vectors, 127,
    183
  literal, 44
Cofactor, 94
Cofactor expansion
  determinants by, 93–98
  elementary row operations and, 104
  of 2×2 matrices, 95
Collinear vectors, 121–122
Column matrices, 26
Column-matrix form of vectors, 128, 225
Column space, 225, 226, 228, 243
  basis for, 229–230, 231
  equal dimensions of row and column
    space, 237
Column vectors, 26, 27, 39

Columns, cofactor expansion and choice
  of column, 97
Combustion, linear systems to analyze
  combustion equation for methane,
  78–79
Comma-delimited form of vectors, 128,
  225
Common initial point, 122
Commutative law for addition, 38
Commutative law for multiplication, 40,
  46
Complete reaction (chemical), 78
Complex vector spaces, 172
Components (of a vector)
  algebraic operations using, 126–127
  calculating dot products using, 35
  complex n-tuples, 316
  finding, 123–124
  in R^2 and R^3, 122–123
  vector components of $\mathbf{u}$ along $\mathbf{a}$,
    147–148
Composition
  of matrix transformations, 263–267
  non-commutative nature of, 264
  of reflections, 265, 277
  of rotations, 264, 275
  of three transformations, 265–267
Compression operator, 257, 277
Computer graphics, 350–355
  rotation, 353–355
  scaling, 352–353
  translation, 353
  visualization of three-dimensional
    object, 350–352
Condensation, 96
Conic sections (conics)
  through five points, 297–298
Consistency, determining by elimination,
  64–65
Consistent linear system, 3–4, 227
Constraint, 304, 309
Consumption matrix, 86, 586
Consumption vectors, 87, 88
Continuous derivatives, functions with,
  182
Contraction, 256, 257
Coordinate map, 217
Coordinate matrices, 205
  three-dimensional computer graphic
    views, 357
Coordinate systems, 200
  "basis vectors" for, 201
  units of measurement, 201